Advances in Oil and Gas Exploration & Production

Series editor

Rudy Swennen, Department of Earth and Environmental Sciences,
K.U. Leuven, Heverlee, Belgium

The book series Advances in Oil and Gas Exploration & Production publishes scientific monographs on a broad range of topics concerning geophysical and geological research on conventional and unconventional oil and gas systems, and approaching those topics from both an exploration and a production standpoint. The series is intended to form a diverse library of reference works by describing the current state of research on selected themes, such as certain techniques used in the petroleum geoscience business or regional aspects. All books in the series are written and edited by leading experts actively engaged in the respective field.

The Advances in Oil and Gas Exploration & Production series includes both single and multi-authored books, as well as edited volumes. The Series Editor, Dr. Rudy Swennen (KU Leuven, Belgium), is currently accepting proposals and a proposal form can be obtained from our representative at Springer, Dr. Alexis Vizcaino (Alexis.Vizcaino@springer.com).

More information about this series at http://www.springer.com/series/15228

Shell International and The Development
Research Center (DRC) of the State
Council of the People's Republic
of China
Editors

China's Gas
Development Strategies

Editors
Shell International
Shell Centre
London
UK

The Development Research Center
(DRC) of the State Council of the
People's Republic of China
Beijing
China

ISSN 2509-372X   ISSN 2509-3738   (electronic)
Advances in Oil and Gas Exploration & Production
ISBN 978-3-319-86690-1   ISBN 978-3-319-59734-8   (eBook)
DOI 10.1007/978-3-319-59734-8

Printed on acid-free paper

This Springer imprint is published by Springer Nature
The registered company is Springer International Publishing AG
The registered company address is: Gewerbestrasse 11, 6330 Cham, Switzerland

Foreword I

How to Encourage the Development of Natural Gas as a Major Energy Source in China

Widespread development of natural gas is a textbook approach adopted by developed countries and in emerging economies during optimisation and updating of energy systems. It is also widely recognised as being a key transitional route in the move towards future sustainable energy source systems. From the perspective of China, whether or not to follow this general pattern and the challenges, opportunities and policies required to achieve the development of natural gas are issues of major concern to policy makers, academics and the business world in general. In order to respond to these questions, following from the previous instalment of the *China Mid- to Long-term Energy Development Strategic Joint Study*, the State Council Development Research Center and Shell organised a number of international and local institutions to join forces in conducting further strategic research into the development of natural gas in China. This has received extensive support from both the Chinese and British governments, and on December 2, 2013, Premier Li Keqiang and Prime Minister David Cameron witnessed the signing of bilateral cooperation documents in respect of this. After a year and a half of research, the results of this research have been collated, and form the basis for the following primary viewpoints and judgements.

Global Natural Gas Development Trends Against the Backdrop of the Shale Natural Gas Revolution

1. **There are plentiful global natural gas resources, and these will go on to take a more significant role in future energy systems**. There are sufficient global natural gas resources to allow for an additional 200 years of extraction, conventional natural gas and unconventional natural gas, each accounting for about a half. There are 50 years of viable confirmed reserves, while the rate at which further reserves are confirmed is greater than the consumption. In the last 10 years, there has been 2.7% growth in natural gas consumption, making it the fastest growing fossil energy type. In 2014, natural gas accounted for 23.8% of global energy consumption,

and it is predicted that by 2035 natural gas will have become the largest global energy source.

2. **Natural gas trade and pricing approaches will undergo major changes, and there will be an increase in terms of the interconnectedness of the global natural gas market**. The increase in importance of LNG in addition to the rapid increase in the number of LNG exporting and importing nations will increase the fluidity of the global natural gas market. Natural gas pricing mechanisms will automatically adjust. Currently, the price of 43% of global wholesale natural gas is established by market competition and is not pegged against oil prices. As both trade and pricing models change, there will be an increase in the interconnectedness of the three main regional markets of the Americas, Europe and Asia, with the difference in prices shrinking, and developing towards integration.

3. **In the short term, there will be a large drop in natural gas prices, but in the long term the price of imported natural gas in China will not be excessively low**. There has recently been a relaxation in terms of natural gas supplies, with a large drop in natural gas prices, as the price of Japanese LNG has dropped from $20/MMBtu in 2014 to the current price of around $7. But in the long term, the price of natural gas when it arrives at Chinese coastal regions from Australia, the United States and Canada, the main newly developed sources of LNG, will not remain below the $10/MMBtu level for long due to development, liquefaction and transport costs being fairly high.

4. **China's share of the global natural gas market is gradually rising, making it necessary to become more actively involved in the international natural gas market**. In 2014, the share of global natural gas consumption accounted for by China was less than 5%, but by 2030 this will have grown to between 9% and 12%. There are abundant global natural gas resources in China, and China also has fairly large natural gas resources. Increases in natural gas consumption in China will not result in a pronounced increase in the global market price. Apart from this, China should not always be on the receiving end of fluctuations in the international natural gas price, and it should be possible for the nation to take a more active role in the international natural gas market via trade and investment.

Changes in the Mode of Supply of China's Natural Gas in the Future and Main Sources of Conflict

1. **China's economic development is entering a stage of a "new normal", with energy demand entering a stage of medium rate increases**. The rate at which China's economy has been developing has already transitioned from a high rate of growth towards a medium-high rate of growth. There will be major changes in economic structures, with the tertiary industry proportion growing while the secondary industry proportion drops. A peak in steel, cement and other heavy/chemical industries will

occur in the period surrounding the 13th Five-Year Plan. Energy demand will revert from a high growth rate to a medium growth rate, while it is predicted that by 2020 total energy consumption may reach 5 billion tonnes of standard coal. Between 2016 and 2020, average annual growth will be approximately 3%, while by 2030 it is possible that energy consumption may have reached 5.7 billion tonnes of standard coal.

2. **If appropriate policies are put in place, China's natural gas demand will still continue to grow relatively quickly**. Development of the service industry, expanding urbanisation and atmospheric pollution control will drive continuous growth in natural gas consumption. Quantitative analysis indicates that, under the current policy climate, natural gas consumption will have reached 300 billion cubic metres by 2020, reaching 450 billion cubic metres by 2030, respectively accounting for 8.0% and 11.0% of energy consumption. With enhanced environmental measures and imposition of a carbon tax, natural gas consumption would reach 350 billion cubic metres by 2020 and 580 billion cubic metres by 2030, respectively accounting for 10.8% and 15.4% of energy consumption. This is a significant increase when compared to the 5.8% proportion in 2014. Residential gas, natural gas heating, transport gas, industrial gas and power generation gas will be fields in which a relatively rapid rate of growth of natural gas use occurs.

3. **There is significant potential in terms of natural gas supplies within China, natural gas import capacity will increase by a large amount**. China actually has fairly plentiful natural gas resources. Based on Ministry of Land and Resources data, there are 40 trillion cubic metres of accessible natural gas reserves (dynamic appraisal of national oil and gas reserves in 2013), while there are 25 trillion cubic metres of shale natural gas and 10 trillion cubic metres of coalbed methane. Annual detection of new reserves is increasing rapidly, reaching 1.1 trillion cubic metres in 2014. However, extraction is both difficult and costly, requiring both technical and institutional innovation, before potential resources can be converted to actual output. Under the current institutional arrangements, the natural gas output for 2020 and 2030 will be 230 billion cubic metres and 380 billion cubic metres respectively, while shale natural gas will account for 40 billion cubic metres and 80 billion cubic metres respectively. If institutional systems were reformed, the 2020 and 2030 natural gas output quantities would be 270 billion cubic metres and 470 billion cubic metres respectively, shale natural gas accounting for 60 billion cubic metres and 150 billion cubic metres respectively. In terms of imports from abroad, pipeline gas import capacity will reach between 135 and 165 billion cubic metres by the end of 2020, with LNG imports reaching approximately 65 billion cubic metres. If one then adds to that the coal gasification output capacity currently planned for construction, in 2020, there will be a supply capability of between 415 billion and 495 billion cubic metres. By 2030, between 665 billion and 800 billion cubic metres of supply capacity will exist.

4. **Pronounced changes will occur in supply and demand relationships, requiring correct handling of the challenges faced in the development of natural gas.** From the above supply and demand analysis, supply and demand of natural gas will begin to exhibit a wider balance over the next 5 to 10 years. Supply quantities will no longer be a major factor restricting the development of natural gas. Low efficiency, high prices, difficulties in transport and inflexible institutional arrangements will gradually become more significant, while the main sources of conflict in the development of the natural gas industry will undergo a transformation. If appropriate adjustments are not immediately made to policy, the momentum necessary to achieve a rapid rate of natural gas growth will be lost and this will have an enduring effect on the development of the natural gas industry.

Strategy for the Development of Natural Gas Over the Next 15 Years

1. **Great efforts should be made towards developing natural gas, in order to develop natural gas into becoming a major energy source.** Natural gas is clean, highly efficient and convenient. Regardless of whether it is the issues of controlling atmospheric pollution or increasing the efficiency of energy systems that concern us, or the need to provide peak regulation that will be necessary with the future development of renewable energy on a huge scale, each of these requires large-scale development of natural gas. Natural gas presents effective investment opportunities, due to its potential for massive growth allied with the fact that there is no excess output capacity. If access to markets were relaxed, there is the potential for annual investment of as much as CNY 400 billion, which would result in a 0.6% increase in GDP. It is safe to say that expansion of effort in the development of natural gas, in order to develop it into a major energy source, conforms to the general patterns encountered in the evolution and updating of global energy systems and the urgent needs for economic and social development encountered in China.

2. **Natural gas should be developed in a safe, highly efficient and sustainable manner.** By safe, we are really referring to two layers of safety involving natural gas resource assurances and manufacturing and operations, as these are prerequisites for the development of the natural gas industry. Increasing local supplies in addition to expanding the sources of imported gas and increasing modes of import are both essential to this process. At the same time, national natural gas network connections need to be improved, which should be accompanied by construction of multi-level natural gas storage in order to ensure stable operation. High efficiency relates to increasing efficiency in natural gas exploration, transport and usage, as this will form the foundation of the development of natural gas in China. Sustainability relates to the balancing of ecological and environmental needs that the development of natural gas allows, this being a major aim of developing natural gas. At the same time as extraction, delivery and usage occur, special attention needs to be paid

to environmental protection and the sustainability of resource supplies. Sustainability concepts must be applied to the design and planning of the overall energy systems. In particular, this can be achieved by ensuring that the external costs of environmental pollution are fully reflected, in order to allow for the full adoption of natural gas.

3. **Clarification of the strategic aims and pathways to be adopted in the development of natural gas**. The aim is that by 2020, the share of primary energy consumption accounted for by natural gas should reach 10%, with consumption increasing to 350 billion cubic metres. By 2030, the share of primary energy consumption accounted for by natural gas should reach 15%, with consumption reaching 580 billion cubic metres. The main pathways are: guiding and nurturing the natural gas consumption market, accelerating the diversification of upstream natural gas supplies, creating a safe, fair and transparent natural gas pipeline transport and storage network, relying on technical innovation to improve the technical standards of the industry, actively promoting natural gas pricing reforms, and creating a modern natural gas market system and government regulatory system.

Guiding and Nurturing the Natural Gas-consuming Market, Expanding the Scope of Natural Gas Usage

1. **Key to expanding the use of natural gas is improving its economic competitiveness**. Central heating, natural gas power generation and industrial fuels have a low natural gas price tolerance and, under the current pricing system, natural gas is not competitive. To increase natural gas competitiveness would, first, require the internalisation of external costs, the effects on the environment and health of energy use would therefore need to be reflected, which would allow greater advantage to be taken of natural gas environmental strengths. Second, natural gas usage efficiency needs to be improved, an improvement in efficiency helping to restrict the increasing costs of energy usage.

2. **Increase environmental monitoring and improve energy usage efficiency, with the replacement of dispersed coal usage**. The replacement of dispersed coal usage is one of the most important areas in which natural gas should be developed. First, environmental monitoring would be increased, with online coal boiler pollution and emissions reduction detection systems, capable of real-time monitoring of waste gas emissions. Second, more effort would be put into energy-saving upgrades, improving energy usage efficiency, with the government providing a certain amount of fiscal support. Third, in China's eastern and central regions, where there are major issues in terms of controlling atmospheric pollution, no additional new coal-fired boilers should be permitted, with gas use becoming obligatory. It is predicted that by 2030, the replacement of dispersed coal usage would require 280 billion cubic metres of natural gas.

3. **Enhance planning and technical standards, accelerating the development of gas use in transport**. In transport, natural gas is capable of replacing oil. It reduces emissions and has major significance in terms of energy safety and environmental protection, apart from which it is both economical and viable. The construction of gas filling stations and infrastructure should be accelerated. Industry standards that encompass the complete production chain should be created so as to adopt suitable safety and regulatory standards. Full advantage should be taken of the power of government purchasing to lead this development, increasing the usage of natural gas-powered buses. It is predicted that by 2030, gas used in transport will reach 75 billion cubic metres.

4. **Improve pricing mechanisms, encouraging appropriate development of gas power generation**. Where base load generation is concerned, natural gas does not present any advantages compared to coal. But natural gas is well suited to peak regulation and integrated thermoelectric cooling, both of which are essential aspects of power systems. First, there is a need to establish peak and trough electricity and gas prices, which will allow better usage of the peak regulation effect, thereby allowing low price gas to be used to generate high-price electricity. Second, the localised manufacture of key technology needs to be achieved, which will reduce equipment and maintenance costs. Third, the introduction of environmental taxes is needed, accompanied by carbon trading, which then allows advantage to be taken of the benefits natural gas power generation presents in terms of being clean and environmentally friendly. It is predicted that by 2030, peak regulation and base load generation will account for between 60 and 70 billion cubic metres of natural gas, while gas consumption for cogeneration and distributed type energy source generation will be approximately 100 billion cubic metres.

5. **Peak regulation, residential and chemical industry gas pricing, reduction of cross-subsidising between different users**. Residential use gas and chemical industry gas have been priced below the cost of supply for a long period of time, the subsidies for this having been borne by other users. Residential use gas prices should be adjusted up to above the gas supply cost, with most residential users being perfectly capable of withstanding price increases, while special case social groups could be provided with subsidies. From the point of view of chemical industry users, establishing market mechanisms such as peak and trough gas pricing and interruptible gas pricing would still provide them with price discounts. However, maintaining prices at less than the cost of supply in the long term is not sustainable and there is a need to establish a timeframe for bringing chemical industry gas prices into line.

Introduction of New Market Entities, Large-scale Increases in Natural Gas Supplies

1. **Introduction of new development modes to the fields of conventional oil and gas**. China has ample sources of gas, but key to supply is the issue of institutional regulation. New development modes need to be introduced

to the fields of conventional oil and gas. One of these is the relaxation of market access, which would involve allowing state-owned energy companies that already own shale natural gas rights, or overseas energy companies already active in the oil and gas industry access to conventional oil and gas mineral rights. The second would be adjustment of the minimum prospecting investment: the minimum prospecting investment should be increased by a factor of between three and five, which would change the current phenomena of areas being occupied without any prospecting taking place. The third would be the creation of confirmed natural gas reserve trading mechanisms and a trading platform, which would then allow the development of latent reserves by allowing the circulation of mineral rights.

2. **More extensive and bolder regulatory reform, establishing regional trials in the fields of unconventional oil and gas**. Not only does unconventional gas require new technology, it has an even greater need for institutional reform. It would be possible to establish a regional trial in the Sichuan basin and surrounding areas, which would explore the following areas. First, regarding mineral rights management, resources other than those currently under production by one of the three main oil companies, or those which have been confirmed and where extraction work is currently taking place, should be uniformly put out to tender. Second, regarding market access, it should be provided not only to other state-owned enterprises, but also to privately owned enterprises and overseas businesses. With respect to foreign investors, independently owned foreign enterprises that satisfy technical and environmental standards and that have passed state safety assessments should be allowed access. Apart from this, new breakthroughs could be made in terms of distribution of benefits accruing from such resources, reform of mixed ownership structures and environmental monitoring.

3. **Diversification of imported gas sources and importing modes**. In general terms, China's natural gas imports are secure. Of greater importance is diversification in terms of sources of imports, entities involved, contract durations and pricing mechanisms in order to spread risks out more evenly. First, this would require increased LNG import from Indonesia, Australia, North America and East Africa, which would allow further diversification of the sources of imported gas. Second would be the introduction of long-term, mid-term and short-term spot trading contract portfolios, with pricing based on a combination of oil and typical natural gas centre pricing indices. Third would be the introduction of new purchasing entities, encouraging large-scale end users to buy natural gas from the international market.

4. **Development of natural gas trade and overseas investment from a commercial perspective**. Encourage overseas investment in natural gas, including regional investment and investment in individual nations which are major sources of gas imports. Investment should satisfy the strategic direction of China's policies, in addition to ensuring that such activity is commercially profitable to the greatest extent, basing project cooperation on economic considerations. Of course, expectations should be realistic

where it comes to high-price contracts that have been signed in the past and an approach should be sought to ensure that the historical costs therein are shared by both commercial enterprises and the nation.

Enhanced Planning and Construction of Pipeline Networks and Gas Storage Facilities, in Order to Ensure Interconnected Networks and Fair Access

1. **Accelerate the construction of basic infrastructure, to overcome storage and transport bottlenecks**. Efforts need to be made to ensure that by 2020 the pipeline network extends to 150,000 km, with a gas delivery capacity of 400 billion cubic metres. This should be extended to 250,000 km by 2030, and gas delivery capacity should exceed 690 billion cubic metres, establishing the "western gas piped east, northern gas going south, maritime gas reaching the land, supplies being based on whichever is nearest" approach to supply. The working capacity of gas storage facilities should reach between 35 and 40 billion cubic metres by 2020, reaching 65 billion cubic metres by 2030. Large-scale gas storage banks and LNG tanks capable of providing seasonal peak regulation, small-scale LNG daily urban peak regulation tanks, CNG spherical tanks and associated storage installations need to be constructed.

2. **Planning and realisation of interconnected networks needs to be enhanced**. The interconnectedness of basic infrastructure is key to ensuring the safety and stability of natural gas operations. The national pipeline backbone requires interconnected junction stations and branch lines, provincial pipeline network backbones connecting to distribution hubs, thus allowing flexible distribution, with subterranean gas storage and LNG receiving stations connecting to the state-owned long-distance pipeline backbone, in conjunction with the creation of regional networks based on adjacent provinces and cities. During the planning stage this will require effective cooperation, which will allow connection between different networks; government departments will need to adopt a unified approach to the construction of infrastructure, in addition to providing back-up during both implementation and subsequent follow-up.

3. **Introduction of societal capital, encouraging the diversification of investment entities**. Under the premise of a unified plan, involvement by all types of capital in natural gas infrastructure investment should be encouraged. The three main oil companies should open up construction of the national pipeline backbone to outside investment, allowing the entry of societal capital. The planned Western No. 4 line, the Western No. 5 line, the new Guangdong–Zhejiang line and other such major national natural gas backbone pipelines should be constructed as joint ventures using societal capital. Removal of the monopoly of provincial gas companies for the construction of provincial infrastructure, completely opening provincial branch pipeline construction rights to outside investment, will allow companies that are appropriately qualified to take part in construction and operation in a compliant manner.

4. **By clarifying and establishing functional roles and regulatory operations, it is possible to achieve the separation of pipeline service and sales businesses and to allow third-party access**. Promote the division of pipeline services and sales businesses in terms of business operations, finances and legal status. Regulate the scope of business operations, costing rules, operational modes and openness of information. Introduce third-party access to services as soon as possible. Ensure effective monitoring of access contracts, service prices and service quality, in order to ensure that operators provide non-discriminatory services.

5. **Clarify responsibility for gas storage and improve pricing mechanisms in order to ensure gas supply security**. Implement a tiered gas storage management regime. The state can be responsible for strategic natural gas storage facilities, upstream natural gas companies can bear the responsibility and duties of providing seasonal peak regulation and emergency storage, and individual provincial gas companies can bear the responsibilities and duties of providing daily and hourly peak regulation. Implement peak and trough pricing for various types of users, in order to encourage sensible consumption, finally acting to compensate for troughs and peaks. Establish a real-time, sensitive advance warning emergency response system.

Accelerate Development of Pricing Mechanism and Government Regulatory System Institutional Reforms, Creating a Modern Natural Gas Market System

1. **Optimise current pricing approaches, in order to resolve conflicting pricing**. First, the netback method needs to be implemented as soon as possible. The reference oil price level and pricing conversion indices need to be adjusted, as these should reflect changes in the cost price of alternative energy sources and should not be based on the previous price of high cost gas imported by the three large oil companies. At the same time, the pricing adjustment frequency needs to be changed, with the aim that seasonal pricing adjustments should be introduced. Second, the transport costs and prices applied to long-distance pipelines, branch pipelines, provincial pipelines, municipal pipelines and distribution pipework should be established in a scientific and appropriate manner. Greater regulation should be applied to this, in order to ensure appropriate returns on pipeline investment. Third, a revision of metering standards and valuation methods is necessary, with conversion from flow quantity-based or mass-based pricing to calorific-based pricing. It will also be necessary to accelerate elimination of inappropriate charging activities by local government.

2. **Encourage market-set natural gas prices, accelerating the creation of natural gas trading centres**. In locations where natural gas sources are relatively diversified and there are fairly abundant supplies, such as Shanghai, Guangdong, Zhejiang and Jiangsu, remove regulation of the provincial and municipal gate prices, the prices then being set directly by upstream and downstream users. Encouraging the creation of regional

trading centres would then accelerate the creation of Shanghai, Beijing, Guangdong, Hubei and Xinjiang natural gas trading centres. Efforts should be made to establish Shanghai as the international trade and pricing centre. The creation of natural gas trading centres must commence with the introduction of spot trading, at which point futures trading can then be developed.

3. **Improve overall energy regulation and establish a single, independent, specialist regulatory system**. Increase efforts to establish an overall energy regulator, which will be responsible for establishing energy strategy, planning and policy; it would also be responsible for adjustments in the balance of total energy sources, thus ensuring energy security, while regulating energy structures, being responsible for both energy conservation and energy efficiency. This would also be responsible for greater data collection and analysis, energy-related scientific and technological innovation and international energy cooperation. Create a unified energy regulator, which in addition to having environmental and territorial resource regulatory functions would gradually implement regulation of the economic and social aspects of the complete energy production chain, creating an independent, unified, specialist energy regulatory body and a comprehensive regulatory system.

4. **Deepen natural gas-related legislation, establishing a complete legal system**. Establish a comprehensive legal framework, centred on an "Oil and Natural Gas Law", which sets out natural gas-specific legislation. One of the main objectives is to perfect resource ownership rights and exploration and extraction contracts. Another is to apply production safety and environmental subsidies, transnational investment, import and export trade and other such legislation to the regulation of infrastructure construction and operation, storage, sale and usage, safety and advance warning and emergency response. The relationship between the Oil and Natural Gas Pipeline Protection Law and other laws must be coordinated according to legal interpretation.

The above topics have been formulated based on many different suggestions after extensive discussion. It would be right to say that these topics consist of an open, collaborative and inclusive study, with the members of the various teams coming from many different institutions, with Chinese representatives coming from the government, research institutes and state energy companies as well as private enterprises. The achievements of this project are based on the assimilated knowledge of all of these parties and involve an ongoing knowledge-sharing process. In addition, when it came to establishing the details pertinent to each area and the conclusions drawn, this was achieved through constant discussions between a number of consultants for each topic as well as being based on the suggestions of specialists in each field, and involved field trips to Shanghai, Zhejiang, Sichuan and Chongqing to properly assess each aspect of the production chain. During the course of researching this project, the research teams submitted many different internal reports, on topics including: negotiation of the China–Russia natural gas east pipeline project; natural gas pricing reform; oil and gas storage; and shale natural gas development. They have also attracted comments from the State Council on a number of occasions in relation to support for instigation of the related reforms and policies. After the research into the topics in question was complete, the main conclusions and a summary report were submitted to the leaders of the State Council and attracted further comments and a positive response. Of course, as open research, inaccuracies and oversights are difficult to avoid, and experts and friends are invited to kindly provide their feedback and corrections where deemed appropriate.

Li Wei
Director of the State Council Development
Research Center

Foreword II

Joining Hands in Dealing with the Challenges Presented by Energy and the Climate

On behalf of Royal Dutch Shell, it has been an honour to work with the Development Research Center (DRC) of the State Council on China's energy evolution. In our previous joint research, we looked broadly at the country's energy system. In this project, we have focused on China's gas development.

The aim was to provide insights to help shape China's 13th Five-Year Plan, and this meant facing a very tight timescale. The DRC and Shell teams have risen to the challenge. The standard of work is high, especially seeing the pace at which the teams had to work, and the results are very rich and insightful.

The teams have worked together exceptionally well, with each side contributing their own specific expertise. DRC brought a deep understanding of China's energy system and the development challenges to be addressed, and this was complemented by Shell's international experience and knowledge of global gas markets, regulatory mechanisms and the drivers of gas demand.

Why is gas important? One of the world's toughest challenges is meeting greater long-term demand for energy. According to the IEA, global demand is expected to be 37% higher in 2040. Population growth, economic development and the need to provide access to modern energy to more people are critical factors in bolstering demand. China is no exception to these trends.

China will need all sources of energy to meet demand, including natural gas. But this is only one part of the picture. Another is the role the energy sector can play in building a sustainable future if it reduces the impact of energy production and use on the environment. When burned for power, gas produces around half the CO_2 and one-tenth of the air pollutants that coal does. Using more gas will therefore help address China's environmental challenges.

Natural gas is also flexible. A gas-fired plant takes less time to start and stop than a coal-fired plant. This makes gas an ideal partner to intermittent energy sources like wind and solar, which are increasingly crucial to the future of the China's energy system.

The 12th Five-Year Plan (2011–2015) showed a clear will to diversify China's energy sources with a view to ensuring a cleaner, more sustainable energy system. The promotion of natural gas was one of its key elements. The 13th Five-Year Plan (2016–2020) is expected to keep natural gas as a priority in China's energy planning.

This report offers insights on how best to increase the share of gas in the energy mix while balancing the objectives of energy security, energy affordability and the environment. The DRC and Shell worked hard together to map out the different factors which drive the increasing share of gas in the global energy system; and how to apply the insights to the local context to determine the potential for gas in China. Globally, a key factor has been the changing level and mix of economic activity in various countries, such as a switch from industry to services. The move to natural gas from other energy sources, in particular oil, was another key factor.

Another important element of this study was grappling with complex economic and technical factors to develop credible outlooks for natural gas production in China. The teams have considered many sources of gas, including conventional gas, unconventional shale gas, coal bed methane and synthetic natural gas production from coal.

This study was also aimed at building confidence that at expected levels of gas imports, including both pipeline and liquefied natural gas (LNG), there is sufficient diversity of supply that energy security will not be an issue for China. Finally, the study looked at the international experiences of market liberalisation and diversifying gas infrastructure and applied them to the China context. This was done in a way which will benefit China but still respects its important political principles and values.

The work has resulted in robust and pragmatic policy recommendations. It includes supply-side measures to accelerate domestic gas production and demand measures to promote gas usage in China. It also considers the reforms needed in the midstream.

The DRC's collaboration with Shell demonstrates how governments and companies can collaborate effectively over energy and climate challenges, and I believe other countries will be keen to learn from this.

The DRC and Shell have established and demonstrated the ability to work together. I now look forward to further collaboration on topics where we can combine Shell's international energy market experience with DRC's deep understanding and thought leadership on China's development.

Ben van Beurden
CEO, Royal Dutch Shell

Acknowledgements

Project Chair	Minister Wei Li, Development Research Center of the State Council of China	Ben van Beurden, CEO, Royal Dutch Shell
Project Executive	Vice President (Vice Minister) Shijin Liu, Development Research Center of the State Council of China	Xinsheng Zhang, Executive Chairman, Shell Companies in China
Project Sponsor	Changwen Zhao, Director-General, Research Department of Industrial Economy, Development Research Center of the State Council	Jeremy Bentham, VP Global Business Environment, Shell International B.V.
Project Team Lead	Jinzhao Wang, Deputy Director General of Research Department of Foreign Economic Relations, Development Research Center of the State Council	William King, Group Strategy Advisor Shell International B.V.

Shell International project team:

- Mallika Ishwaran, Senior Economist, Shell International Limited
- Martin Haigh, Senior Energy Advisor, Shell International Limited
- Taoliang Lee, LNG Markets Advisor, Shell Eastern Petroleum Private Limited
- Shangyou Nie, Competitive Intelligence Advisor, Shell International Exploration and Production B.V.
- Cindy Wang, Government Affairs-Policy and Integrated Gas & New Energy Business Support, Shell (China) Limited

DRC project team:

- Zhaoyuan Xu, Director of Research, Research Department of Industrial Economy Development Research Center of the State Council
- Xiaoming Wang, Director of Research, Research Department of Industrial Economy, Development Research Center of the State Council

- Jigang Wei, Director of Research, Development Research Center of the State Council
- Zhonghong Wang, Secretary of the Party Committee and Vice-Director-General of China Economic Times, Development Research Center of the State Council

With special thanks to Vivid Economics, Aurora Energy Research and Oxford Institute for Energy Studies.

Contents

Abbreviations

Bod	Barrels of oil per day
CCHP	Combined cooling, heating and power
CHP	Combined heat and power
CNG	Compressed natural gas
CNY	Chinese yuan
EIA	US Energy Information Administration
GDP	Gross domestic product
IEA	International Energy Agency
LNG	Liquefied natural gas
LPG	Liquefied petroleum gas
MMBtu	Million British thermal units
mtpa	Million tonnes per annum
NOC	National Oil Company
OECD	Organisation for Economic Co-operation and Development
OPEC	Organization of the Petroleum Exporting Countries
TSCE	Tonnes of standard coal equivalent

List of Figures

List of Tables

General Summary

1

Natural gas is an efficient, easy to use fuel. There are abundant resources, and advances in technology have brought increased supply and lower costs. The golden age of natural gas has arrived, and in the future it will play an even greater role in the global energy system. In China, the development of the natural gas industry must be promoted as part of the optimisation of energy resources to control atmospheric pollution while meeting the energy needs for economic growth. Efficient, safe and sustainable development of the natural gas industry requires the rationalisation of the structure of the industry, greater demand, increased supply, improved safety, higher efficiency and the creation of a modern natural gas industrial system and government management system. This will lead to natural gas becoming one of the main energy sources in China.

1.1 Global Trends in the Natural Gas Industry

1.1.1 Abundant Resources, and a Greater Role for Natural Gas in the Future

According to International Energy Agency (IEA) statistics, global current natural gas resources stand at 751 trillion m^3, and at the present levels of production, these can be exploited for more than 200 years. These resources include 420 trillion m^3 of conventional gas reserves and 331 trillion m^3 of unconventional gas reserves, with the latter including 206 trillion m^3 of shale natural gas, 76 trillion m^3 of tight natural gas and 47 trillion m^3 of coalbed methane. The proven reserves (resources which can be extracted under current economic and technological conditions) are estimated at 185.7 trillion m^3 and can be exploited for more than 50 years. Moreover, the new resources discovered every year exceed consumption, and natural gas resource to production ratio of 55.1 is showing a steady upward trend. It is, therefore, safe to conclude that in the future gas reserves will remain abundant.

The importance of natural gas in the global energy structure is growing all the time. The global consumption of natural has grown by approximately 2.7% between 2005 and 2014, the fastest of any fossil fuel (BP 2014). Assuming continued strong consumption growth in the future, natural gas consumption may increase at

* This chapter was compiled by Jinzhao Wang and Changwen Zhao, with important guidance and final vetting provided by Wei Li, Shijin Liu, Junkuo Zhang, Laiming Zhang, Guoqiang Long and Bin Yu from the Development Research Center of the State Council. The study group consultants and core experts provided important suggestions for revision. International team members and Chinese team members also offered important contributions to the drafting of the report.

Shell International and The Development Research Center (Eds.), *China's Gas Development Strategies*, Advances in Oil and Gas Exploration & Production, DOI 10.1007/978-3-319-59734-8_1

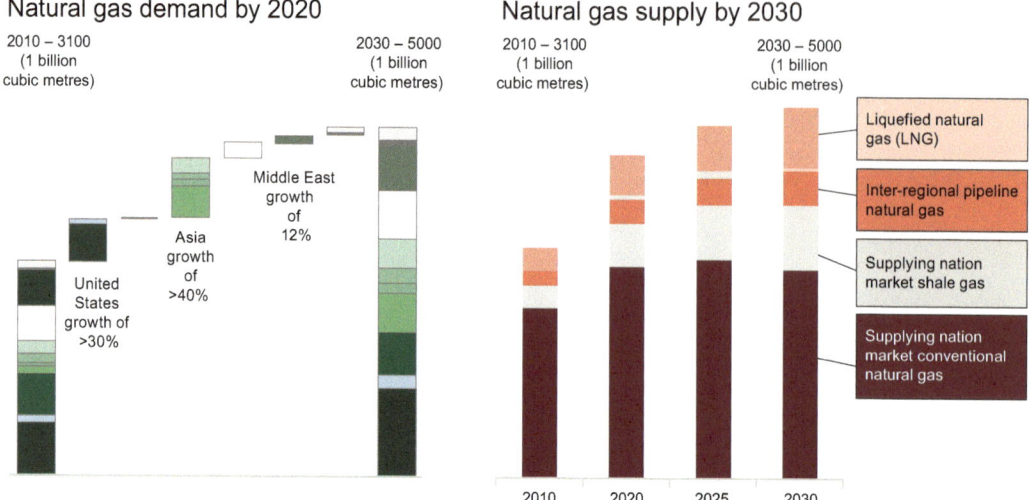

Fig. 1.1 Global natural gas supply and demand

approximately 2% per annum until 2030, and by 2035 it will become the largest primary energy source (IEA, Current Policies Outlook). The total supply and demand for natural gas is expected to have grown from 3.1 trillion m^3 in 2010 to approximately 5 trillion in 2030 (Fig. 1.1).

Recently, due to factors such as sluggish economic recovery, the low price of coal and large subsidies received by the renewable energy sources industry, the growth in the demand for natural gas in developed countries such as the EU, the USA and Japan has been limited. Demand in non-OECD nations is much higher, and is expected to continue to be so: in the period up to 2030, non-OECD countries are expected to account for 80% of the increase in global natural gas consumption.

1.1.2 Changes to the Global Trade and Pricing, and a More Interconnected Market

Advances in liquefied natural gas (LNG) technology have greatly changed the global natural gas trade, and the proportion of LNG in global trade is gradually growing. In 2013, the volume of the international gas trade stood at 1 trillion m^3, of which 320 billion (32%) was LNG (IGU 2014). This share has grown and diversified dramatically over the last few decades. In 1990 there were eight LNG-exporting countries and nine LNG-importing countries. The numbers have now grown to approximately 30 and 20 respectively, and future estimates are that this will continue to grow, creating a much more liquid global market. There have also been extensive changes in which nations export LNG. Australia's production capacity has been growing rapidly, and recent investments into seven LNG projects there will bring its total annual capacity to 85 billion m^3 (95 billion m^3 if projects under construction in neighbouring Papua New Guinea are included). The success of the shale natural gas industry in the USA has turned the country into another potential exporter. The capacity of LNG projects now under construction worldwide is now estimated at 130 billion m^3, and if these projects become operational as planned, total global annual production capacity of LNG is expected to rise from 390 billion m^3 in 2012 to approximately 520 billion by 2017. Exports from Australia, the United States and Canada are expected to affect the global LNG trade structure

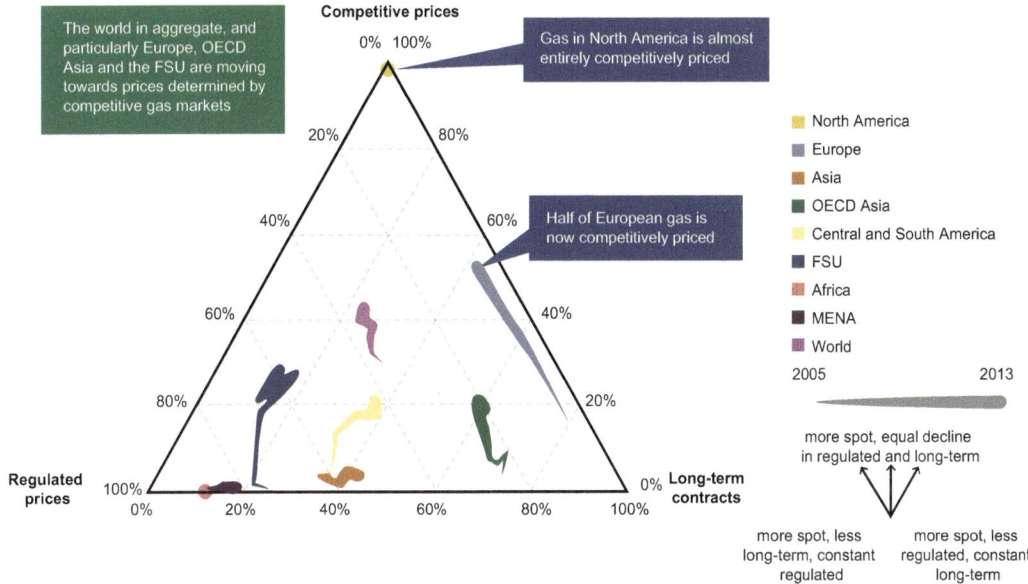

Fig. 1.2 Evolution of global natural gas pricing mechanism. *Note* The width of the line increases with time; if a line moves towards the corner of the triangle, this indicates a tendency towards that pricing method

the most, and with this change come both benefits and risks to China's LNG imports.[1]

The pricing of natural gas has also been changing. Formerly dominated by linkage to oil price and by long-term supply agreements, now there is a diversity of methods in modern markets for setting the price for natural gas, and there is also regional variation in prices.[2] The 2014 report by International Gas Union (IGU) showed that just 19% of natural gas was linked to oil price (Fig. 1.2).

Changes in the natural gas trade and pricing methods have increased the interconnectivity of natural gas markets. Previously, gas markets were regional—Europe, Asia and North America were the three main regional markets, and the was a large price differential between them, caused by the regional differences in resources, transportation methods and pricing methods.

North American prices were the lowest, followed by Europe and then Asia. In extreme cases, Asian prices were double those in Europe and quadruple those in North America. With the changes in trade and pricing mechanisms, the interaction between the markets will strengthen, reducing the price gap, and putting the global market on the road to greater integration.

1.1.3   Falling Oil Prices and Their Effect on Natural Gas Markets—but no Great in the Price of Imported Gas in China in the Medium to Long Term

International oil prices, which in June 2014 stood at over 100 $/bbl (Brent), have more than halved, and in October 2015 were below 50 $/bbl. Where gas price is linked to oil price, this has meant a substantial fall in gas price. In Japan, LNG prices fell from 20 $/MMBtu in 2014 to the current level of 7 $/MMBtu. Gas-importing countries whose supply agreements are based mainly on spot prices or short-term contracts

---

[1]For Australia, the United States and Canada, LNG development has an influence on Chinese imports. Further discussion is provided in Sect. 8.3.

[2]Contracts linked to oil prices can also be competitive, but it is fixed-price gas contracts that are generally more competitive, and are able to more rapidly respond to changes in the market.

have also seen a fall in the cost of imports. As for the gas supplier countries, the largest impact was on current projects under construction, as many of those had been launched with the expectation of high oil prices, and the profitability of such projects will have reduced, with some perhaps becoming no longer profitable. Falling oil prices have also greatly affected planned natural gas projects in the long term. The total capacity of projects where final investment decision has been made is 30 billion m^3, and for those at the Front-End Engineering Design phase is 370 billion, while the capacity of projects at even earlier phases has reached 500 billion (IGU 2014). Falling oil prices have affected the investment process for several of these projects. The final investment decision has been postponed, for example, for the Browse gas project in Australia and for Canada's Pacific Northwest Project. The global fall in oil prices is caused adjustments in natural gas markets. The falling oil prices have resulted in natural gas supply being lower than predicted, and this will, to a certain extent, provide a limit to how low international natural gas prices can sustainably fall.

Causes of the Current Fall in Oil Prices and Future Trends

The shale oil and gas revolution in the United States has changed the global oil and gas supply structure and changed the international balance of power as regards price control.

The breakthrough in the shale oil industry stem originated in advances in the US hydraulic fracturing technology, and in 2011 the supply of natural gas overtook demand, leading to a fall in prices. The effects of the shale revolution the made themselves felt in oil, and there was a large increase in US oil output in 2011, with supply rising from 5.5 million barrels per day to 9.3 million in 2014, an increase of 3.8 million barrels per day in just three years. This increase allowed the

United States to overtake the output of both Iran and Iraq, the number two and number three OPEC producers, the equivalent of the United States having become, in the space of three years, a major OPEC nation.

More importantly, the shale oil and gas revolution has changed the international balance of power in price control. Previously, oil supply capacity and demand were such that Saudi Arabia was able to regulate market balance using its spare capacity of 2–3 million barrels per day—they were effectively able to set oil prices. Currently US shale oil output is 4.2 million barrels per day and production is able to respond rapidly to higher market prices. The result is a flattening of the global oil supply curve and a reduction in the ability for Saudi Arabia to use its capacity to manage international oil prices.

The slowdown in Chinese demand has also greatly affected the global demand for oil. Demand in China grew by an average of 6.5% per year between 2004 and 2013, an annual increase of 500 million barrels per day, accounting for 45% of the annual global increase of 1.1 million barrels per day. In some years the increase in demand for oil in China exceeded 1 million barrels per day, and the oil bull market of the past 10 years is to a certain extent attributable to this rapid growth of Chinese demand. However, recently there has been a slowdown in the growth of China's oil demand, with growth falling to 3.7% or 0.39 million barrels per day. In 2014, growth fell further, to 2.3% or 0.22 million barrels per day, slipping below the past 10-year average. The slowdown in the growth of the Chinese economy might have become the norm, and the situation of demand exceeding supply caused by high-speed Chinese economic growth might never return.

The medium- to long-term view is that the oil prices may recover to 70–80 $/bbl, with the average cost of US shale oil at 60–70 $/bbl. As global demand gradually grows, it will absorb the spare capacity we see now, and international oil prices will again exceed the cost of supply of shale and other unconventional oil.

Therefore, in the short and medium term, natural gas supply will exceed demand. Asian natural gas spot prices may slide to below 7–8 $/MMBtu. This can be an opportunity for China to expand natural gas usage, provided there is timely liberalisation of imports. In the medium to long term, if there is to be increased use of natural gas this will require improvements in the efficiency of usage and a rationalised pricing system. Also, the costs of natural gas exploitation, liquefaction and transportation in Australia, the United States and East Africa are high, and, once delivered to Chinese ports, the price is expected to exceed 10 $/MMBtu.

1.1.4 China's Share of the Global Natural Gas Market Is Slowly Growing—China Must Become an Active Player in the International Natural Gas Market

In 2014, China's consumption of natural gas stood at 183 billion cubic metres—less than 5% of the world's total—but it is expected to rise quickly in the future. According to several different development strategies, by 2030 China's consumption of natural gas will rise to 450–570 billion m^3 (9–12% of total global consumption) and account for 15–25% of the global increase in natural gas demand.

The increase in China's natural gas consumption is not expected to lead to a marked increase in global prices. First, China has abundant resources of natural gas, and domestic production can cover part of the increase in consumption. Second, global natural gas resources are plentiful. If China sends out a clear demand signal to global markets, this can stimulate upstream investment to expand future supply capacity, without having a significant impact on global gas price. Third, ending the national reliance on coal will cause Chinese coal prices to fall, and, as China accounts for over 50% of global coal consumption, this could have a material impact on international coal prices. Developing countries, more sensitive to economic costs, may increase their coal consumption and reduce their consumption of natural gas, thus triggering a global crowding-out effect and, to a certain extent, hedging against the rise in price caused by the increase in Chinese demand. Quantitative model analysis shows that active measures to increase the consumption of national gas would only lead to an approximately 4% increase in international gas prices.

China must stop being a natural gas consumer nation, passively accepting natural gas market prices, and should, using trade and investment, play an active role in the international natural gas market. Chinese enterprises could play this role via equity investment, joint development and provision of technical services. By implementing the 'One Belt, One Road' development strategy, China will have an opportunity to establish an in-depth partnership with major natural gas-producing countries such as Russia, Turkmenistan, Qatar and Iran. This is undoubtedly a great opportunity for China to connect to the global natural gas industry. Moreover, China can actively move forward the core natural gas construction projects, raising its status both in the domestic market and in the East Asian market. By 2030, the demand for natural gas from developing countries, such as India, is expected to enter a phase of explosive growth and China would be able to be proactive, yet operate from a position of security, in the face of global market competition.

1.2 China's Natural Gas Supply and Demand and Its Primary Challenges

1.2.1 The Chinese Economy's New Normal

The characteristics of China's new economic normality are as follows. First, the speed of economic growth has shifted from fast to medium. According to forecasts, GDP growth during the 13th Five-Year Plan period is expected to be in the region of 6.66%, followed by 5.56% between 2021 and 2025, and 4.64% in 2026–2030. It is hoped that in 2020 GDP per capita will reach $10,000 (using 2013 fixed prices and basing the calculation on an annual appreciation rate of yuan renminbi of 0.5%), rising to 20,000 in 2030. The second characteristic is the structural change in the Chinese economy. Currently the Chinese economy is in the late stage of industrialisation, which is expected to be completed by 2020. The industrial structure is expected to change—the proportion of the service industry in the economy is expected to rise to 53.1% in 2020 and 61.2% in 2030, while the contribution of secondary industry will shrink to 39.8% and 33.4%, respectively. In the industrial sector, the contribution of heavy chemical industry will decrease, with the demand for steel, cement and other heavy chemical industry peaking around the period of the 13th Five-Year Plan. As far as energy is concerned, China's demand will go back to medium growth from fast growth. Model predictions tell us that in 2020 the total energy consumption will reach 5 billion tons, a 3.4% average annual increase between 2010 and 2020, and in 2030 the energy consumption will increase further to 5.7 billion tons.

1.2.2 Growth Natural Gas Demand in China

Even though there has been a slowdown in economic growth and energy demand, the demand for natural gas will continue to increase for a period of time in the future. The first reason is the low contribution of natural gas to total energy consumption. In 2014 Chinese consumption of natural gas stood at 183 billion m^3, which, according to the latest statistics, accounted for 5.8% of total energy consumption that year. Globally, that proportion stands at 23.8%, with more than half of countries at 20–40%. During the 30-year period between 1982 and 2012, the United States, Japan, Germany and the United Kingdom saw the part played by natural gas in their energy consumption grow from 23, 9, 12 and 27%, respectively, to 40, 33, 29 and 45%. The same statistics for the emerging economies of Turkey, Malaysia and Egypt show an increase from 1, 2 and 9% to 41, 48 and 63% respectively, a huge increase for both the developed and the emerging economies.

The second reason is that the expansion of the services industry, growing urbanisation and the control of atmospheric pollution all spur on the continuing growth of natural gas consumption. The data for the developed countries shows that for each 1% increase in the service industry there is a corresponding 0.84% increase in the contribution of natural gas to the total energy consumption. One cause for such a close link between the growth of the service industry and the increase in natural gas consumption is that the demand for high-quality energy sources rises with economic development, and especially with the growth of the services industry. The second cause, and an important one, is the people's demand for clean air, a consequence of growing urbanisation, accompanies the growth of the services industry. The experience of developed countries shows a close relationship between natural gas consumption and fine particulate atmospheric pollution—as the role of natural gas in energy consumption rises, it is followed by a large drop in the airborne concentration of fine particles.

We have carried out a quantitative scenario analysis on the future demand for natural gas. The results of the analysis are as follows. Under the "business as usual" scenario, in 2015 the demand for natural gas approaches 200 billion m^3. In 2020 it will break the 300 billion m^3 barrier, and in 2030 it will exceed

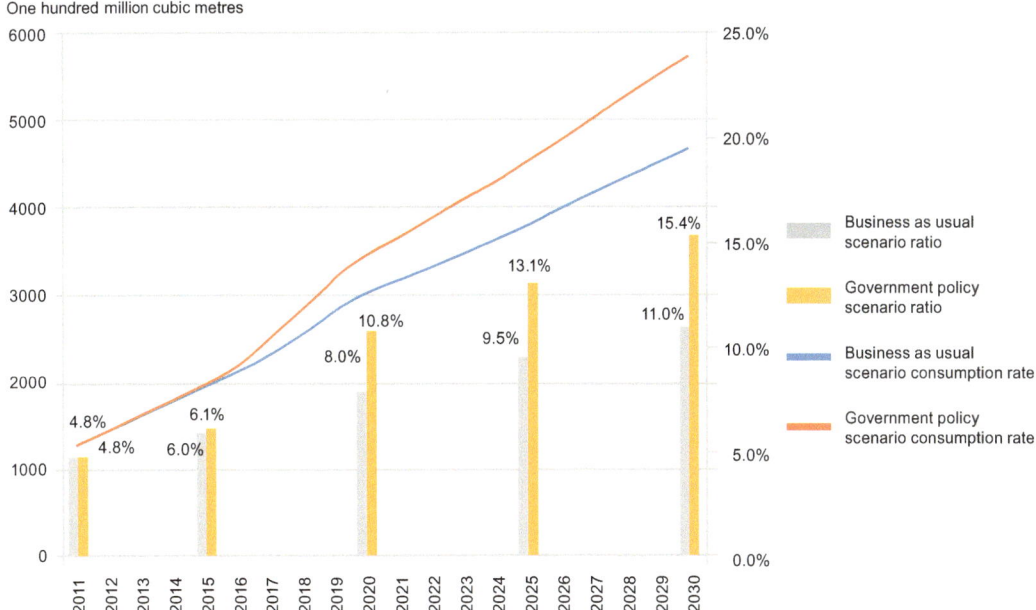

One hundred million cubic metres

Fig. 1.3 Natural gas demand volume and its role in energy consumption under the different scenarios

450 billion m³. The fastest growth is expected to occur in residential usage: gas, heating, transport and electricity generation. Despite the predicted fast growth in natural gas consumption under the "business as usual" scenario, the targets for natural gas industry development stipulated in the 12th Five-Year Plan and the Medium- to Long-Term Energy Resources Development Strategy will not be met. This clearly illustrates that, in order to meet these targets, more effective policies must be adopted to further boost the growth of the demand for natural gas.

Using model testing we have concluded that the most effective economic measures are the strengthening of environmental regulation and the setting of carbon prices by introducing carbon trading and carbon tax. Under the conditions of stronger environmental regulation and carbon pricing, the Chinese demand for natural gas may indeed reach the 12th Five-Year Plan targets of consumption of 350 billion m³ by 2020 and 580 billion by 2030. These policies will lead to the proportions of natural gas in national energy consumption of 10.8 and 15.4% respectively, a large increase on the 2014 figure of 5.8% (Fig. 1.3).

Under the policy scenario, the areas where the increase in natural gas consumption is expected to be the fastest are those sensitive to changes in the coal prices and in the environmental costs. These include gas heating, gas electricity generation and transport. Compared with the "business as usual" scenario, in 2030 the consumption of natural gas for electricity generation will rise to 45 billion m³. Gas heating is another area with great demand potential, which is also very sensitive to cost changes, and under the policy scenario, the consumption of natural gas for heating is expected to rise to 30 billion m³. For the other areas, such as the residential use of gas, the changes in demand are relatively small (Table 1.1).

1.2.3 Potential to Expand Domestic Natural Gas Supply and Import Capacity in China

The statistics tell us that China is rich in coal, lacks oil and is poor in natural gas—but in fact China has rather abundant natural gas resources.

Table 1.1 Natural gas consumption under the different scenarios

	Business as usual scenario				Policy scenario			
	2015	2020	2025	2030	2015	2020	2025	2030
Extractive industry	151	169	138	115	150	188	161	141
Petroleum processing and coking industry	143	216	286	356	144	228	304	384
Non-metal mineral industry	132	179	179	180	137	236	251	265
Metallurgical industry	79	101	97	98	82	138	140	148
Natural gas power generation	249	499	766	1120	257	673	1057	1574
Natural gas heating	57	252	415	589	59	349	601	887
Transport	269	435	623	787	268	446	652	842
Chemical industry	316	402	420	456	314	396	415	454
Other industries	168	177	205	233	176	246	287	326
Residential	405	576	683	742	397	548	662	734
Total	1968	3005	3811	4678	1985	3447	4531	5756

According to Ministry of Land and Resources oil and gas data for 2013, China's conventional gas resources stand at 68 trillion m^3, of which 40 trillion m^3 are recoverable. China also possesses extensive unconventional gas resources, comparable in size with its conventional resources—25 trillion m^3 of recoverable shale natural gas and 11 trillion m^3 of recoverable coalbed methane. As for newly discovered proven resources, in 1986–1990 less than 40 billion m^3 of natural gas deposits were confirmed, in 1991–1995 this figure rose to almost 100–150 billion m^3 in 1996–2000, to 300 billion m^3 in 2001–2005, to 320 billion m^3 in 2006–2010, and in the three-year period between 2011 and 2013 the proven geological reserves exceeded 600 billion m^3, reaching the 1.1 trillion m^3 mark in 2014. The volume of the recoverable newly confirmed reserves exceeds 500 billion m^3, and, based on the data from the previous years, we can expect proven resources to continue to grow in the future. China's production of conventional natural gas billion cubic metres in 2014 was 128 billion m^3 with, in addition, 2.6 billion m^3 of coal seam gas and 1.3 billion m^3 of shale natural gas.

China might possess abundant natural gas resources, but even so, natural gas extraction in China is very difficult and the costs high. One third of all the recoverable natural gas in China is tight natural gas, deemed to be an unconventional resource outside China, and its extraction is more difficult and expensive than that of the common natural gas. Many of the conventional gas deposits in China lie very deep, are rich in sulphur and have high extraction costs. China has very rich shale natural gas resources, but, apart from few locations such as Fuelling in Chongqing, most of the deposits of shale natural gas are situated at great depth with difficult surface conditions, and the exploration and exploitation costs are thus higher than in the United States. Innovation, both in technology and in organisation, is needed to exploit the full potential of these resources.

Analysis has shown that, with things organised as they currently are, by 2020 conventional gas output may reach 180 billion m^3, shale natural gas 40 billion m^3 and coal seam gas 10 billion m^3, bringing total natural gas output to 230 billion m^3. In 2030, conventional natural gas output will rise to 260 billion m^3. The total output of shale natural gas is expected to rise to 80 billion m^3, as the output from Southern Sichuan and East Sichuan's Longmaxi shale formations is expected to increase, and output breakthroughs are expected at other shale natural gas fields. Rough estimates predict that the output of coal seam gas will reach 40 billion m^3, with coal seam gas contributing 20 billion m^3, and coal bed tight natural gas and shale natural gas another 20 billion m^3. The total gas output in China will thus reach 380 billion m^3. If an organisational reform is carried out, more players are attracted to open up and exploit the resource, and further technological breakthroughs are achieved, then in 2020 the output of shale natural gas could reach 60 billion m^3, coal seam gas 15 billion m^3, and coal bed tight gas and shale natural gas 15 billion m^3, bringing the total to 270 billion m^3. By 2030, conventional gas output could rise to 280 billion m^3. An output of 150 billion m^3 of shale natural gas can be achieved by 2030 if the new system used in Southern and East Sichuan production proves to be effective, and is used successfully at other shale formations. Rough estimates predict the coal seam gas output to reach 40 billion m^3, thus pushing the total natural gas output to 470 billion m^3. Moreover, the coal gas industry can also grow, and may reach production capacity of 30 billion m^3 by 2020 and 50 billion by 2030. The methane hydrate industry is expected to remain at the research and development and experimental and demonstration stages in 2030, as large-scale commercialisation of this energy source remains very difficult.

Turning to consider imports from overseas, China's import capacity is undergoing rapid growth. Central Asia–China Gas Pipeline Lines A, B and C are already operational, with total capacity of 55 billion m^3/year. The capacity of Line D of the same pipeline is 30 billion m^3/year and this is estimated to become operational in 2016. The Myanmar–China Pipeline became operational in 2013 and has a capacity of 12 billion m^3/year. The contracts have been signed for the construction of China–Russia East Route pipeline, which is estimated to start operations around 2018, and have capacity of up to 38 billion m^3/year. The Memorandum of Understanding was signed in November 2014 for the western China–Russia pipeline route, planned to start operations before the year 2020, and its capacity is expected to reach 30–60 billion m^3/year. In total, by 2020 China's import capacity via pipelines will reach approximately

100 billion m^3/year, rising to 135–165 billion m^3 by 2030.

As far as LNG imports are concerned, according to the already signed contracts, including sales and purchase contracts, letters of intent and memorandums of understanding, the import volume will rise from the current 25 billion m^3 to approximately 57 billion in 2020. The Australian contribution to LNG imports will increase greatly, from the current 5 billion m^3 to approximately 25 billion in 2015. Other major contributors to this import growth are Canada (8 billion m^3), Russia (4 billion m^3), Papua New Guinea (3 billion m^3) and other third parties (9 billion m^3). Regarding LNG terminal construction, the capacity of China's LNG terminals may reach 85–225 billion m^3/year, greatly surpassing the expected future volume of LNG imports. Based on these facts, and considering the long-term LNG and pipeline contracts that China has already signed, in the coming few years China's capacity for natural gas imports is expected to grow quickly, from 53 billion m^3 in 2013 to over 160 billion m^3 in 2020 and is expected to reach 210–240 billion m^3 in 2030.

If both domestic supply and imports are added up, the estimation is that the total supply capacity will reach 425–495 billion m^3 by 2020, rising to 640–800 billion m^3 in 2030. It is important to make it clear that these capacities can only be reached if breakthroughs are made in the development of the domestic natural gas, and especially non-conventional gas, resources, and if the international contracts are fulfilled. The annual capacity and import capacity depend on growing demand, opening up of domestic resources, progress made in the construction of international projects, and changes in the domestic and the international supply prices.

1.2.4 Key Problems Facing the Development of the Natural Gas Industry

The past 10 years saw a rapid growth of natural gas demand, with demand exceeding supply, but in the coming five to ten years the situation is expected to be different. As domestic output grows and import capacity increases, the supply and demand for natural gas will have reached a broad general equilibrium. According to model analysis, even under the scenario of strengthened environmental regulation and levying of carbon tax, the demand for natural gas will reach approximately 350 billion m^3 in 2020 and 580 billion in 2030, but the supply capacity will exceed 400 and 600 billion m^3 in the same years. Supply is no longer the key factor in inhibiting the growth of natural gas industry, as the problems facing the industry have changed greatly. The increase in natural gas consumption that we saw over the past few years has slowed down greatly, and the current problems that have gradually emerged are low efficiency, high prices, problematic transportation and organisational inflexibility. The key problems at all the links in the industrial chain—resource extraction, transportation, sale and usage—must be systematically identified understood and addressed.

First, the environmental benefits and the economic value of natural gas must be realised in order to boost its consumption. Natural gas is an ideal substitute for other fuels—it is a clean, environmentally friendly, efficient and easily regulated energy source. However, the problem when comparing it with coal is its persistently higher price. Burning of natural gas does not produce SO_2 and soot, but this environmental benefit is not factored into the cost equation. When used in electricity generation, natural gas has a great advantage—its capacity for regulation and response—but its potential to be used for peak shaving has not yet been exploited either. The thermal efficiency of natural gas when used for CCHP generation is twice that of coal, but we are having difficulties realising this potential to save energy and reduce greenhouse gas emissions. Failure to externalise the economic and environmental benefits of using natural gas will make the maintenance of the previously fast consumption growth problematic. Now that total Chinese resource consumption has entered a medium growth period of 3–4% per annum, an increase in the natural gas consumption cannot

come solely from the demand for new energy sources, but must also, to a great extent, come from coal and oil substitution. Currently, the surplus production of coal in China is extremely severe, with prices at their historical minimum, and, if we cannot exploit the environmental and economic benefits of natural gas, it will have difficulties competing with coal and the slow-down of its consumption growth will continue.

Second, we can, using a rational price, achieve a sufficiently stable supply. According to our modelling, compared to current levels, by the years 2020 and 2030 China's consumption of natural gas will double and treble respectively. Failure of domestic production to keep up will lead to gas imports quickly rising to over 50%, and supply security issues will then impede the sustainable development of China's natural gas industry. In addition to the protection of natural resources, price levels and market stability are also critical. Currently, the upstream exploration and exploitation of national resources are not competitive with imports. This results in irrational resource allocation and rapidly rising upstream natural gas costs. The price in the contracts signed recently is relatively high. There is an objective reason for this—currently demand is higher than supply on the international natural gas market—but another reason is the misplaced belief by the three largest players in the oil industry that their market superiority allows them to pass the high price downstream. Allowing the upstream costs of natural gas to grow further may result in exceeding the potential for end user switching and thus running the risk of making natural gas unsellable at this high price.

Third, we must achieve safe, fair and transparent interoperability of natural gas transmission and distribution. In recent years China's natural gas pipeline network has grown greatly and the construction of the unified pipeline network framework is already at preliminary stage. The length of the network stands at 80,000 km, three times the figure for 2004. However, taking the medium- to long-term view of natural gas consumption and demand, the transmission and distribution capacity of the network are far from sufficient, and we need to mobilise and apply the capacity of all forces for its development, investment and construction. Looking at the market structure, there is a lack of overall planning, pipeline transmission and sales services are bundled together and there is a severe problem of market monopolisation. There are also a whole series of problems, from insufficient branch lines and peak storage facilities to pipeline network operational safety.

Fourth, we must standardise markets and pricing. Pricing mechanisms are key if we want to co-ordinate the benefits which extend down, up and mid-stream on the industrial chain, and they will have direct relationship with healthy sustainable economic development in China. There is currently a great deal of government intervention in price setting in China, and with the pricing management arrangements as they stand, the government is over-regulating in some areas, while other areas are left completely disregarded. Currently, in pricing, the net present value method is being gradually replaced by the "cost plus" method, but the situation still exists when only the upstream urban gate price can be set, without a way to solve the problem of disconnecting oil and gas prices. The midstream pipeline transmission services are bundled with terminal sale services, there is a lack of standardised fees for pipeline transmission and, due to lack of supervision, provincial and metropolitan networks artificially inflate their prices. The downstream sales market is monopolised by local gas companies, and also there is a cross-subsidy issue between industrial clients and private residential users. The inverted situation has arisen, where the price for industrial clients is higher than that for private individuals, which absurdly contradicts the international principle of large customers enjoying lower prices. This is also detrimental to coal substitution and environmental protection.

Fifth, a modern natural gas industrial system and government regulatory mechanism must be formed. A reconstruction of Chinese natural gas industry setup and market management system must be carried out in order to support and benefit the growth of Chinese natural gas market. Looking at the market situation, it is clear that the market is currently dominated by single large players—state-owned enterprises. No matter where you look, at the ownership of resources or the pipeline construction and operation, the market share of the three large oil companies exceeds 90%, with the share of PetroChina alone at over 70%. There is currently virtually no market competition in the industry and there is no sign of rising efficiency. It will, therefore, be extremely difficult to sustain the predicted future output of 580 billion m^3. In the past, the government relied primarily on administrative command when it came to managing the natural gas industry, which is not suited to a diversified, competitive and flexible industrial system. We must gradually explore the transition from the "full control" to the "standardised regulation" method, set clear regulation content and methodology, and assess its cost and feasibility.

1.3 China's Natural Gas Industry Development Strategy for the Next 15 Years

1.3.1 National Strategy and Policy Are Crucial for Natural Gas Industry Development in China

We have developed two scenarios for the future development of China's natural gas industry. According to the "business as usual" scenario, influenced by expensive natural gas, cheap coal and oil and weak environmental policies, the growth of the demand for natural gas will slow down, and natural gas will remain a niche energy source. In the event of there being no breakthroughs in non-conventional gas sectors such as shale and coal seam gas, growth in supply will also slow. With natural gas being currently widely available on the international market, imports into China will rise rapidly, but volatility in international markets will start raising problems of supply security. Without extensive restructuring of natural gas production and management systems, the industry will lack dynamism, have low operational efficiency and, as the price for both imported and domestically produced gas remains relatively high, growth in the natural gas market will be restricted, and the entire development of the natural gas industry in China will enter a bottleneck.

The other scenario is a "transformation" or "revolution" scenario. On the demand side, the relevant policies must be formulated and implemented, levying environmental and carbon taxes to reflect external environmental costs. Concurrently, stronger competition will drive the cost markedly down, leading to rapid expansion of the market, with natural gas gradually becoming the main energy source. For supply, as new players are attracted to the market, the output of shale and coal seam gas will rise greatly, with most of the supply becoming domestic, and the domestic market, via the influence of the international market, will undergo optimisation and adjustment. The new players on the market will bring with them a greater degree of competition, markedly driving down supply costs, and pushing development into an optimal cycle. A multifaceted, highly competitive market will then emerge, together with a modern, rationally structured industrial system. The industrial system will be adapted and transformed—the role of the government, which previously gave priority to planning and direct instructions, will give way to regulation fundamentally by the market itself. A modern regulatory system will emerge—the role of the government will be reduced to ensuring environmental protection, production safety and fair competition in the market.

It must be emphasised, however, that the two scenarios described above do not represent the only possible outcomes. The natural gas industry is still at the initial stages of its development, and its future development is, therefore, very flexible. China is endowed with comparatively abundant gas resources, has a great development potential, and innovation—and especially structural reform—will allow China ample operational space to expand its gas consumption and use global natural gas resources. Therefore, if natural gas is to become a principal energy source, the crux of the matter is the strategy and policy choices that China makes.

Different development strategies do not only have a deep effect on the future development of China's natural gas industry and energy resource structure. Robustly developing the natural gas sector would also be an important method to increase effective investment and promote economic growth.

Robust Development of the Natural Gas Industry Stimulates Economic Growth and Plays a Role in Protecting the Environment

There is great opportunity for natural gas industry growth in China. Currently, annual consumption of natural gas in China does not exceed 200 billion m^3, possibly rising to 350 billion m^3 in 2020 and rising further to 580 billion m^3 by 2030. The end market alone is expected to double. Taking into account the potential for growth in the upstream, midstream and downstream segments of the natural gas industry, the investment demand is very considerable. It is estimated that in the period between 2015 and 2060, cumulative total investment to the value of almost CNY 6 trillion will be created. Equivalent to CNY 400 billion a year, this can drive up the annual GDP growth by 0.6%.

Regarding the upstream production of gas, China's production potential may reach 470 billion m^3 in 2020, a figure which includes 280 billion m^3 of conventional gas and 190 billion of shale and coal seam gas. To build these capacities will require cumulative investment of CNY 1.6 trillion. Just one shale natural gas field project in Sichuan basin with annual capacity of over 100 billion m^3 requires an investment of 0.8 to CNY 1 billion.

As regards midstream pipeline investment, China needs to expand its pipeline network to over 150,000 km by 2030, and, based on the investment budget of CNY 10–15 million per km of pipeline, this could result in a total investment of approximately CNY 2 trillion. The total investment required for the construction of natural gas peak and reserve facilities and LNG terminals exceeds CNY 500 billion.

Regarding urban gas pipeline networks, here the total investment must be raised to CNY 1.2–1.6 trillion, based on previous calculations of CNY 3–4 invested for each additional cubic metre of gas.

With regard to sources of investment, the Chinese natural gas industry is still at an early, fast-growth stage, and therefore investment risks are low and returns are stable, making it very attractive to many different types of capital. In recent years, after the launching of coastal LNG terminals, investment of over 50 billion followed within a short period of time, a very positive market response. Even without government funds, the deregulation of relevant areas brings a great deal of social capital.

There are also clear environmental and social benefits to actively developing the natural gas industry. Estimates suggest that by 2030 natural gas will replace the use of 300–400 million tons of coal, and rough calculations show that this will lead to a 10% reduction in $PM_{2.5}$ concentration, which will be of great benefit to the health and the quality of life of China's citizens.

1.3.2 Strategic Objectives of High Efficiency, Safety and Sustainability

Development of the natural gas industry must be carried out with a fundamental emphasis on high efficiency, safety and sustainability. The specific implications are:

- Achieving efficiency means working to raise efficiency at all stages of the process—exploration and production, transport and utilisation. Efficiency is the foundation of the development of natural gas industry in China.

China is endowed with gas resources, but their quality cannot compare with that in the Middle East, Central Asia and Russia, and the conditions are unfavourable when compared with those in the United States. To make the extraction of natural gas economically viable, both extraction and transmission cost-effectiveness must improve. Moreover, the end-user price for natural gas in China is kept high due to cheap alternative energy sources and the long distances involved in gas imports. To make gas competitive, therefore, it must be used with greater efficiency so that its qualities of being an efficient and clean fuel can be fully realised.

- Safety, both natural gas resource security and production safety, is a prerequisite for development. The importance of energy security is self-evident as far as leading nations are concerned; the energy security of a leading nation starts within its borders, and is based on robust regulation. It is on this basis that resource security can be achieved under the open economic conditions of global resource allocation. In order to strengthen resource and energy security, domestic supply must be boosted, and gas import sources and import mechanisms diversified. At the production and operation level, the peak and trough differences in the demand for natural gas are very large, and we must strengthen the national natural gas network and boost national gas storage reserves in order to ensure the security of supply of natural gas. Moreover, the safety of gas usage remains important, as it is used by a great number of households nationwide.

- Sustainability is the relationship between the effective management of the development of natural gas industry and the environment and ecology, and is a goal of natural gas industry development. First, in order to improve the environment rather than damage it, the natural gas industry must emphasise environmental protection at production, transmission and usage stages, and also pay attention to the sustainability of resource supply. Second, the concept of sustainable development must be used throughout the natural resources industry. In order to protect the environment, the external cost of pollution must be fully reflected, and enlarging the usage of natural gas will raise the efficiency of natural resource industry, lowering the environmental pollution caused by the utilisation of natural resources.

1.3.3 Clearly Defined Channels and Targets of the Natural Gas Industry

China's target for the next 15 years is to vigorously develop the natural gas industry. Efforts must be made for natural gas to reach 10% of all primary energy source consumption by 2020, increasing consumption to 350 billion m^3, and more than 15% and 580 billion m^3 respectively by 2030. The main channels to achieve this are outlined below.

First, end consumer markets must be developed. In order to turn natural gas into a principal energy source, multiple measures must be applied simultaneously to enlarge natural gas consumption:

- Environmental regulation must be strengthened, and there must be an emphasis on using natural gas a substitute for coal, including in rural/urban residential heating and fuel usage in light industries such as textiles and paper mills. This will be critical in reducing smog and improving the environment.
- Natural gas transport must be developed. The emphasis here will fall onto road passenger transport and waterway freight transport. The promotion of gas as a replacement for oil must be achieved by strengthening infrastructure and perfecting the relevant technological standards.
- In electricity generation and heavy industry, relying on competitive markets for energy resources, green finance and taxation reform must ensure that the environmental and

economic benefits of using natural gas (such as energy saving and peak storage capacity) are valued and realised. We must raise the economic benefits generated by, and thus the competitive power of, natural gas in peak electricity reserves, trigeneration and the natural gas chemical industry.

Second, the upstream natural gas supply must be diversified. Using a two-pronged approach involving increasing domestic production and using foreign trade, relying on the dynamic complementarity and boosting the competition between the two can raise the efficiency of resource allocation, and achieve market liberalisation and diversification of the natural gas supply. Domestic production of natural gas resources must be strengthened, and the following three-step strategy regarding conventional gas, tight natural gas and shale coal seam bed gas promoted:

- In the near term, growth must be centred on the technological reserves for the production of conventional gas and tight natural gas, shale and coal seam gas.
- The medium-term aim is to simultaneously advance conventional gas, tight and shale natural gas industry. China must also actively participate in foreign markets, and, using commercial principles, promote natural gas trade and investment, and strengthen control and bargaining power capacity for imports coming into China.
- Sustainable development of trade with Russia, the Middle East, South East Asia (Myanmar) and the maritime trade must be promoted, partnership with the Shanghai Co-operation Organization, ASEAN and Canada must be enlarged further, and import methods and pricing mechanisms must be gradually diversified. China must combine its "One Belt, One Road" strategy with the active implementation of the "Go Out" policy, and play an active role in the international markets by means of long-term agreements, equity investment, joint development, pipeline network construction and ocean shipping.

Third, a gas pipeline and storage network must be constructed consistent with the principles of safety, fairness and transparency.

- The system must have unified planning, be interconnected and integrate the national gas producers, imports and consumption. The top-level design of the pipeline network must be carried out using a scientific approach, ensuring the interconnectivity of different gas-producing areas and markets in various locations.
- There must be a focus on diversifying investment in the construction of the pipeline network, encouraging the entry of non-state-owned enterprises and capital. The security risk issues can be avoided by demarcating pipeline hierarchy and differentiating the investment accordingly. Long-distance pipeline must be constructed in combination with national-level planning; construction of provincial-level trunk lines must be performed under the guidance of local government, using both state-owned enterprises and social capital. The city- and regional-level pipeline construction market must be gradually liberalised.
- Access to the pipeline network must be fair. National pipeline access standards must be unified. In order for gas from different locations to have access to the pipeline network and reach the end customers, the only requirement must be that the correct safety standards are complied with.
- The reserve capacity must be strengthened, and the obligation to maintain reserve capacity must be backed up by legislation. The gas reserve price must be fixed, and all parties must be mobilised to boost national peak regulation reserves and the emergency capacity.

Fourth, innovation must be applied to raise the technological levels in the industry. Vigorous efforts must be made to promote technological innovation at each segment of the industrial chain, from exploration and production, transmission and distribution to conversion and utilisation, allowing the technological revolution to achieve technological and economic efficiency at every stage.

- The strategic layout of natural gas production and conversion key technology will be launched in advance: exploitation of shale, deep sea, coal seam bed and methane hydrate reserves launched, key problems in areas such as gas engines and combined cycle electricity generation facilities tackled, both human talent and financial investment boosted, and technological innovation in institutional mechanisms activated.
- Research and development and application of complete technological sets and solutions must be explored in depth, focusing on the integration of the equipment systems, reduction of costs, environmental protection and green development. Efforts must be made to ensure that, by 2020, China achieves a clear increase in the technological level of the equipment, that great strides forward are achieved in technical autonomy and domesticising of shale and sea gas prospecting and exploitation technology, and that the efficiency of the industry's functioning, production and operation is raised. By 2030, the industry's technological level and production and operation efficiency will achieve leading international standards, and domestic technological innovation capacity be raised greatly.

Fifth, natural gas price reform must be vigorously promoted. As gas markets and infrastructure in different locations mature, liberalisation of gas price setting must be pushed on in waves, gradually reducing the direct intervention by the government, actively perfecting the pricing mechanism and market environment, and finally realising a complete liberalisation of natural gas prices. Full advantage must be taken of the multiple import channels and mature infrastructure in South-Eastern coastal locations such as Shanghai, Guangdong and Zhejiang.

- Take the initiative in upstream market diversification, abolishing price setting at gas terminals, and promoting independent pricing at upstream markets.
- Building on this, the separation of the midstream pipeline transmission and downstream sales must be pushed on, third-party access to pipeline network transmission allowed, direct access provided to large customers and liberalisation of the end market gradually promoted. In central and western China, where infrastructure is not yet mature, an upper cap must be placed on gate price and transmission and distribution, and the end-market pricing mechanism must be further improved, with price adjustment frequency increased. Prices for transmission and distribution and storage must be formulated rationally, and measures implemented to improve the situation as regards the seasonal price gap, the peak and trough price gap and the end price. By 2020, the situation will be ripe for gradually pushing for cross-regional price merger using methods such as third-party access to the pipeline network, and finally achieving liberalisation of the prices of natural gas.

Sixth, a natural gas industry system and a government regulatory system must be built. In order to have a healthy, sustainable modern natural gas industry, we need to allow the market to play a decisive role in resource allocation, and at the same time take advantage of the role of the government. When designing the system, we must:

- completely understand the crucial junctions in the reform of the natural resources system;
- promote the liberalisation of the natural gas industry;
- design the organisation and modes of operation in upstream, midstream and downstream segments of the production chain;
- diversify resources and subjects;
- create standardised, legitimate competition and co-operation; and
- apply the utmost effort to building an efficient, open, orderly market.

Regarding government regulation, it is necessary to carry out, as soon as possible, top-level design and division of responsibility in natural gas market regulation; regulation must place an emphasis on environmental protection, fair access to the pipeline network and low-stream separation of pipeline network transmission and sales. Also, unified pipeline transmission price control measures must be formulated. The aim is to achieve, by 2020, multifaceted competition in Chinese natural gas market, breakthroughs in exploration and production and imports, price liberalisation in some regions, an efficient government regulation system, and industry growth that shows greater vitality and efficiency. By 2030, the aim is to have a system of robust, fair markets with multiple players, where the market itself plays the decisive role in price setting. Also, by the same year, a unified, independent, professional regulatory system should be in place. The dynamic, efficient new industry will enjoy much greater influence internationally, and its output will be internationally competitive.

1.4 Raising Efficiency and Supporting Improved Policies to Expand Natural Gas Use

In order for natural gas to become a main energy source, natural gas use must be expanded, and raising its competitiveness is the key. Natural gas price sensitivity for residential use, industrial services, vehicular transport and hydrogen production is low, while that of central heating, electricity generation, usage for industrial fuel, production of synthetic ammonia and methanol is higher, and under current pricing conditions natural gas lacks competitiveness. One way of raising this competitiveness is to internalise the external costs, reflecting the impact that resource utilisation has on the environment and people's health, and this will enable us to express the advantage of natural gas as a clean resource. Another way is to raise the efficiency of natural gas utilisation, and this higher efficiency will prevent the end cost of using the expensive

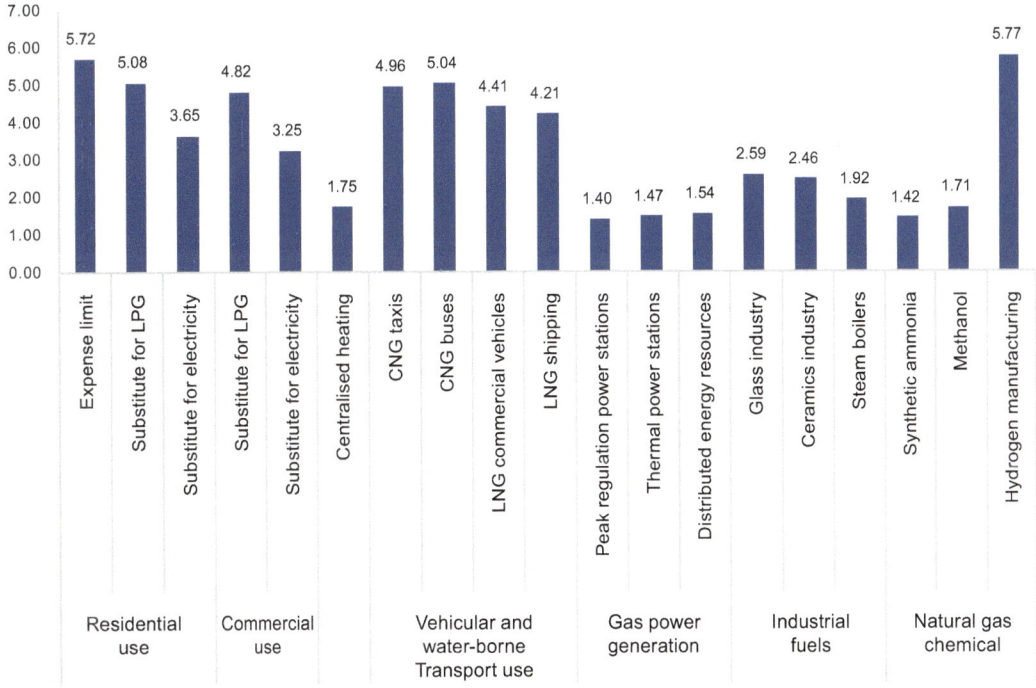

Fig. 1.4 Natural gas price adaptability of different user categories (in CNY/m³). *Note* The user adaptability described above is based on the international crude prices at $80 per barrel, and TCE at CNY 600 per tonne

natural gas from increasing markedly. Specific areas and policies are as follows (Fig. 1.4).

1.4.1 Strengthen Environmental Regulation, Improve Resource Utilisation Efficiency, Substitute Natural Gas for Coal

Substituting natural gas for the coal used in decentralised energy generation is one of the priorities in the development of the natural gas industry, and there is a great potential for development in this area. Decentralised use of coal is inefficient and highly polluting, leading to high pollution control costs. Currently approximately half of the coal in China is used for centralised electricity generation, and the other half is used in a decentralised manner. Half of the latter is accounted for by areas such as metallurgy and cement firing, and the rest is made up of public heating, textiles, ceramics, glass and

paper production. In all these areas there is potential to substitute gas for coal. Looking at the experience of developed countries, the proportion of coal used for electricity generation stands at 78% in OECD countries, and the same figure exceeds 90% in the United States. There is also very little use of coal for decentralised electricity generation.

There are three main channels for coal to gas substitution:

- Catering for heating users in small and medium-sized northern cities and some industry clients located in the industrial zone, coal-fired boilers can be replaced with coal-fired CHP plants. This is an economy-based, environmentally friendly substitution route.
- The same approach (replacement of coal burning with CHP plants in public heating) is difficult to implement for large and very large northern cities due to limited environmental capacity, the existing limitations on the land

available for power plants and problems with coal transportation. Moreover, it is very difficult, using coal-powered CHP plants, to satisfy the requirements of industrial clients who require extremely high temperatures and those users whose locations are widely dispersed. Their energy requirements can only be satisfied by using natural gas and expanding natural-gas fired CHP.

- For the urban fringe and in the countryside, where coal is used in decentralised manner for heating and cooking purposes, apart from substituting LPG and LNG for coal, there is also an option of using biogas produced from the organic waste generated in the villages. The expectation is that, using the above methods, by 2030 the use of gas will rise in the areas where it is most used—to 160 billion m^3 among the industrial users, and to approximately 120 billion in urban heating, and residential and commercial use.

Regarding policy, one option is to strengthen environmental regulations, increasing the cost of coal when used in decentralised energy generation. For example, an online monitoring system for reduction of atmospheric pollution produced by coal-burning boilers could be set up, as well as emissions monitoring systems. Second, energy-saving reform could be strengthened in order to improve energy resource usage efficiency, for example boosting the construction of energy-saving buildings, and optimising industrial processes and their methods of using energy. For high-quality, expensive energy resources, the economic burden must not be passed down to the end user, and the government can provide financial support for the replacement of the energy facilities and to aid with energy-saving reform. Third, to address the addition of new coal-burning boilers in areas such as northern and central China where atmospheric pollution management comes under the greatest pressure, enforcing compulsory usage of natural gas can be considered, to avoid adding new boilers in the future.

1.4.2 Improve Planning and Technological Standards, and Promote the Development of Natural Gas Transportation

Expanding the use of natural gas in transport would be a highly significant move and have important implications for safeguarding national energy security. Natural gas is a clean and efficient fuel, but using it as a substitute fuel has several specific economic and technological implications. Pollution and carbon dioxide emissions from gas-powered vehicles are on average lower than those from traditional internal combustion vehicles. From the economic point of view, over the past few years, the large difference in price between gas and oil, as well as sales tax exemption, has made gas-powered vehicles economically viable and propelled fast growth in their use, especially as CNG urban taxi fleets and public buses.

There is enormous potential for the growth and development of gas-fuelled transport. In China the total gas volume used by vehicles reached approximately 12 billion m^3 in 2013, of which CNG vehicles contributed 10 billion m^3 and LNG vehicles 2 billion. The CNG market for vehicles is gradually entering a period of fast growth. It is expected that by 2020 annual consumption will reach 20 billion m^3, then 23 billion in 2025, and that this will be followed by a slowdown. The market for LNG vehicles is expected to maintain fast growth in the future; the forecast is that in 2020 the LNG market will exceed 20 billion m^3, with expectations of exceeding 40 billion by 2030. The use of LNG as ship fuel is another important potential market, but, due to the more rigorous environmental policies, growth here is expected to be slow initially, and changes in environmental policy will dictate the outcome of the later stages of development. According to optimistic calculations, the market for LNG as ship fuel may exceed 2 billion m^3 in 2020 and 10 billion in 2030.

However, there are barriers obstructing the development of the natural gas transport fuel industry. Unlike the fixed-source users in power generation, and residential and industrial users, transport users are mobile, and need an external network. While the change from coal to gas in electric power generation can be called a "point" substitution, the change from petrol or diesel to gas in transport is more of a "network" or "system" substitution. Large-scale development of the industry does not just depend on the technical and economic qualities of one segment of the industrial chain or a certain industry, but rather on the network effect of the entire system. Only if a sufficient number of service facilities, such as refuelling stations and repair shops, join the network can the customers' mobile, network-based demand be satisfied. The network can then reach a tipping point, and the old "chicken and egg" problem can be solved. This process of establishing a network will require a great deal of time if left to the market's "invisible hand". Markets can also malfunction from time to time, so the government can be extremely helpful here. These are the recommendations for promoting gas for petrol substitution:

- The construction of filling station infrastructure must be accelerated. The first step is to formulate a filling station network construction plan for China's important freight transport road routes, water transport routes and key areas. In order to benefit vehicle users, conflicts, fragmentation, duplicate construction and harmful competition must be avoided. Second, suitable tax reduction and exemption policies must be formulated to make investment in gas fuelling facilities at filling stations attractive, thus accelerating the availability of gas at filling stations until a tipping point is reached. Having reached the tipping point, a gradual withdrawal of the incentives may be considered. Moreover, any obsolete regulations must be identified and amended, for example any regulations based on the premise of using gasoline fuels must

be revised, the approval process for filling stations must be simplified, and the ban on LNG transport on the railways must be lifted.

- Industrial standards must be formulated as quickly as possible for the entire industrial chain, for example regarding fuel quality, measuring fuel calorific value, construction of filling stations and gas vehicles' component parts. Accelerating the standardisation of the entire industrial chain will allow economies of scale to be achieved.

- The government must take the lead, with the industry associations playing the role of co-ordinators. Full advantage must be taken of the government's leading role in procurement; funds for gas-fuelled vehicles must be allocated a clear proportion in the financial support and subsidies issued by the municipal administration, public transport and government departments for vehicle procurement. In addition, the co-ordinating role of the industry associations must be given free rein. Co-ordination is particularly necessary in building the natural gas transport industry network, and, since transportation has already been fully liberalised, the government must not interfere directly in the markets, and so the role of associations as co-ordinating intermediaries is vital.

1.4.3 Improve the Pricing Mechanism, Making the Development of Natural Gas Electrical Power Generation Justifiable

Using natural gas to generate electricity has many benefits compared with coal. It is cleaner, provides flexibility to respond to demand peaks and requires a much smaller land area. It is highly suitable for peak-time electricity generation and CCHP usage, and these qualities make natural gas indispensable in electricity generation. Looking at the international experience, in 2010 the global contribution of natural gas to electric power generation stood at 30% of all fossil fuels, and in 2030 it is estimated to rise to 37%. In developed countries, electricity generation is a major portion of natural gas usage, accounting for 30–40% of total natural gas use.

China has abundant coal resources and coal-based electricity generation can be easily made clean and efficient. Coal is an excellent, competitive fuel when used for baseline electricity generation. For this reason, in China, the use of natural gas to generate electricity should centre on peak-time electricity generation and CHP generation as well as decentralised power-generating, heating and cooling. In the decentralised electricity generation sector, we must take into account the economies of scale in asset utilisation and assign priority in the order: region, building, household. The liberalisation of the electricity market, especially in electricity sales, provides an important opportunity to promote the development of the decentralised energy sources. It is estimated that by 2030 the volume of gas used in peak and baseline electricity generation will reach 60–70 billion m^3, whereas CHP generation and decentralised resource electricity generation will use approximately 100 billion m^3. The key measures to be implemented in order to boost the competitive power of electricity generation from natural gas are as follows:

- Peak and trough electricity and gas prices must be set. As regards electricity generation, gas is a flexible energy source for this purpose, and has high peak regulation and reserve values. Its value for peak generation is even more significant against the background of the increased use of renewable and intermittent energy sources. Looking at the natural gas part of the system, there is seasonal variation in the use of electricity and gas. For example, during winter a lot of gas is used for heating, but electricity is used relatively little, while in summer a great deal of electricity is used for refrigeration, but there is little heating being used. The use of natural gas in electricity generation, therefore, can reduce the difference between the peaks and the troughs in natural gas demand. Using cheap gas to produce expensive electricity is a competitive advantage of electricity generation using natural gas. Currently the grid tariffs are higher for electricity generated using gas than for electricity generated using coal, which to a certain extent reflects its peak regulation and reserve usage value. The next step may be to introduce time-of-use pricing to further express the benefits of using natural gas for peak shaving. Time-of-use pricing for natural gas should be formulated in such a way as to fully express the competitive power of gas-generated electricity.

- The external environmental cost must be reflected. The external cost of electricity generation can be separated into two categories. The first is the costs of normal pollution such as dust and sulphur and nitrogen oxides. Increasing investment can lead the emissions produced by coal-based electricity generation to approach the levels produced by gas-based electricity generation, as the new coal electricity generating units now basically have the same pollutant concentration limits as gas units (using 6% oxygen content as reference, the concentration of dust, sulphur dioxide and nitrogen oxide emissions does not exceed 10, 35 and 50 mg per m^3 respectively). Boosting investment in environmentally friendly facilities and equipment and factoring this into the budget increases the cost of electricity produced by large generator units by 0.01–0.02 CNY/kWh, and this figure is slightly higher for older, smaller generators which use inferior-quality coal. Overall, the measures to control general pollution do contribute to the gap in the cost of electricity production between gas and coal, but this effect is not large. The other aspect, a key one, is the cost of coal emissions. Using coal for generating electricity produces 0.8 kg/kWh of emissions, while using gas produces only 0.37 kg/kWh. As the price of coal increases, the corresponding costs of pollution from generating electricity using coal are also showing an upward tendency. According to an example from the Shenzhen Emissions Exchange, on June 18, 2013 the online opening price for one ton of coal stood at CNY 30 per tonne, and the cost of coal-produced electricity at 0.013 CNY/kWh. When the coal price peaked at 143.99 CNY/tonne on October 18, 2013, the cost of coal-produced electricity rose to 0.06 CNY/kWh.

- The core equipment and facilities must be produced domestically. Currently, core technology such as turbines and computerised turbine equipment is a monopoly of foreign companies, which creates high long-term costs and keeps transportation and long-term preventive maintenance fees high. Solving this problem will lower the fixed costs of gas-generated electricity. Lowering investment per unit by 15% can lower the electricity cost by approximately 0.014 CNY/kWh. At the same time, changing to the domestic production of key equipment can lower total lifecycle costs. If the total repair fees are lowered by 50%, then the corresponding reduction in electricity costs will be reduced by approximately 0.01 CNY/kWh. Changing to the domestic production of key equipment can lower the electricity cost to the approximately 0.024 CNY/kWh.

1.4.4   Adjust Gas Prices for Residential and Chemical Industry Use, and Reduce Cross-Subsidies for Different Users

There are large subsidies in place to provide low-cost natural gas to residential and chemical industry customers in the long term. For example, in 2013, the average price of gas for end urban residential users nationwide was 2.15 CNY/m^3, which was 1.5 CNY/m^3 lower than the national average marginal residential supply cost. The volume of gas used by

residential users is approximately 18 billion m^3, which means that the subsidy from the gas suppliers to urban residents reached approximately CNY 27 billion. For fertiliser producers, the gas price is now 1.80 CNY/m^3 on average, which is 1.10 CNY/m^3 lower than the national average marginal commercial supply cost. The volume of gas used by the fertiliser industry is approximately 20 billion m^3, implying that the subsidy by the gas suppliers for this industry's users is approximately CNY 22 billion. It appears on the surface that these large subsidies are provided by the gas suppliers, but, in actual fact, it is other users that bear the cost. The resulting higher price for these users is detrimental to the expansion of natural gas use in their sectors.

The price for the residential users must be gradually raised until it exceeds the cost for the gas suppliers. The first reason is that the volume of gas consumed by residential users (including for heating purposes) is expected to grow greatly, from 28.8 billion m^3 in 2012 to 54.8 billion in 2020 and 73.4 billion in 2030. Such increases will make the continuation of subsidies highly problematic. The second reason is that the price adaptability of residential users is high, and they can be regarded as top-quality downstream customers. Looking at the competition, and comparing with oil, LPG or electricity, as long as gas prices remain below 4 CNY/m^3, gas remains competitive. Regarding the adaptability of the consumer, the average amount of gas used by one person each year is approximately 60 cubic m, and a hypothetical price increase of 1 CNY/m^3 results in additional CNY 60 per year spent on gas. In 2013 the disposable income of the low-income stratum (the bottom 20%) of the rural/urban population stood at CNY 11,434, which means that the increased spending on gas amounts to a mere 0.5% of that figure. Moreover, the price of heating must also be adjusted, otherwise it will be difficult to continue with the use of natural gas for this purpose. Naturally, the use of gas for cooking and heating is absolutely essential in people's lives, and for this reason a subsidy can be given to the part of the population living in poverty when the prices are adjusted. Choosing the right opportunity and the right method to raise the prices for the residential users is crucial, as is the importance of communication with the public. Leaving the decision-making regarding the extent of the price adjustment range to the lower levels of authority, such as local government, can also be considered.

For chemical industry users, certain discounts may be given using market measures, such as setting peak and trough prices and interruptible prices, but the situation so far, with prices below cost, can no longer continue. The chemical industry must no longer rely on cheap gas, but has to optimise its industrial process and strengthen its production management to be able to fully take advantage of time of day prices and discounts for interruptible prices. In addition, it must continue with industrial upgrading and develop high value-added industries. Regarding policy, a clear timeline must be set for the price merger between residential users, the chemical industry and other users, to provide local government with ample room for adjustment.

1.5 Measures to Ensure a Safe and Efficient Gas Supply

At the same time as expanding the use of natural gas, we must also raise gas supply capacity, as well as ensuring the safety and efficiency of the gas supply. The key to ensuring a safe and efficient gas supply is attracting new players to the exploration and exploitation of gas resources, thus greatly boosting the national supply capacity. In addition, import sources and methods must be diversified further, to ensure the security of natural gas imports. The advance of pipeline network construction must be moderated, and the interconnectivity of the pipeline network boosted. Construction of gas reserve facilities and reform of the relevant institutional mechanisms must both be consolidated to ensure the smooth operation of natural gas industry. We must also raise the technological level of the equipment and facilities and of the innovation capacity of

the industry, strengthening the very foundation of its long-term development.

1.5.1 Attracting New Players to Resource Exploitation to Expand Natural Gas Production

China is richly endowed with gas resources and has the fundamental technology it needs to exploit them. Priority is given to conventional gas, tight natural gas, shale and coal seam gas, and to the appropriate development of coal gas industry. The key problems lie with the organisation of the industry. As mentioned above, under the current system, the output volume of natural gas in China will reach 230 billion m^3, possibly reaching 380 billion in 2030. If the system is reformed, the same figures may rise to 270 and 470 billion m^3 respectively. This sections outlines the measures that should be taken in order to attract new players to resource exploitation and thus expand national natural gas production volume.

1. Development model

A new model of development must be applied. Exploration and production and the pipeline transmission of natural gas are subject to the regulations of the State Council, and are operated by the state-owned oil companies. The advantage of this system is that it allows intensive development of the natural gas industry, and makes it convenient to undertake large-scale exploitation, but it also faces the problem of high costs. The other development route is to reform the current monopolised control system and strengthen the role of market forces, attracting businesses with different models of ownership to the exploration and production of natural gas, and achieving diversified development. The resulting market competition, "the survival of the fittest", will create a new structure of exploration and production in the natural gas industry. It is recommended that the current institutional mechanisms be adjusted appropriately, and conventional oil and gas operational authority be moderately liberalised. Before anything else, though, conventional oil and gas mining rights must be opened up to the state-owned energy enterprises which have already obtained shale natural gas mining rights, as well as to those energy enterprises which are engaged in overseas imports. This change will attract these enterprises to the domestic exploration and production of conventional resources.

2. Minimum investment level

The minimum exploration investment must be adjusted. The 1996 Chinese Mining Law and the follow-up supporting laws and regulations state that the minimum investment for oil and gas prospecting rights is 5000 CNY/km^2 in the first year, CNY 7000 in the second and CNY 10,000 in the third. However, it fails to consider other factors, such as currency inflation. Due to the costs having risen greatly and the fixed-price budget, these unchanged minimum prospecting and surveying investment regulations have led to a decrease, year after year, in the exploration workload per unit area. In the past, three exploration wells could be completed with a budget of CNY 100 million, but currently it is only possible to complete one. Physical exploration investment per unit area has is currently barely between one third and one fifth of its level 15 years ago and there is a severe lack of physical exploration work done in the regions favourable to gas production. According to the investment standards and regulatory requirements for shale natural gas block tenders, the basic exploration investment required for conventional oil and gas blocks is 3–5 times greater than the current levels, with a minimum investment of 30–50,000 CNY/km^2. Raising minimum exploration investment, raising the prospecting efficiency of the blocks, encouraging block withdrawal and reducing the surface area of the blocks will transform the current prospecting stalemate. At the same time, the current advanced

block application system must be reformed, and, as with shale natural gas block sales, it should be possible to sell conventional oil and gas blocks through a competitive process. In addition, mechanisms for block exit that are widely practised internationally can be used. Monitoring teams must also be set up to oversee the implementation of all the rules and regulations.

3. Trading mechanism for reserves

Trading mechanisms and platforms for proven natural gas reserves need to be established. In recent years, the average annual increase in proven geological natural gas reserves has exceeded 600 billion m^3, and this tendency to fast growth is expected to continue in the coming years. However, there is investment and opening up of the newly proven reserves is not promptly achieved, with parts of the reserves remaining unexploited for prolonged periods of time. The reason for this is the difficulty of making the exploitation of some of these reserves economically viable. The recommendation, therefore, is to establish a trading mechanism for the hard-to-exploit proven reserves, allowing them to enter the market, where they can be transferred to a different party, and can eventually come into possession of an enterprise capable of conducting more expeditious exploitation and making it economically viable. The exploration enterprises can thus recover their exploration costs and the producer enterprises, by purchasing these reserves, gain a production opportunity. To ensure that these reserves are developed promptly, the situation where rights transfer is not followed by development must be avoided.

Recommendations for Setting up a 100 billion m^3 Annual Capacity Shale Natural Gas Integrated Experimental Area in Sichuan Basin

The development of the shale natural gas industry touches upon many areas—the conditions of the natural resources, technology, equipment, construction, investment, utilisation and environmental costs, for example—and these must all be considered. We estimate that with appropriate organisation and policy support, Sichuan Basin can become a leader and focal point for shale natural gas development in China, and the potential production of shale natural gas in this region could be as high as 100 billion m^3 by 2025. The reasons for this are as follows:

1. **Guaranteed resource base**: According to PetroChina evaluation data, the total reserves of shale natural gas in the Longmaxi and Qiongzhusi formations approach 40 trillion m^3. These include Longmaxi marine facies shale natural gas, where production breakthroughs have already been made, with a total area of 75,000 km^2, containing reserves of 25 trillion m^3 of shale natural gas, with recoverable resources of 3.7 trillion m^3. Of these 75,000 km^2, 35,000 have been found to have conditions that are especially favourable to shale natural gas extraction, containing almost 14 trillion m^3 of shale natural gas, of which 2.8 billion m^3 is recoverable. In a scenario where the 20,000 km^2 area of Longmaxi shale coal formation is used, development is started in 2015, then the exploration and well distribution can be completed and 10,000 wells can be drilled by 2025, with the potential production of shale natural gas reaching 100 billion m^3 in 2025. Following that, drilling approximately 800 wells annually, production can remain stable for 20 years.

2. **Mining technology and equipment**: Currently China is at already in the initial stages of mastering shale natural gas geophysics, well drilling and completion, and fracturing technology. China currently possesses 3500 m capability of horizontal drilling and stage fracturing. Chinese non-seismic geophysical prediction and identification technology, fully mobile tracked drilling rigs, large fracturing vehicles and construction environmental protection technology all comply with advanced international standards, and there has also been a breakthrough in domestic bridge plug production. Currently the only fields where there is a gap between Chinese technology and advanced foreign technology are well drilling geological guidance, measurement while drilling (MWD), micro- and nano-structures and component analysis. However, there are already many foreign shale oil companies in China taking part in shale natural gas development, and these can supply the key technology required. The fact that Sinopec's Chongqing Fooling shale natural gas field has already entered the commercial exploitation phase proves that the development of Chinese shale oil industry has passed an important benchmark.

3. **Exploitation is productive and economic viability guaranteed**: By November 30, 2014 (when the research group was on-site), Sinopec, in Fooling, Sichuan, completed 69 fracturing test wells, all of which produced medium and high shale natural gas flow. The average flow at a single well was measured at 32,000 m^3 per day. On average, a single well, under conditions of uninterrupted flow, can produce over 21.6 million m^3 of gas, with the average cost of single well standing at CNY 80 million (including exploration, production and transmission costs). Adding taxes and other simple management fees, the well cost is below 1.5 CNY/m^3. Currently the wellhead price of shale natural gas stands at 2.78 CNY/m^3, and adding the 0.5 CNY/m^3 national financial subsidy gives us the actual gas price of 3.18 CNY/m^3, and a single well can,

therefore, dependably generate an annual income of up to CNY 68.69 million. Other areas in Sichuan, such as Changning and Weyuan, also have shale deposits, situated at great depth, but here shale natural gas fields are layered with conventional gas deposits and the economic value lies in their combined extraction, while shale natural gas production in Exi, in the east of Chongqing, is made highly economically viable due to the presence of associated light oil. These facts confirm the assured economic viability of shale natural gas production in Sichuan Basin.

4. **Existing environmental protection precedents to draw lessons from and guaranteed water resources required for gas extraction**: The United States currently has over 100,000 shale natural gas wells, and under effective government supervision and public scrutiny, there have so far been no environmental accidents which might have been detrimental to society. In Sichuan Basin, the extractable layers of shale natural gas are situated at the depths of over 2000 m, and measures such as adequate well casing and increased safeguards (adding an extra casing) can effectively prevent the water used for fracturing from permeating to the surface. As far as water use is concerned, the production of 100 billion m^3 of gas consumes approximately 400 million m^3 of water, which amounts to 1.6% of the current water consumption in Sichuan province. Therefore, Sichuan Province, richly endowed with water resources, can guarantee sufficient water reserves for the industry.

In order to meet the 100 billion m^3 annual mark, CNY 800 billion of investment is required to build the capacity (to cover the drilling and completion of approximately 10,000 wells between 2015 and 2025). Under the current system the project would rely solely on PetroChina Southwest Oil and Gasfield Company. However, this company's investment track record is insufficiently dynamic—as evidenced by the Laongangmiao gas field (with over 400 billion m^3 of capacity), discovered by PetroChina Southwest Oil and Gasfield Company in central Sichuan (these conventional gas fields also require large exploration investment). Thus a new way of thinking is needed to develop the shale natural gas industry, and we propose establishing an integrated shale natural gas exploitation experimental zone in Sichuan Basin and its periphery. First, preliminary testing will be conducted under safe conditions and strict environmental management, and the consequent exploitation must then proceed using new, remodelled methods, mechanisms and regulations. As well as playing a role in bolstering shale natural gas exploitation, it will become a trailblazer for China's oil and gas reform, and the experience gained from it can be replicated and promoted. Important initiatives include:

1. **Innovative mining rights management and market access management**: A unified tender can be put out for resources, with the exception of those areas where the three main enterprises have started already production, are verifying resources or are carrying out exploration. Market access must not be restricted to other state-owned enterprises but must also be opened to privately run companies and foreign-owned enterprises. It is not only the large transnational foreign enterprises that need to be attracted to take part, but also small and medium-sized companies with extensive experience and strong capacity for innovation. It is these companies that played a key role in shale oil and gas

revolution in the United States. If the environmental targets for technology are met and national security clearance is obtained, foreign enterprises can enter the Chinese market as wholly foreign-owned enterprises.

2. **The full benefits of the mixed ownership system must be obtained**: This practice has already been implemented by PetroChina and Sinopec. In the Changning block PetroChina has set up Sichuan Changning Gas Development Co., Ltd., while Sinopec established Chongqing Shalegas Exploration and Development Co., Ltd. in Chongqing, both of which are under mixed ownership. A mixed system of ownership not only grants an investment opportunity to local state capital, private capital and foreign capital to invest, but is also conducive for central enterprises' usage of social capital in order to bolster resource exploitation. The presence of local state capital is beneficial with regard to demolition and resettlement, road building and the immediate usage of natural gas. The participation of private and foreign capital is beneficial in reforming corporate governance and raising extraction efficiency and the level of services. In order to deepen the mixed ownership reform in shale natural gas industry, it is necessary to apply great effort in the areas of corporate governance, strategic synergy and the division of labour. We must maintain the initiative across the board, and avoid misappropriation of funds or situations where the mixed system of ownership exists only on paper, and the external investors are denied participation in management and operations.

3. **Effective regulation methods for shale natural gas industry**: Appropriate policy system, management standards, regulation system and rules for shale natural gas prospecting and

exploitation, as well a platform to gather and share information, can be set up and applied for a period of 3–4 years in the experimental area. Regarding the environmental regulations, the environmental regulation will be organised and implemented locally following the standards and policies set by the central government. The Environment Ministry is currently researching and formulating the environmental impact assessment guidelines for the shale natural gas industry, and these can be put to a pilot test in the experimental area, bringing out and implementing, for the first time, the national standards and regulations regarding land use, vegetation restoration, water resource utilisation, waste water treatment, waste disposal, gas emissions, exploratory well drilling and well completion. Regarding sharing geological data and information, it is possible to follow the American example and introduce legislation to enforce the obligatory compilation of geological shale natural gas data, building it up into a geological engineering information management platform, with the government controlling the natural resources information, which will then be shared, boosting the efficiency of shale natural gas exploitation.

4. **Fiscal and taxation policy**

Fiscal and taxation policy regarding natural resource extraction must be improved. Conventional gas projects generate high income after price liberalisation, and the natural resource taxation rate should be raised. However, for tight natural gas, shale natural gas and coal seam gas, in order to encourage development of the unconventional gas industry, it is necessary to consider reforming the financial and taxation system, adjusting the relationship between local and central taxation, lowering the natural resource tax rate and mobilising the role of local government. Regarding shale natural gas, during the 2012–2015 period, the central financial administration provided a subsidy of 0.4 CNY/m^3 to shale natural gas-producing enterprises meeting the relevant conditions. The 13th Five-Year Plan period is a crucial time in the development of China's shale natural gas industry, as it moves from the investment stage to growth stage. Even in the current situation of very low international oil prices, with the expectation that the price of a barrel of oil is to remain around the 60–80 $/bbl in the next 3–5 years, and with global investment in shale natural gas shrinking, it is especially important to encourage active exploration and production of shale natural gas. For this reason, the subsidies for shale natural gas industry must be extended and continued. Currently the Finance Ministry and the Natural Resources Ministry have already decided that the subsidies will be fixed at the level of 0.3 CNY/m^3 for the period of 2016–2018 and 0.2 CNY/m^3 for 2019–2020.

Coalbed gas exploration must also be supported. First, in coal seam gas exploitation, we must, with safety and environmental protection as guiding principles, closely integrate the development of mines and wells within coal mining areas. Ground drainage must be performed five years in advance, pumping water from the well must be advanced moderately, and old reservoir coal seam gas (methane) must be actively extracted. Independent enterprises must be encouraged to purchase coal seam gas (methane) from different mine and well locations, large-scale usage must be realised, and subsidies extended to the enterprises directly or partially using this gas. Second, commercial exploitation of non-coal mine coal seam gas should be promoted using targeted policy. Coal seam bed gas production must be commercialised in partially exploited coal bed gas areas situated in the locations outside the regions with coal mining programmes, as well as in those mines which for 10 or 20 years have not engaged in coal production. The current categorised incentive policy does not allow the extension of these incentives to the coal mining locations which do not address the safety and environmental issues.

1.5.2   Achieving Import Diversification, Safeguarding the Security of Natural Gas Imports

The general orientation of support and encouragement for natural gas imports is as follows. With regard to natural gas import policy, China needs to reconcile the conflict between the following three aspects: the need to address the problem of environmental pollution, a continuous increase of natural gas consumption caused by economic development and a rise in people's incomes, and the need to ensure the security of natural gas supply. The domestic consumption of natural gas must be expanded both as a proportion of the total energy resource consumption (currently at 5.8%) and as regards the penetration rate in residential use (16%). Preventing air pollution has already become urgently necessary and is a strong motivating factor in the expansion of the use of natural gas. In summary, to solve the fundamental contradictions facing the issue of how to expand domestic natural gas consumption, the formulation and issuing of all the relevant natural gas policies and regulations in China must be subordinate to the need to effectively resolve these contradictions. The orientation of China's import policy in the near future (from the present day to 2020 and then 2030) is as follows: imports must be cautiously encouraged, but at the same time efforts must be made to achieve diversification of the import sources and import partners, as well as of contracts and pricing methods, in order to guarantee overall control of natural gas imports.

Analysis of Chinese Natural Gas Supply Security

The degree of dependence of China on foreign natural gas is rising fast, reaching 32% in 2013. For this reason, some take the view that restrictions should be placed on the proportion of natural gas imports. Our analysis shows that China is capable of maintaining overall control of the security of natural gas supply, without the need for restrictions being placed on the proportion of natural gas imports. The reasons for this are as follows:

- Natural gas is a substitutable resource and can be easily substituted using oil, coal or renewable energy sources, which makes the problem of its supply security easily solvable.
- The security of natural gas supply in China is markedly better than that of oil. One reason is the diversity of sources of natural gas imports—from Central Asia in the North West, Russia from the North East, Myanmar in the South West, and on the Eastern seaboard the LNG arriving via maritime routes from different countries. The situation is clearly different from oil, which relies on the supplies from the Middle East, and over-relies on the supplies delivered via the Strait of Malacca. The second reason is that, compared with other countries, the current ratio of natural gas in its energy structure and the ratio of its imports, make China's diversified imports comparable with the European Union, and in the future will be comparable with Japan and South Korea.
- The national reserves of conventional gas, shale natural gas and coal seam gas have very large development potential, and, provided the institutional mechanisms are rationalised and enough time is allowed, production output can be increased greatly.

It could be said that the key and the incentive to achieve national gas supply security are firmly in China's own hands. For this reason, in the context of China's ample global natural gas supply and its pressing need to control emissions,

increasing gas imports will allow China to simultaneously increase natural gas usage and reduce emissions. Naturally, while encouraging imports, the operation of gas stocks and reserves as well as the interconnectivity of the pipeline network must be increased, in order to mitigate against potential short-term market fluctuations (Fig. 1.5).

First, the sources of imports must be diversified. According to the current LNG and pipeline contracts signed by China, in the coming years natural gas imports are expected to undergo fast growth, from 53 billion m^3 in 2013 to approximately 192–230 billion m^3 in 2020. Generally speaking, the diversification of China's natural gas imports is growing. The main importers into China are Turkmenistan and Russia, with each country accounting for roughly 30% of the total, while Australia's imports come second, at 12%. The remaining 30% or so is contributed by eight different countries—including Myanmar, Qatar, Uzbekistan and Papua New Guinea—and also by centralised natural gas buyers. These figures represent a significant change from the proportion of natural gas imports in 2013: the imports from the Middle East into China fell the most, from 17.7% in 2013 to just 3%, Turkmenistan's fell from 47 to 30%, but imports from Russia grew from zero to 32%, with a slight increase in Australian imports also.

Second, the import methods must be diversified. Regarding contracts and price setting:

- The contracts must include long-, medium- and short-term ones, as well as merchandise on hand contracts, in order to facilitate greater flexibility and also to better respond to and control market changes.
- When setting prices, the oil price index as well as the standard diversified price index must be included.
- When estimating the competitiveness of the price, it is necessary to consider factors such as the risks involved in the project and the

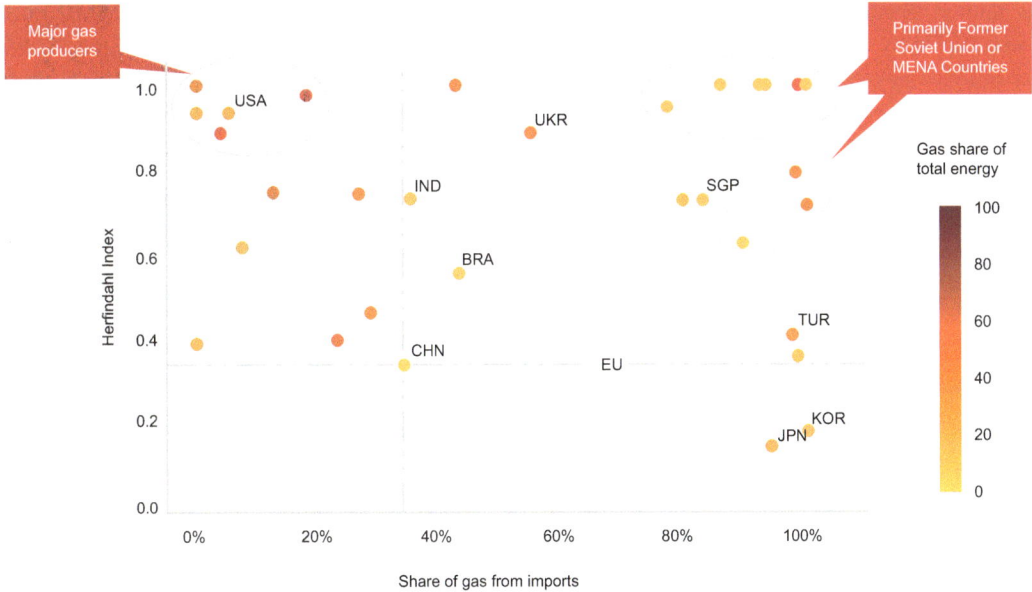

Fig. 1.5 Proportion of natural gas imports and import diversity in selected countries. *Note* The Herfindahl index is used to measure and describe varied indicators. The lower the value, the higher the diversification

commercial structure; risks can be mitigated by using rolling contracts for purchasing natural gas, which are also suitable for balancing domestic supply and demand.

The natural gas supply structure must involve diverse geographic sources. Also we must attract new participants in the import industry, encouraging more large-scale end users to purchase natural gas on the international LNG market. Such new customers, both electricity companies and large-scale urban heating enterprises, will bring great benefits to the market, and by reducing the intermediate links this can improve the efficiency of the natural gas value chain. Supervision and regulations must be strengthened, and access by third parties to the relevant facilities, such as receiving stations and the pipeline network, must be effectively promoted to ensure an efficient, unimpeded flow of natural gas imports into the country.

Third, natural gas trade and foreign investment must be bolstered using commercial principles. Encouraging foreign investment in natural gas, especially in the "One Belt, One Road" region and by the major gas-producing countries,

is particularly beneficial for safeguarding the security of natural gas supply, as well as being beneficial to extending the commercial cover to more segments in the natural gas value chain. It also helps the chain to withstand the risks of price fluctuations. When developing foreign trade and investment in natural gas, in order to consistently maintain commercial principles, we must not adhere only to national strategic guidance, but must also make the utmost effort to obtain commercial benefits, and make economic viability a foundation of project partnership. Naturally, in order to realistically tackle the problem of contracts signed at the time of high prices, a new mechanism of sharing the historical cost between enterprises and the state can be explored.

1.5.3 Accelerating the Construction of Pipeline Network, Promoting Network Interconnectivity

Pipeline network is key infrastructure and its construction must be advanced moderately, and its unified planning, interconnectivity and dispatch flexibility strengthened. Specific measures are outlined below.

First, the construction of pipeline network infrastructure must be accelerated and transmission capacity increased. China's 2020 natural gas market demand will be approximately 350 billion m^3, and 580 billion in 2030. Knowing that resource supply to demand ratio must be 1:1.1 and the transfer capability to demand ratio 1:1.15 in order to ensure proper distribution and allocation within the market, we can see that the capacity of the pipeline network must reach 400 billion m^3/year in 2020, and 670 billion in 2030. At the end of 2014 the transmission capacity of China's pipeline network stood at 240 billion m^3, capable of satisfying just 60% of the 2020 transmission demand and 36% of the 2030 transmission demand. Therefore, the construction of pipeline network infrastructure must be accelerated in order to meet the demands of the growing market. Efforts must be made to extend the total pipeline network length to 15,000 km by 2020, achieving transmission capability of over 400 billion m^3, and 26,000 km and 690 billion m^3 respectively by 2030. The aim is to achieve a pattern of "western gas to the East, northern gas goes South, eastern gas arrives by sea, using the nearest supply". In other words, Central Asia imports of gas supply the North West of China, Myanmar supplies gas to the South West, the North of China will be supplied by Russia, and Eastern China by maritime imports of LNG, forming four great import pathways. Between 2020 and 2030 Lines A, B, C, and D of the Central Asia–China Gas Pipeline will achieve transmission capacity of 85 billion m^3, the Myanmar–China Pipeline a minimum capacity of 10 billion, the Russia–China pipeline 68–69 billion, while LNG imports are expected to realise a reception capability of 100 million tons.

Second, the construction of pipeline network must achieve optimal layout and interconnectivity. The fundamental idea is to create interconnectivity by constructing a backbone trunk of the long-distance national pipeline, equipped with hubs and tie-lines. The flexibility of allocation

control will be achieved by provincial-level backbone pipeline trunks running through different hubs. Underground reservoirs and LNG reception terminals will be connected to the national pipeline network and the adjacent provincial and metropolitan markets will form a regional network. Specific measures required are:

* Achieve interconnectivity of the main trunk of the national pipeline network. The interconnectivity of the main trunk relies first on the key hubs at the junctions of the network. Such hubs have been formed in Zhongwei in Ningxia, Yongqing in Hebei and also in Shanghai and Guangdong. Among these, Ninxia's Zhongwei is the junction on the route connecting Western China to the East, and is connected to the system transporting Western gas eastwards, linking up the Xinjiang coal methane pipeline, the Shaanxi–Beijing pipeline, the Zhongwei–Jingbian pipeline and the Zhongwei–Guizhou pipeline. Zhongwei brings together the gas from Central Asia, from Xinjiang's and Changqing's gas fields as well as the Xinjiang coal methane. Yongqing in Hebei is northern China's regional hub; it connects the Shaanxi–Beijing Pipeline, the Russia–China Natural Gas Pipeline, the gas storage reservoirs in northern China and the Tianjin–Hebei resources pipeline. It is the northern region's centre for the regulation of the natural gas flow, it optimises natural gas allocation and mitigates peak demand. Many lines of the backbone national pipeline traverse Hubei Province, which is a key hub in South Central China, a key junction connecting the supplies of conventional gas and coal methane from Xinjiang, the imported gas from Central Asia and Russia, as well as shale natural gas from Sichuan and Chongqing. Shanghai, an important natural gas market centre, is not only an important hub for China's West to East Pipeline's Pipelines 1 and 2, and the Sichuan to East Pipeline, it also receives gas from the Sea of Japan (East Sea) and also a large quantity of LNG imports. Shanghai is a gas trading hub and is also a national pricing

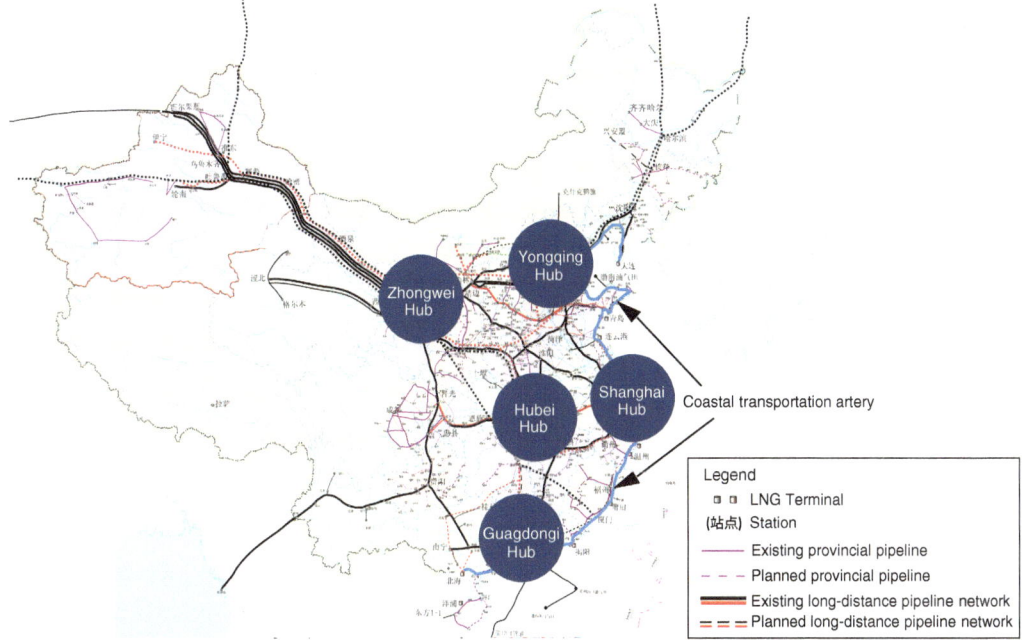

Fig. 1.6 National natural gas pipeline concept

benchmark, and the city is destined to become a national hub in the future. Guangdong Province is an important natural resources hub, and in the future will form a staging area for different types of gas. Its supply structure is varied, with pipeline gas, sea gas and LNG, and its supply pipelines include China's West to East Pipeline's Pipelines 2 and 3, the Xinjiang–Guangdong–Zhejiang pipeline, sea gas and also LNG terminals. Guangdong is the most diversified province as far as sources of gas are concerned, and several different categories of gas resources are pooled here, including conventional gas, coal methane, sea gas and LNG (Fig. 1.6).

- Unified planning, interconnectedness and flexible deployment of resources of provincial-level pipelines are a priority—in order to avoid replication of infrastructure and loss of efficiency, the infrastructure construction must take place under a unified planning programme and proceed step by step according to the degree of market development. The provincial natural gas companies must be in charge of the construction and

operation of the provincial pipelines and peak regulation facilities. However, provincial gas companies must not purchase natural resources. This is done by the downstream customers negotiating directly with upstream suppliers.

- Because of factors such as administrative divisions and examination and approval red tape, the structure and development of regional pipeline networks, apart from the trans-provincial long-distance pipeline and its tie-lines, is largely concentrated within the borders of provinces, creating large natural gas consumption gaps between neighbouring provinces or economically similar provinces. In the future, relying on government guidance and support, pipeline network tie-lines must be built to connect the backbone provincial pipelines, achieving resource complementarity, bolstering regional natural gas industry development and improving the regional pipeline network system.

Third, overall planning of infrastructure construction is necessary, strengthening the supervision of the schedules of planned projects. Currently, infrastructure construction in China is mainly carried out by the three major oil and gas companies, with no effective links between them. As the infrastructure keeps improving, the inter-connectedness of the infrastructure layout will become crucial in ensuring the security and sta-bility of the functioning of China's natural gas industry. This will become especially significant in the future following the construction of the Xinjiang–Guangdong–Zhejiang Pipeline, West to East Pipeline Lines 4 and 5, and the Russia–China Pipeline, as well as of coastal terminals and underground storage reserves. Effective infrastructure interconnection can only be achieved if effective co-ordination is included in the planning stage. For this reason, the govern-ment departments must become involved in infrastructure construction, carrying out com-prehensive co-ordination and strengthening the supervision, regulation and control of the con-struction projects. In order for the planning tar-gets to be implemented, the progress and the actual status of the projects must also be tracked.

Fourth, we must diversify the investors. China encourages and supports all categories of capital to participate in investment into the construction of infrastructure under the umbrella of unified planning. The three major oil and gas companies must open up their investment rights for the construction of the national backbone pipeline, and private capital must be attracted to this area. The monopoly of the provincial gas companies on infrastructure construction must be broken. By attracting private capital to participate in the infrastructure construction, we will bolster infrastructure construction and raise construction capacity. Attracting new shareholders, such as the National Council for Social Security Fund, urban infrastructure industrial investment funds and Baosteel, will break the established pattern of 100% investment by the three major oil and gas companies, and establish private capital in the construction of the national backbone long-distance natural gas pipeline. The planned West to East Pipeline Lines 4 and 5 and the Xinjiang–Guangdong–Zhejiang Pipeline must, using the method applied for West to East

Pipeline Line 3, create a joint capital venture using private capital. Investor diversification in long-distance pipeline infrastructure will be achieved by bringing in investors other than one of the three major oil and gas companies. To accelerate provincial infrastructure construction, those provinces with their own gas companies must attract private capital with a desire for investment and the necessary investment eligi-bility to form investment partnerships. The con-struction rights for the provincial branches of the main pipeline network must be completely lib-eralised and the eligible enterprises will be allowed to take part in infrastructure construction and operation, in accordance with national and local government's planning.

1.5.4 Accelerate the Construction of Natural Gas Reserve Facilities and Implement Corresponding Institutional Reform

Natural gas storage and peak regulation are extremely important in ensuring the smooth operation of natural gas supply. Currently, Chi-na's effective storage capacity stands at approx-imately 4 billion m^3, merely 2.2% of the consumption volume, very different to the 15–20% observed in developed countries. The sys-tems of organisation are also inadequate. The specific recommendations are as follows.

- Clear targets must be set for the construction of gas storage facilities. Natural gas storage facilities in China must be built in accordance with the scale of China's future natural gas consumption, taking into account the current peak regulation capacity, the conditions of the underground storage reservoirs and the advancements in the construction of LNG terminals, as well as drawing on the experi-ence of major foreign gas consumer countries. The integrated system of the underground storage facilities and LNG storage facilities will, by 2020, have a storage capacity of 30 billion m^3 of natural gas and

5–10 billion m^3 of LNG. With a working storage capacity of 35–40 billion m^3, this provides a storage equal to 10–11% of the total demand and will transform the current peak regulation method of reducing customer demand. At the same time, in order to guarantee supply security of key regions, the national natural gas emergency response system must be set up as quickly as possible, specific emergency plans perfected, and an early warming emergency response system built. During the period leading up 2030, the scale and capacity of the storage system for peak regulation must continue to be gradually expanded, and the working capacity of storage facilities expanded to 65 billion m^3, or 12% of the total demand, converging with the global average. Around 2030, the strategic storage capacity must reach approximately 5% of the import volume in order to create an emergency reserve storage system that is capable of dealing effectively with demand and supply fluctuations and able to cover the full national demand.

- Emergency peak regulation reserves storage construction must be accelerated and its layout optimised. The first step is to boost the pace of the construction of large-scale natural gas storage reservoirs and LNG tanks for winter peak emergency. In northern, north-eastern and north-western regions of China, where the seasonal peak/trough difference in gas consumption is very large, and where there is no shortage of locations to build the storage facilities, priority must be given to the construction of underground storage facilities, as well as supplementary small and medium-scale liquefying facilities and LNG terminal tanks. For peak regulation in Central and South Western China, with its specific geological conditions and the proximity of oil- and gas-producing areas, it is possible to use depleted oil and gas pools to construct underground storage reservoirs, and at the same time use the upstream gas fields, supplementing this with small and medium-sized LNG installations. In eastern and southern China and other regions with

rather inadequate conditions for the construction of underground storage reservoirs, priority must be given to the construction of LNG tanks, while underground reservoirs and small and medium-sized tanks can be used as a supplementary system for peak regulation. The second step is to prioritise the construction of small-scale gas storage facilities in order to satisfy daily peak regulation needs. In cities with high natural gas consumption, in order to solve the important problem of daily peak time demand, free hand must be given to gas companies, and the construction of small-scale LNG tanks and CNG spherical storage vessels must be accelerated. Furthermore, it is worth considering using the newly discovered large-scale gas fields as storage facilities (with strong regulation capacity), but the relevant technology and incentive measures must be researched first.

- Formulating the laws and regulations on gas reserve storage facilities, and thus clearly defining the obligations of the operating natural gas companies, is of critical importance. The national policies and regulations must be improved as quickly as possible so as to ensure the optimal external policy environment for both the structure and the institutional framework. Enterprises must receive encouragement to become engaged in the construction of storage facilities and research must be launched into innovative storage facility operational models. National rules and laws on the management of natural gas storage facilities must draw on international experience, and clearly state the organisational and managerial roles and their duties and obligations, as well as create a graded management system for storage facilities. While strategic gas storage reserves remain a responsibility of the state, the responsibility and duty for the daily and hourly peak regulation must be shouldered by the provincial and municipal gas companies, together with the operation of, and all the investment required by, the regulation storage facilities.

- Pricing and taxation policy for gas storage must be improved. The gap between the peak and the trough prices is a reflection of the value of the natural gas at peak hours. We recommend that China introduces peak-trough price differences for all types of customers, which will lead to rational consumption, and the reduction in peak and trough differences this causes will the raise the operating efficiency of the system. In the United States and France, the difference between peak and trough prices is usually a factor between 1.2 and 1.5. January–February and November–December will be designated as peak gas usage season, when peak pricing will be implemented, with the price rising above the gas prices at that time, establishing a rational mechanism for pricing. The tariffs can then be calculated on the basis of the operating costs of storage reservoirs and storage facilities, and on the basis of the internal return rate approved by the government. The government approves the price level, but periodic evaluation and adjustment of storage reservoir prices must be carried out. The formulation of preferential policies and measures will further encourage enterprises to become involved in the construction of storage facilities. The recommendation is for the government to apply preferential policies to the costs of the construction of storage facilities. Here the government can use the law regarding commercial oil reserves as a reference, where the upstream supplier can offset 30% of the cost of LNG used in storage and of the total cost of the investment into the construction of LNG emergency peak regulation storage faculties against tax. This will encourage upstream suppliers to set up an appropriate amount of LNG storage, enabling them to counter the problem of temporary supply disruptions or excessive peak demand.

- Reform of the management system of gas storage must be accelerated. Taking into account the Chinese gas industry's special characteristics, future gas storage operations in China may at first adopt an operating model that is not fully market-based, with independent storage operating companies set up within companies managing pipeline transmission. Gas storage facilities will thus become independent profit-making bodies, which will not only help the gas storage industry to operate in a more professional and market-oriented way, but will also support independent market price setting. As China's gas storage services industry develops at great speed, there will be a developing tendency for storage services to separate from pipeline transmission services, becoming an independently operating link. The diversification of peak regulation storage facilities construction must be encouraged. For this purpose, an independent state-owned enterprise, municipal gas companies and independent storage operating companies must be formed to create a diversified system, which will help to ensure natural gas supply security and fast growth in gas storage capacity.

- An effective early warning emergency response system must be set up. The first step will be to establish a forecasting and early warning system. Statistical data monitoring and evaluation and information dissemination mechanisms must be established as soon as possible, communication channels must be fully integrated, and the system responsible for information and natural resources statistics collection must be continuously upgraded. Next, an emergency response system covering all the aspects of the natural gas industrial chain—production, transportation and sales—must be set up. Different levels of contingency plans must be decided upon in accordance with the specific enterprise and regional characteristics. In order to guarantee timely, rigorous and authoritative emergency response, overall co-ordination must be entrusted to the National Energy Commission, establishing an assessment system and a decision-making mechanism capable of handling major events in the natural resource sphere, as well as providing feedback and information disclosure.

- Finally, research must be conducted into the appropriate scale of strategic natural gas reserves. On a global perspective, the gas industry has not yet seen an international supply crisis like the oil industry has, and there are no prominent security issues. However, compared to the oil industry, the gas industry harbours a great potential risk—it has a high degree of integration between the downstream, midstream and upstream. A problem at any of these points can have an effect on supply security. The Russia–Ukraine gas war of 2009 is one such example. For this reason, China must, as soon as possible, on the basis of its commercial reserves, start researching the issue of strategic gas storage reserves, evaluating the future mid- and long-term trends in national and international natural gas security, commencing preliminary planning, and formulating a strategic reserves scope and model suitable for China.

1.5.5 Improve Technology Capacity and Strengthen the Long-Term Development Capability

By expanding joint research and co-operation with foreign partners in science and technology, innovative, advanced technology must be developed and deployed. The technology of exploration, extraction and production of shale natural gas, suitable for the specific nature of Chinese geology and subterranean structure, must be clearly understood, and a China-specific, cost-effective, environmentally friendly and scalable solution developed:

- **Equipment**: To begin with, there should be a switch to more powerful equipment, to reduce the quantity of equipment and the scale of well sites and to facilitate operations in mountainous terrain. Following that, a switch should be made to module-based, smaller-sized, portable equipment for those operations in areas with geologically complex terrain.
- **Cost reduction**: In order to achieve large-scale cost reduction, we must actively work on research on the crucial and complementary areas of technologically optimal design and low-cost fracturing fluids.
- **Environmental protection**: The safety and environment-friendliness of the additives used must be guaranteed. In the next three to five years, the aim is to achieve, through innovation and the introduction of advanced technology, an economically viable system for shale natural gas that addresses China's specific needs.

We must strive to achieve, after 2020, a mature and complementary shale natural gas industry capable of developing its own technology. We must bolster research into shale natural gas geology, and explore advanced yet workable methods for shale natural gas exploration and commercialisation.

The intensity of technological innovation must be increased. There must be a focus on making key technological breakthroughs, and greater input must be applied to the research in key common technology in areas such as unconventional gas extraction, deep sea gas industry development, and LNG storage and transportation facilities. Strategic planning must be carried out, as soon as possible, with regard to research and development of key technologies in eight major areas—shale natural gas extraction, sea gas exploitation, coal seam gas extraction, methane hydrate extraction, coal gas, gas-powered vehicles (ships) engine manufacturing, combined cycle gas turbine generator units, and carbon capture and storage. While focusing on achieving breakthroughs in key technologies, we must also carry out, in advance, fundamental research in the relevant fields.

Finally, innovative energy technologies require an innovative management system. The national energy authorities must take the lead, and organise and co-ordinate the work of the ministries of technology, national resources, industry and information technology, finance, environmental protection and of the Standardization Administration. Every organisation's role and responsibility in supporting natural gas technological innovation should be clearly defined. The supporting policies must be formulated according to the special characteristics of generic technologies in order to strengthen the support system for innovative natural gas technologies. A national unconventional natural gas research laboratory must be set up, bringing together human, material and financial resources to achieve breakthroughs in the key areas of technology.

1.6 Constructing a Modern Natural Gas Market Mechanism and Management System

Expanding supply and demand, raising efficiency and safeguarding supply safety all require a reform of the institutional mechanisms. The orientation of such reforms is identical with the reform spirit and the orientation of the Third Plenary Session of the 18th Central Committee of the Communist Party of China. Namely, while the decisive role must be given to market and resource allocation, the government role must also be applied to the greatest possible extent. In order to rationalise the comparative prices for natural gas and alternative energy sources, a pricing system and a green finance system must be set up which reflect the degree of scarcity of natural resources, the supply and demand relationship, and the external cost of environmental damage. Diversified competition and greater openness to foreign involvement must be boosted at all the stages of the natural gas industrial chain, but we must also at the same time strengthen the regulations regarding market access fairness and general services in order to establish a diversified, opened, orderly and modern market system. Specific measures are as follows.

1.6.1 Liberalisation of the Upstream Sector and True Import Liberalisation

In order to have competition in the natural gas market, it must have a certain number of participants. The United States and Europe are examples of classic competitive markets, with many players and vigorous competition at every stage, from exploration and production, and storage and transportation, to wholesale and retail and market hubs. Resource acquisition is the prerequisite of natural gas usage, but in this industry it is often the upstream producers that control the operations and the profit allocation along the entire industrial chain. For this reason, the liberalisation of access to upstream exploration and development is crucial for the growth of the natural gas market.

Market Characteristics of Major Countries

The characteristics of a competitive, responsive, fluid natural gas market are:

- The number of participants in the market must exceed a certain number: a greater variety of services is made available to consumers by greater market participation and midstream competition.
- Competitive prices at wholesale and retail levels.
- Non-discriminating liberalised market access, including access to infrastructure facilities such as pipeline network, storage reservoirs and LNG terminals (Table 1.2).

Currently the oil and gas concessions in China are almost entirely concentrated in the hands of large oil companies. Nationwide, the total area of registered oil and gas exploration concessions stands at 4 million km^2, and the area granted to the three main oil companies—Sinopec, PetroChina and China National Offshore Oil Corporation—stands at 3.9 million km^2 or 97% of the total. The area of registered mining concession is approximately 118,000 km^2, of which 117,000 km^2 (99%) belongs to the three main oil companies. Exploration and development of unconventional gas resources, such as shale and

Table 1.2 Natural gas market characteristics of major countries and regions

Natural gas indicators	Category	USA	EU	UK	Japan	South Korea	China
Supply (billion m³/year)	National production	689	269	38	3	0.5	115
	Imports	37	231	39	123	53	49
Consumption (percentage of total amount)	Electrical power	40	30	30	65	50	15
	Industry	20	20	10	5	20	45
Pipeline (km)		500k	200k	8k	5k	4k	50k
Wholesale market competition		✓	Constrained (oligopoly)	✓	Constrained (oligopoly)	X	X
Access openness	Upstream	✓	✓	✓	✓	X	X
	Transmission	✓	✓	✓	✓	X	X
	Allocation	diversified	diversified	✓	X	X	X
Unbundling of transportation and sales rights		✓	diversified	✓	X	X	X
Independent (federal) market power		✓	✓	✓	X	X	X
Liquidity market centres		✓	✓	✓	X	X	X

Note The above data is from 2013, but the data for China's natural gas consumption for electricity generation and by industry is given at 2011 levels
Source Vivid Economics, and based on IEA, EIA, Chinese Government and ENTSOG data

coal bed gas, has already been opened up to a variety of different enterprises, but the number of exploration blocks is limited. Furthermore, the three main oil companies still enjoy the leading positions in the development of the unconventional gas resources due to the fact that under many circumstances the degree of the overlap between non-conventional gas (mainly shale natural gas) and conventional gas blocks is very high. Moreover, in the import segment of the industrial chain, even though policy allows a greater number of enterprises to participate in the natural gas import trade, currently, because of the limiting factors such as reception terminals, the number of players in the domestic natural gas trade market is still very limited, being restricted to a few enterprises such Huadian Group, Jiufeng Energy and ENN Energy.

Solving the problems which plague the issue of mining concessions, such as over-concentration, "enclosure without prospecting" and "control without development" is a crucial part of the comprehensive, deep upstream reform of the industry. For this reason, starting with the existing laws, regulations and management methods, the maximum effort must be exerted to promote investment diversification, and improve market access standards. The tender bid system must be comprehensively applied to natural gas concessions, and we must push forward, in an orderly and efficient manner, the construction of primary and secondary prospecting and mining rights markets, while at the same time putting fundamental natural gas resource data to work for the public benefit.

Regarding imports, we must, on the one hand, expand the standardisation, public disclosure and transparency of the approval process for built facilities, such as receiving terminals, and, on the other hand, draw on the international experience, producing detailed policy regarding third-party access, and effectively strengthening the necessary regulation. We must compel those enterprises that meet the necessary conditions to turn towards the natural gas import industry (pipeline transmission and LNG) and make independent choices to engage in natural gas imports.

1.6.2 Step-by-Step Natural Gas Price Reform, Accelerated Construction of Natural Gas Trade Hubs

Natural gas reform is key to resolving the numerous conflicts in the natural gas sector. Such reform must not only focus on adjusting natural gas price levels. The focus must also be on the reform of the pricing mechanism.

1. The netback method must be improved further, as quickly as possible. First, price regulation frequency must be adjusted, and efforts must be made to achieve seasonal regulation as soon as possible. Second, the oil price levels and depreciation indices used as a reference for price adjustment must, to the greatest possible extent, reflect the changes in the price of alternative energy sources, rather than be based on the high-price import costs paid previously by the three large gas and oil companies.

2. The cost and the prices for gas transmission using long-distance pipelines, pipeline branches, provincial and metropolitan pipelines, and distribution pipelines must be determined on a scientific and rational basis. In addition, the regulation in this sector must be strengthened and the return on pipeline investment rate adjusted from 12% down to 8–10%. The current situation of inflated transmission and distribution prices must be reformed in order to create a more optimal environment for price reform and for the development of the natural gas reform industry.

3. Improvements must be made in the implementation of seasonal price differences, the

peak and trough price differences, intermittent supply prices and storage prices. Measurement standards and valuation methods must be revised (from being based on flow rate or quality, to a valuation based on calorific value) and at the same time we must strengthen, both in technological and administrative areas, the supervision and checks of pricing violations. The problem of the lack of organisation in tax collection by local governments must be solved as quickly as possible. Currently, as part of the natural gas price adjustment fund, the local government levies a fee of between 0.3 and 1 CNY/m^3 of natural gas sold, which adds to the burden on the consumer. This practice must be abolished. The aim is to gradually rationalise the price of natural gas for residential use.

4. Price liberalisation must be carried out in areas which comply with the necessary conditions. In areas with multiple sources of natural gas and abundant supply, such as the eastern regions of Shanghai, Guangdong, Zhejiang and Jiangsu, the imports of natural gas must be liberalised by allowing third-party access to infrastructure; liberalisation must also extend to the gate price of every province, with prices decided directly by the up- and downstream users.

5. The construction of regional trading hubs should be promoted. The construction of the Shanghai and Guangdong hubs can be pushed forward in the near future. In the medium and long term, new regional hubs can be set up in Beijing, Sichuan, Hubei, Ningxia and Xinjiang. Efforts must also be made to make the Shanghai hub into an Asian, and even international, natural gas trade hub. Eventually, the Shanghai hub will develop into a trading hub system where its international pricing hub capability can interact with that of a multi-regional hub, acting as a link and comprehensively connecting the domestic and the foreign demand and supply. The

trading hubs must first engage in spot trading, then expand its trade volume and scope, and only then commence developing trading in futures, increasing the market depth.

The Function, the Conditions and the Implementation Phases of Trading Hubs

Spot trading provides greater flexibility and fluidity to long-term contracts, and its role in international natural gas trade is continuing to grow. Trading hubs have two important roles. The first is to physically bring together the sellers and the buyers, and the second is to allow for the competition to decide the prices. For this reason, as the signal of market-oriented pricing raises the economic efficiency of the trading and investment decisions, correspondingly lowering trading costs, market participants will benefit. Moreover, the trading hubs can, by using the pricing mechanisms, also solve the problem of supply and demand imbalance, safeguarding the security of gas supply. The potential shortcomings are that the prices might experience intense short-term fluctuations, since they are decided entirely by competition, and led entirely by non-governmental market forces. The data shows that price fluctuations are often greater when prices are set using the trade hub method than when set using oil price as a standard.

The success of trading hubs depends on three prerequisites. First, there must be non-discriminatory access to an improved and open natural gas transmission pipeline network for the market participants to operate in. Liberalisation of the pipeline network is the key to successfully liberalising prices, and failure to meet this condition can cause the natural gas supply, regardless of whether the gas is domestically produced or imported, or transported by pipeline or as LNG, to be unable to link up with local demand. Second, many of the independent buyers and sellers actively take part in arbitrage trade, and none have strong market allocation capacity. For this reason, the trading hubs must be guided by competitive prices, thereby avoiding the problem of the market forces influencing and distorting prices and trade volume. Third, the government must support the liberalisation of prices on the wholesale gas market, along with providing stable, transparent and reliable regulations. The trading hubs must place priority on arbitrage trade, which will raise market operational efficiency, but the arbitration trade must rely on accurate and highly transparent reports of natural gas prices.

China has many of the conditions necessary to become Asia's regional natural gas trading hub: the scale of its natural gas production, the growth of its pipeline network and the volume of its LNG imports. Moreover, other Asian countries cannot match China on the Asian natural resources market. However, the Chinese market is still comparatively centralised, third-party access to the pipeline network is still substandard, and the prices remain under the government control. These factors will hinder the establishment of trading hubs. A roadmap for the next 5–10 year development of China's natural gas trading hubs must be formulated. First, pilot projects must be launched. A mechanism comprising third-party access to pipeline network and transparent prices can be set up in the Shanghai region, creating the conditions of competitive natural gas demand and supply. Then the area allocated to pilot projects can be expanded, new categories of sellers and buyers added, while at the same time strengthening market supervision. Gradually, as the effectiveness of the pilot project grows, it will attract a growing number of producers and consumers. As more and more different organisations join the hub trade, the network effect can be achieved, increasing the advantages of the hub. Both the hub's scale and its role in setting prices will continue to grow until the hub becomes limited by the size of the land area allocated to it.

With price liberalisation action as a foundation, it is also necessary to use the regulatory role of taxation policy, and to establish and improve the policy system of environmental taxation and carbon trading. Environmental tax and carbon tax can be used as an incentive mechanism to expand the natural gas industry, and will also send out a powerful signal to the investors. The period of the 13th Five-Year Plan must witness the formulation and appearance of the environmental tax, the expansion of carbon trading and the establishment of a robust national market, expanding across the entire economy, including primary, secondary and tertiary markets, and integrated into the global market.

1.6.3   Clear Allocation of Functions and Robust Supervision to Support Unbundling and Third-Party Access

In order to ensure the fairness of market transactions, we must, in accordance with the demands of the Third Plenum, push forward with the separation of pipeline network from natural gas transmission services. The detailed methods are outlined below.

First, we must push forward with the separation of pipeline network from natural gas transmission services. Both international experience and the practices established in China indicate that unbundling of services, accounting and legal affairs alone is not sufficient; the key is to achieve the unbundling of functions and structure. If it is successful, and is backed up with robust regulation, the splitting of ownership rights is not required. More specifically, in order to complete these three tasks, we must first, as fast as possible, push forward the unbundling of services,

accounting and legal affairs; pipelines already built can remain in the possession of the three main oil and gas companies, but their company branches must be turned into subsidiaries. As for the newly built pipelines, here the three main companies can take the lead, but a diversity of investors must be brought in. The three major companies must be encouraged to form consortiums with up- and downstream enterprises, especially those engaged in natural gas services, for joint investment, construction and operations. Even more important is to provide the gas transportation companies with business scope, cost regulations, operational methods and information disclosure methods. The transportation companies must have independent management rights, and the information disclosure by the parent company, the affiliated companies and other companies must be synchronised.

Unbundling of the Natural Gas Pipeline Network and Natural Gas Transmission

The unbundling of services in itself is not very effective. In the United Kingdom, in the early stages of this process, the dominant position of the established brand, British Gas, hindered competition. In Japan, where the degree of unbundling of accounting and functioning was not strong enough at the early stages, the unbundling of services did not have much value. The European experience also shows that the unbundling of legal affairs on its own does not have much use either, and must be supplemented by deeper structural unbundling using measures such as functional unbundling.

Functional unbundling is a crucial factor in the entire unbundling process. Functional unbundling often involves a series of restrictions and requirements which affect the operations of midstream services (regardless of whether or not they are an independent legal entity). Unbundling includes the following: establishment of an independent trademark and brand;

Table 1.3 Five methods for the unbundling natural gas pipeline network and natural gas transmission

Method	Changes	Aim	Comparison with prior to unbundling
Services unbundling	Midstream business (most importantly pipeline services) are separated from the gas retail businesses, becoming independent suppliers	If transmission businesses do not independently supply natural gas retailers, the competition becomes distorted as the supplier is not able to sell the natural gas to the pipeline users apart from the pipeline company	N/A
Accounting unbundling	Midstream businesses must independently consider their profits and costs, and be separated from the up- and downstream services	Prevention of midstream cross-subsidies or creation of favourable conditions for the relevant up- or downstream business	If midstream businesses collect tariffs independently without being a separate service, it makes separate bookkeeping impossible
Legal affairs unbundling	Midstream business become a separate legal entity, but remain vertically integrated into the company structure, for example as a wholly-owned subsidiary	A legal duty for the subsidiary company to become credited as a legal person. Management unbundling can be more effective if done as an independent legal entity	A separate legal entity must manage separate accounts and supply independent services
Structural unbundling	Effective and extensive separation of the midstream business operations policy incentive mechanism from the incentive mechanism of up- and downstream enterprises	Separation of the midstream business operations incentive mechanism from the relevant entities, creation of a more effective market by eliminating the motivation for anti-competitive behaviour	Midstream business must be a legal entity with clearly defined actions and duties
Ownership unbundling	All the midstream businesses must be separated from the vertically integrated company, and transferred to entities with independent ownership rights	Comprehensive unbundling of ownership rights must achieve the fullest possible separation of midstream business from the benefits of low and upstream sector participants	The independently owned companies must be independent legal entities, with separate accounts and services

data processing restrictions; unbundling of auxiliary services such as legal affairs, accounting and IT; management unbundling; and compliance and reporting planning (Table 1.3).

If the regulatory mechanism is sufficiently robust, there is no need for ownership rights to be unbundled. Only the UK and The Netherlands have carried out ownership rights unbundling. The French and German policy makers advocate structural unbundling without the unbundling of ownership rights, and the latest European Union natural gas directives

have stipulated robust regulatory measures in order to boost structural unbundling. According to this new reform plan, the large-scale electricity and gas enterprises must choose one of the following three reform programmes:

- unbundling of ownership rights and sale of the pipeline transportation network, achieving a complete separation of pipeline network and transmission;
- unbundling of operational rights while preserving the pipeline network ownership rights, but with the establishment

of an independent company (independent system operator) to run the pipeline network transmission operations;

- management rights unbundling while retaining the possession and operation of pipeline transmission, but the management of the pipeline transmission must be handed over to an independent subsidiary endowed with management and decision-making rights (independent transmission operator).

The reform programme also requires all member states to set up an independent regulatory mechanism in order to ensure that the large enterprises carry out effective "separation of production and supply" and truly implement the principles of free competition in the energy market. The American experience also shows that a competitive gas market can be sustained without the compulsory unbundling of ownership rights, provided that strict regulations are imposed on the operations of transmission and supply companies. Under normal circumstances a robust regulatory system on its own is sufficient to achieve effective structural unbundling.

Third-party access services must be provided as soon as possible. Robust regulation must be applied to the terms of access, service prices and service quality, ensuring that the main operators provide fair, impartial services, and creating the conditions to cultivate a competitive market. In order to establish detailed regulations regarding third-party access, and thus ensure third-party participation, we need to promote information disclosure and strengthen the regulatory capacity. An open and transparent oil and gas management and operations information platform is required, covering all the information relevant to industrial operators, such as current excess production

capacity infrastructure and the conditions and costs of third-party access.

Moreover, the unbundling of the pipeline transmission business and sales business and third-party access reform are complementary when it comes to liberalising the free choices of downstream users. The graded (according to the volume of annual natural gas consumption) liberalisation of large users' direct choice of gas suppliers must be pushed forward as soon as possible. For this purpose, large users include urban gas companies, electricity generating stations and CCHP stations with 20,000 kW capacity, large enterprises (including raw materials and fuels) and LNG/CNG suppliers. The storage and city distribution link of the natural gas industrial chain must also gradually introduce the third-party access mechanism for customers at different consumption scales. Starting with the largest annual non-residential users, an annual schedule and the list of gas consumption volume must be made, and the large non-residential users can be presented with supplier choices based on their own time schedule and operations. The users may also choose to avoid the urban gas distribution networks completely or chose solely the services of the city distribution networks where the price is fixed by the government.

1.6.4 Strengthen Energy Resource Management, Establish a Regulatory System and Build Regulatory Capacity

When promoting marked-oriented reform of the natural gas market, we need to strengthen comprehensive energy resource management and professional regulation. The component parts of this are: building a comprehensive energy management regulatory system; formulating a natural resource strategy, regulations and policies; controlling and regulating the overall balance of energy resources; ensuring energy resource security; adjusting the energy resource structure; adopting energy-saving and efficient management practices; strengthening information collection and analysis; supporting technological innovation and international co-operation in energy; and ensuring co-ordination with the relevant departments.

We must strengthen the energy resource regulatory mechanism and maintain order in the energy market, promote market competition in the industry, and solve market disputes. The authority and the position of the regulators must be independent, and the regulators must make correct decisions while extending equal treatment to all players in the markets. The specific steps are as follows:

- In the near future, we must further improve the regulatory and co-ordinating mechanism. The relevant departments must exercise better communication, co-ordination and co-operation when formulating regulatory laws, rules and standards, national energy development planning and strategies and energy industry policies. The departments must also improve their information sharing and co-ordination during regulatory operations. On the local level, purely provincial-level energy regulation departments must be established which will directly send out professional regulation teams to any locations required. Moreover, the central regulatory authorities must carry out robust inspection and supervision of the local regulatory departments.
- In the medium and long term, we need to set up a unified energy regulatory mechanism. This mechanism, apart from monitoring the environment and national resources, must be gradually given an economic and social regulatory function across the entire energy

industry chain. It will explicitly co-ordinate its work with the relevant departments, with a detailed breakdown of its duties and responsibilities, and gradually become an independent regulatory body. The final outcome must be an independent, unified and professional energy regulation mechanism, and a top-down regulatory organisation system. According to the convention of Chinese administrative reform, in 2018 the administrative system will enter into an adjustment phase and at that point these plans to build a unified integrated energy management mechanism and an energy regulation mechanism can be considered.

1.6.5 Support the Development of Professional Services and Technological Companies; Use Market Forces to Promote the Expansion of the Division of Labour

The experience of the American shale revolution demonstrates that a development style that involves diverse participants and the division of labour can mobilise progress in many areas, including venture capital, research and development, upstream resource extraction, infrastructure, market development and terminal applications. In a free competition environment, highly specialised, technologically advanced technical services companies can provide engineering services such as horizontal drilling, well completion, well cementing and multi-stage fracturing, as well as technological services, such as well logging and experimental testing, to mining rights holders. Some companies can simply withdraw after having completed their services at a certain stage of the industrial chain, and then be replaced by an enterprise from the next stage of the chain. High division of labour has therefore turned the extraction of shale natural gas, at each stage, into a low-investment, short-operating-cycle process with fast investment return. It has also attracted a great deal of venture capital and private capital into the extraction of shale natural gas. We must therefore exploit the role of market forces to the full and strongly encourage the development of specialised technological services companies and advanced technology companies in the oil and gas industry. In accordance with the principle of "demand for production, advances in technology, excellence in reputation", we should use methods such as market forces and quality control to regulate these service enterprises and their operations. We must organise specialised construction teams, and establish the system of rules for each segment, including exploration, production, cost, safety and environmental protection, with the aim of achieving an orderly, standardised operation of the production process, as well as ensuring orderly exploration and production.

The policy measures needed to encourage the development of specialised companies may include tax breaks in the first instance. Another method is to split off smaller auxiliary companies from the large state-owned enterprises as part of the reform of state-owned enterprises. These venture exploration, oilfield and engineering technology auxiliaries can, after becoming independent, form specialised companies and independently enter the market. There are several advantages to this. First, the burden on the main company will be lowered and operational efficiency improved. Second, the development and growth of the service companies will be accelerated.

Despite the concerns expressed by some over the survival of such auxiliary companies after separation, experience has shown that many such auxiliary companies have demonstrated excellent growth as specialised operations. For example,

Bureau of Geophysical Prospecting Inc., China National Petroleum Corporation, after leaving the management of CNPCC and forming a specialised company, grew rapidly and became strongly competitive internationally. It is now ranked third globally in overall strength, and is in a firm first place for ground operations. After successful example is that of the subsidiaries of China National Offshore Oil Corporation—China National Oilfield Services and Offshore Oil Engineering Company—which have shown excellent progress.

A third possibility is to form a rational and scientifically-based oil industrial chain—after the split off the auxiliaries, the main company will have a greater choice of specialist companies that can provide services, and the auxiliaries will also have more choices over their co-operation with the main companies. In addition, expansion into the international market can also follow. Even more important is that specialised companies can provide technology services to the new entrants into the competitive market, which can break the current technology monopoly.

1.7 Deepen the Laws and Regulations System of the Gas Sector, and Establish a Sound Legal System

Although the formulation of the legal framework for the gas sector in China is already in its initial stages, the problems of lack of co-ordination and systematisation and of inconsistency persist. For example, China still lacks a specialised, comprehensive Oil and Gas Law to cover all the links on the industrial chain. In the area of exploration and production there is a legislation base from the Mineral Resources Law and Solid Ore Mining Law, and there is a common problem with specification of mining resources, but it fails to address the issues specific to the special properties of gas as a gaseous mining resource. A great deal of the current legislation retains the particularities of a command economy, with restrictions on the acquisition and transfer of mining rights. As a result, such legislation is unsuitable for the requirements of market reform and the special characteristics of natural gas. The regulatory legislation and rules covering gas transmission, storage and distribution at each segment of the industrial chain from pipeline network to end consumption, as well as those covering environmental protection and other aspects, remain unsound. The duties and the obligations of the local government and the relevant enterprises are not clearly defined. The technology specifications and standards for a modern natural gas industry require improvement and standardisation, and the legal framework of the natural gas sector must be overhauled as soon as possible.

The key measure necessary is the establishment of a sound Natural Gas Law, using a specialised gas law to support for the entire legal framework. The laws and regulations must be made more practical, adapting them to the special characteristics of natural gas exploration, production, transmission, storage and distribution, and usage, and improving the drafting of the specific rules and supporting regulations that are needed. The natural gas-specific laws and regulations must be formulated as quickly as possible. When formulating and revising the Oil and Gas Mining Rights Regulations, the Mid–and Downstream Natural Gas Management Regulations, the Natural Gas Extraction Environmental Protection Regulations, the Sea Oil and Gas Pipeline Management and Protection Regulations, the Natural Gas Storage Regulations and other such administrative regulations, the focus must be on the legal system of the resources' property rights, exploration and development contracts, infrastructure construction, operational management, storage, sales and usage, safety and emergency warning and response system, production safety and ecological benefit compensation, cross-border investment and imports, exports and trade.

Conflicts between the Oil and Natural Gas Pipeline Management and Protection Regulations and other laws must be solved by resorting to the judiciary. There is an urgent need to clarify many areas: the co-ordination of pipeline development planning and other specialised programmes, the convergence of pipeline construction planning and urban-rural planning, the legal definition of the pipeline underground rights of passage, the restrictions imposed by pipeline safety on land use, and the conflict between pipeline safety and the safety requirements of highways and railways.

PART I
Analysis of Natural Gas Demand

Developments in Global Natural Gas Consumption

2

China is a country rich in coal but lacking in oil. For many years, coal has held the leading position in the country's energy mix, while petroleum and natural gas consumption has been low. Recently, with economic growth and increasing energy consumption, the increasing pressure on the environment brought about by the extensive use of coal has led to unprecedented attention being paid to the development of clean energy by the state. On the supply side, the development of shale natural gas, coalbed methane and other unconventional gas sources has also been accompanied by greater development of conventional gas sources. Natural gas import volumes are also expected to grow. If China's natural gas supply capabilities continue to grow at the current rapid rates, then in the future supply will no longer be the primary obstacle to the development of natural gas. Given these circumstances, research into applications for natural gas and its development potential, allowing targeted measures to support the growth of natural gas consumption, is of major significance in the promotion of the adoption of natural gas.

* This chapter was overseen by Zhaoyuan Xu from the Development Research Center of the State Council and Martin Haigh from Shell International, with contributions from Baosheng Zhang and Shouhai Chen from the China University of Petroleum, Lianzeng Zhao from the China Petroleum Planning Research Institute, Linji Qiao from ENN and Juan Han from Shell China. Other members of the research group participated in discussions and revisions.

2.1 Major Factors Affecting Natural Gas Consumption Growth in Other Countries

In recent years, natural gas consumption grown rapidly in many countries, so that it accounts for an increasing proportion of energy used. China can learn crucial lessons by analysing natural gas consumption growth in these countries.

2.1.1 Proportion of Energy Consumption of Natural Gas in Various Countries

The horizontal axis in Fig. 2.1 indicates the proportion of energy consumption represented by natural gas (0–100%, divided equally into 10 groups) while the vertical axis indicates the proportion of natural gas consumption of each country compared to global natural gas consumption. It is clear that, out of 103 countries (or regions), there are 29 countries whose natural gas proportion is between 0–10% of total energy consumption and these 29 countries account for 8.34% of global natural gas consumption. There are 14 countries whose natural gas consumption is between 20–30% of total energy consumption and these 14 countries account for around 32.7% of global natural gas consumption. There are another 19 countries whose natural gas consumption is between 30–40% of total energy consumption, accounting for around 19.64% of

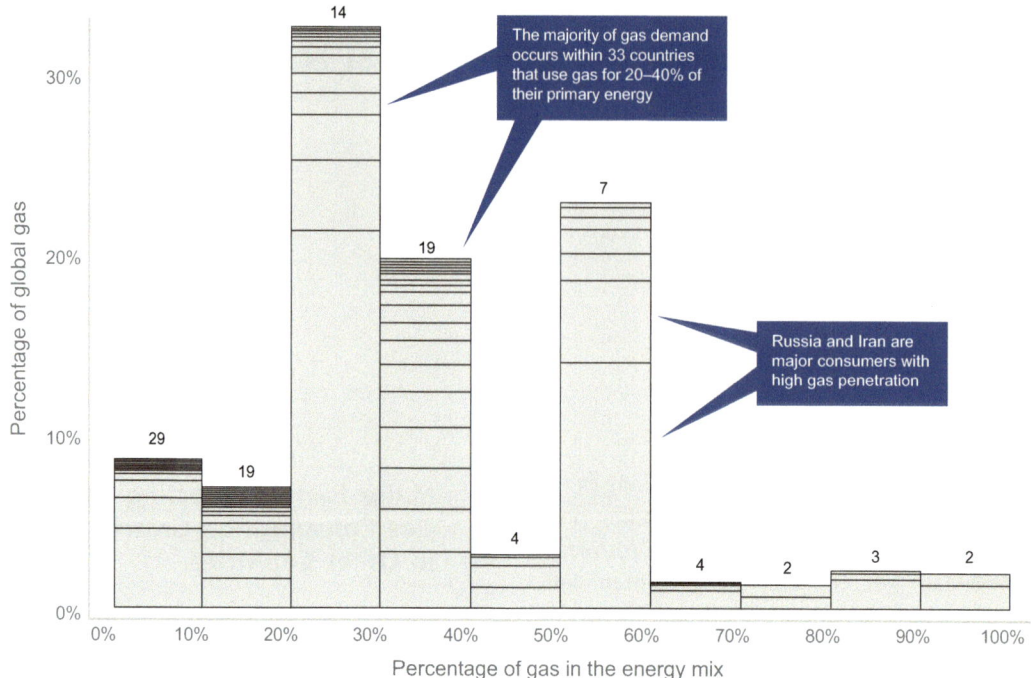

Fig. 2.1 Proportion of energy consumption of each country accounted for by natural gas and proportion of global natural gas consumption of each country. *Source* Formulated by Vivid Economics based on IEA and EIA data

global natural gas consumption. There are few countries where natural gas usage accounts for a proportion greater than 40%, and these countries often possess large quantities of natural gas reserves. However, with the exception of Iran and Russia, these countries are not major natural gas consumers in terms of quantity. The majority of global natural gas demand comes from countries where natural gas usage accounts for between 20–30% and 30–40% of total energy structure.

In order to analyse the prospective growth of China's natural gas consumption, we chose seven countries to act as "benchmark countries" for China, and carried out detailed analysis of the growth in consumption of natural gas in these seven countries. Each of these seven countries is a major consumer of natural gas, and the proportion of energy consumption accounted for by natural gas has increased relatively quickly in each case. Of these, four countries are developed countries: the United States, Japan, Germany and

the United Kingdom. There are also three emerging countries: Malaysia, Turkey and Egypt. There are other major natural gas-consuming nations where the proportion of energy usage accounted for by natural gas is quite high, but these have large natural gas reserves, excluding them from the comparison due to China's relatively small reserves. In addition, there are countries where growth in natural gas consumption has been slow over a number of decades, which are therefore not suitable for adoption as "benchmarks" for the analysis of China's future natural gas demand (Fig. 2.2).

Since 1982, the proportion of natural gas consumption of these benchmark countries exhibited major increases, especially in the three emerging countries of Turkey, Malaysia and Egypt, where natural gas proportions have risen from very low levels to extremely high levels over the past 30 years. During this same period, the proportion of energy sources that natural gas

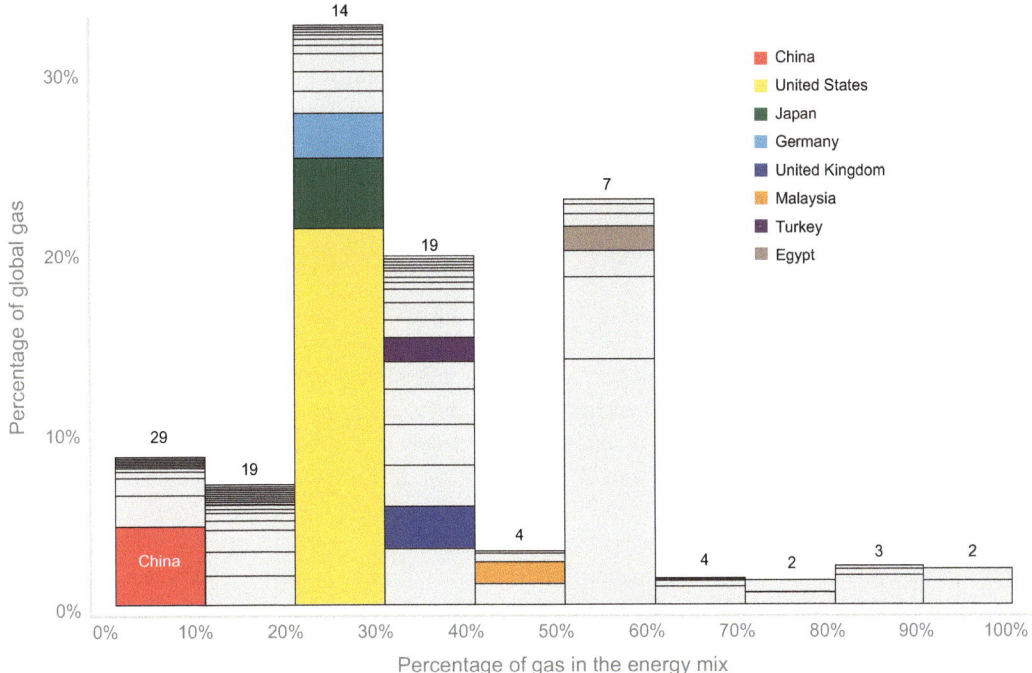

Fig. 2.2 Natural gas consumption in seven benchmark countries. *Source* Formulated by Vivid Economics based on IEA and EIA data

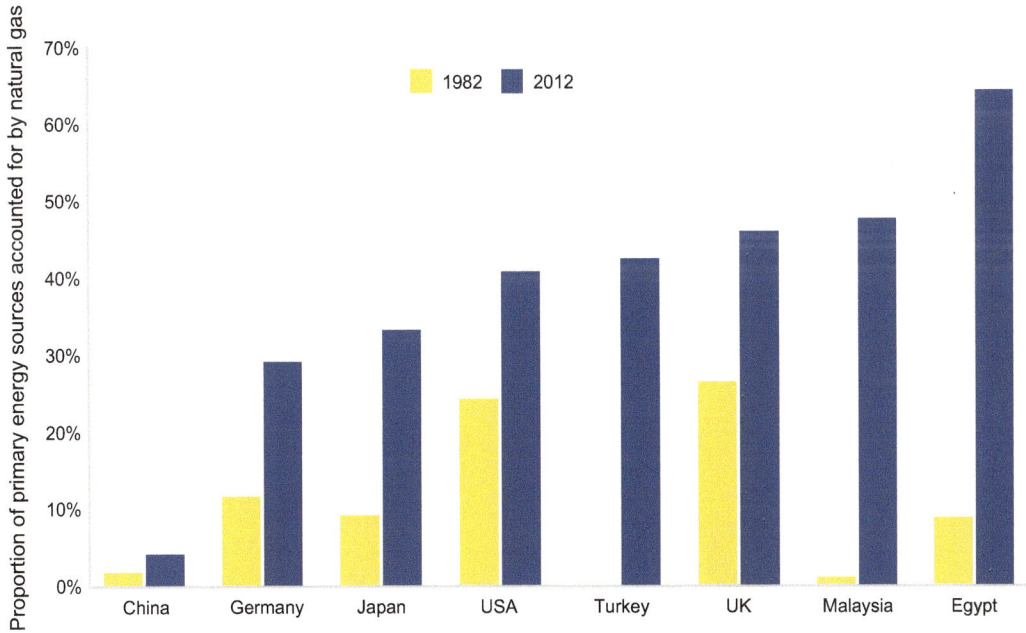

Fig. 2.3 Changes between 1982 and 2012 in the proportion of natural gas consumption of seven benchmark countries. *Source* Formulated by Vivid Economics based on IEA data

accounted for in China rose from 2% in 1982 to 4% in 2012 (Fig. 2.3). This indicates that an increase in the proportion of energy sources accounted for by natural gas could occur.

2.1.2 Breakdown of Natural Gas Consumption Growth in Seven Benchmark Countries

To carry out analysis of the natural gas consumption growth in these seven benchmark countries, we used index breakdown analysis methods, carrying out quantified analysis of the most important factors driving natural gas demand, allowing the degree to which various factors contributed to the growth in natural gas demand to be calculated. The main factors include economic activity, energy density and fuel conversion.

1. Method of analysis

In different economic sectors, natural gas demand changes are related to changes in economic activity, energy density and proportion of natural gas usage. Each economic sector's natural gas demand can be expressed as a function that includes a series of drivers (namely, economic activity, energy density and proportion of energy usage represented by natural gas), as shown in the equation below:

$$\begin{matrix} Gas \\ demand \end{matrix} = \sum_i^{sectors} \begin{matrix} economic \\ activity \end{matrix}_i \times \begin{matrix} energy \\ intensity \end{matrix}_i \\ \times \begin{matrix} share\ of\ energy \\ from\ gas \end{matrix}_i$$

In the equation above, the total natural gas demand G is broken down into the sums of the natural gas demand of various sectors G_i ($i = 1$, …,n). The natural gas demand of each sector is then broken down into three parts, namely each sector's economic total GVA_i, energy intensity $\frac{E_i}{GVA_i}$ and natural gas share of total energy consumption in the sector $\frac{G_i}{E_i}$. Based on natural gas

demand functions, indexes can be set for each driver, and these indexes can be broken down again to quantify the change that occurs to natural gas demand due to changes in each driver over a period of time. This kind of index breakdown approach is called a logarithmic mean Divisia index (LMDI). LMDI is a common analytical method found in energy literature. Based on the relative changes of each variable driving factor, it breaks changes in overall natural gas demand into the effects of three factors. This allows the analysis of natural gas growth prospects from the perspective of each factor's development trends. Due to the limitations in the data, the influence of fuel price changes has not been eliminated from the LMDI used in this analysis.

$$G = \sum_{i=1}^{n} GVA_i \times \frac{E_i}{GVA_i} \times \frac{G_i}{E_i} \rightarrow \sum_{i=1}^{n} GVA_i \times \frac{E_i}{GVA_i} \times \frac{G_i}{E_i} \rightarrow \sum_{i=1}^{n} G_i$$

Theoretically speaking, these three driving factors will affect natural gas demand in different ways:

- **Economic activity**: As economic activity grows, the energy input required by such economic activity grows as well, so if all other factors remain equal, natural gas demand will also exhibit a corresponding increase.
- **Energy intensity**: When the level of economic activity remains unchanged, then as energy consumption intensity drops, the amount of energy required to create the same amount of economic activity is reduced, and thus energy demand will drop. If all other factors remain equal, total natural gas demand will exhibit a corresponding drop.
- **Conversion to natural gas**: If a given economic sector converts from coal or other fuels to natural gas, then natural gas demand will increase.

2. Breakdown of factors affecting international natural gas demand

This section presents a breakdown of natural gas demand changes in the seven benchmark

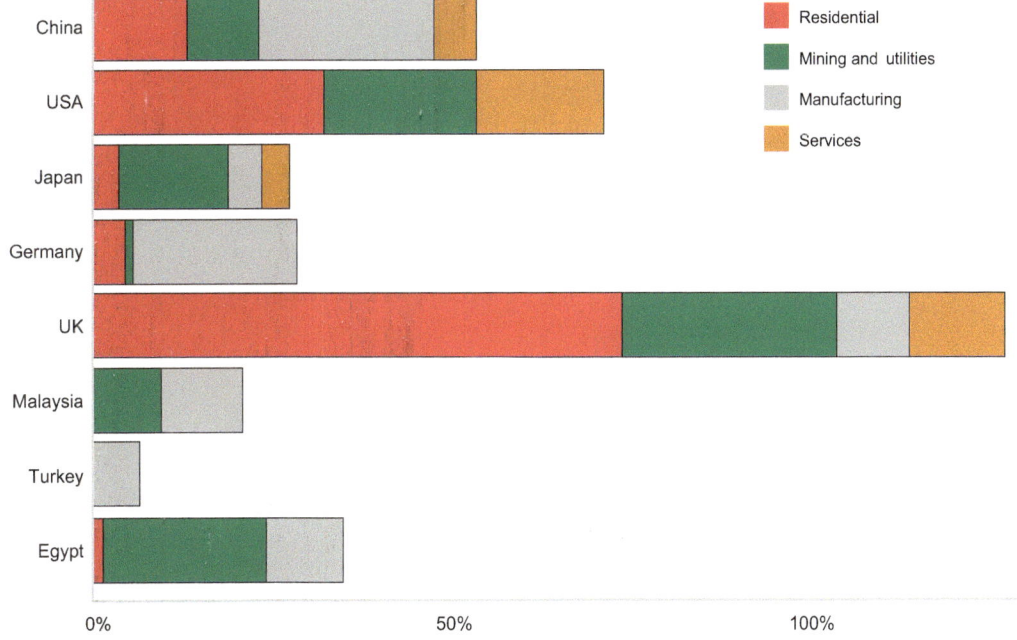

Fig. 2.4 Contribution of increased economic activity in increasing demand for natural gas. *Source* Formulated by Vivid Economics based on UN and IEA data

countries between 1982 and 2012, and for China between 2005 and 2012, based on LMDI.[1]

(I) Influence of economic growth on natural gas consumption

As economic activity increases, each sector's natural gas demand will increase. Figure 2.4 shows the proportion that the increase in natural gas demand as a result of economic growth encountered in the seven benchmark countries between 1982 and 2012 represents in terms of actual growth in demand (treating actual total natural gas growth in consumption as 100%). It is clear that in these seven benchmark countries the economic activities in all economic sectors grew, and each sector's natural gas demand also rose. However, the most notable increases were in the residential and public utility sectors. The

possible cause of this may have been that as incomes increased, household heating expenditure rose, and this was accompanied by increased usage of natural gas in power systems to satisfy the increased power demands. In addition, there was a growth in natural gas demand for manufacturing in each country. This is an indication that as economic activity increases, more natural gas is needed to satisfy the ever-growing demand for energy.

(II) Influence of energy intensity on natural gas consumption

Reduced economic sector energy intensity implies that the amount of natural gas required to generate the same output has dropped. Natural gas demand in China, the United States, Germany and the United Kingdom has seen a net reduction, which is primarily due to a decline in residential and manufacturing energy intensity (Fig. 2.5). There are also some countries, such as the United States, Japan and Egypt, where the mining industry and public utility sector energy intensity

---

[1]The period over which analysis is possible for China is relatively short, mainly due to the difficulties encountered in obtaining the pre-2005 data required to carry out this analysis.

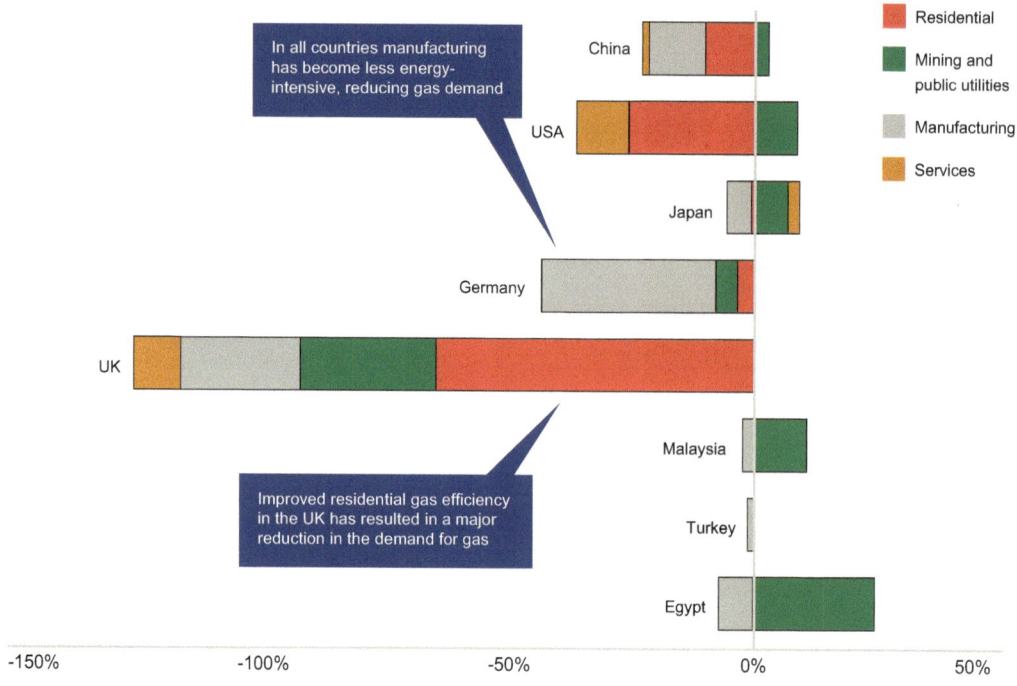

Fig. 2.5 Influence of energy intensity changes on natural gas demand. *Source* Formulated by Vivid Economics based on UN and IEA data

have seen increases. However, since it is not possible to break down mining industry, public utility and heating plant data, the underlying causes of these increases is difficult to analyse. In addition, over the same period of time, the non-gas-consuming sectors of the economy of any particular country (such as economic activity that requires electrical power) may also exhibit reduced energy intensity, resulting in an overall improvement in total energy intensity of these countries, such as that encountered in Japan. The results of this breakdown indicate that China, the United States, Germany and the United Kingdom all show reduced demand for natural gas as a result of reduced energy intensity.

(III) **Influence of fuel conversion on natural gas consumption**

Among the seven benchmark countries, all economic sectors chose to reduce usage of other fuels and to increase the proportion of natural gas. In other words, they achieved fuel

switching. This is a major factor behind the increase in natural gas demand (Fig. 2.6). The effect was particularly marked in the mining and public utility sectors, and their fuel conversion and the corresponding natural gas demand increase accounted for a large proportion of growth in consumption. In addition, China, Germany, the United Kingdom and Turkey, among other countries, also saw residential usage fuel conversion, which resulted in an increase in demand for natural gas. In the United States and Malaysia, manufacturing exhibited a relatively large-scale transition towards natural gas. In Japan, Germany and the United Kingdom, the service industry also showed similar trends.

2.1.3 The Importance of Fuel Switching in Natural Gas Demand Growth

Figure 2.7 is a summary of each country's natural gas consumption growth factors for

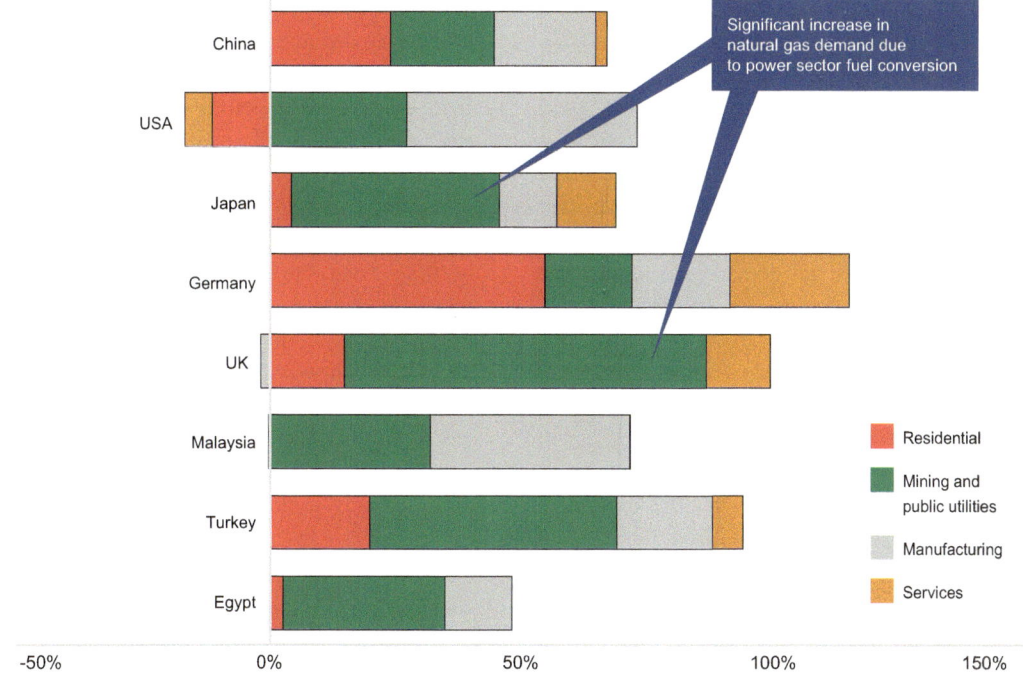

Fig. 2.6 Influence of fuel switching on natural gas demand. *Source* Formulated by Vivid Economics based on UN and IEA data

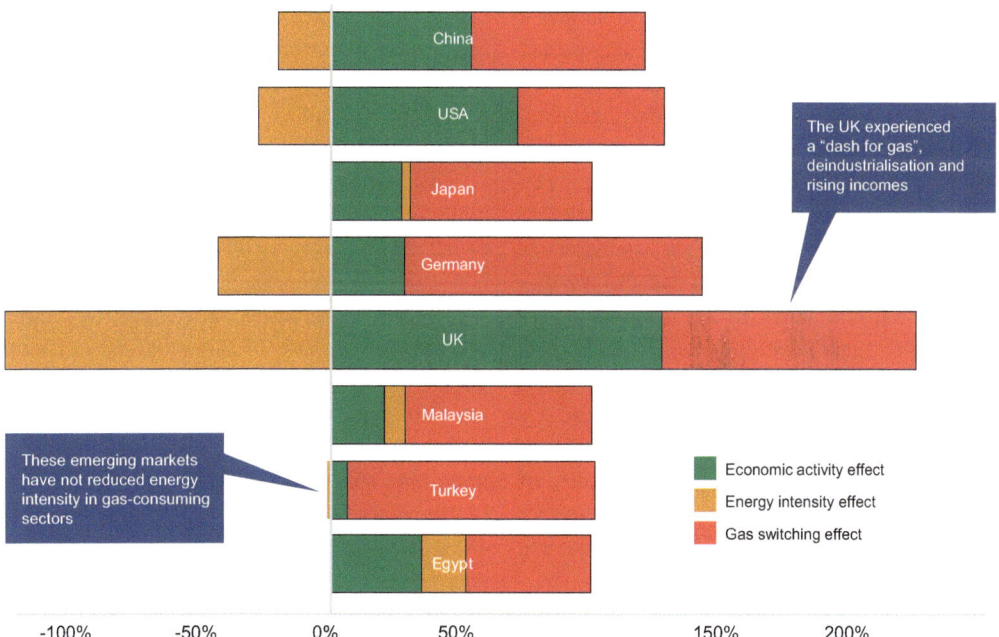

Fig. 2.7 Factor breakdown of benchmark country natural gas demand: overall results. *Source* Formulated by Vivid Economics based on UN and IEA data

1982–2012 (China 2005–2012) (each country's natural gas demand total change is set at 100%), giving an indication of the relative importance of each factor. From Fig. 2.7 we can see that, apart from the United States, in all other countries the influence of switching from other fuels to natural gas on natural gas demand was greater than that of economic activity. This shows that the main driver behind natural gas demand increases is the choice to adopt natural gas as a replacement for other fuels.

Analysis of this breakdown found that the nature of natural gas demand in the United Kingdom was different from other countries. In fact, all three drivers have a major influence in the United Kingdom. Between 1982 and 2012 the United Kingdom was primarily influenced by three major development trends:

- the "dash for gas"—many people converted from coal to natural gas, which played a major role in the overall fuel conversion figures;
- growth in GDP, which made the effects of economic activity more apparent;
- de-industrialisation and schemes to improve household energy efficiency, which resulted in a major increase in energy efficiency and had an opposite effect on natural gas demand.

Even though energy intensity factors can cause demand for natural gas to decline, the other two factors caused natural gas demand to increase. Each of these factors had a very significant result, equivalent to the total change in natural gas demand during the same period. The experience of the United Kingdom shows that, given the circumstances soon to be faced by China, for instance rapid development of domestic natural gas resources, high rates of economic growth and the economy becoming more service industry-oriented, factors that drive up natural gas demand will continue to be very significant.

Generally speaking, the seven benchmark countries analysed each experienced a large increase in natural gas demand. A detailed analysis of the results based on economic sector

shows that each of the three factors has the effect of influencing demand for natural gas as follows:

- Increased economic activity will cause natural gas demand in all economic sectors to increase, especially the power-hungry sectors of housing, mining and public utilities.
- Reduced energy intensity in residential housing and manufacturing will cause reduced natural gas demand.
- Fuel conversion is an important driver, in which the primary driver is the conversion of power plants and heating facilities to natural gas, while residential housing and manufacturing sector fuel conversion was also a major factor in most countries.

2.2 Primary Factors Motivating Switching from Other Fuels to Natural Gas

Based on international experience, natural gas is generally uncompetitive on pricing alone, compared to other fuels. It follows that there are other factors that contribute to the decision to switch from other fuels to natural gas and it is therefore necessary to explore these less-obvious factors more deeply.

This research employs two methods to explore the motivating factors behind fuel switching. The first consists of cross-sectional regression analysis conducted on the 2012 data for OECD countries, focusing on factors including: the share of the economy accounted for by the service industry; whether or not there are natural gas reserves; total natural gas reserves (adjusted by GDP); and degree of openness to trade. The second involves a descriptive analysis of the majority of the world's countries from 1980–2013 in terms of natural gas share of the energy mix and data on motivating factors towards natural gas, with a focus on several variables, including service industry proportion, manufacturing industry proportion, urbanisation and atmospheric pollution (including PM_{10} and SO_2).

2.2.1 Various Approaches to Natural Gas Replacement

1. **The seven benchmark countries primarily switched from oil to natural gas**

Each country uses oil, natural gas and coal in various proportions to satisfy its fossil fuel needs. As time passes, energy source structures gradually adjust. Figure 2.8 shows the fossil fuel energy mix changes over time in China and the benchmark countries. Each country's energy source structure change is shown as a pathway. The thin end represents the 1980 energy mix, and the thick end represents the 2012 energy mix. The energy mix change path follows the timeline from the thin end to the thick end.

If a country's path moves to the top of the triangle, this indicates that the share of natural gas increased in that country. If a country's curve moves away from "Oil" in the bottom left towards the top of the triangle, this indicates that natural gas replaced oil, as in Egypt (pink),

Turkey (purple) and Japan (green). If a country's path moves from away from "Coal" in the bottom right towards the top of the triangle, this indicates that coal had been replaced with natural gas in that country, as in United States (yellow) and the United Kingdom (blue). If the country's curve goes straight up toward the top of the triangle, then this indicates that natural gas has replaced both oil and coal in that country, such as in Germany over the past few years (grey). China's energy mix (red) has always been focused on coal, with only slight changes so far.

From Fig. 2.8 it is apparent that the countries with the fastest growth in demand for natural gas are those switching from oil to natural gas, whereas China is switching from coal to natural gas. To allow a conclusion to be reached that is of relevance to China, a large range of countries is required for comparison, in particular those that have converted from coal to natural gas, even if the scope of conversion to natural gas in these countries is smaller than that encountered in the seven benchmark countries.

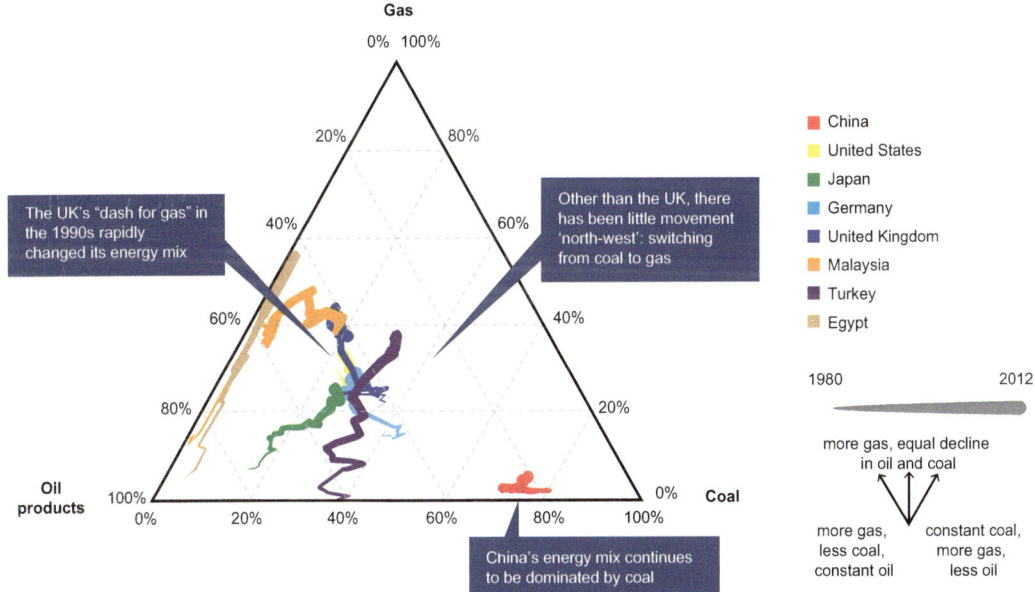

Fig. 2.8 Natural gas replacement pathways in China and benchmark countries (1980–2012). *Source* Formulated by Vivid Economics based on IEA data

2. **Main replacement pathways in countries transitioning from coal to gas**

Conversion from coal to gas is mainly encountered in Europe, Australia and the United States. Figure 2.9 shows the changes in energy structure in terms of coal and natural gas for major countries that have transitioned from coal to gas. Curves running from the bottom right-hand corner represent a high percentage of coal, while movement toward the top left corner is toward a high percentage of natural gas. Developed countries, primarily in Europe, and especially the United Kingdom, have seen a high proportion of coal usage transition toward a high proportion of natural gas. From the perspective of overall global trends (orange pathway) in recent years relating to fossil fuels, coal proportions have seen some increase, primarily due to declines in oil rather than natural gas. The motivating factors for European countries to switch to natural gas may

also be of relevance when considering switching from coal to gas in China.

2.2.2 Analysis of Driving Factors in OECD Member Countries Switching to Natural Gas

We conducted regression analysis on natural gas consumption proportions and other factors for OECD countries for 2012. The aim of this analysis was to identify the primary motivating factors behind OECD countries switching to natural gas.

Based on the regression results, there were four factors—service industry share, normalised GDP-adjusted total natural gas reserves, degree of openness to trade (a ratio between the import/export totals and GDP) and current natural gas reserves—that acted as major variables in

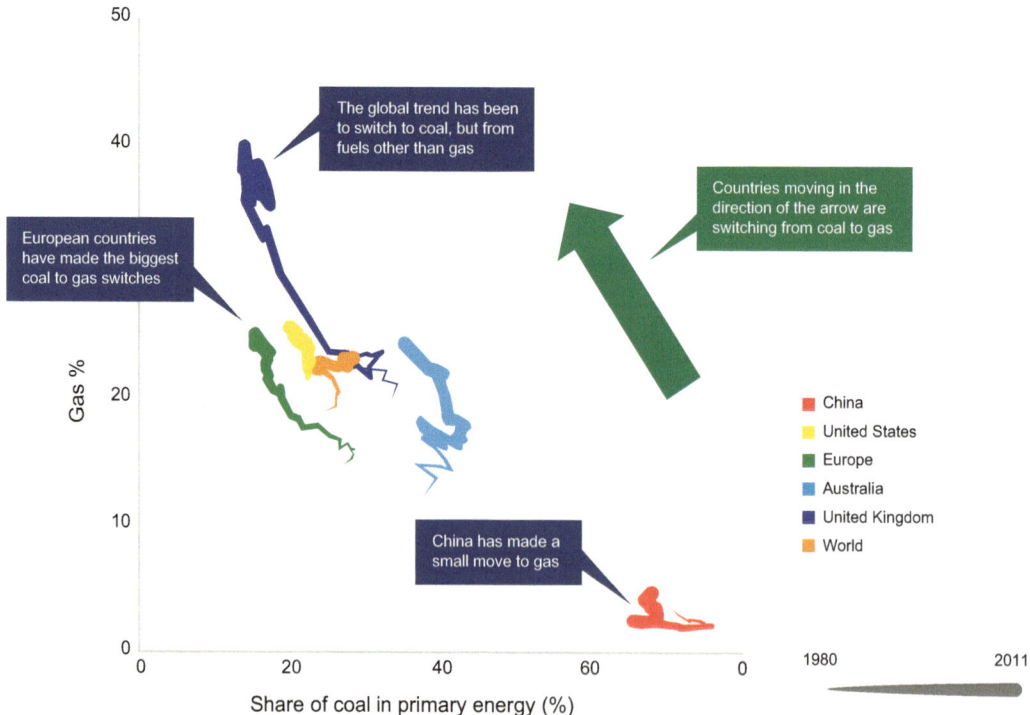

Fig. 2.9 Changes in energy usage from coal to natural gas. *Note* The data represents non-renewable energy source percentage; the data only shows countries that have experienced major transitions to gas. *Source* Formulated by Vivid Economics based on IEA data

natural gas share in OECD countries. Changes in these four variables have a pronounced influence on natural gas share. For example, if a member country's service industry share changes by 1%, then that country's natural gas proportion will change by 0.86% (Table 2.1). By conducting further computation, it can be deduced that these four variables are responsible for 53% of the changes in OECD member country natural gas share. That is to say, differences in OECD member countries' natural gas shares can mainly be traced back to these four variables.

When all factors are considered, service industry share is the factor that has the greatest effect on proportion of natural gas in OECD member countries. Figure 2.10 shows standard values for regression coefficients. By employing standard regression coefficients, one can compare the importance of such variables. The higher the standard value, the greater the importance.

Service industry share is the most important variable, followed by GDP-adjusted natural gas reserve levels. Based on these results, it is clear that service industry share is not only the most important variable, but is also an index that can be used as a yardstick for other driving factors. For example, countries with a larger service industry share could be at the stage of introducing increasingly strict controls on air quality, and this will further increase the market share of natural gas. However, this research does not incorporate air quality monitoring data; other variable factors that could explain natural gas share, such as urbanisation, were incorporated into the factors we considered, but did not affect the results. This is because by 2012 the majority of OECD member countries had already achieved a high degree of urbanisation, and thus there were no significant differences in urbanisation between them.

Table 2.1 Role of four variables in OECD country natural gas share

Variable	An increase of 1% of any variable will cause:
Share of service industry Gross value added in services sector ($GVA_{services}$/GVA)	Natural gas share rises by 0.86%
Natural gas reserve levels (adjusted for normalised GDP) Reserve levels/GDP	Natural gas share rises by 0.49%
Degree of openness to trade (Imports + Exports)/GDP	Natural gas share rises by 0.09%
Whether there are natural gas reserves or not 1 represents entirely no reserves, 0 represents the existence of reserves	If a country does not have natural gas reserves, then natural gas share drops 0.15%

Note This cross-sectional regression analysis uses 2012 data
Source Vivid Economics

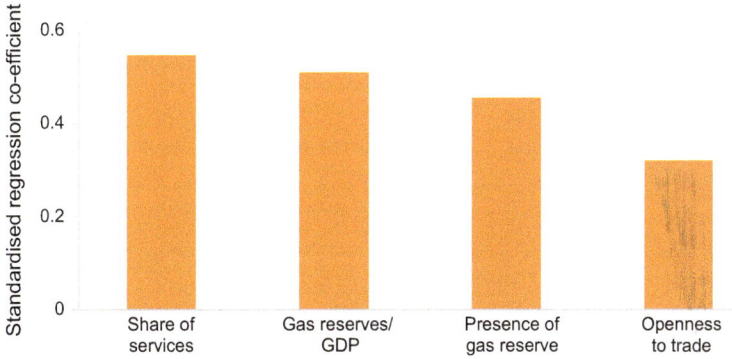

Fig. 2.10 Regression results for 2012 natural gas share in OECD countries

The OECD country analysis results show that when a country has natural gas reserves and has already entered the stage of development in which services are the primary focus, natural gas accounts for a very large share of the energy mix. Domestic reserves are undoubtedly important, because those reserves are often a country's cheapest source of natural gas and, regardless of how extensive they are, they will influence the natural gas share. This implies that countries without natural gas reserves might not make natural gas a primary energy source, that natural gas would not be used for infrastructure and by institutions, and that the import of natural gas would not be encouraged. The OECD countries South Korea and Japan are an exception to this, though; their domestic natural gas reserves are negligible, they have a high degree of openness to trade and they restrict other energy sources, which results in natural gas having a relatively high share in their energy structures.

The analytical data shows the influence of factors driving changes in the use of natural gas in developed countries, but does not consider the effects of urbanisation and atmospheric pollution. These factors are of major relevance in China, but they are less influential in OECD member countries. The next section will therefore expand the range of countries looked at so that a more comprehensive range of driving factors can be examined.

2.2.3 Further Analysis of Motivating Factors for a Country to Switch to Natural Gas

To further analyse the factors motivating a country to switch to natural gas, we carried out a comparison of data on natural gas share and natural gas motivating factors for most of the world's countries for the period 1980–2013. This permitted a more comprehensive analysis of how changes in each country's natural gas share correlated with changes in the variables. A global dataset was built for the years 1980–2013,

including natural gas share, service industry share, manufacturing share, urbanisation and atmospheric pollution including PM_{10}, and SO_2 for each country. The changing trends are outlined in Figs. 2.11, 2.12, 2.13, 2.14, 2.15 and 2.16.

1. **Apart from China, service industry share is often closely connected to natural gas share**

Changes in service industry share and natural gas share exhibit the same trends. In other words, when one variable changes, the other variable changes along with it. This kind of common trend suggests that a connection exists between the two and all that remains is for it to be confirmed by statistical analysis. There is a pronounced pattern in the United Kingdom, Japan and South Korea, and many other countries also exhibit such characteristics. In Fig. 2.11, only the data from China fails to demonstrate a relationship between service industry share and natural gas share. Nonetheless, China's service industry share is similar to levels in Japan and South Korea in 1980, so it is likely that this trend will manifest in the future.

2. **The relationship between the manufacturing industry share and natural gas share is not pronounced**

Only in a minority of countries is there a connection between manufacturing share and natural gas share; in the majority of countries the two do not have a pronounced relationship. Figure 2.12 shows that, with the exception of China, Asia's emerging markets all use natural gas to develop their manufacturing industries. In Europe, the natural gas share continues to rise as deindustrialisation proceeds. This implies that the relationship between manufacturing industry share and natural gas share could change depending on industrial and energy mix policies, as well as changing as a result of economic development.

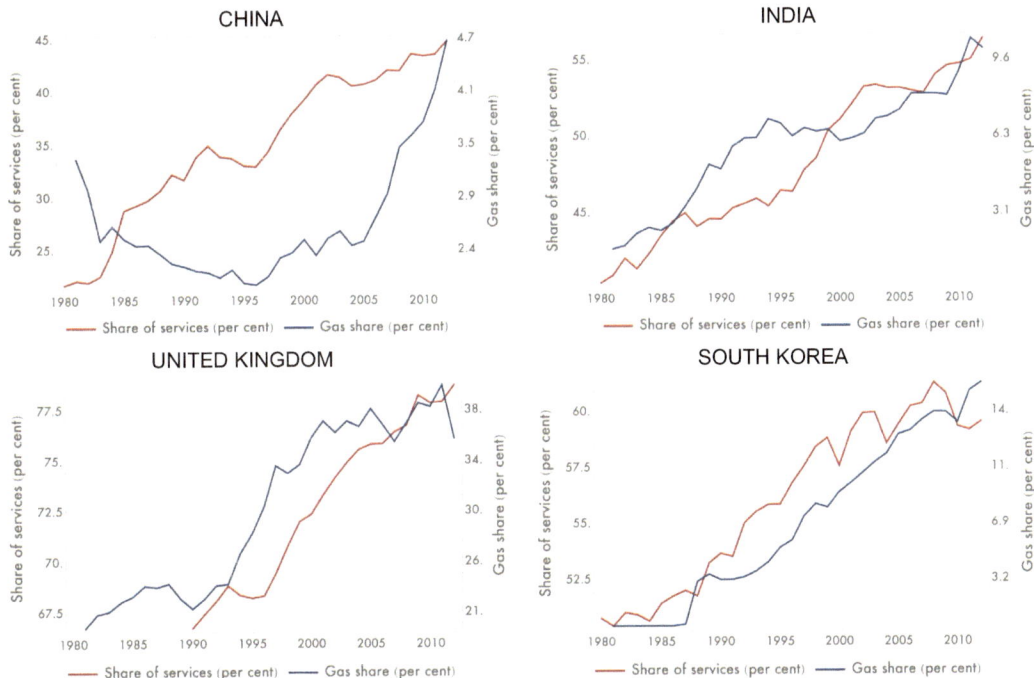

Fig. 2.11 Size of service industry and natural gas share of energy mix, 1980–2010. *Note* Natural gas share refers to the share of natural gas within total non-renewable energy sources. *Source* Vivid Economics, based on EIA and World Bank data

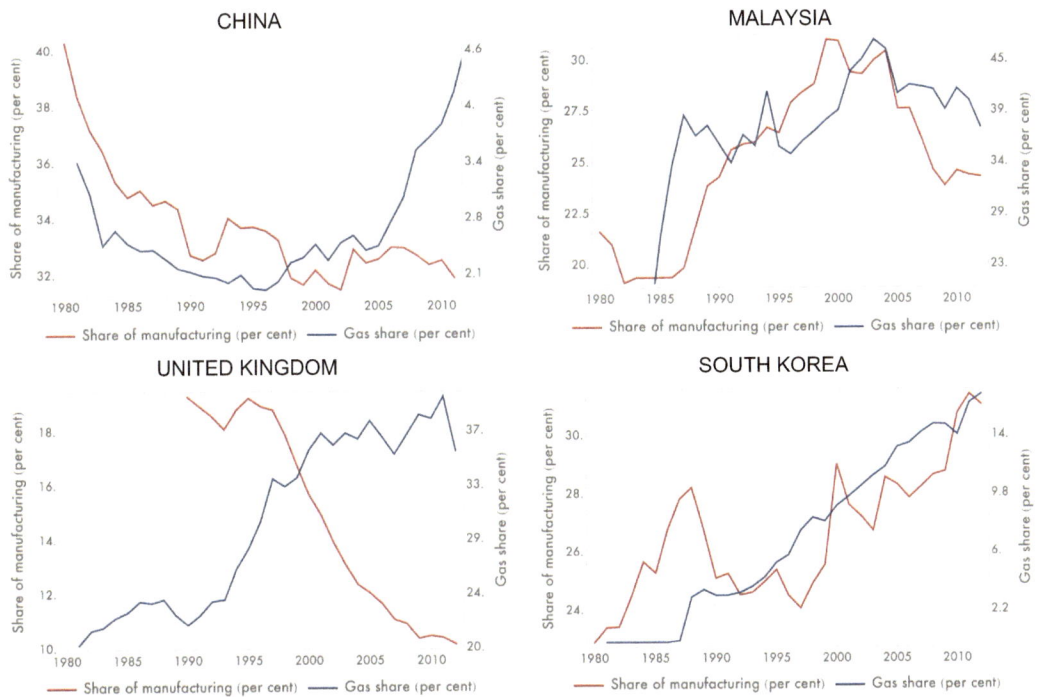

Fig. 2.12 Size of manufacturing industry and natural gas share of energy mix, 1980–2010. *Note* Data is for share of gas in primary energy. *Source* Vivid Economics, based on EIA and World Bank data

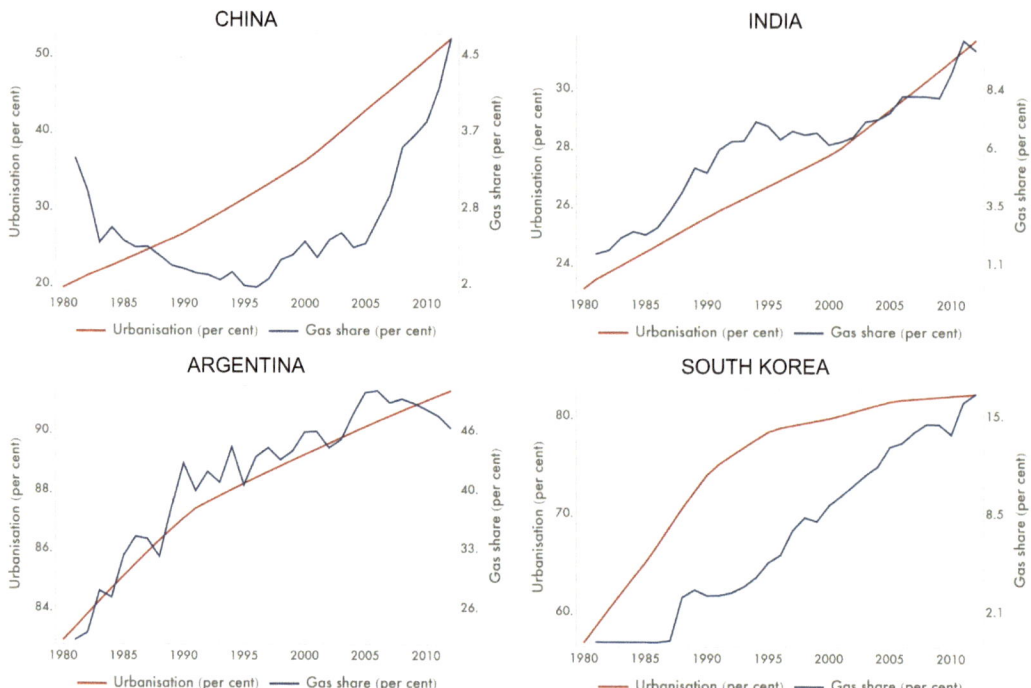

Fig. 2.13 Urbanisation and natural gas share of energy mix, 1980–2010. *Note* Data is for share of gas in primary energy. *Source* Vivid Economics, based on EIA and World Bank data

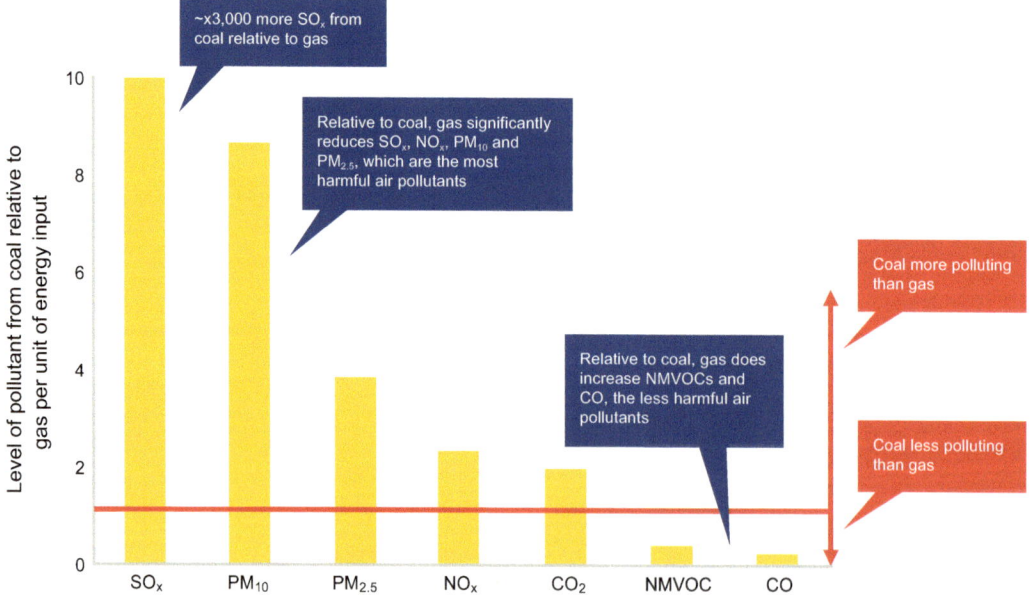

Fig. 2.14 Comparison of pollution resulting from natural gas power generation and from coal power generation, per unit calorific value. *Source* Formulated by Vivid Economics based on EEA and UNFCCC data

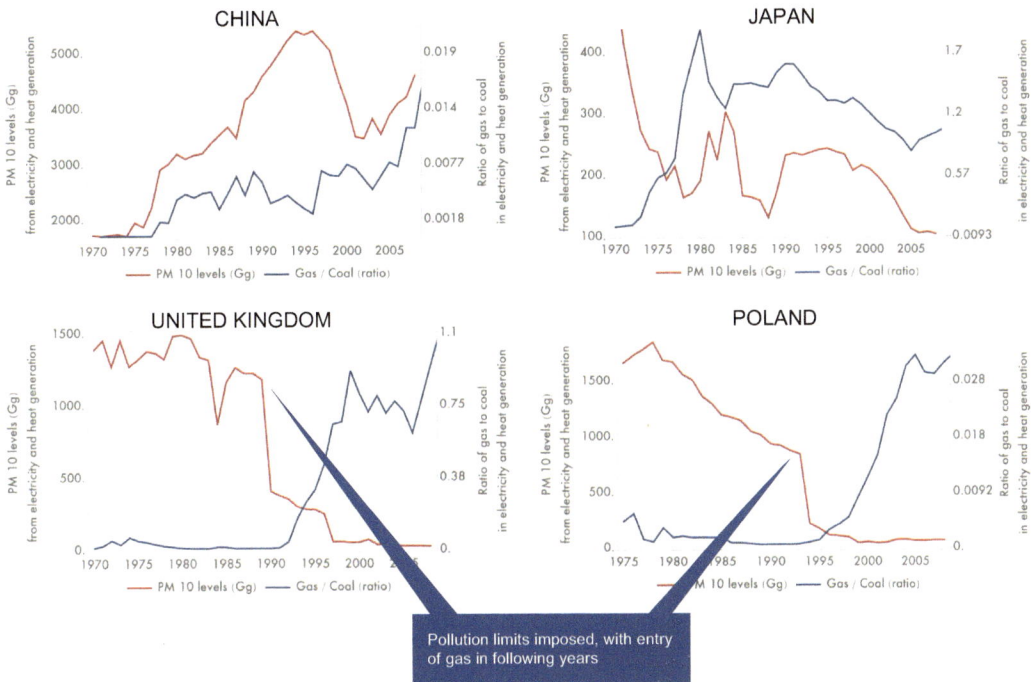

Fig. 2.15 PM_{10} emission levels and natural gas share of heat and power generation, 1970–2010. *Source* Formulated by Vivid Economics based on EEA and UNFCCC data

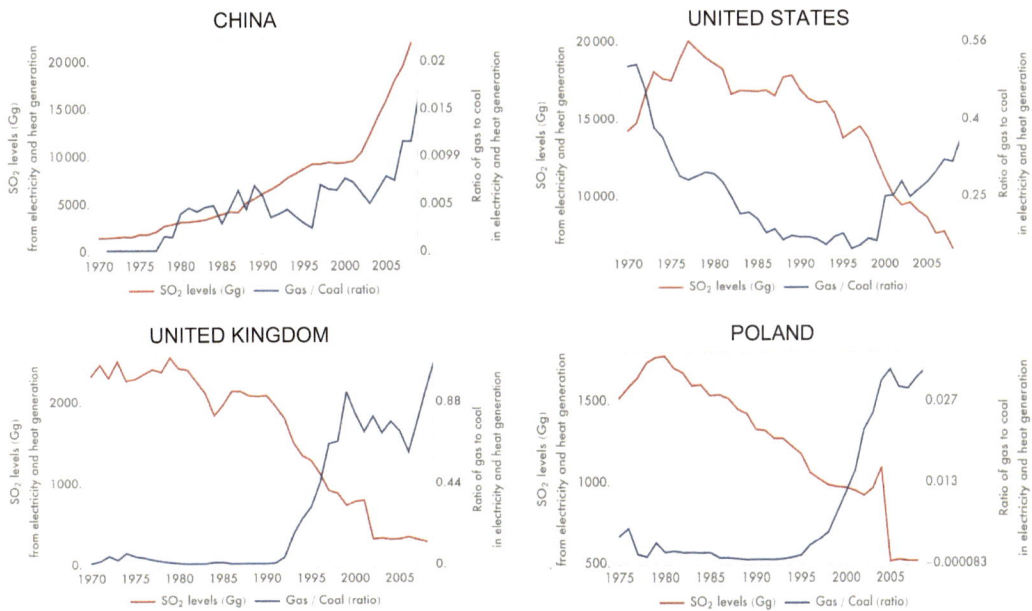

Fig. 2.16 Relation between power generation, SO_2 emission levels and natural gas share of heat and power generation, 1970–2010. *Source* Formulated by Vivid Economics based on EEA and UNFCCC data

3. **Urbanisation and natural gas share are closely related**

Urbanisation is closely related to natural gas usage, and China is now in the early stages of urbanisation. From Fig. 2.13 it can be seen that urbanisation and natural gas share are closely related in many countries. This is very likely due to the fact that natural gas is particularly well suited to use as a fuel in urban environments, since it is easily transported via networks of pipes, its combustion is easily controlled and it generates minimal atmospheric pollution. This is very important to an area with a high population density. So far, China does not exhibit this trend in the same way as other countries, but because China is in the early stages of urbanisation, greater demand for natural gas may result as urbanisation progresses.

4. **Many countries switch from coal to natural gas after enacting air quality legislation**

Compared to coal, natural gas is a comparatively clean fuel. For example, when power generation with natural gas is compared to power generation with coal, natural gas only results in small amounts of the most harmful air pollutants (Fig. 2.14). Many countries have therefore switched from coal to gas after air quality legislation is introduced.

5. **In countries where there are competitive power markets, the natural gas power generation share increases after PM_{10} restrictions are put in place**

Figure 2.15 shows that in the United Kingdom and Poland, PM_{10} emissions generated by power generation dropped significantly in the year in which air quality legislation was introduced. Furthermore, the share of natural gas power generation then increased rapidly. This shows that in the completely competitive power markets of the United Kingdom and Poland, even though efforts were made to restrict PM_{10} emissions by other courses of action, the cost of coal power generation eventually became prohibitive and power generation began its transition to natural

gas. Figure 2.15 also shows that in Japan, where the power generation market is more regulated, the increase in the share of natural gas power generation has not lagged behind changes in PM_{10}, though there was also a marked direct effect in the 1970s, which shows that natural gas can be used directly to control atmospheric pollution. In China, even though the share of natural gas power generation is increasing, the total remains relatively small, and coal power generation's overwhelming growth suggests that the PM_{10} levels are set to remain elevated.

Figure 2.16 shows that in the United Kingdom, the United States, Poland and other countries, SO_2 emission levels improved significantly after air quality legislation was enacted. The direction taken by changes to SO_2 emissions may be similar to that for PM_{10} emissions, but the rate of reduction is slower. For coal-fired power stations, compared to satisfying the restrictions on PM_{10}, SO_2 restrictions will only be met successfully if they are adopted at a more gradual pace. In China, the enormous expansion of coal-fired power stations means that SO_2 levels are set to remain high.

Many countries have implemented strategies to improve air quality by switching to natural gas. Analysis reveals that the approaches to implementing such strategies differ from country to country. Some countries rely on market forces and pollution control measures which force up the cost of coal, causing a transition to gas. Other countries adopt natural gas development planning. The ultimate result of these different approaches to controlling air quality is essentially the same, but the results in terms of controlling atmospheric pollution become apparent at different stages.

2.2.4 Summary

The main conclusions that can be drawn from our analysis of the drivers for transition to natural gas are that, in the course of development, an increase in the size of the service sector, the pressures of urbanisation and an urgent need to control atmospheric pollution result in all countries switching gradually to natural gas. Major adjustments have been made in Europe, the United States and Australia in switching from coal to natural gas—and this now also needs to occur in China. At first glance, these adjustments are a result of a growing service sector share driving an increase in demand for natural gas, but there is also evidence that other trends drive demand for natural gas, such as urbanisation and demands for cleaner air. These two factors play a major role in the motivating the transition in many countries. As urbanisation in China proceeds, an increasing amount of attention is being paid to air quality, and this is likely to encourage greater use of natural gas.

2.3 The Influence of Natural Gas Price on Demand

As well as factors such as service sector growth, urbanisation and air quality controls, natural gas price also has a direct effect on energy costs, and can play an important role in affecting demand. This section carries out an in-depth analysis of the relationship between natural gas price and demand. The analysis mainly concentrates on the experiences of OECD countries, because as they are the primary natural gas-consuming countries, there is relatively good data available. Furthermore, in the majority of cases, these countries have liberalised energy price markets, making it easier to follow the changes in natural gas pricing, and easier to analyse the relationship between pricing and demand.

2.3.1 The Relationship Between Natural Gas Price Changes and Changes in Demand in OECD Countries Is not Pronounced

1. **Natural gas demand in OECD countries grew rapidly from relatively low levels between 1987 and 2000**

Figures 2.17, 2.18, 2.19 and 2.20 show the evolutionary progress of natural gas prices and the demand of main sectors in the United Kingdom, the United States, Japan and Germany. Each point represents data for a year. The colours of the points represent annual rates of change in natural gas demand. The vertical axis plots the natural gas price to industry, with prices expressed as 2010 PPP (purchasing power parity —that is, including inflation and relative price change adjustments). This price thus represents changes in natural gas affordability. There is a clear increase in natural gas prices during the early 1980s, with a drop in 1987. Since then they have remained fairly stable until 2000. Subsequently, and in particular from 2005 until the Global Financial Crisis and the shale natural gas revolution in the United States, prices have risen continuously. Although industry sector gas prices are considered, it seems that they have increased at the same rate—the price of residential gas is higher than that used by industry.

Based on the experience of four countries, natural gas prices between 1987 and 2000 were low and stable and yet, compared to other periods, natural gas demand continued to grow rapidly. During this period, the number of years when there was higher growth in sector demand for natural gas was greater than in the period when the price was higher. For example, Japan and the United Kingdom saw more years with increased demand during this period than at other times (tending towards red on the scale).

2. **Natural gas demand rose in many countries when natural gas prices were high**

Even though natural gas prices began to rise in 2000, in the United Kingdom, Japan and Germany, demand for natural gas remained consistently steady or grew. From 2000 to 2008, natural gas prices in OECD countries actually rose by 60%. Despite this, during the overlapping period of 2000–2012, industrial and residential natural gas demand in OECD countries only dropped by 8%. This indicates that, once demand stabilises, it is capable of withstanding wide economic fluctuations. This could be due to the fact that natural gas demand is not sensitive to price changes (i.e. it is non-elastic), at least in

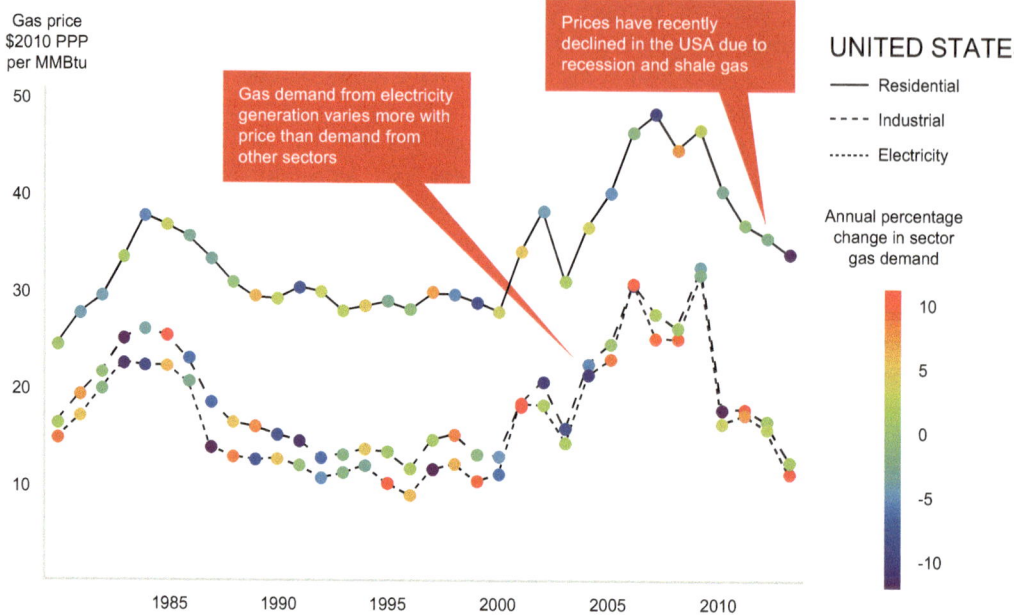

Fig. 2.17 Relationship between growth in demand for natural gas and pricing in three sectors in the United States. *Source* Vivid Economics based on IEA data

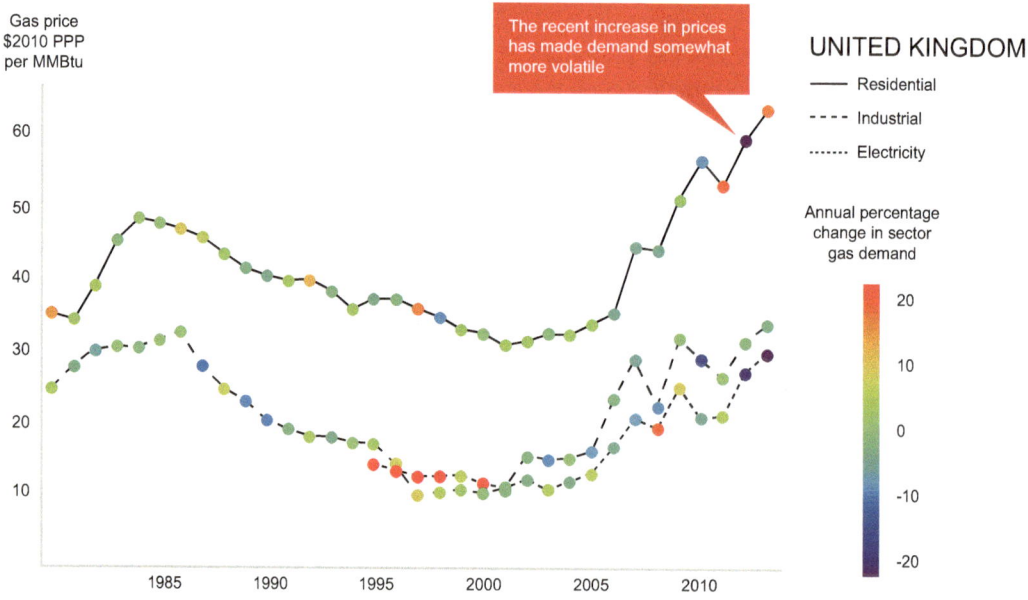

Fig. 2.18 Relationship between growth in demand for natural gas and pricing in three sectors in the United Kingdom. *Source* Vivid Economics, based on IEA data

terms of residential usage. Moreover, once natural gas is adopted, residents are not willing to give up natural gas for other fuels. As a clean fuel, allowing controllable heating and with easily accessible exothermic properties, when natural gas is compared to other fuel types, such as coal, it becomes apparent that due to its individual properties, as soon as the necessary

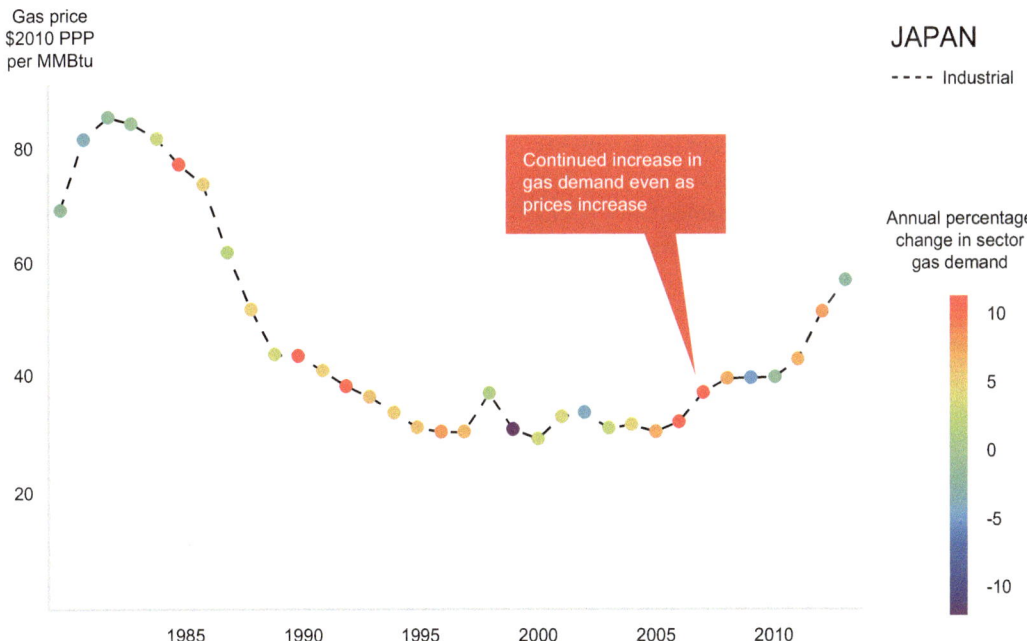

Fig. 2.19 Relationship between growth in demand for natural gas and pricing in industry in Japan. *Source* Vivid Economics, based on IEA data

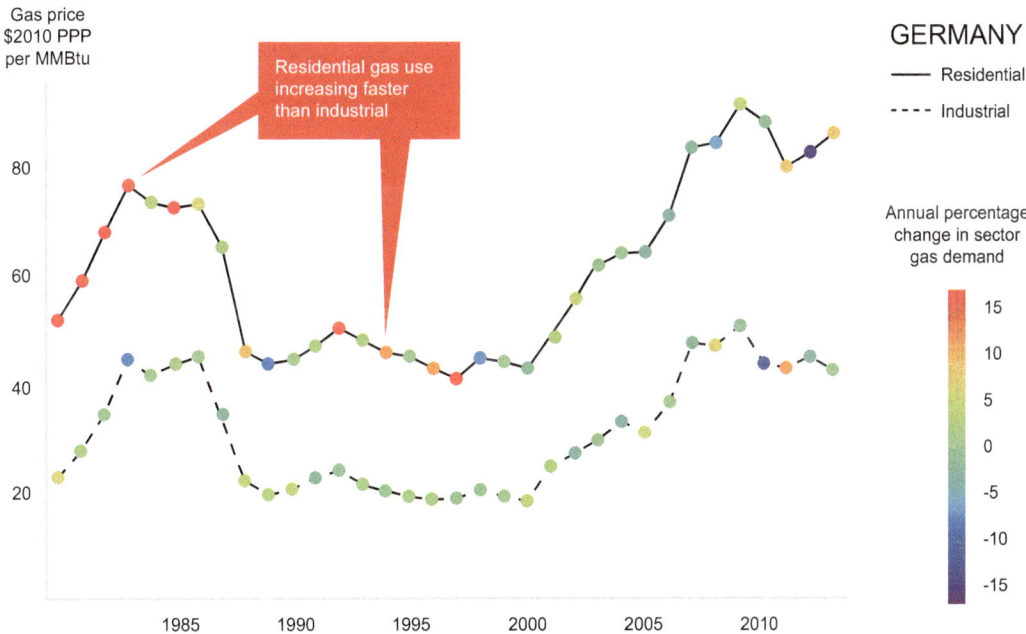

Fig. 2.20 Relationship between growth in demand for natural gas and pricing in the residential and industrial sectors in Germany. *Source* Vivid Economics, based on IEA data

infrastructure is in place, residents and manufacturers are less concerned about price. From
these observations we can see a kind of "ratchet
effect": once natural gas demand reaches a certain level, the likelihood of switching to another
energy source is lower, therefore there is less
likely to be a reduction in demand. Only scientific innovations such as electric heat pumps, that
may be capable of offering even higher-quality
heat and efficiency, will change the current
"non-elastic" state of affairs.

On the whole, during periods when natural
gas prices are rising, increases in demand for
natural gas from manufacturing, residential and
power generation sectors seem to keep
step. Despite a lack of data covering all industries
in all countries, the data that is available shows
that demand for natural gas generally increases
across the board in all sectors. Although this is
not applicable to the United States, the United
Kingdom, Japan and Germany, it is still applicable to other major natural gas-consuming
countries such as The Netherlands and Spain.

3. **No pronounced relationship between sector gas price and demand share**

Figure 2.21 shows natural gas PPP price against
overall demand share in various sectors for OECD
countries in 2012. This shows that there is no pronounced relationship between sector gas price and
demand share in each country; although the
industrial sector gas price in some of these countries
is relatively low, the industrial sector gas demand
share in those countries is not necessarily high.

From Fig. 2.21 it is also possible to see that the
price of residential sector gas in OECD countries is
relatively high, with residents willing to pay higher
gas charges. This could be because residential
natural gas usage requires the economies of scale,
since the transportation and delivery costs are much
higher. Practically speaking, there is no pronounced relationship between residential demand
and gas price, since the primary replacement energy
source for residential usage is electricity. Electricity
is relatively more expensive, and natural gas is
more convenient. In fact, the high price of natural
gas doesn't seem to have had a negative impact on
residential demand for natural gas. Over recent
years there has been a greater increase in demand
for gas for residential use than for industrial use in
OECD countries.

2.3.2 Limited Effect on Demand of Difference in Price Between Natural Gas and Other Energy Sources

Economically speaking, the demand for a particular item is not only determined by its price, but
also by the price of alternatives. This suggests that

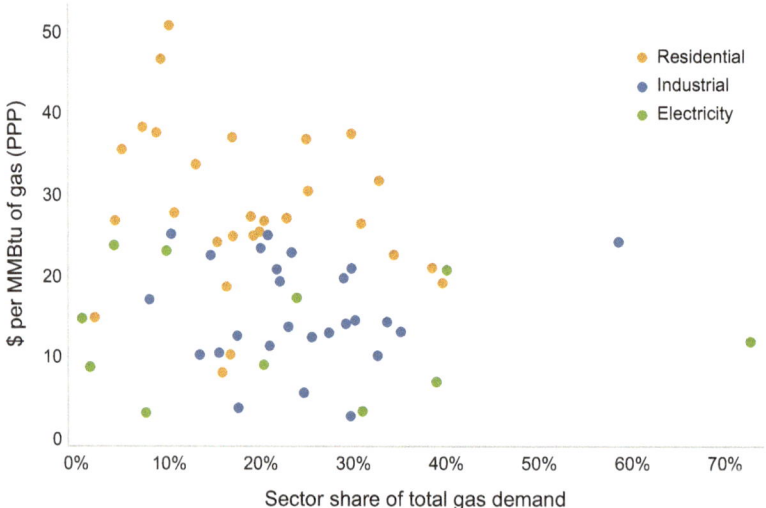

Fig. 2.21 Relationship between sector gas price and share in OECD countries (2012). *Source* Vivid Economics, based
on IEA data

natural gas demand will also be affected by the difference in price between natural gas and other energy sources.

1. The difference in price between natural gas and electricity in OECD countries has not affected the demand for gas in the industrial and residential sectors

In OECD countries, the natural gas price is lower than that of electricity in both residential and manufacturing sectors. Compared to countries where the price of electricity is lower than that of natural gas, natural gas has a slightly higher share of primary energy sources, though this effect is still not all that pronounced. Figures 2.22 and 2.23 show the relationship between the pricing of natural gas and electricity in different countries and the natural gas share in the industrial and residential sectors. The dots in the diagram represent different countries, while the colour of the dots indicates natural gas share.

Figure 2.22 shows the residential sector, while Fig. 2.23 shows the manufacturing sector. These two diagrams demonstrate that pricing competition between the residential and

manufacturing sectors has very little influence on demand for natural gas or electricity.

2. Rate of growth in demand for natural gas in OECD countries was especially rapid during a period of stable relative energy source prices

From 1987 to 2000, demand for natural gas in OECD countries grew rapidly. However, during that period the prices of natural gas, electricity and coal remained relatively stable. Figures 2.24 and 2.25 both indicate this situation. For example, during this period, looking at the actual unit price of energy in the manufacturing sector, the mean gas price was only 22% of the electricity price, or around one quarter. However, compared to the price of coal, the price of natural gas was over three times higher, at around 320%. When relative prices remain stable, the competitiveness of each energy source remains unchanged, and thus demand will not change as a result of pricing competition between energy sources. However, it was under precisely such circumstances that natural gas demand nevertheless grew steeply, an indication that factors other than price play a

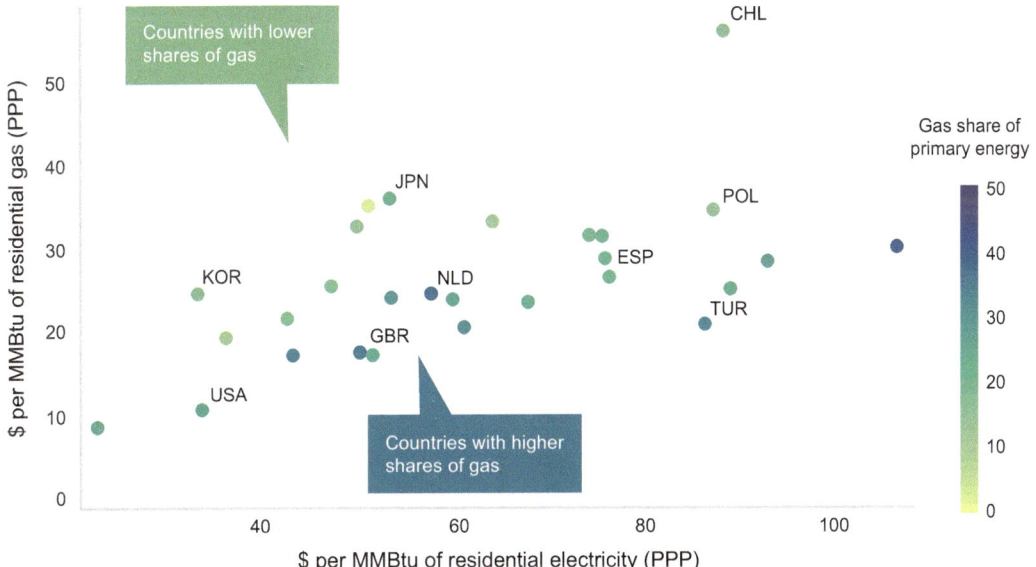

Fig. 2.22 Residential gas share and pricing differences between gas and electricity. *Note* Data is for 2012, electrical units have been converted from MWh to MMBtu. *Source* Vivid Economics, based on IEA data

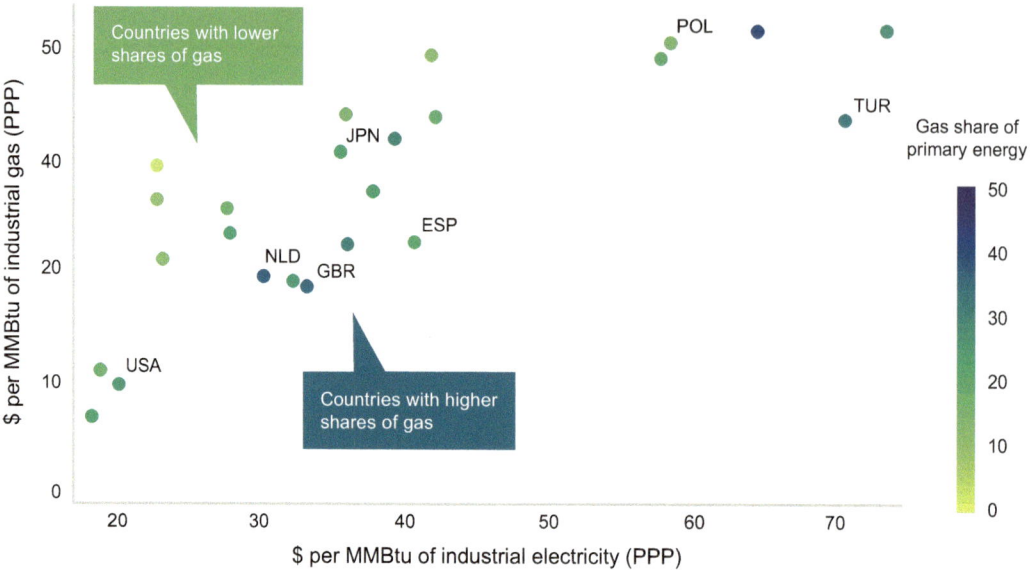

Fig. 2.23 Industrial sector natural gas proportion and variance in natural gas and electricity price. *Note* Data is for 2012, electrical units have been converted from MWh to MMBtu. *Source* Vivid Economics, based on IEA data

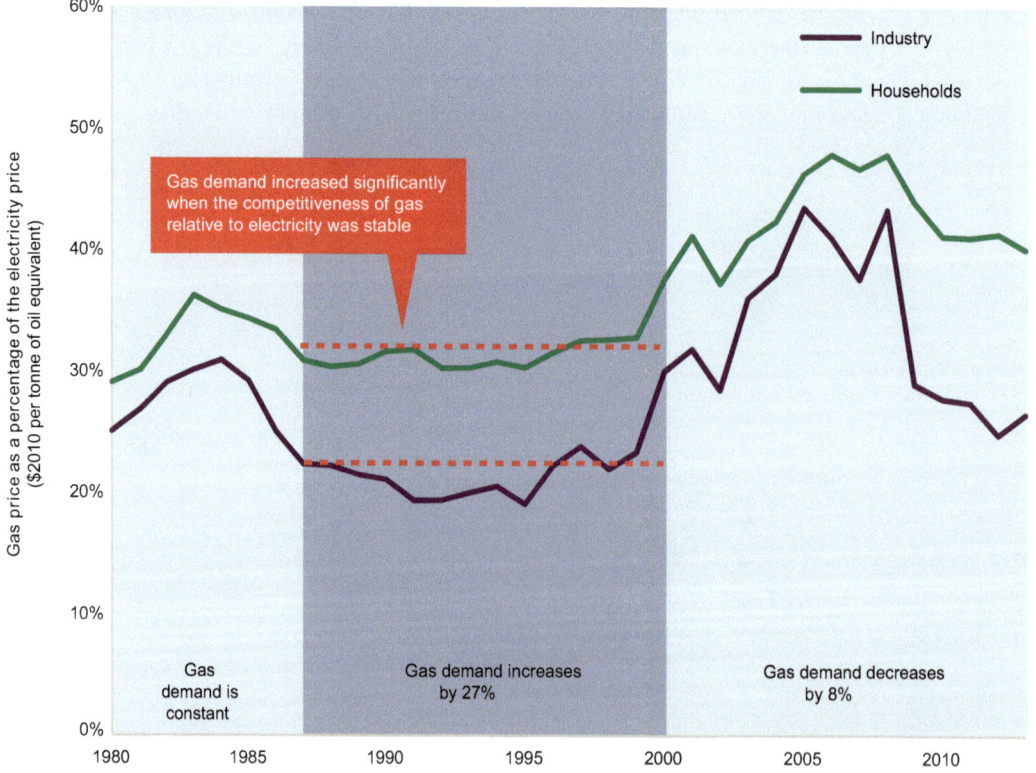

Fig. 2.24 Natural gas and electricity price difference and demand growth. *Data source* Vivid Economics quoting the IEA

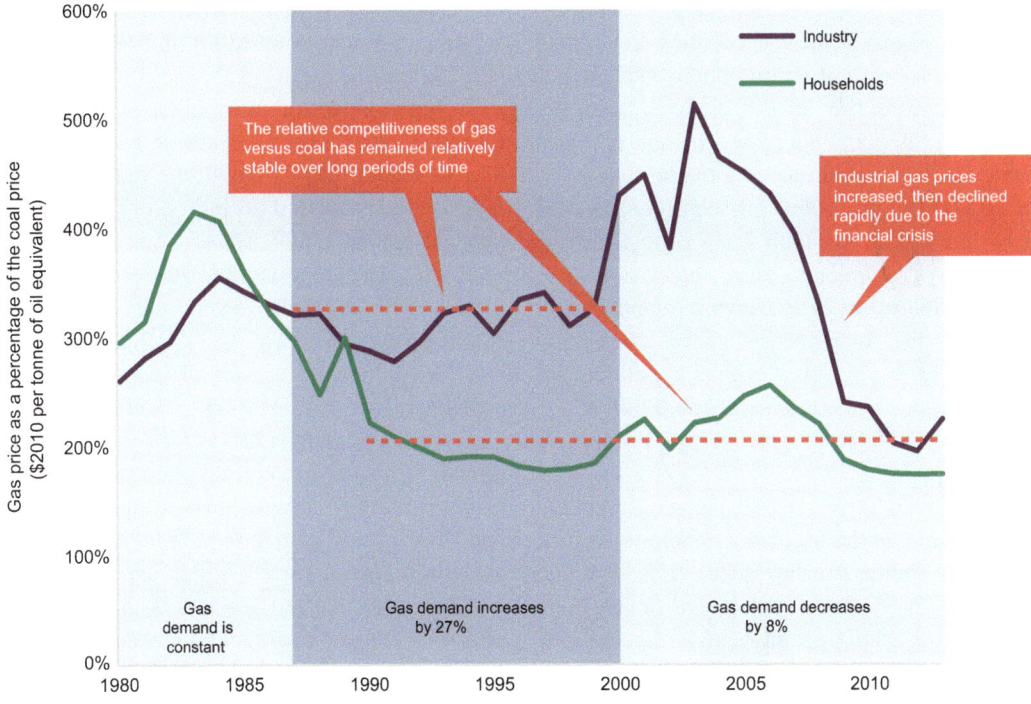

Fig. 2.25 Natural gas and coal price difference and demand growth. *Note* Natural gas demand is for industrial and residential usage. The red dotted line is the average level during the same period of time. *Data source* Vivid Economics quoting the IEA

significant role. Another conclusion that can be drawn is that pricing competition between energy sources is not actually all that pronounced.

Among OECD countries, competition between oil and natural gas is only on a small scale. In OECD countries, natural gas is generally not used as a transportation fuel, so it rarely competes with oil. In addition, oil is now less likely to be used for power generation in OECD countries. Due to this, competition between oil and natural gas is not pronounced in these countries.

In summary, natural gas demand seems to have no pronounced relationship to the price of natural gas. So, even if the price of natural gas in China remained high, it is not that likely to reduce demand for gas. Even though in OECD countries the price of gas is more competitive and gas has a higher overall share, these countries have not seen an increased demand for natural gas due to pricing competition between different energy sources. This is a direct result of

the important non-pricing-related characteristics of gas. These results show that even though gas prices in China have been high in the past, other factors such as urbanisation and atmospheric pollution that are now confronting China will most likely result in major growth in demand for natural gas.

2.4   Current State of Natural Gas Use in China and Future Trends

2.4.1   Sector Distribution and Total Gas Consumption in China Since 2000

1. **China's natural gas industry has grown rapidly, but overall share is still low**

Due to its relative abundance, coal has always played a major role in China's energy mix—its share of total energy consumption has never been less than 65%. After coal, oil is the second most important energy source, accounting for around 20% of total energy consumption. Natural gas consumption is the lowest, with a share not exceeding 6%, and the low amount of clean energy in the overall mix is a major deficiency. In 2000, China's natural gas consumption was only 24.5 billion m^3, with 2.2% of total energy source consumption. This had risen to 106.9 billion m^3 by 2010, having tripled in size. However, the total energy source share only reached 4.4%. In 2013, China's natural gas consumption had reached 166.0 billion m^3, accounting for 5.8% of total energy consumption (Fig. 2.26).

By looking at production and consumption, we can arrive at the following conclusions: first, an energy source structure focused on coal and oil will be hard-pressed to achieve fundamental change over a long period of time. Second, the importance of natural gas in China's energy source system is continually rising, but it is far from occupying a central position. To increase natural gas production and consumption would still require major government policy support.

2. Natural gas is primarily used in manufacturing, power generation and residential

In terms of sector distribution in China, the manufacturing industry has always been the main user of natural gas. In 2000, the manufacturing industry consumed 11.8 billion m^3 of natural gas, accounting for 48.5% of China's total natural gas consumption. In 2003, this figure reached 50%, the highest ever. Subsequently, the share began to decrease, and 2010 consumption was 57.3 billion m^3 (33.4% of total natural gas consumption).

Power generation is the sector with the fastest growth in natural gas consumption. In 2000, the power industry consumed 800 million cubic metres of natural gas, only 3.3% of total natural gas consumption. By 2006, this had doubled to 7.0%, and in 2012 it reached 23.5 billion m^3, 16.0% of China's total natural gas consumption.

Transport and residential are two other major sectors using gas. In 2000, transport and shipping used 880 million m^3, accounting for 3.6% of total consumption. By 2012, this had risen to 15.5 billion m^3, or 10.6% of consumption.

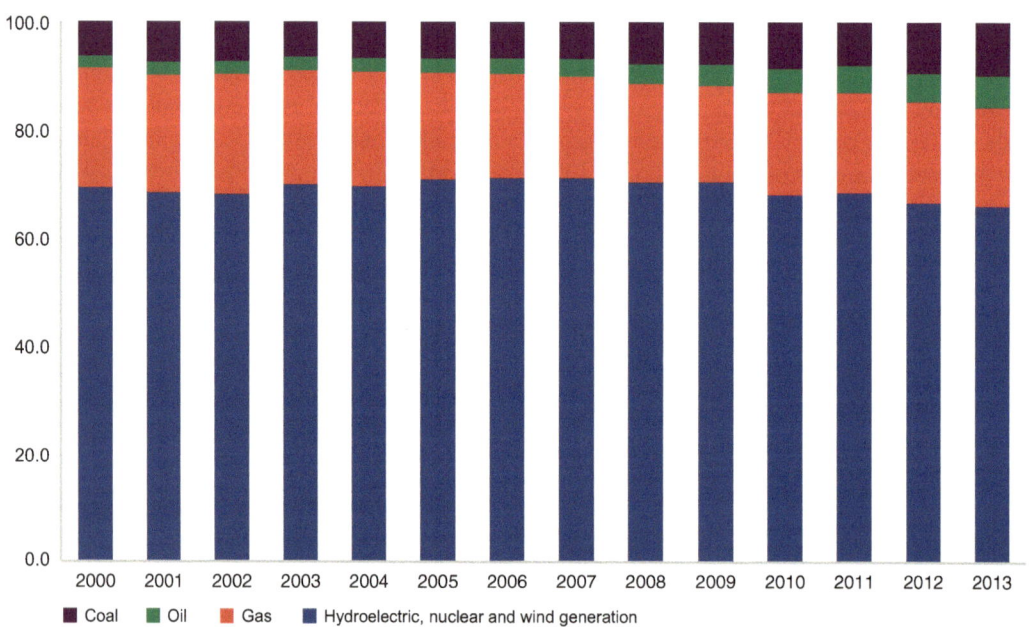

Fig. 2.26 Chinese energy source consumption structure (%)

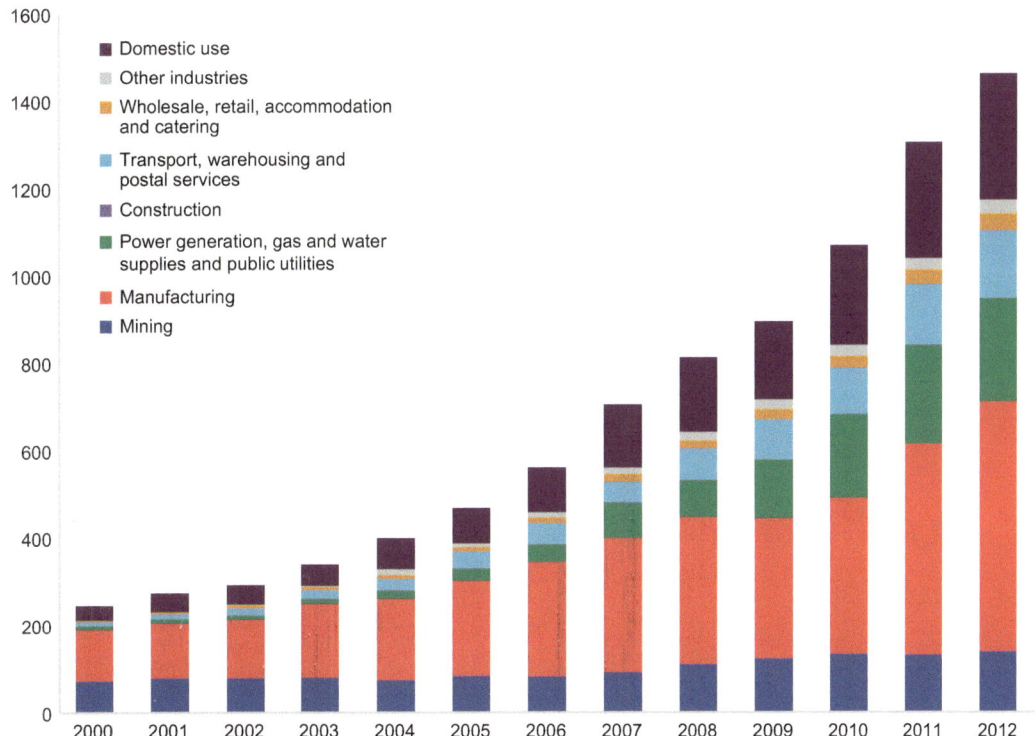

Fig. 2.27 Natural gas consumption in China by sector

Relatively speaking, the residential natural gas consumption share has been relatively stable. In 2000, residential consumption accounted for 13.2% of overall consumption, and this had risen to 19.7% by 2012 (Fig. 2.27).

3. **Compared to other countries, China's natural gas prices are relatively high**

In recent years, China's natural gas prices have increased relatively quickly. In the late 1980s, China's natural gas price was relatively low and stable. However, since 2000 the price of China's imported natural gas has risen, just as it has in the United Kingdom and Japan, as can be seen in Fig. 2.28. There was a decline in 2008 as a result of the Global Financial Crisis, but the rise continued after 2009.

Compared to other countries, at the same time as China experienced rapid growth in demand for natural gas it also had a relatively high natural gas price. Figure 2.29 shows rapid growth in demand for natural gas in OECD countries in the 1990s, but at that time prices were relatively low. China now experiences relatively high prices, a pronounced characteristic of the development of natural gas. In addition, China's natural gas market also exhibits problems, including price differences between users: compared with the situation elsewhere in the world, in China natural gas production prices are lower, residential natural gas prices are lower, but natural gas prices for industrial use are higher.

2.4.2 Mid- to Long-Term Energy and Natural Gas Development Plans in China

In 2013, the State Council published the Energy Source Development 12th Five-Year Plan. In June 2014, the State Council issued the Energy Source Development Strategy Action Plan

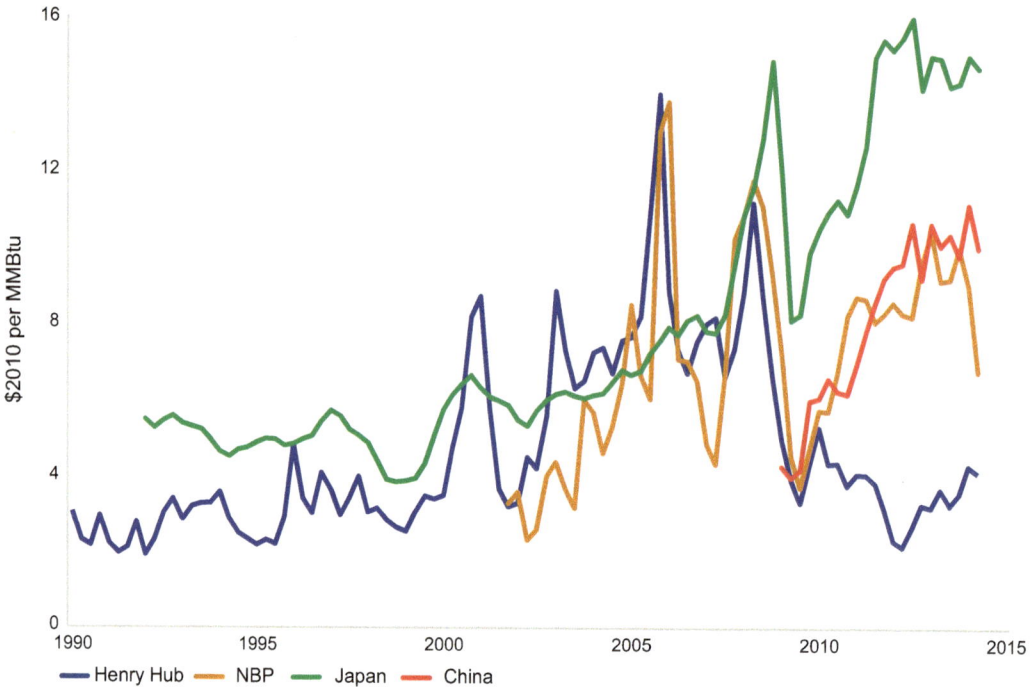

Fig. 2.28 Natural gas prices in China and other countries. *Data source* Vivid Economics

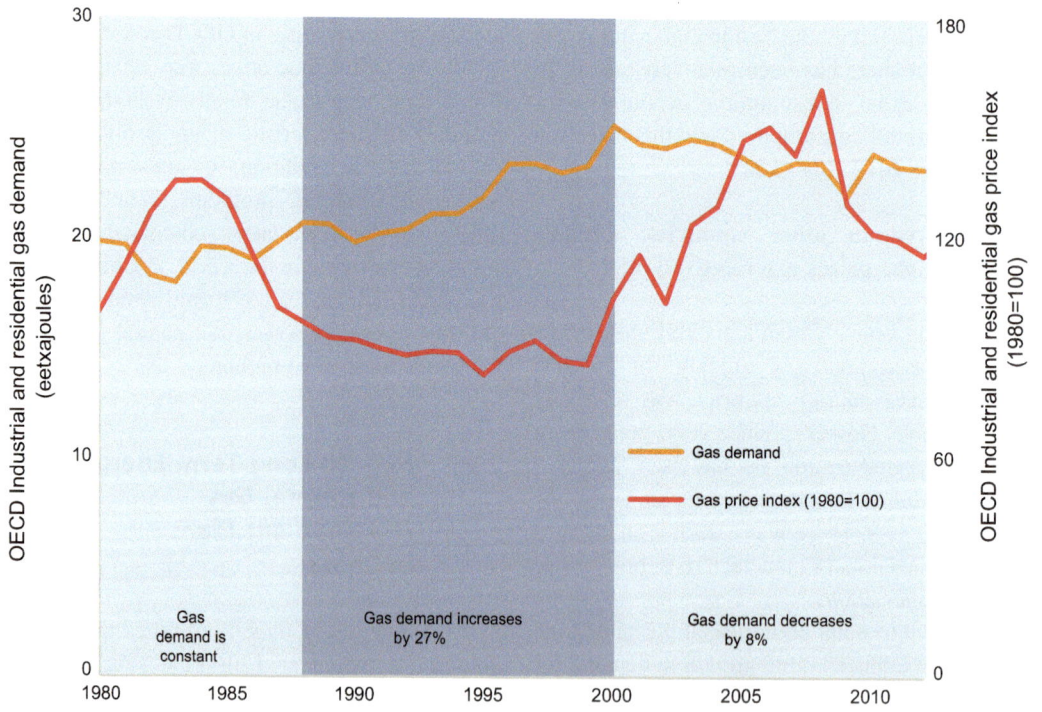

Fig. 2.29 Long-term movements in natural gas price and demand in OECD countries

(2014–2020) and in November 2014 China and the United States published the Sino-American Joint Declaration on Climate Change. These important documents laid out planning and outlines for China's energy source development up to 2030. Based on these documents, China's energy source development plan includes several important landmark tasks.

1. **Overall energy consumption and energy supply capabilities**

Prior to updating its energy statistical data, China set objectives for total volume of energy development.

In 2015, China's energy consumption will reach 4 billion tons of standard coal. By 2020, non-renewable energy consumption will be controlled at around 4.8 billion tons of standard coal, with total standard coal consumption controlled at around 3.0 billion tons, a drop of 62.5% in the proportion of total energy consumption.

By 2020, a relatively comprehensive energy security system will have been established. Domestic non-renewable energy source output will have reached 4.2 billion tons of standard coal, with energy self-sufficiency maintained at around 85%. In addition, the oil production to storage ratio will have increased to 14–15. By 2020, installed generating capacity of nuclear energy will have reached 58 GW, with another 30 GW of capacity under construction.

2. **Energy source structure**

By 2015, non-fossil fuel's share of energy source consumption will have risen to 11.4%, and non-fossil fuel's share of installed power generation capacity will have reached 30%. The natural gas primary energy source consumption share will have risen to 7.5%, while coal's share of consumption will have dropped to around 65%. By 2020, non-fossil fuel energy sources will account for 15% of primary energy consumption, natural gas's share will have reached 10% and coal's share will be controlled to within 62%. Newly added natural gas will be earmarked

to support residential usage and replace dispersed coal usage, while a programme will be introduced to encourage urban residents to adopt clean energy sources. By 2020, urban residents will essentially be using natural gas. By 2020, an attempt will have been made to ensure that standard hydroelectric power reaches around 350 million kW of installed capacity, while wind power will reach 200 million kW of installed capacity. The feed-in price of wind-generated and coal-generated electricity should be the same. Photovoltaic installed capacity will reach around 100 million kW. The price of photovoltaic electricity should be the same as the power grid electricity sale price.

By 2020, accumulated newly added proven reserves of conventional gas will be 0.55 trillion m^3, with annual production of standard natural gas of 185 billion m^3. By 2020, shale natural gas production will aim to exceed 30 billion m^3, while coalbed methane production volumes will aim to reach 30 billion m^3.

3. **Long-term objectives of the Chinese government to be achieved by 2030**

In November 2014, during the APEC summit, China and the United States released their Sino-American Joint Declaration on Climate Change, which revealed the long-term objectives of China up to 2030. China's intentions are that CO_2 emissions should peak by around 2030 and work should be carried out to try to achieve this peak sooner. Planning outlines include 2030 non-fossil fuel energy sources rising to account for around 20% of primary energy source consumption.

2.4.3 Mid- to Long-Term Natural Gas Development Trends for China Based on International Experiences

Compared to other countries, the share of China's current energy source consumption accounted for by natural gas is at a relatively low level. This is partly due to the available natural resources, but as the economy and society develop, the effects of conflicts in supply and demand and of environmental restrictions on resources in relation to energy consumption will alter. As a result, the overall structure of China's natural gas consumption will differ from that seen in the past.

Based on the experiences of seven countries with similar natural environments and availability of natural gas resources, natural gas development is primarily influenced by growth in the service sector as a share of economic activity, degree of urbanisation and the urgency of controlling atmospheric pollution. Europe, the United States and Australia have all made significant adjustments to allow for the transition from coal to natural gas.

China has a potentially enormous demand for natural gas. With suitable policies, the natural gas share of energy consumption will rise considerably. As China's current economic development enters a "new normal", especially with the service sector share already exceeding that of secondary industry, and as further urbanisation continues to take place, and in recent years much greater attention has been paid to air quality, it can be predicted that China's natural gas consumption will see increasingly rapid growth.

In 2000, China's natural gas consumption was only 24.5 billion m^3, accounting for 2.2% of total energy consumption. In 2014, China's natural gas consumption was 183 billion m^3, 5.8% of total energy consumption. Looking at the breakdown of consumption, the manufacturing industry is showing a declining trend, dropping from a high of 50% of total natural gas consumption down to 33.4% currently. Power generation has risen from its 2000 share of 3.3% up to the current 16.0%. Transport and residential usage of natural gas have grown relatively rapidly, with shares rising from 3.6 and 13.2% in 2000 to 10.6 and 19.7% respectively in 2012. Looking at the long term, demand for natural gas in China will continue to show a trend for rapid growth.

Potential for Natural Gas to Act as a Substitute Fuel in China

3

Unlike coal and oil, natural gas does not have an exclusive niche. Analysis of international experiences has found that growth in natural gas consumption in other countries has come about largely as a result of substituting gas for other fuels. It is only substituted for other fuels when its price is competitive compared to other energy sources, incentivising users to adopt it. The competitive forces that drive such changes derive from users' desire to have their requirements for energy services met while reducing their energy source costs to a minimum.

This section therefore examines the price of natural gas that would make it competitive in different application sectors, relative to the currently preferred fuel used in that sector.

The primary application sectors for natural gas are power generation, transport and shipping, urban gas, industrial fuel and natural gas chemical engineering. In the power generation sector, natural gas competes primarily with coal. In the transport and shipping industry, it primarily competes with diesel. In the urban fuels gas sector, competition in residential usage is primarily with liquefied gas, artificial gas and electricity. In the heating and small-scale boiler sector, natural gas primarily competes with coal. In the industrial fuel sector, competition is primarily with fuel oil, artificial coal gas and coal. In the field of natural gas chemical engineering, competition is mainly with coal and naphtha. This section mostly consists of an analysis of price tolerance in these primary markets.

3.1 Power Generation: Cost Comparison of Gas as a Substitute for Coal in Power Generation

International experiences of development suggest that power generation is an important sector for natural gas. Natural gas power generation accounts for 39.4% of total installed capacity in the United States and 29% in Japan, whereas this number is only about 3% in China. China's natural gas power generation therefore has plenty of space to grow. In terms of its specific characteristics as a commodity, electricity is a classic homogenised product. Whether it is coal, hydroelectric, gas, wind or photovoltaic power, once it enters the grid, from the point of view of the end user it is essentially the same. It is primarily cost that affects the competitiveness of such commodities.

* This chapter was overseen by Zhaoyuan Xu from the Development Research Center of the State Council and Martin Haigh from Shell International, with contributions from Baosheng Zhang and Shouhai Chen from the China University of Petroleum, Lianzeng Zhao from the China Petroleum Planning Research Institute, Linji Qiao from ENN and Juan Han from Shell China. Other members of the research group participated in discussions and revisions.

Shell International and The Development Research Center (Eds.), *China's Gas Development Strategies*, Advances in Oil and Gas Exploration & Production, DOI 10.1007/978-3-319-59734-8_3

3.1.1 Direct Cost Comparison Between Gas Generation and Coal Generation

In order to better compare and analyse gas power generation competitiveness and other factors affecting it, we adopted base-load coal power generation as the object of reference. Based on this, coal-fired units rely on the currently widely adopted ultra-super critical unit, with an installed scale of 600,000/1,000,000 kW. For gas-fired units, the widely adopted 9F-grade CCGT unit was adopted, with an installed scale of 400,000 kW. Coal power generation baseline price was set at 550 CNY/ton (price of 5500 kilocalorie power coal delivered to power station), which converts to a standard coal price of 700 CNY/ton.

1. **Coal-fired power generation kWh total cost calculation**

    The coal-fired power generation 1 kWh cost primarily includes capital costs, non-fuel operational costs and fuel costs. Referring to national energy department reports and coal-fired power station feasibility reports, and supposing that a coal-fired power generating unit operating cycle is 20 years, the discount rate would be 9%. Coal-fired generating units initially require a unit investment of 3900 CNY/kW, with annual operating hours of 5000. Therefore, the capital cost of

1 kWh is 0.085 CNY/kWh. Other costs of a kWh come to around 0.045 CNY/kWh. When these are combined, the total is 0.13 CNY/kWh, a figure close to the fixed cost of a kWh. As far as fuel costs are concerned, 300 g of standard power coal is consumed on average, therefore the fuel cost of a kWh is around 0.21 CNY/kWh. Combining these three together, the overall cost of a kWh is 0.34 CNY/kWh (see Fig. 3.1).

2. **Gas-fired power generation kWh total cost**

    Combining research data and feasibility research report data, the initial unit investment for a gas-fired power unit was established to be 3000 CNY/kW. Assuming that annual operating hours are 3500 h (at peak load and intermediate load), after conversion the capital cost of 1 kWh is 0.094 CNY/kWh. Based on research data, other operating costs for 1 kWh are 0.065 CNY/kWh. As regards fuel cost, depending on region and construction date, the price of gas supplied to power stations can differ significantly. This report uses south-eastern coastal areas as the main reference region, and adopts a price of 2.8 CNY/m³ as the basis for calculation. In terms of the gas required to generate 1 kWh, in actual gas-fired generation the differences in gas source result in fuels having different calorific values. There is quite a wide variation in the amount of gas consumed in power

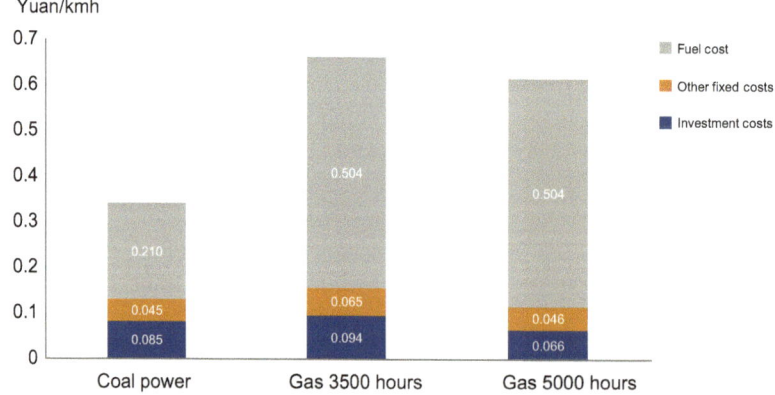

Fig. 3.1 Preliminary comparison of coal power generation and natural gas power generation costs

generation. By combining research data, a figure for gas consumption related to power generation can be set at 0.18 m^3/kWh, which converts to a kWh fuel cost of 0.504 CNY/kWh. Combining these three together, the overall cost of a kWh is 0.663 CNY/kWh. If annual operating hours were raised to 5000 h, the 1 kWh fixed cost including capital costs and non-fuel related operating costs would drop, causing the cost of 1 kWh to be reduced to 0.615 CNY/kWh.

It is apparent from this comparison that, based on current energy source prices, the 1 kWh cost for gas power generation is still far greater than coal, regardless of whether 3500-h or 5000-h operations are concerned. Looking at the direct costs, gas-fired generating units are not competitive. Even the fuel cost of gas-fired power generation alone (that is, the variable cost, or marginal cost) is much higher than total coal-fired power generation costs. This makes gas-fired generating units uncompetitive compared with coal-fired power generation, even if one ignores the fixed costs or where fixed costs have been covered by subsidies.

3. **Price tolerance calculations for gas-fired power generation**

Without considering changes to other factors, if gas-fired generating units were competitive with coal-fired power generation, then what level of gas price would we be talking about? Sensitivity analysis shows that when the gas kWh fuel cost is equivalent to the total cost of coal fuel, the natural gas price would drop to around 1.9 CNY/m^3, which is equivalent to European gas price levels. At this price level, due to the external benefits of gas-fired power generation, the fixed cost would attract a subsidy and gas would become competitive with coal. The sensitivity analysis shows that if the total fuel gas cost for 1 kWh was required to be the same as the total cost of 1 kWh generated with coal, then for an annual operation of 5000 h, natural gas prices would need to drop to around 1.3 CNY/m^3, while for 3500 h of operations it would need to drop to around 1.0 CNY/m^3, which is trending closer to US gas prices.

3.1.2   Other Factors Influencing the Competitiveness of Gas-Fired Power Generation

Apart from the actual gas price, there are other factors that affect the competitiveness of gas-fired power generation, including those that have an effect on the internal costs of gas power generation and its associated external benefits.

1. **Localised production of core equipment for natural gas power generation**

In terms of core equipment, the overseas monopolies in turbine equipment and process control have led to long-term elevated equipment costs, while LTP, operating and maintenance fees are very high. An improvement in this situation would help to reduce the fixed kWh costs for gas power generation. If unit investment could be reduced by 15%, it would also be possible to reduce the kWh cost by around 0.014 CNY/kWh. At the same time, localisation of manufacture would reduce long-term maintenance and inspection costs. If overall maintenance fees could be reduced by 50%, this would also bring about a reduction in the kWh cost of around 0.01 CNY/kWh. By localising the manufacture of core equipment, the associated kWh cost could be reduced to a level around the 0.024 CNY/kWh mark.

2. **Operating hours**

By increasing the annual operating hours of generating units, it is possible to reduce the kWh fixed costs that need to be absorbed, thereby reducing gas power generation costs to a certain degree. However, the benefits of this decrease progressively. Based on the configuration data in this report, and by adopting sensitivity calculations, when generating unit operating times are respectively increased from 3500 h to 4000, 4500 and 5000 h, the respective cost reduction will be 0.02, 0.035 and 0.048 CNY/kWh (see Fig. 3.2).

There are many factors restricting natural gas installation operating hour increases. The first is a

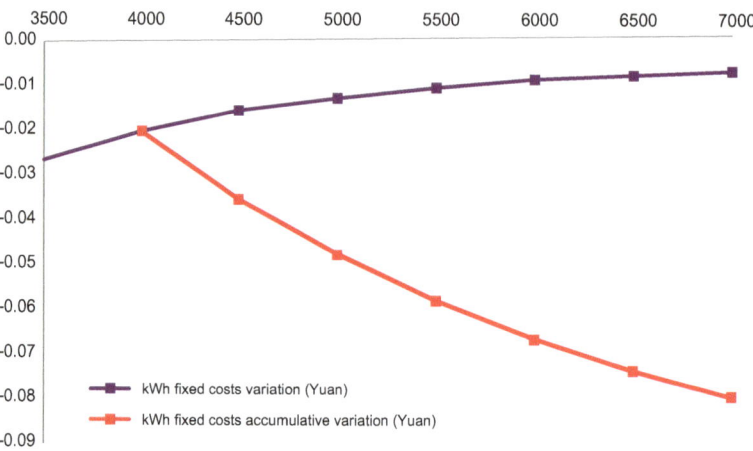

Fig. 3.2 Power generation cost analysis

high natural gas price and limited government subsidies, which greatly shorten the hours of power generation. Some new generating units that form part of this research were only capable of annual power generation of 1000 h. In some locations, in order to cope with high gas prices, an approach had been adopted whereby coal-fired power stations were used as a proxy in power generation, but no solution was sought to resolve the fundamental issue of gas power generation facilities sitting idle. The second is that in south-eastern coastal areas, due to the lack of hydroelectric and other externally generated power for use in peak regulation, when major fluctuations in power demand occur, the effect of this is also felt on the operating hours of gas power generation. At the same time, frequent starting and stopping of operations also shortens the useful life of generating units and increases maintenance costs, none of which is beneficial to the long-term operation of gas power generating units.

3. Emissions standards

China has introduced relatively strict emissions standards with regard to coal-fired power generation. In September 2014, the three committees forming the National Development and Reform Commission issued the Coal Power Energy Savings and Emissions Reductions Upgrade and Reconditioning Action Plan (2014–2020), which makes clear pronouncements concerning strict control over emissions of atmospheric pollutants and which requires the atmospheric pollutant emissions concentrations ratings of newly built coal-fired generating units in eastern regions to conform to the restrictions applicable to gas turbine emissions (namely that with a baseline oxygen content of 6%, smoke and dust, SO_2 and nitrogen oxide concentration may not exceed 10, 35 and 50 mg/m^3). Emissions ratings of newly built coal-fired generating units in central regions must approach or achieve the restrictions applied to gas turbine emissions, while emissions ratings of newly built coal-fired generating units in western regions are encouraged to approach or achieve the restrictions applied to gas turbine emissions. Apart from the addition of environmental installations, the plant operating modes were not changed. With regard to additional investment and the operational costs of such environmental installations, pricing conversions can be made with reference to the upgrades and modifications applied to the installations of certain current power stations and additional environmental investment encountered in new projects. After conversion, it was discovered that in order to comply with the strictest emissions standards, coal-fired power generation kWh costs rise by between around 0.01 and 0.02 CNY/kWh. Large-scale generating unit costs increase slightly less, while cost increases are slightly higher for older and smaller units or where coal quality is poorer.

4. **Carbon tax**

The high carbon emissions of coal power represent an important advantage for gas power. The price of carbon or the level of carbon tax is a critical factor influencing the competitiveness of gas power costs. Collection of carbon taxes will cause coal power and gas power kWh absolute costs to both rise. However, because coal power absolute cost increases will be greater, the difference in costs will result in the relative cost of carbon emissions associated with 1 kWh for coal power being greater than for gas power. Based on various energy source emission product coefficients provided by the Shenzhen Carbon Exchange as well as IPCC report coefficients, it is possible to calculate the coal power carbon emissions as 0.8 kg/kWh and gas power carbon emissions as 0.37 kg/kWh. As carbon prices rises, so does the carbon emission cost for coal power (Fig. 3.3). Taking as an example the Shenzhen Carbon Exchange June 18, 2013 online trading opening price of 30 CNY/ton, the total increase in the relative cost of coal power was around 0.013 CNY/kWh. Taking as an example the highest price listed on October 18, 2013 of 143.99 CNY/ton, this affects the relative cost by around 0.06 CNY/kWh.

5. **Auxiliary services**

Gas power auxiliary services refer to the system value of gas power in terms of adjustments during peak periods and as a back-up. The term also highlights the advantages of gas power over coal power as regards energy quality. Currently, with there being no auxiliary services market and no time-based power pricing mechanisms, capacity pricing based on a two-tier electricity price would take advantage of the benefits of auxiliary services. The determination of the capacity electricity price takes into consideration all aspects covered by gas power fixed costs, and can also be understood as including the additional investment in power capacity required to guarantee the reliability and stability of power supplies. Shanghai currently has power stations that implement a two-tier system for electricity pricing, with a gas power capacity price of 45.83 CNY/kW/month. In large cities such as Shanghai, Beijing and Shenzhen, where a two-tier system has been implemented, the industrial power price is between 40–44 CNY/kW/month. In overall terms, the influence of auxiliary services on competitiveness is fairly small.

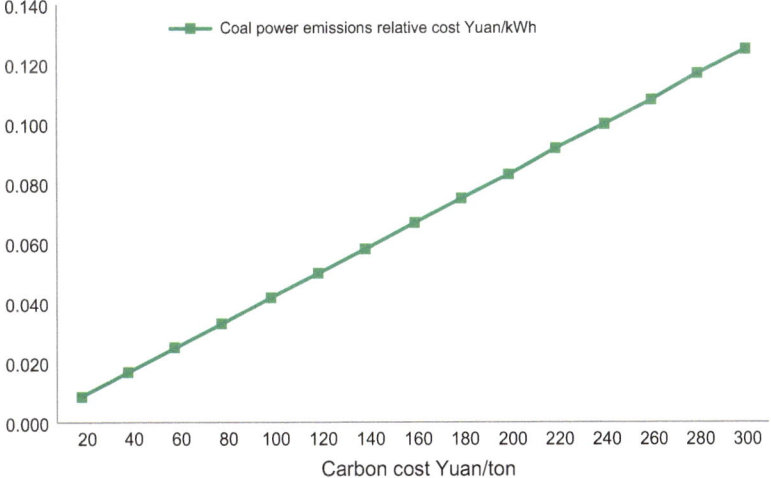

Fig. 3.3 Coal power emissions relative cost (CNY/kWh)

3.1.3 Conclusions Regarding the Prospects of Gas Replacing Coal in Power Generation

Natural gas power generation competitiveness is determined by various relative costs. Here, the coal price adopted is 550 CNY/ton. Based on the overall trends in coal prices since 2014, this is unlikely to increase in the short term. Regarding the natural gas price, considering that natural gas will gradually form a globalised market, with relatively large amounts of natural gas being imported by China, China's natural gas price will be inseparable from international gas prices. As things stand, increased marginal supply affecting the international natural gas market primarily comes from North American LNG, while East Asia is mainly responsible for marginal increases in consumption. Therefore, North American natural gas prices will have the greatest influence on LNG prices in East Asia. In fact, the biggest influence will be felt in China's south-eastern coastal regions, where there has been the greatest growth in natural gas power generation. Summarising forecast data from various research institutions, assuming a future US Henry Hub natural gas price of 5 $/MMBtu, total LNG transportation fees of 5 $/MMBtu and a CIF price in China of 10 $/MMBtu, it would be necessary to reduce or exempt import duty on LNG. At the same time, for large power plant users near LNG receiving stations, the gas reception and piping costs can be estimated at 0.2 CNY/m^3. This would result in the price of natural gas delivered to the power plant being 2.39 CNY, which can be converted to a kWh cost of 0.43 CNY/kWh.

1. **Comprehensive cost comparison for base loading power generation**

For a base loading generating unit (5000 h), the fuel cost based on an international gas price of 5 $/MMBtu would be 0.43 CNY/kWh, giving a kWh total cost of 0.541 CNY. By introducing localisation of equipment manufacturing, it would be possible to reduce kWh costs by 0.024 CNY. By increasing annual use to 6000 h,

costs could be reduced by 0.0185 CNY/kWh, reducing the kWh cost of gas power to 0.499 CNY/kWh. After the cost of emissions treatment has been applied to coal, the cost is still only 0.36 CNY/kWh. If the carbon tax price is 120 CNY/ton, then the cost of coal power would rise by 0.05 CNY/kWh, reaching an overall cost of 0.41 CNY/kWh. Therefore, gas power costs would still be higher than coal power by around 0.088 CNY/kWh. As a result of this, gas power would not be competitive in China.

2. **General cost comparison of peak regulation**

When gas-fired peak regulation (annual operation of 3500 h) is compared to base loading power generation, the greatest difference is that it allows the advantages of auxiliary generation services to be exploited. For example, by adopting capacity power pricing, once localised manufacturing has been taken into account, the compensatory level is around 39 CNY/kW/month, which converts to an electricity price of 0.134 CNY/kWh. At the same time, if coal power emissions standards were very strict, the additional pollution processing costs would rise by 0.02 CNY/kWh. Coal power kWh total costs would rise from 0.34 CNY/kWh to 0.36 CNY/kWh. If the carbon tax rate was 120 CNY/ton, then coal power costs would rise by a further 0.05 CNY/kWh. This would result in total coal power costs of 0.41 CNY/kWh, while the gas power total cost would be 0.43 CNY/kWh. Unless the carbon tax price is increased further, or government subsidies are increased, gas power will continue to lack a competitive advantage.

In overall terms, gas power lacks competitiveness with coal, not only at current gas price levels and in the current market environment, but even when the CIF price of international LNG is at the $10 levels, under which circumstances gas power would still need a series of combined measures in order to achieve market competitiveness. Due to this, for a market to arise where gas power was able to develop would not only require changes in the global natural gas market, but also require the implementation of a series of wider-reaching policy measures. This would include providing encouragement for major scientific breakthroughs, gaining control over gas pricing mechanisms, streamlining the administration of the natural gas industry, the creation of liberalised gas and electricity pricing mechanisms and the internalisation of environmental costs and effects, particularly in terms of sensible carbon emissions pricing. Only then would growth in gas power generation become possible.

3.2 Transport: Analysis of Natural Gas as a Substitute for Diesel

Starting in the 1930s, when Italy promoted usage of natural gas in cars due to the petrol shortage at that time, there have been over 70 years of history of gas being used as vehicle fuel. However, transport is different from power generation in that its use in transport relies to some extent on user perception, energy source availability and transportation safety, while also being affected by external networks. The replacement of diesel by natural gas in the field of transportation would require system-wide adjustments and involve a relatively complex process.

3.2.1 Primary Motivators for Natural Gas to Replace Diesel

Looking at the development history of natural gas vehicles, safety, cleanliness and economy have all been primary motivators in promoting the development of natural gas cars. After the 1973 oil crisis, national worries about the safety of oil supplies resulted in greater attention being paid to the development of natural gas cars. In the 1980s, urban air quality problems resulting from vehicle exhaust pollution attracted increasing attention in Europe and North America, while the fact that levels of gas and emissions were much lower than petrol or diesel led to western countries, led by Italy, paying more attention to the promotion of natural gas cars. However, since the 1990s, as oil product quality and emissions control technology have improved, petrol and diesel vehicle pollution emissions have been greatly reduced and the advantages of

natural gas in terms of cleanliness are no longer so pronounced. The pace of development of natural gas-powered vehicles in Europe and North America has now slowed. At the same time, up until around 2005, as international oil prices rose, the economic attraction of natural gas became more pronounced, and there was a rapid growth in the use of natural gas-powered cars in developing countries with relatively abundant gas resources.

Currently, in developed countries, oil supplies are relatively stable and there are fairly good oil reserves. Unless a major incident occurs, oil security has long since stopped being a driving factor for the development of natural gas vehicles. Moreover, with advances in engine technology, the advantages of natural gas cars as regards cleanliness have been reduced, reducing interest in the development of natural gas cars in developed countries. As regards economic factors, the energy consumption cost expenditure of European and North American users is relatively low, with more attention being paid to the driving experience, therefore without enhanced policies to drive forward the transition, the difference in price between diesel and gas would have no effect in terms of encouraging users to switch.

These factors exist to some extent in China as well, and in particular cleanliness and economy still play a major role in the rapid development and adoption of natural gas-powered vehicles. In terms of cleanliness, natural gas has the advantage of having a single component ingredient, allowing the easy removal of impurities, in addition to being of stable quality and having excellent physical characteristics, making it easier to improve cleanliness and meet emissions requirements. Of course, in developed countries, where oil quality is satisfactory and emissions control technologies are in place, petrol and diesel vehicles can meet the emissions levels of natural gas vehicles. Aside from this, vehicle emissions control requires the integration of fuel processing, engines controls, exhaust handling and other related technological systems and appropriately enhanced emissions standards. In China, due to the supply network and its locked-in nature, high levels of new investment in systems improvements and upgrades with a long-term strategy would be necessary to resolve problems relating to crude oil quality, refining processes, the vehicle industry and industrial standards that currently exist; natural gas, however, offers many advantages due to its cleanliness, as well as a good tolerance towards it in petrol engines, so its adoption would show quicker results in terms of environmental requirements.

The fact that China is a developing country means that economical fuel is still a major driver behind the development of natural gas as a transportation fuel. Fuel costs account for around 40% of operating costs, so reducing fuel costs is a major factor in making transport businesses more competitive. In the past few years, the significant difference in the price of diesel and gas and exemption from consumer taxes has resulted in the rapid development of natural gas vehicles in China. Taking CNG in taxis as an example, for a period of time the CNG price was just half that of petrol, meaning that the conversion cost could be recovered within a few months, resulting in rapid growth in the market for CNG in cars.

3.2.2 Natural Gas Oil Replacement Price Tolerance in the Urban Transport Sector

1. **Taxi natural gas petrol replacement price tolerance**

Urban CNG taxis and public transport are currently the largest markets for natural gas. CNG taxis are mainly being adopted in second- and third-tier cities, as well as in first-tier cities where natural gas was introduced relatively early, such as the cities in the Sichuan-Chongqing basin and in Chengdu. In some cities, adoption of natural gas fuel is almost universal. However, after rapid development in the past, this market has slowed in recent years. In first-tier cities where taxis did not invest in CNG early on, the high demand for land for urban development has made it difficult to site natural gas filling stations, which restricts the development of CNG taxis.

City taxis can use CNG to replace petrol, the cost of taxi conversion being around CNY 5000. Based on average distances travelled of 100,000 km, and the cost of conversion being absorbed over a year, the conversion cost is 5 CNY/100 km. The price tolerance of gas used as a replacement, based on the running costs for gas taxis being the same as for petrol taxis, is 5.58 CNY/m^3 for an international crude oil price of 100 $/bbl, and 4.34 CNY/m^3 when the international crude oil price is 60 $/bbl (Table 3.1).

2. **Bus natural gas diesel replacement price tolerance**

Urban public transport is also an important submarket for CNG and LNG applications. Because public transport companies normally have their own bus stations, it is possible for them to construct LNG/CNG stations even in first-tier cities. As a result, there is greater potential for development.

Urban public transport vehicles can use CNG to replace diesel, and vehicle conversion costs around CNY 12,000. Basing calculations on

Table 3.1 Taxi price tolerance of natural gas as replacement for petrol

International price of crude oil ($/bbl)	Petrol		Fuel cost (CNY/100 km)	Natural gas		
	Price (CNY/L)	Oil consumption (L/100 km)		Gas consumption (m^3/100 km)	Conversion cost (CNY/100 km)	Price tolerance (CNY/m^3)
100	6.77	8	54	8.8	5	5.58
80	6.08		49			4.96
60	5.39		43			4.34

Data source Results of calculation

Table 3.2 Bus price tolerance of natural gas as replacement for diesel

International price of crude oil ($/bbl)	Diesel		Fuel cost (CNY/km)	Natural gas		
	Price (CNY/L)	Oil consumption (L/100 km)		Gas consumption (m³/100 km)	Conversion cost (CNY/L)	Price tolerance (CNY/m³)
100	6.72	35	235	38	15	5.80
80	5.90		206			5.04
60	5.08		178			4.28

Data source Results of calculation

recovering the cost of conversion over a year, with a distance travelled of 80,000 km, the cost is the equivalent of an additional 15 CNY/100 km. The price tolerance of gas used as a replacement, based on the running costs for gas buses being the same as for diesel buses, is 5.80 CNY/m³ when the international crude oil price is 100 $/bbl, and 4.28 CNY/m³ when the international crude oil price is 60 $/bbl (Table 3.2).

CNG private cars and light commercial vehicles also account for a share of the market, primarily concentrated in second- and third-tier cities where natural gas taxis are relatively well developed. Private cars are normally relatively economical low-end compact cars, such as those costing less than CNY 100,000. In the next 10 years, the market will remain relatively stable, and as income levels rise, new users will continue to enter the economy car market. However, at the same time, some of the original economy vehicle users will upgrade to mid-range and high-end vehicles. As for penetration of electric private vehicles, it is mainly in first-tier cities affected by vehicle policies, and especially in cities where license plate restrictions are in place, that there is a considerable variance in terms of market share when compared to CNG private vehicles, and as far as overall numbers go, this does not represent a major influence on the CNG private vehicle market at the present time. There has been relatively rapid growth in light commercial vehicles in the past two years, but numbers still remain low, as conversion of diesel engines presents problems, resulting in low levels of interest and indicating minimal market growth in the future.

3.2.3 Commercial Vehicle Natural Gas Diesel Replacement Price Tolerance

The commercial vehicle LNG market is the sub-market with the greatest growth, and is primarily composed of heavy lorries, inter-city buses, medium-sized lorries, works vehicles and other vehicles. In terms of economic incentives or market scale, heavy lorries are the application market with the greatest growth potential. Moreover, long-distance inter-city buses, if travelling over 400 km, also have a marked economic effect, but are affected in turn by the rapid development of high-speed rail and the quickness of air travel. It is expected that this market will tend to shrink. However, medium-sized lorries, works vehicles and other vehicles only make economic sense in regions where gas sources are relatively inexpensive, due to their relatively low fuel consumption. Since 2011, there has been a rapid increase in the construction of LNG filling stations, and the expectation is that by 2020 filling station coverage will include major ports, industrial parks and main logistics routes. Moreover, as networks are optimised, the growth in numbers of LNG vehicles will outrun the increase in LNG filling stations, resulting in increased operational efficiency and profitability of LNG filling stations, and the emergence of a self-sustaining cycle in this industry.

We calculated the commercial vehicle price tolerance taking inter-city long-distance heavy lorries as an example. Inter-city long-distance lorries can use liquefied natural gas (LNG) instead of diesel. The purchase price of LNG vehicles is

Table 3.3 Transport lorry price tolerance of natural gas as replacement for diesel

International price of crude oil ($/bbl)	Diesel		Fuel cost (CNY/100 km)	Natural gas		
	Price (CNY/L)	Oil consumption (L/100 km)		Gas consumption (m³/100 km)	Price (CNY/L)	Price tolerance (CNY/m³)
100	6.72	45	302	50	45	5.15
80	5.90		265			4.41
60	5.08		229			3.67

Data source Results of calculation

higher by roughly CNY 90,000 than that of diesel vehicles. Basing calculations on annual travel of 100,000 km, and recovering the cost of conversion over 50% of the vehicle scrapping distance, the calculated cost is the equivalent of an additional 45 (37.5) CNY/100 km. The price tolerance of gas used as a replacement for diesel, based on the running costs for gas commercial vehicles being the same as for diesel commercial vehicles, is 5.15 CNY/m³ when the international crude oil price is 100 $/bbl, and 3.67 CNY/m³ when the international crude oil price is 60 $/bbl (Table 3.3).

3.2.4   Ship Transport Natural Gas Diesel Replacement Price Tolerance

Waterborne transport relying on LNG is a submarket that is just starting to develop. Ships consume considerable amounts of fuel, and shipping routes are relatively fixed and generally linear in their distribution. There are low external requirements for filling station networks, and with mature technology and commercial models, rapid development is likely. Currently, ship fuels are divided into light oil and heavy oil. Because heavy oil prices are low, LNG is not competitive. The main market for LNG is as a replacement for light oil. Light oil use is primarily concentrated in interior river systems and used in river transport, dredging and works vessels, as well as being used by some coastal fishing boats. Interior river goods vessels have stable annual oil consumption, have fixed routes and can normally fill

up at service areas. There are not that many dredgers, but they have high fuel consumption, refill frequently, have stable fuel consumption and have relatively dense concentrations, making it possible to construct dedicated filling stations. These two types of user are submarkets that could be reasonably easily developed. Even though coastal fishing boats are numerous, they are affected by factors such as the fishing season and have some seasonal variances with non-fixed routes and irregular operation, accompanied by rapid deterioration, therefore short-term development will be slow. Currently, LNG shipping applications are being promoted heavily by central and local government. Because single unit fuel consumption is considerable, economy of scale is particularly pronounced. Compared to constructing an LNG filling station network, adding filling stations along shipping routes is much simpler and more convenient. Once conditions develop, this market will easily pick up, and it requires only a short development cycle.

Price tolerance for shipping to use LNG instead of oil is relatively high. Basing calculations on a 2000 ton grade dual-fuel (30% diesel + 70% LNG) powered ship, the conversion cost is CNY 750,000. Based on 10 journeys per year, each voyage being 1250 km, and offsetting the conversion costs over 5 years of voyages, the conversion cost works out at 1200 CNY/100 km. The price tolerance of gas used as a replacement for diesel, based on the running costs for gas-powered vessels being the same as for diesel vessels, is 4.96 CNY/m³ when the international crude oil price is 100 $/bbl, and the price tolerance is 3.47 CNY/m³ when the international

Table 3.4 Shipping price tolerance of natural gas as replacement for diesel

International price of crude oil ($/bbl)	Diesel		Fuel cost (CNY/100 km)	Natural gas		
	Price (CNY/L)	Oil consumption (L/100 km)		Gas consumption (m³/100 km)	Price (CNY/L)	Price tolerance (CNY/m³)
100	6.72	950	6384	1045	1200	4.96
80	5.90		5605			4.21
60	5.08		4825			3.47

Data source Results of calculation

crude oil price is 60 $/bbl. Pricing parity for natural gas versus diesel is around 0.70 (Table 3.4).

3.2.5   General Factors in the Transportation Market Relating to Gas Replacement of Diesel

Even though natural gas is currently the most realistic alternative transportation fuel, and its clean, low-carbon and economical qualities mean that it has great potential in China, the realisation of this presents many challenges. In addition to economy, factors affecting the switch from oil to natural gas include natural gas supply, development of natural gas transport modes, natural gas filling stations, consumer expectations and the development of electrically powered cars.

1. **Assurances related to natural gas and transportation usage**

   In terms of natural gas fuel, users primarily focus on assurances concerning gas supplies, price and quality. Currently, as transitions take place in the layout of global natural gas supplies and demand, and China's natural gas supplies diversify, it can be expected that natural gas supplies will be fully guaranteed in the future. It is therefore to be expected that concerns about natural gas supply guarantees will no longer play a major role. Natural gas prices, pricing mechanisms and price development trends play a more significant role. In terms of quality, unlike in power generation, consumers focus more on the calorific value of natural gas.

2. **Extent of deployment of natural gas filling stations**

   The extent of deployment of natural gas filling stations is an important factor affecting the convenience of vehicle operation. Transportation is affected by network externalities, and as more filling stations come on line, this will attract new users and create wider network benefits. This will, in turn, increase the value of new and existing filling stations, creating a virtuous circle that increases the use of gas in transport.

3. **Development of natural gas transportation**

   Transportation using natural gas currently includes CNG vehicles, LNG vehicles and LNG shipping, and involves manufacturing or conversion, operation, maintenance and scrapping. CNG vehicles are primarily compact cars, light goods vehicles and buses. Because CNG and petrol both use a combustion engine, the manufacture, conversion and maintenance processes are simple. Currently the main problem encountered is that public transport departments are worried that converted vehicles present a safety hazard. LNG vehicles are primarily used as heavy lorries, light commercial vehicles, inter-city passenger vehicles and in other long-distance uses where vehicles consume large quantities of fuel. They can also be used in mining, engineering and other sectors. Vehicle conversion requires return to the factory, but the technology is mature. The cost of switching from diesel to LNG is relatively high due to the small scale on which it occurs. In addition, LNG vehicle repair shops are few and far between. If a malfunction occurs, it is often necessary to return the vehicle to the factory for repair, resulting in inconvenience and financial losses being incurred by the user. LNG ships are currently in the trial adoption stage. Depending on the type of combustion engine that vessels rely on, conversion to LNG may require a return to the factory in order to convert the engine to the spark ignition type. Alternatively, conversion may be to a mixed diesel/LNG type engine, and appropriate conversion schemes still require improvement through further trials.

4. User perception also influences replacement of oil-based fuels with natural gas

User perception is the subjective experience of the user relating to the use of natural gas in transportation, and this has an important influence on the development of natural gas transportation and in the analysis of submarkets. User perception includes three aspects: performance, user sensitivity and usage environment.

In terms of performance, natural gas vehicles vary somewhat from petrol vehicles. Compared to petrol, the burn rate of natural gas is slower, the theoretical highest burn temperature is lower, and the amount of air required during the burn process is greater. At the same time, to accommodate burning of both fuels, the engine is often not configured to the optimal state to use natural gas. These factors each influence the drive output of natural gas engines, fuel efficiency and other aspects. In terms of user sensitivity, this involves fuel economy, power output, usage space, convenience and many other factors. For example, commercial vehicles focus more on economy and factors related to economy, whereas private vehicles focus more on output, usage space, cleanliness etc. In terms of usage environment, electric cars have also been labelled as innovative, technologically advanced, fashionable and environmentally friendly alternatives to both gas and petrol vehicles, whereas gas-fuelled private vehicles tend to have an image of being more affordable.

5. Other threats posed by replacement with electric cars

Compared to natural gas vehicles, electric cars offer some competitive advantages, and therefore could present fierce competition to natural gas vehicles. First, electric motor efficiency is inherently higher than that of the internal combustion engine, there being less mechanical wear in the electric car drive mechanism; as a result of this, less energy is consumed than with internal combustion engine cars. Moreover, electric motors emit essentially zero emissions. Second, in terms of output and performance, output in electric cars can be increased through simple changes to voltage, current and frequency. Their structure is more compact, while they have a flexible driving dynamic. Finally, in terms of economy, even though current purchase costs are high, they become more economical as driving distance increases.

Currently, the greatest obstacle to the development of electric vehicles comes from insufficient battery recharging facilities, in addition to which battery costs are high and there are challenges when it comes to extending voyage distances. The potential threat that electric cars pose to natural gas-powered cars and the space their development occupies depends on the technological development of electric vehicles, and the speed at which battery exchange and charging facilities can be extended.

3.2.6 Potential Prospects for Transportation Market Natural Gas Demand

In 2013, natural gas vehicles used around 12 billion m^3 nationally. Of this, CNG vehicles used around 10 billion m^3, and LNG vehicles used 2 billion m^3. Because natural gas is still the most practical alternative fuel for transportation compared to other alternative fuels, it has a relatively large potential for development.

Looking at the larger picture, markets for vehicles using CNG will gradually enter a period of slower development, with expectations that this growth will be slightly higher or equivalent to natural gas consumption market growth rates. By 2020, consumption will reach 20 billion m^3, and by 2025 it will reach 25 billion m^3, after which growth will stagnate.

LNG vehicle markets in the next 10 years will maintain high rates of growth, with expectations that in 2020 vehicular use of LNG will exceed 20 billion m^3. By 2030, that number is expected to reach 40 billion m^3.

Shipping LNG is also an important potential market, but conversion schemes and commercial models have yet to be further optimised. Moreover, interior river system markets using light oil only have small capacity, whereas coastal shipping still depends on the degree of implementation of environmental policy. Early development is therefore expected to be slow, with later development depending on changes in environmental protection policies. Optimistic projections for ship usage of LNG envisage usage exceeding 2 billion m^3 by 2020, and over 10 billion m^3 by 2030.

3.3 Urban Use: Assessment of Price Tolerance for Natural Gas to Replace Other Fuels

3.3.1 Price Tolerance is Relatively High for Residential Usage of Natural Gas

Urban residential usage of natural gas is primarily for cooking, hot water and heating systems. Gas use indices are affected largely by lifestyle, infrastructure and price. Based on calculations in the *Urban and Rural Construction Statistical Yearbook 2012*, per capita natural gas usage across China in 2012 was 73 m^3. However, in regions such as Sichuan and Chongqing, where there are more abundant resources, the price is lower, and where urban introduction occurred early on, this figure reaches 120 m^3. In major cities there is a high quality of life required

and they exhibit a higher range of 70–80 m^3. Other cities generally have a range of 50–60 m^3.

Residential options for energy sources include electricity, bottled liquefied gas and natural gas. Therefore, natural gas, liquefied gas and electricity compete to replace each other in urban residential usage. Since electricity replaced coal, urban residential usage of natural gas has an indirect relationship of acting as a substitute for coal.

1. **Price tolerance for natural gas to replace liquefied gas**

Liquefied gas is a by-product of the oil refining process, and its price is very much related to the price of crude oil. When the international price of crude oil is 100 $/bbl, household bottled liquefied gas costs 8.23 CNY/kg, and when the international price of crude oil is 60 $/bbl, household bottled liquefied gas costs 6.64 CNY/kg.

The heat efficiency of urban residential piped natural gas and bottled liquefied gas is around 60%. Basing calculations on the principle of equivalent effective calorific value costs, the price tolerance for natural gas as a replacement for liquefied gas can be calculated as 5.62 CNY/m^3 when the international price of crude oil is 100 $/bbl and 4.54 CNY/$m^3$ when the international price of crude oil is 60 $/bbl (Table 3.5).

2. **Price tolerance for natural gas to replace electricity**

Electricity is a clean and efficient household energy source. When natural gas prices are not competitive with those of electricity, residents will switch to using electrical appliances, for example switching from a natural gas hot water heater to an electrical hot water heater. Based on the available information, residential natural gas heat efficiency is around 75%, whereas electric hot water heat efficiency is around 98%. Calculating the price tolerance of natural gas to replace electricity based on the principle of equivalent effective calorific value cost, with current residential electricity

Table 3.5 Price tolerance for residential usage natural gas replacement of liquefied gas

International crude oil price ($/bbl)	Small bottle liquefied gas			Effective calorific value cost (CNY/GJ)	Natural gas		
	Price (CNY/kg)	Calorific value (MJ/kg)	Calorific efficiency		Calorific efficiency	Calorific value (MJ/m^3)	Price tolerance (CNY/m^3)
100	8.23	50.25	60%	273	60%	34.34	5.62
80	7.43			246			5.08
60	6.64			220			4.54

Data source Results of calculation

Table 3.6 Price tolerance for residential usage natural gas replacement of electricity

Residential electricity			Effective calorific value cost (CNY/GJ)	Natural gas		
Price (CNY/kWh)	Calorific value (MJ/kWh)	Calorific efficiency		Calorific efficiency	Calorific value (MJ/m^3)	Price tolerance (CNY/m^3)
0.6	3.60	98%	170	75%	34.34	4.38
0.5			142			3.65
0.4			113			2.92

Data source Results of calculation

Table 3.7 Residential disposable income price tolerance for natural gas

Income group	Per-capita disposable income (CNY/year)	Upper limit of residential fuel expenditure (CNY/year)	Resident average gas usage (m^3/year)	Acceptable natural gas price (CNY/m^3)
High income	56,389	1692	60	28.2
Mid- to high income	32,415	972		16.2
Middle income	24,518	736		12.3
Mid- to low income	18,483	554		9.2
Low income	11,434	343		5.7

Data source Results of calculation

prices being 0.40–0.60 CNY/kWh, natural gas price tolerance as a replacement for electricity is 2.92–4.38 CNY/m^3 (Table 3.6).

3. **Natural gas price tolerance in terms of disposable income of urban residents**

Compared to liquefied gas, natural gas is convenient, does not require bottling or transport and avoids the inconvenience of frequent replacement. It is symbolic of quality of life, and thus even if natural gas use has a higher cost than liquefied gas, residents will still opt to use natural gas, so long as the cost is within the range tolerable to their income level.

According to Price Law, residential use natural gas is a public commodity and is included in the scope of government price setting. Government procedures must be followed, and the most important factor from the point of view of government price setting is the disposable income level of urban residents. A level of average residential usage of 3% of the disposable income of low-income residents is used as a reference, resulting in an acceptable residential natural gas price of 5.7 CNY/m^3 (Table 3.7).

3.3.2   Price Tolerance is also Quite Resilient in Commercial Service Natural Gas Use

Commercial service natural gas users primarily include airports, government institutions, employee canteens, kindergartens, schools, hotels, restaurants, the catering industry, shopping centres and office buildings. Individual unit usage is far higher than that of residential users, with usage primarily for cooking and for cooling and heating systems. Alternatives include liquefied gas and electricity.

1. **Price tolerance of natural gas replacing liquefied gas**

   Unlike residential users of liquefied gas, commercial service users generally use large tanks of liquefied gas, the price of which is lower than that of the smaller bottles of liquefied gas used residentially. Basing calculations on the principle of equivalent effective calorific value costs, the price tolerance for natural gas replacing liquefied gas can be calculated as 5.34 CNY/m^3 when the international price of crude oil is 100 $/bbl and 4.31 CNY/m^3 when the international price of crude oil is 60 $/bbl (Table 3.8).

2. **Price tolerance of natural gas replacing electricity**

   Among commercial service and institutional users, there is a competitive alternative relationship between natural gas and electricity in the promotion of natural gas direct combustion air conditioning units (Table 3.9). Basing calculations on the principle of equivalent effective calorific value costs, the price tolerance for natural gas replacing electricity can be calculated as 4.52 CNY/m^3 when the international price of crude oil is 100 $/bbl and 3.01 CNY/m^3 when the international price of crude oil is 60 $/bbl.

3.3.3   Price Tolerance is Weak for Centralised Urban Heating Using Natural Gas

China's centralised heating is primarily focused in the northern regions. The traditional areas relying on winter season heating are the three north-eastern

Table 3.8 Price tolerance for commercial service usage of natural gas as replacement for liquefied gas

International crude oil price ($/bbl)	Large tank liquefied gas			Effective calorific value cost (CNY/GJ)	Natural gas		
	Price (CNY/kg)	Calorific value (MJ/kg)	Calorific efficiency		Calorific efficiency	Calorific value (MJ/m^3)	Price tolerance (CNY/m^3)
100	7.82	50.25	50%	311	50%	34.34	5.34
80	7.06			281			4.82
60	6.31			251			4.31

Data source Results of calculation

Table 3.9 Price tolerance for commercial service usage of natural gas as a replacement for electricity

Commercial usage of electricity				Effective calorific value cost (CNY/GJ)	Natural gas			
Price (CNY/kWh)	Calorific value (MJ/kWh)	Conversion efficiency	Conversion cost (CNY/GJ)		Conversion cost (CNY/GJ)	Conversion efficiency	Calorific value (MJ/m^3)	Price tolerance (CNY/m^3)
0.6	3.60	95%	300	475	320	75%	34.34	4.00
0.5				446				3.25
0.4				417				2.50

Data source Results of calculation

Table 3.10 Urban centralised heating price tolerance of natural gas as a replacement for coal

Coal				Heating cost (CNY/m²)	Natural gas			
Price (CNY/ton)	Calorific value (MJ/kg)	Consumed amount (kg/m²)	Conversion cost (CNY/m²)		Conversion cost (CNY/m²)	Consumed amount (m³/m²)	Calorific value (MJ/m³)	Price tolerance (CNY/m³)
800	20.94	30	12	36	9	12	34.34	2.25
600				30				1.75
400				24				35

Data source Results of calculation

provinces, five north-western provinces, five north China provinces and Shandong and Henan, giving a total of 15 provinces (regions, cities), most of which rely on coal-fired boilers. Conversion from coal to natural gas has been carried out based on local environmental policy requirements. Basing calculations on the equivalent unit cost per heated area, when the coal price is 600 CNY/ton, natural gas price tolerance as a replacement for coal is 1.75 CNY/m³ (Table 3.10).

3.4 Industrial Use: Cost Comparison of Natural Gas Replacing Other Energy Sources

Natural gas as an industrial fuel is primarily used in smelting furnaces, heating furnaces, hot treatment furnaces, roasting furnaces, drying furnaces and other industrial furnaces (construction materials, electromechanical, metals and other industries) and industrial furnaces used to generate steam for process requirements, and is used to replace fuel oil, coal gas and coal.

3.4.1 Price Tolerance is Very Low for Replacement of Fuel Oil by Natural Gas in the Glass Industry

Glass is primarily divided into flat panel glass and special type glass. Flat panel glass is divided into general flat panel glass and floating glass. The general process for the production of glass is: raw material–melting–forming–cooling–cutting–packing–placement into inventory–sale. Natural gas is primarily used at the melting stage, the main natural gas-consuming equipment being a glass furnace. Based on glass raw material melting principles, glass enterprises generally use high calorific value fuels such as fuel oil. Natural gas may therefore act as a substitute for fuel oil in the glass industry.

Basing calculations on the principle of equivalent effective calorific value costs, the price tolerance for natural gas replacing fuel oil can be calculated as 3.25 CNY/m³ when the international price of crude oil is 100 $/bbl and 1.94 CNY/m³ when the international price of crude oil is 60 $/bbl (Table 3.11).

Table 3.11 Glass industry price tolerance of natural gas as replacement for fuel oil

International crude oil price ($/bbl)	Fuel oil			Effective calorific value cost (CNY/GJ)	Natural gas		
	Price (CNY/kg)	Calorific value (MJ/kg)	Heating efficiency		Heating efficiency	Calorific value (MJ/m³)	Price tolerance (CNY/m³)
100	3.74	41.88	85%	105	90%	34.34	3.25
80	2.98			84			2.59
60	2.24			63			1.94

Data source Results of calculation

3.4.2 Price Tolerance is Relatively Weak for the Ceramics Industry for Replacement of Coal Gas by Natural Gas

The ceramics industry can be broken down into four sub-industries: hygiene ceramics, articles for everyday use, art and other ceramics. According to the production processes for ceramic products, the overall process is: raw material moulding and forming—firing—cooling and application of enamel—re-firing and other steps, with natural gas primarily used in kiln heating.

In order to guarantee the quality of the ceramic products, the fuel needs to be pure and without contaminating matter and the supply must be stable. Normally coal gas is used, but natural gas has the potential to be a substitute. Basing calculations on the principle of equivalent effective calorific value costs, the price tolerance

for natural gas replacing coal gas can be calculated as 2.46 CNY/m^3 when the price of coal is 600 CNY/ton (Table 3.12).

3.4.3 Natural Gas Price Tolerance is Relatively Poor for Steam Production as a Replacement for Coal

Steam is broadly used in industrial manufacturing processes, with steam generated by industrial boilers. Industrial boiler steam production does require much fuel, and normally low-priced coal is used. Therefore, in the industrial production of steam, natural gas has the potential to compete as a substitute for coal. Basing calculations on the principle of equivalent effective calorific value costs, the price tolerance for natural gas replacing coal can be calculated as 1.92 CNY/m^3 when the price of coal is 600 CNY/ton (Table 3.13).

Table 3.12 Ceramics industry natural gas price tolerance as a replacement for coal gas

Coal				Coal gas calorific value (MJ/m^3)	Effective calorific value cost (CNY/GJ)	Natural gas	
Price (CNY/ton)	Calorific value (MJ/kg)	Conversion efficiency (m^3/kg)	Conversion cost (CNY/m^3)			Calorific value (MJ/m^3)	Price tolerance (CNY/m^3)
800	20.94	3	0.7	16.75	82	34.34	2.80
600					72		2.46
400					62		2.12

Data source Results of calculation

Table 3.13 Industrial boiler steam production price tolerance of natural gas as a replacement for coal

Coal				Steam cost (CNY/ton)	Natural gas			
Price (CNY/ton)	Calorific value (MJ/kg)	Consumed amount (kg/ton)	Conversion cost (CNY/ton)		Conversion cost (CNY/ton)	Consumed amount (m^3/ton)	Calorific value (MJ/m^3)	Price tolerance (CNY/m^3)
800	20.94	200	80	240	50	78	34.34	2.44
600				200				1.92
400				160				1.41

Data source Results of calculation

3.5 Chemicals: Potential for Increased Natural Gas Use in the Production Process

Natural gas chemical engineering refers to the use of methane, the primary component of natural gas, in manufacturing processes, the products of which are primarily synthetic ammonia, methanol and hydrogen.

3.5.1 Price Tolerance of Natural Gas is Extremely Low for the Manufacture of Synthetic Ammonia

Production capacity and production volume in China's synthetic ammonia industry are the highest in the world, and coal and natural gas are the main raw materials, accounting for 76 and 22% respectively. The price of the final urea product is essentially guided by the production cost of the coal raw material. Thus, from the perspective of industry development, natural gas competes as an alternative to coal in the manufacture of synthetic ammonia.

Basing calculations on the principle of equivalent finished product costs, the price tolerance for natural gas replacing coal in the manufacture of synthetic ammonia can be calculated as 1.42 CNY/m^3 when the price of coal is 600 CNY/ton (Table 3.14).

3.5.2 Price Tolerance of Natural Gas is Very Weak for the Manufacture of Methanol

Production of methanol in China relies on coal and natural gas as the main raw materials, accounting for 63 and 28%, respectively. The price of the urea end product is essentially guided by the cost of the coal raw material. Thus, from the perspective of sector development, natural gas has a relationship as a substitute for coal in the manufacture of methanol.

Basing calculations on the principle of equivalent finished product costs, the price tolerance for natural gas replacing coal in the manufacture of methanol can be calculated as 1.71 CNY/m^3 when the price of coal is 600 CNY/ton (Table 3.15).

Table 3.14 Natural gas tolerance in production of synthetic ammonia as replacement for coal

Coal			Urea finished product cost (CNY/ton)	Natural gas		
Price (CNY/ton)	Consumed amount (kg/t)	Conversion cost (CNY/ton)		Conversion cost (CNY/ton)	Consumed amount (m^3/ton)	Price tolerance (CNY/m^3)
800	1080	1000	1738	800	600	1.77
600			1554			1.42
400			1369			1.07

Data source Results of calculation

Table 3.15 Methanol production price tolerance of natural gas as replacement for coal-based production

Coal			Methanol finished product cost (CNY/ton)	Natural gas		
Price (CNY/ton)	Consumed amount (kg/ton)	Conversion cost (CNY/ton)		Conversion cost (CNY/ton)	Consumed amount (m^3/ton)	Price tolerance (CNY/m^3)
800	1400	1500	2457	900	870	2.02
600			2218			1.71
400			1979			1.40

Data source Results of calculation

Table 3.16 Hydrogen production price tolerance of natural gas as replacement for coal-based production

International crude oil price ($/bbl)	Naphtha			Hydrogen finished product cost (CNY/ton)	Natural gas		
	Price (CNY/kg)	Consumed amount (kg/ton)	Conversion cost (CNY/ton)		Conversion cost (CNY/ton)	Consumed amount (m³/ton)	Price tolerance (CNY/m³)
100	10.64	3600	1000	33,731	650	5200	7.19
80	8.51			27,185			5.77
60	6.38			20,638			4.34

Data source Results of calculation

3.5.3 Price Tolerance of Natural Gas is Very Strong for the Manufacture of Hydrogen

Natural gas has a high methane content, and when used as the raw material in the production of hydrogen it results in a high hydrogen yield in addition to reducing fuel consumption, making it the ideal raw material. The majority of the world's hydrogen is manufactured using natural gas as a raw material. However, in China the raw material most used in the manufacture of hydrogen is coal, although petrochemical enterprises manufacture hydrogen using mainly naphtha. Natural gas is primarily used in the manufacture of hydrogen by petrochemical enterprises as an alternative to naphtha. Basing calculations on the principle of equivalent finished product costs, the price tolerance for natural gas replacing naphtha can be calculated as 7.19 CNY/m³ when the international price of crude oil is 100 $/bbl and 4.34 CNY/m³ when the international price of crude oil is 60 $/bbl (Table 3.16).

3.6 China's Natural Gas Demand Curve and Ways to Increase Natural Gas Consumption

3.6.1 China's Natural Gas Prices in 2013

1. **Shanghai as the baseline**

   In July 2013, the National Development and Reform Commission issued the *Notice Regarding Natural Gas Prices* (FGJG 2013, No. 1246), in which a reform plan was proposed for natural gas prices. Considering China's natural gas market resource trends, consumption and management distribution as a whole, Shanghai was chosen as the pricing baseline location. Station prices across the country in all provinces as well as all user terminal natural gas prices were tied to the Shanghai station price.

2. **Individual subsidies for each provincial station**

   The price difference between each province's station price and the pricing baseline of the Shanghai station price was the subsidy. The main factor affecting provincial subsidies was transportation cost (Table 3.17).

3. **End price calculations for provincial users**

Provincial station price = Shanghai station price - subsidy
City station price = provincial station price + provincial pipeline network pipeline transport fee
City natural gas user terminal price = city station price + city distribution fee
Large industrial user delivery price = city station price + provincial pipeline network pipeline transport fee
CNG filling sales price = city station price + CNG compression and distribution fee
LNG filling sales price = city station price + LNG liquid filling fee
Depending on the provincial pipeline network pipeline transport fee for each province, the city

Table 3.17 Station price subsidies in each province

No.	Province	Subsidy	No.	Province	Subsidy	No.	Province	Subsidy
1	Shanghai	0.00	11	Shandong	0.20	21	Sichuan	0.53
2	Guangdong	0.00	12	Liaoning	0.20	22	Chongqing	0.54
3	Zhejiang	0.01	13	Hubei	0.22	23	Hainan	0.54
4	Jiangsu	0.02	14	Jiangxi	0.22	24	Ningxia	0.67
5	Anhui	0.09	15	Hunan	0.22	25	Gansu	0.75
6	Henan	0.17	16	Shanxi	0.27	26	Shaanxi	0.84
7	Guangxi	0.17	17	Heilongjiang	0.42	27	Inner Mongolia	0.84
8	Beijing	0.18	18	Jilin	0.42	28	Qinghai	0.91
9	Tianjin	0.18	19	Yunnan	0.47	29	Xinjiang	1.03
10	Hebei	0.20	20	Guizhou	0.47	30	Tibet	1.50

distribution fee, CNG compression and distribution fee and LNG liquid filling fee, the average values are 0.2 , 0.75 , 1.5 and 2.0 CNY/m^3, respectively.

3.6.2 Natural Gas Demand Curve for China in 2013

1. **Summary of price tolerance for different users**

Based on the calculations in this chapter, when the international price of crude oil is 80 $/bbl, and when raw coal price is 600 CNY/ton, the terminal price tolerance for natural gas usage by users in various sectors is 1.4–5.77 CNY/m^3. Within this range, the price tolerance is relatively high for residential, commercial services, vehicle and ship transport, and natural gas manufacturing of hydrogen, but relatively low for centralised heating, natural gas power generation, industrial fuels, synthetic ammonia and methanol (Fig. 3.4).

During the optimisation of primary energy consumption structures, for natural gas replacement of oil as the fuel in industrial processes, the natural gas price tolerance is 2.59 CNY/m^3, while for natural gas chemicals, the price tolerance of natural gas for use in the production of hydrogen as a replacement for naphtha is highest, with a maximum natural gas price tolerance of 5.77 CNY/m^3.

In terms of natural gas as a replacement secondary energy source for electricity in residential and commercial use, natural gas price tolerance

ranks only second to natural gas substituting petroleum, at around 3.7 CNY/m^3 (Fig. 3.5).

During the process of natural gas replacing coal, natural gas price tolerance is generally low, and only when it is used as a substitute for syngas is the natural gas price tolerance relatively high at 2.46 CNY/m^3. Price tolerance for other coal use, whether for heating, synthetic ammonia, methanol production, steam boilers or power generation, is only around 1.5 CNY/m^3.

2. **Construction of the natural gas demand curve**

Step 1: The total consumption of natural gas in China of 166 billion m^3 was divided between 30 provinces, 5 categories and 16 types of user.

Step 2: The prices of alternative energies such as coal, fuel oil, naphtha, #93 petrol, #0 diesel, LPG, electric power and data on urban residents' disposable income was collated, and an evaluation of the terminal price tolerance of 16 kinds of natural gas user carried out.

Step 3: Terminal price tolerance of natural gas users was normalised using natural gas pricing mechanisms based on the Shanghai natural gas benchmark price.

Step 4: The tolerance for the Shanghai benchmark price of the 30 provinces, 5 categories and 16 types of user was organised from high to low, yielding the effective China natural gas market demand curve.

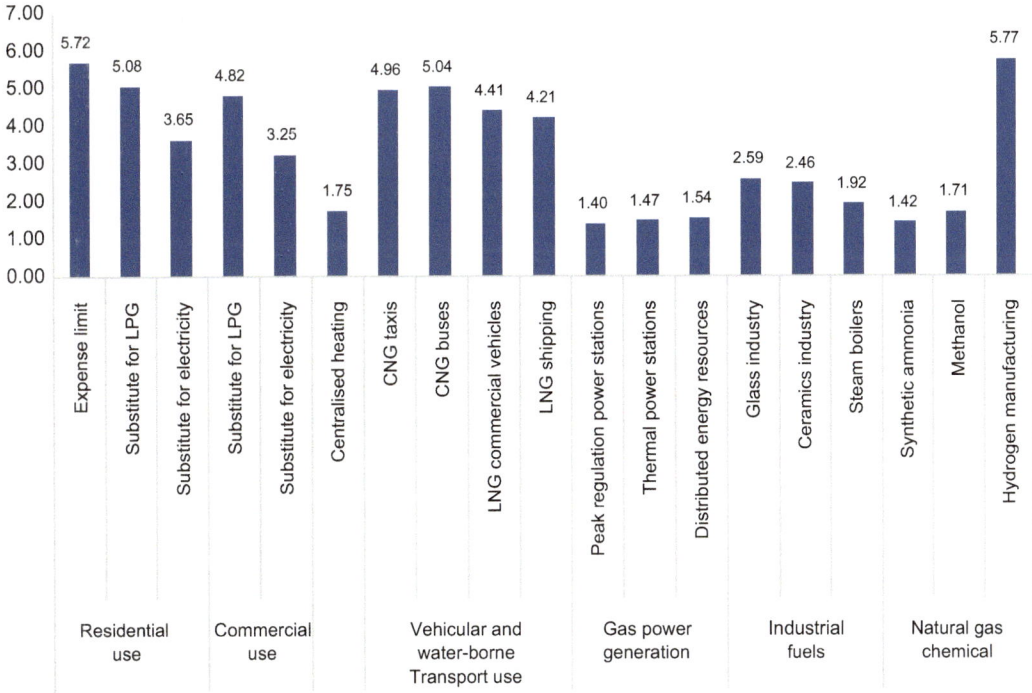

Fig. 3.4 Terminal price tolerance for users in various natural gas sectors

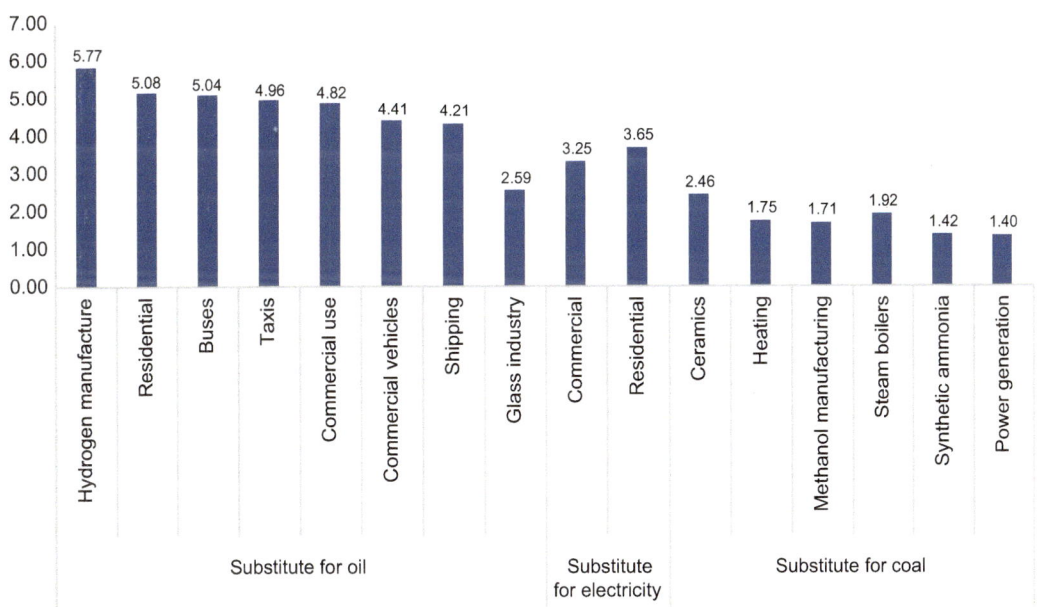

Fig. 3.5 Terminal price tolerance of main gas market users in terms of energy substitution

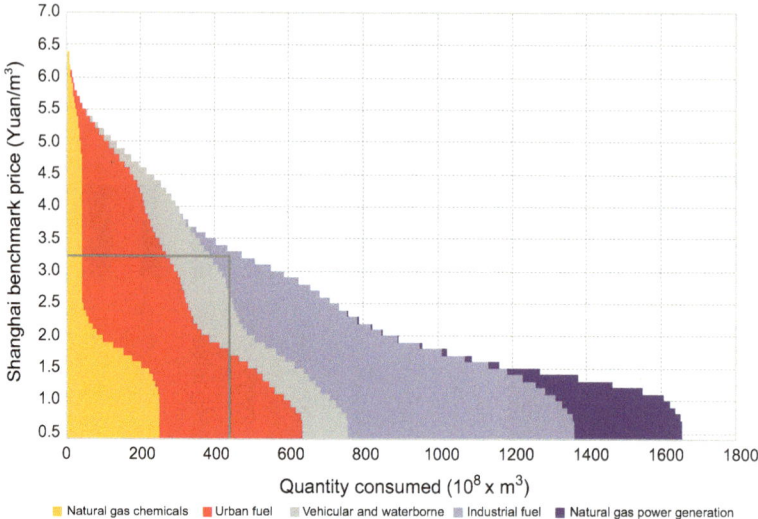

Fig. 3.6 Effective 2013 China natural gas market demand curve

This finally yielded the 2013 China natural gas demand curve, as shown in Fig. 3.6.

3.6.3 Implementation Model of Effective Demand and Actual Natural Gas Consumption Based on Natural Gas Price Reform Targets

In 2013, China proposed a natural gas price reform scheme, which set the Shanghai benchmark price of natural gas at 3.32 CNY/m^3. From the effective demand curve of the natural gas market it can be seen that, in 2013, China's actual consumption of natural gas reached 167.6 billion m^3, and, based on the Shanghai benchmark price of 3.32 CNY/m^3, the corresponding effective demand reaches 50 billion m^3, which only accounts for 30% of actual consumption. The natural gas consumption in urban centralised heating, power generation, ammonia and methanol synthesis and for most industrial fuels cannot be the effective demand based on the Shanghai benchmark price of 3.32 CNY/m^3 (Table 3.18).

In 2013, China's actual natural gas consumption reached 166 billion m^3, which substantially exceeds the effective demand calculated based on price tolerance, being a result of various natural gas supportive policies.

1. **Natural gas price reform transitional policy—gas suppliers providing subsidies of nearly 150 billion CNY to users**

The natural gas price reform scheme implemented by the state specified that the civil natural gas price would not be changed, the fertiliser natural gas price would be slightly adjusted, and non-residential use natural gas prices would have a stored gas and reserve gas two-tier system applied to them. During the natural gas price reform transitional period, an unsatisfactory low price policy was applied to natural gas, with natural gas suppliers providing subsidies to natural gas consumer enterprises to ensure that users could afford gas and that natural gas consumption reached a certain level.

In 2013, the national average urban resident terminal natural gas price was 2.15 CNY/m^3 and the average resident natural gas provincial gate price was 1.37 CNY/m^3, which was 1.50 CNY/m^3 lower than the average national incremental natural gas price. Natural gas suppliers granted around 27 billion CNY of subsidies to urban residential users based on a residential natural gas consumption of 18 billion m^3.

The average fertiliser natural gas consumption price was adjusted slightly to around 1.80 CNY/m^3, being 1.10 CNY/m^3 lower than the average national incremental natural gas price.

Table 3.18 Effective demand of natural gas in China based on the Shanghai benchmark price in 2013

Main purpose		Actual consumption		Effective demand		Growth rate (%)
		Quantity	Proportion (%)	Consumption	Proportion (%)	
Urban fuel gas	Residential	181	10.95	147	30.95	81
	Commercial service	104	6.29	84	17.61	80
	Centralised heating	97	5.87			
Vehicular and waterborne transportation	CNG taxi	59	3.57	59	12.43	100
	CNG bus	32	1.94	32	6.74	100
	LNG lorry	33	2.00	24	4.96	71
	LNG shipping					
Gas-fired power generation	Peak-load regulation power plant	157	9.50			
	Thermal power plant	126	7.62			
	Distributed energy resources	4	0.24			
Industrial fuels		606	36.66	88	17.45	15
Natural gas chemical industry	Synthetic ammonia	152	9.20			
	Methanol	61	3.69			
	Hydrogen production, etc.	41	2.48	41	8.28	100
Total		1653	100	475	100	29

Data source Calculation results
(Units of 100 million m^3)

Natural gas suppliers granted around 22 billion CNY of subsidies to fertiliser users based on a fertiliser natural gas consumption of 20 billion m^3.

In 2013, the stored natural gas price in various provinces in China was 0.88 CNY/m^3 lower than the incremental natural gas. Natural gas suppliers subsidised non-residential stored natural gas users to the tune of around 100 billion CNY based on a non-residential stored natural gas volume of 110 billion m^3.

2. **Special pricing policy for natural gas generation electricity prices**

The *Notice on Relevant Matters for Adjusting Grid Purchase Price of Power Generation Enterprises* issued by the National Development and Reform Commission on September 30, 2013, besides reducing the desulphurised coal power electricity grid price for all provinces (regions and cities), required that the gas-fired power electricity grid price in provinces (regions and cities) such as Shanghai, Jiangsu, Zhejiang, Guangdong, Hainan, Henan, Hubei and Ningxia should be increased, thus easing slightly the conflicting influences on the natural gas price. In China, the grid price of gas power electricity is 40% higher than that of coal power electricity (around 0.20 CNY/kWh). Based on 28.5 billion m^3 residential natural gas consumption, around 570 million CNY of the cost of gas generated electricity is transferred from power suppliers to the user.

3. **Supportive policy for natural gas use**

Generous fiscal subsidies are granted in many locations to encourage the adoption of natural gas centralised heating. Compared with coal-fired centralised heating, the natural gas price tolerance for gas-fired centralised heating only reached 1.75 CNY/m^3. After deducting urban natural gas distribution costs, the provincial gate natural gas price tolerance reached around 1.00 CNY/m^3. If the unit price charged for centralised heating is not adjusted, the economic benefits of gas-fired centralised heating would be almost the same as those of coal-fired central heating. Based on the stored natural gas price of 2.2 CNY/m^3 and a heating natural gas consumption of 10 billion m^3, the fiscal subsidy granted by local government for gas-fired centralised heating reached around CNY 12 billion.

4. Coercive policy for natural gas use

In September 2013, the State Council published the *Action Plan for Atmospheric Contamination Prevention*, which requires the acceleration of coal to gas conversions, with the aim of eliminating small urban coal-fired boilers, establishing coal-free zones etc. to speed up the adoption of natural gas in heating boilers, industrial boilers and thermoelectric projects. Cities at all levels were required to rid themselves of coal-fired boilers with capacity less than 10 tons of steam per hour, and the construction of coal-fired boilers with capacity of 20 tons of steam or less per hour was forbidden. Most industrial fuel users have a low tolerance for natural gas prices, but based on policy requirements, they are now required to absorb the costs of natural gas themselves.

Judging from the natural gas effective demand curve, the demand for stored natural gas that could be supported by industrial fuel users reached 30 billion m^3, while the increased cost of CNY 27 billion associated with the other 30 billion m^3 of industrial fuel consumption had to be absorbed by the enterprises themselves, posing challenges for the sustainable development of natural gas-consuming enterprises.

Environmental and Social Value of Natural Gas

4

Natural gas, as a clean energy source, can reduce the economic losses caused by environmental pollution from coal and petroleum. Based on international experience of natural gas consumption development, it is clear that, while natural gas lacks an obvious price advantage and has even higher direct costs, it is still an effective substitute for other energy sources. The key to this is that restricting pollution reduces the economic and social losses associated with pollution.

4.1 Losses Caused by Atmospheric Pollution in China

4.1.1 Atmospheric Pollution is a Key Cause of Death in China

Based on World Health Organization figures, in 2010 there were 1.2 million people who died prematurely as a result of air pollution, representing a loss of over 25 million years of healthy living. Atmospheric pollution includes indoor and outdoor pollution, among which outdoor pollution is mainly caused by discharge of pollutants through the use of fossil fuels, including coal burning, automobile emissions, dust, etc., while indoor atmospheric pollution is caused by leakage of hazardous substances generated during finishing, and the lack of affordable and clean fuels for residential use; an example of this is the use of direct combustion of large quantities of coal briquettes, which cause significant pollution. In 2010, outdoor atmospheric pollution was ranked fourth as a cause of death for residents in China, while indoor atmospheric pollution was ranked fifth.

From a historical point of view, prevention of indoor atmospheric pollution in China has improved a great deal, but outdoor atmospheric pollution still presents serious challenges. In 1990, indoor atmospheric pollution was ranked second as a cause of death in China. Subsequently, its importance dropped rapidly, and in 2010, it was ranked fifth. However, in recent years, with rapid economic development in China, and a lack of strict energy conservation and emission reduction measures in many energy production and consumption enterprises, the air quality in China has failed to improve significantly, and the threat of atmospheric pollution to health remains significant. During the period 1990–2010, an average of 80–90 persons per 10,000 died of various diseases caused by atmospheric pollution (Fig. 4.1).

* This chapter was overseen by Zhaoyuan Xu from the Development Research Center of the State Council and Martin Haigh from Shell International, with contributions from Baosheng Zhang and Shouhai Chen from the China University of Petroleum, Lianzeng Zhao from the China Petroleum Planning Research Institute, Linji Qiao from ENN and Juan Han from Shell China. Other members of the research group participated in discussions and revisions.

Shell International and The Development Research Center (Eds.), *China's Gas Development Strategies*, Advances in Oil and Gas Exploration & Production, DOI 10.1007/978-3-319-59734-8_4

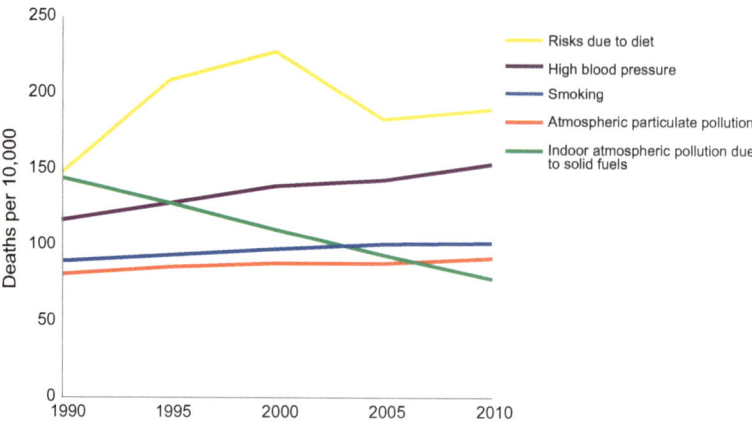

Fig. 4.1 Top five causes of death of residents in China. *Data source* Vivid Economics, referred from Institute for Health Metrics and Evaluation (2014)

4.1.2 Losses from Injury to Health Caused by Atmospheric Pollution in China Account for 3–12% of GDP

Economic activity can improve citizens' income and increase well-being, but the associated atmospheric pollution causes losses in terms of quality of life. A balance must be reached between economic benefits and losses. However, health losses cannot be directly measured in the same way as economic output if there is no quantification of loss of health. Previous experience shows that, if health effects can be economically quantified in a financial index that represents losses caused by the death of a person, this reflects the true cost of economic development, and can act to promote improvements in air quality.[1]

1. Atmospheric pollution losses in the European Union

Based on international standards, the air quality in EU countries is excellent. However, in 2012, the losses from primary pollutants generated by industry and energy generation still accounted for between 0.3% and 1% of GDP

(European Environment Agency, 2014). Figure 4.2 shows estimated hazard costs per ton of pollutants in Europe. In Europe, the hazards of CO_2 cost 9.8–38.1 €/t (2005 prices), while the hazards of PM_{10} cost 23,000–67,000 €/t. Other pollutants besides particulate matter also need to be noted; in particular, the hazard cost of heavy metal and organic pollutants per unit weight far exceeds that of particulate matter pollution, though the quantity of these pollutants is less. The hazard that they represent overall is therefore less than that presented by particulate matter.

2. Estimate of health-related losses caused by atmospheric pollution in China

China faces worse pollution than other countries. In the main cities in China, the average concentration of PM_{10} and sulphur dioxide is respectively 5 times and 12 times the reference value (Nielson and Ho 2013) issued by the WHO in 2009. Significant research has shown that air pollution and energy structure have a direct relationship. In recent years, various methods have been adopted in many studies to assess health-related losses in China.

The *China 2030 Report*, jointly issued by the Development Research Center of the State Council and the World Bank in 2013, estimated that the health-related losses caused by PM_{10} were equivalent to 2.8% of the GNI in 2009,

---

[1]There are also many people who object to using money to measure the value of life, but for the purpose of balancing interests, this measure is still necessary.

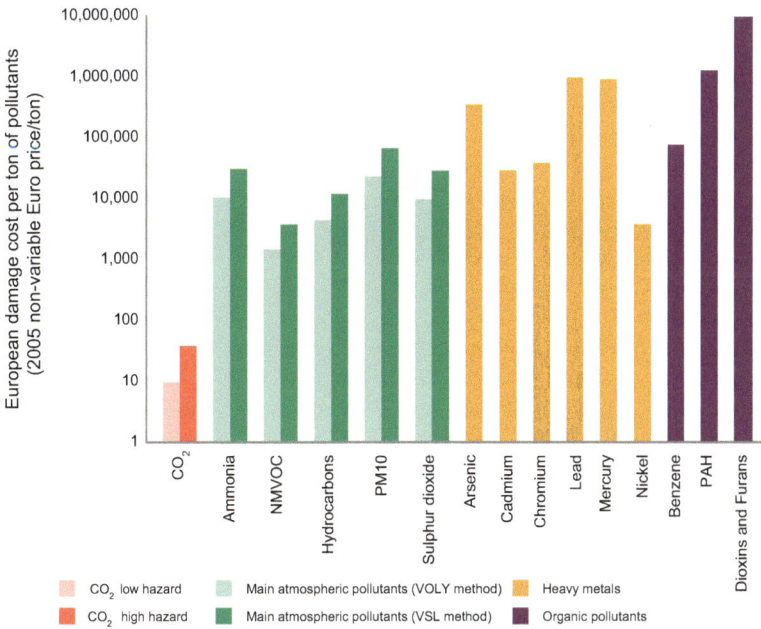

Fig. 4.2 Hazard cost estimation for main pollutants in Europe. *Note* Damage calculations apply for Europe only. *Data source* Vivid Economics, referred from (European Environment Agency, 2014)

calculated based on the two basic parameters of deaths caused by particulate matter discharge and the amount of money that people are willing to pay to reduce the risk of disease.

The *New Climate Economy Report* issued by the Global Commission on Economy and Climate in 2014 estimated that the economic loss caused by deaths of people due to $PM_{2.5}$ in China in 2010 was equivalent to 9.7–13.2% of GDP, with a median value of around 12% of GDP. This result was obtained by multiplying three values: deaths caused by $PM_{2.5}$, economic value of deaths and per capita GDP. The economic value of deaths was obtained by converting the value of statistical life (VSL) used in the European Union, adjusted for per capita income. In other words, the economic value of deaths will grow together with the growth of per capita GDP. Data for deaths caused by $PM_{2.5}$ was sourced from the WHO's 2010 Global Burden of Disease Study.

Matus et al. (2011) estimated the economic loss caused by atmospheric pollution. This paper calculated the health hazard caused by PM_{10} and ozone, and the results show that in 2005 the economic losses caused by health hazards due to

PM_{10} accounted for around 6% of GDP. This paper adopted the multi-department CGE model to calculate the losses due to atmospheric pollution health damage between 1970 and 2005, taking into account the cost of related treatment, reduced working capacity and extra holidays taken. In 2005, around 60% of the losses were due to death, 10% due to the cost of treatment and 30% from other economic losses.

If other pollutants are brought into the equation, the economic losses of atmospheric pollution in China become even higher. These studies only consider particulate matter (PM), including the larger PM_{10} with a diameter of 10 microns and the smaller but more harmful $PM_{2.5}$. In fact, the hazards of other pollutants such as sulphur dioxide and nitrogen dioxide are also severe.

In general, the difficulties encountered in research into the economic quantification of the losses caused by atmospheric pollution in China cause a lot of variation: the research methods, concerns and assessment standards are different, and the research results are also uncertain. Generally speaking, the economic loss caused by atmospheric pollution in China is massive,

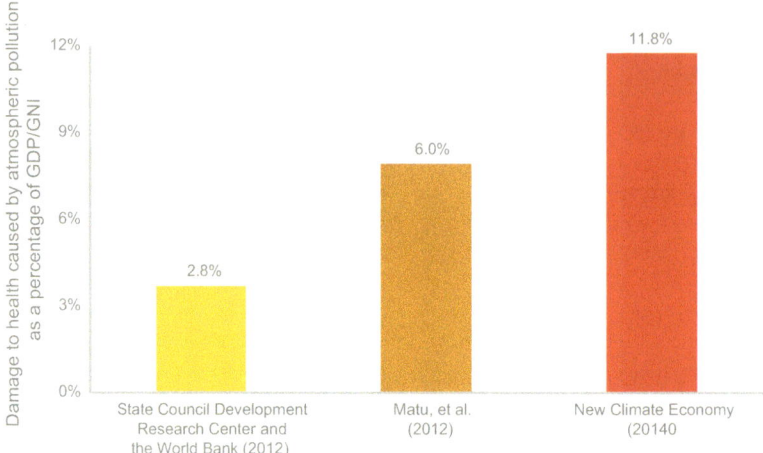

Fig. 4.3 Economic loss caused by atmospheric pollution as a percentage of GDP. *Data source* The State Council Development Research Center and the World Bank 2012, Matus et al. 2012; Global Economy and Climate Commission, 2014

equivalent to between 3% and 12% of GDP. Based on the results of these studies, in 2014, atmospheric pollution caused a loss of between $300 billion and $1200 billion in China. Given that the International Energy Agency estimated that China's energy supply needed $300 billion of investment annually up to 2035, this gives a clear indication that improving air quality would pay for itself (Fig. 4.3).

3. **Other unfavourable effects of atmospheric pollution**

Atmospheric pollution not only causes direct harm to health and economic losses, it also reduces China's capital labour reserves and reduces productivity. This results in a "haircut" being applied in terms of the actual economic benefits of the particularly labour-centred approach to economic activities that China relies on. In addition, the investment attracted by the 12th Five-Year Plan and the political targets put forward during the 3rd Plenary Session of the 18th CPC Central Committee will also be affected. The implementation of policies for expanding urbanisation, developing the service industry, reducing inequality, improving social insurance, etc. will become harder to achieve, as shown in Fig. 4.4.

4.2   The Environmental Value of Natural Gas as a Substitute for Coal

The environmental value of natural gas stems from the fact that it is a clean energy substitute for coal. Coal production, transportation and usage cause serious environmental pollution, and result in tremendous direct and indirect external economic losses. However, as the costs of environmental pollution in China are accounted for as external economic losses, market entities causing environmental pollution consistently fail to pay the price of environmental pollution themselves. Similarly, the environmental benefits of replacing coal and oil with natural gas are not reflected in natural gas's market value, since the costs of failing to make the replacement are not being internalised.

4.2.1   Environmental Pollution and Economic Losses Incurred During Coal Production

Environmental pollution generated during coal production manifests as damage to and

Fig. 4.4 Further effects of atmospheric pollution

appropriation of land, and pollution of water resources and the atmosphere.

Underground mining causes surface subsidence; mining areas suffer from ponding, flooding and salinisation, which accelerate water and soil erosion; and desertification causes damage to surface infrastructure to differing degrees. Open coal mining peels off the topsoil and rock stratum covering the coal bed, destroying land resources, and discharges spoil onto adjoining land. Coal spoil piles up and buries land resources. Based on relevant data, extraction of 10,000 tons of coal will cause surface subsidence of an area of 0.2 hectares, requiring relocation of two residents, while associated spoil occupies 0.01 hectares. The economic losses caused by land resource damage due to extraction of one ton of coal are thus CNY 65.

During the mining process, the scale of extraction grows and even larger quantities of mine water are discharged, causing pollution of the aquatic environment around mines; meanwhile, the water table drops, causing subsidence. The commonly used coal washing process results in large quantities of water that contains high concentrations of sludge, silt and harmful heavy metal ions. Large amounts of harmful substances (especially heavy metal ions) contained in the coal spoil heaps leach out due to rainfall and enter the ecosystem. Based on relevant data, extraction of one ton of coal results in the discharge of 2.3 tons of mine water, 0.35 tons of industrial waste water, 0.05 tons of coal washing water, and 0.04 tons of other waste water, resulting in economic losses estimated at CNY 10 per ton of coal.

The greenhouse effect of marsh gas (the main ingredient of which is CH_4) released during coal extraction is 21 times that of CO_2, making this one of the main gases responsible for global warming. Large quantities of hazardous gases such as SO_2, CO_2, CO etc. produced by spontaneous combustion of coal spoil also affect the atmospheric environment. Based on relevant data, extraction of one ton of coal on average results in the discharge of 10 m^3 of gas, equivalent to a discharge of 138 kg of CO_2, and the estimated economic loss caused by atmospheric pollution produced by extraction of one ton of coal is CNY 15.

Economic losses caused by environmental pollution during coal production per ton of coal extracted are around CNY 90.

4.2.2 Environmental Pollution and Economic Loss During Coal Transportation

The geographical distribution of coal production and consumption in China is highly unbalanced, with coal production bases being distributed mainly in the northern and western regions, while coal consumption is mainly distributed in eastern and coastal areas, determining the basic pattern of transportation: northern coal to the south, and western coal to the east. The average distance coal is transported by rail is 552 km.

Coal in China is transported over long distances, and passes from the coal stores or coal yards of mines to trains (or vehicles or vessels) and then reaches large thermal power generation coal yards, industrial coal consumption enterprise coal yards and fuel company coal yards of various sizes. As coal storage, loading and transportation facilities are usually not integrated, the lack of flow regulation exacerbated by poor management results in serious pollution of the surroundings at all stages of coal storage, handling and transportation.

The spontaneous combustion of mined coal not only happens in coal yards, but also after loading on lorries or vessels. During open stacking, handling and transportation of coal, the coal dust contaminates wide areas and causes serious pollution. Spraying in open coal yards during loading and unloading to cause coal dust to settle results in pollution of bodies of water.

Based on relevant data, there is an annual discharge of between 200,000–300,000 tons of harmful gas due to spontaneous combustion of coal in China, while 10 million tons of coal dust is produced during coal storage and 11 million tons of coal dust is generated during coal transportation. The estimated economic losses caused by pollution from transportation of each ton of coal are thus CNY 20.

4.2.3 Estimates of Environmental Pollution and Economic Losses Arising from Coal Use

Direct combustion of large amounts of coal has caused serious damage to the environment in China, while it is also the main cause of environmental pollution.

SO_2 discharge and acid rain have serious effects on human health and vegetation, and cause deforestation, damage to buildings and metal corrosion. Based on calculations made by various institutions, the economic loss caused by the discharge of one ton of SO_2 is as high as CNY 7000.

CO_2 discharge and the greenhouse effect seriously threaten global ecosystems and the survival of mankind, causing shrinking forests, land salinisation, desertification, extreme weather events, etc. Based on the cost of carbon sequestration, the estimated economic loss caused by the discharge of one ton of CO_2 is CNY 110.

NO_X discharges have serious effects on human health and cause a series of diseases and photochemical reactions, in addition to damaging ecosystems. Based on calculations made by various institutions, the economic loss caused by discharge of one ton of NO_X is as high as CNY 5000.

Dust emissions have an effect on human health and discharge of dust into the atmosphere will reduce atmospheric cleanliness and affect plant photosynthesis etc. Based on calculations made by various institutions, the economic loss caused by discharge of one ton of dust is as high as CNY 200,000.

Based on calculations of coal combustion pollutant discharge quantities, the economic loss caused by the pollution produced by one ton of raw coal is CNY 830 (Table 4.1).

4.2.4 Environmental Value Assessment for Natural Gas Substituting Coal

The process of coal production, transportation and use results in serious environmental pollution. The estimated economic loss incurred is 380–940 CNY/t, of which production accounts for 90 CNY/t, transportation accounts for 20 CNY/t and usage accounts for 295–830 CNY/t.

CO_2 and other micro pollutants are discharged during the natural gas usage process. Calculated based on a discharge of 2.5 kg CO_2 per m^3 natural gas, the economic loss caused by the pollution due to natural gas is 0.3 CNY/m^3.

Based on calorific value and efficiency of coal and natural gas, and calculated for 550 m^3 natural gas replacing one ton of coal, the natural gas environmental value compared to coal which is subjected to similar emissions restrictions is 0.4–1.4 CNY/m^3 and 1.4 CNY/m^3 under conditions of dispersed coal use.

4.3 Social Value of Natural Gas as a Substitute for Coal

Besides environmental losses, coal production, transportation and consumption processes all cause non-economic losses to the population, for instance non-economic losses involving mental

Table 4.1 Calculation table for the economic loss caused by pollution produced by one ton of raw coal

Emission	Discharge quantity (kg)		Unit loss (CNY/kg)	Economic loss (CNY)		Remarks
	Near zero emissions	Distributed combustion		Concentrated combustion	Distributed combustion	
CO_2	2455	2455	0.11	270	270	Calculated based on carbon exchange value
NO_X		5.40	5		27	Calculated based on pollution loss
SO_2		10.40	7		73	Calculated based on pollution loss
Dust		2.30	200		460	Calculated based on pollution loss
Total				295	830	

and physical suffering, such as that due to the pain caused by disease suffered by the victim, and the pain caused to the relatives due to the death of a loved one.

Social losses are calculated based on difference in willingness to pay and human capital. Willingness to pay refers to readiness to pay or a desire for compensation ascertained through research, the calculation of which includes bidding gaming methods, comparison gaming methods and the Delphi method. Human capital refers to the monetary loss connected with health, including increased losses and medical expenses due to premature death, illness or medical leave and other factors.

4.3.1   Social Loss of Coal Production

The coal industry is the industry with the greatest number of fatal accidents in China—annual coal mine fatalities had reached around 6000–7000, but dropped to nearly 4000 by 2007. Accidents in coal mines account for 85% of workplace fatalities in China's mining industry, and coal mining accounts for 50% of all mining fatalities, the highest rate in the country. Apart from this, pneumoconiosis, an occupational disease related to coal production, has very serious consequences on health.

4.3.2   Social Loss of Coal Use

Environmental pollution caused by coal use seriously affects people's health. SO_2 emissions have

obvious effects on the respiratory system, irritating the respiratory mucosa, resulting in acute and chronic inflammation of the respiratory tract. Long-term inhalation of SO_2 will lead to fibre hyperplasia of lung tissue cell walls as well as emphysema and bronchial asthma. Smoke dust discharge is a serious problem too. The dust consists of particulates which can become suspended in the atmosphere for a long time, and at any time it can be directly inhaled into the respiratory tract. These small particulates are a carrier of bacteria, viruses and metallic particles, and bring harmful pathogenic bacteria and toxic particles into the human respiratory system and the alveoli, inducing allergic rhinitis, bronchitis and bronchial asthma, and even malignant tumours.

4.3.3   Social Value Assessment of Natural Gas as a Substitute for Coal

Based on the functional relationships between death rate and the incidence (dose-response function) of certain diseases outlined in *Atmospheric Pollution in Large and Medium-Sized Cities of China* issued by the World Bank, the contribution of coal and natural gas use to atmospheric pollution was evaluated, while considering the loss of life caused by mining accidents and the social losses associated with occupational diseases encountered in the coal production process, providing an estimated external environmental benefit to society of natural gas of around 0.4 CNY/m^3 (Table 4.2).

Table 4.2 Dosage-effect function of inhalable particulate matter PM_{10}

Item	Unit	Value	Unit value orientation (CNY)	
			Human capital method	Willingness to pay (WTP) method
Mortality	Person	6	100,000	500,000
Respiratory disease outpatient rate	Case	12	4000	5200
Emergency cases	Case	235	400	520
Restricted activity days	Day	57,500	40	52
Lower respiratory tract infection/paediatric asthma	Case	23	200	260
Asthma	Case	2068	60	78
Chronic bronchitis	Case	61	80,000	104,000
Respiratory disease symptoms	Case	183,000	10	13

Note The values are the numbers of additional deaths, cases, or days for each increase of 1 $\mu g/m^3$ in PM_{10}

4.4 China's Achievements in Energy Conservation and Emissions Reduction

Thanks to technical progress and greater social awareness, China has significantly enhanced its energy conservation and emissions reduction work in recent years. The results show that energy conservation and emissions reduction resulted in significant direct and indirect benefits being achieved with limited input, providing empirical support for the further development of clean energy in China.

4.4.1 Targets and Main Measures for Sulphur Dioxide Emission Reduction During the Period of the 12th Five-Year Plan

In recent years, China has attached great importance to energy conservation and emissions reduction, and has strengthened the intensity of energy conservation and emissions reduction work, with stated targets becoming binding indexes listed in the Five-Year Plan and government planning. For example, the 11th Five-Year Plan specified that sulphur dioxide emissions between 2006 and 2010 should be 10% lower than those in 2005. Judging from the actual achievements, the implementation of these emission reduction works has seen good results.

In order to achieve the sulphur dioxide emissions reduction targets during the 11th Five-Year Plan period, two main policies were adopted by the Chinese government.

- All newly-built coal-fired power generation plants were required to be fitted with flue gas desulphurisation facilities conforming to specific standards, while most of the old power generation plants were required to undergo retrofitting to meet desulphurisation standards. After the reforms, the coal-fired power station flue gas desulphurisation capacity was enhanced to 83% in 2010 from the 12% adopted in 2005. Flue gas desulphurisation conversion costs are low. The flue gas desulphurisation cost of a 6 million-watt power plant is only around 3.8% of the construction cost, and operational costs are increased by 2.4%—costs that are offset by government subsidies.
- Small, inefficient power generation plants were to be shut down. At the end of 2005, low-use and highly polluting small power plants accounted for one third of national energy production. By replacing these with large power generation plants, costs were reduced by a factor of 2 or 3 per million

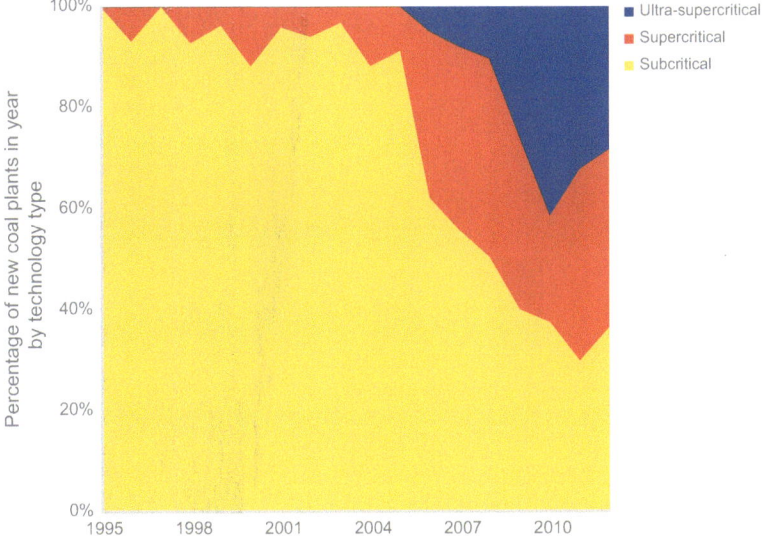

Fig. 4.5 Proportion of new technology use in newly built coal-fired power stations in China. *Source* Vivid Economics, based on PIRA database

watts. Since 2004, the average coverage of new power generation plants has doubled. At the same time, most of the large new power generation plants have adopted more efficient technology, as shown in Fig. 4.5. Based on the shutdown policy for small power plants, 59 billion watts of small power plants were closed during the 11th Five-Year Plan period and replaced by larger power plants.

4.4.2 Emissions Reduction Results Achieved During the 11th Five-Year Plan Period

The goals of reducing sulphur dioxide emissions were achieved. The power sector accounted for 54% and 28% of the total sulphur dioxide emissions respectively in 2005 and in 2011, growth in sulphur dioxide emissions being much slower than that of other pollutants, such as nitrogen oxides and particulate matter. A study conducted by Harvard University and Tsinghua University shows that the sulphur dioxide emission reduction policy achieved a major effect with a low macroeconomic cost, and is expected

to save between 12,000 and 74,000 lives every year, which is equivalent to CNY 8–400 billion (Nielson and Ho, 2013). This policy shows that China can achieve greater benefits (synergically speaking) by improving atmospheric quality, while greater losses result if atmospheric pollution is not controlled (Fig. 4.6).

4.4.3 Enormous Potential for Natural Gas to Substitute for Coal in Industrial Fuel and Residential Heating

Considered from the point of view of pollutant source, industrial and domestic coal use are now major contributors. In recent years, the level of pollution prevention and control technology in coal-fired power generation plants in China has improved significantly, with the proportion of pollution produced during power generation sharply dropping, the proportion of pollution produced by transportation being relatively low, and emissions improving year by year. This is primarily as a result of improved vehicles. For example, new urban vehicles in China must comply with the Europe IV standard, which

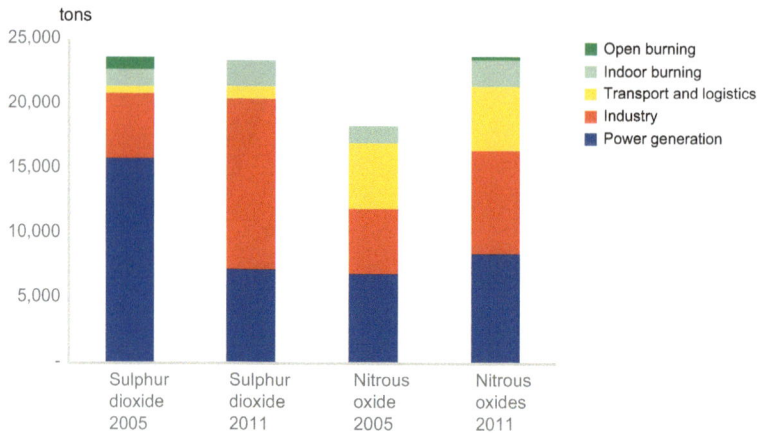

Fig. 4.6 Reduction of sulphur dioxide and nitrogen oxides emission during the 11th Five-Year Plan period. *Data source*: Vivid Economics, referred from Nelson and Ho (2013)

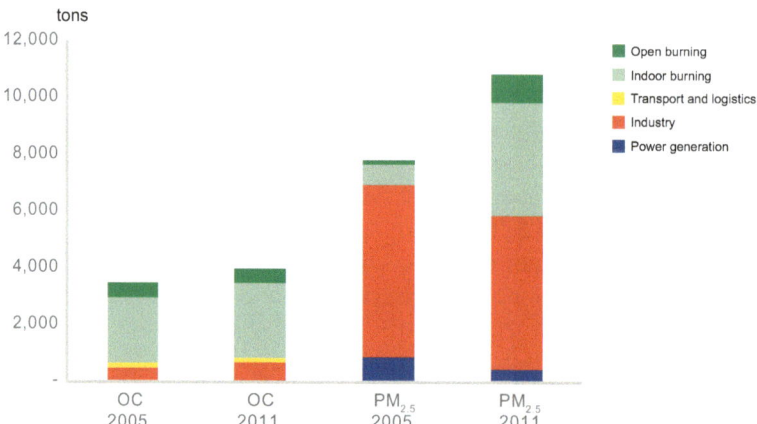

Fig. 4.7 Change in the sources of PM$_{2.5}$ and OC (organic carbon) during the 11th Five-year Plan

greatly reduces automobile exhaust pollutant emissions. In terms of main pollutant sources such as sulphur dioxide, nitrogen oxides, PM$_{2.5}$ and organic carbon (OC), etc., industrial and residential use are the main sources of emissions (Fig. 4.7).

Conversion from coal to natural gas in industrial and residential use is a feasible pollution prevention and control measure. Different methods have been adopted for the control of production-related and residential pollutant emissions and the control of sulphur dioxide emissions during the 11th Five-Year Plan. This is because the dispersal of industrial and residential users and their relatively small scale restricts their ability to adopt the same end technology on a significant scale. In contrast, the average scale of new power generation stations in China has tripled since 2004, and there are now 1425 new power stations. Despite the quantity being significant, it is still small compared to the quantity of industrial and residential user locations in China. Natural gas is suitable for small-scale applications, and this represents an economically valuable approach to reducing coal pollutant emissions from industrial production and residential heating.

Analysis of Medium- to Long-Term Natural Gas Demand and Supply

5

This section explores various aspects of the development of industry and the economy, and the changes in supply and demand of different energy sources, considered in combination with projected changes in economic growth, advances in technology and shifts in industrial structure. The aim is to determine the medium- to long-term demand and supply of natural gas and other energy sources in China.

5.1 The Natural Gas Supply-Demand Model

In order to analyse and predict the future of natural gas demand, a Computable General Equilibrium (CGE) model was devised to reflect medium- to long-term economic growth and changes in industrial structure, in terms of total demand for energy sources and the interchangeability between natural gas and other energy sources. The CGE model relies on general equilibrium theory, using actual economic data as the initial equilibrium, and reflects optimised decision-making by the market components

(manufacturers, consumers, government departments). Compared to general economic models, CGE models are more often used for simulating the effect that policies have on the economy, whether direct or indirect, making them an effective tool for analysing policies. The CGE model used in this study is based on the Computable General Equilibrium model developed and maintained over time by China's State Council's Development Research Center (DRC-CGE) for dynamically simulating the economy of China. The development of the model began in 1997; it has been perfected over the years and applied to studies of the effects of China joining the World Trade Organization, infrastructure construction, energy conservation, emission reduction, urbanisation and other policies.

5.1.1 Basic Characteristics of the Model

The model uses 2010 as the base year and, adopting the Social Accounting Matrix based on 2010 input-output tables of China for basic data, simulates the economy from 2010 to 2025. it includes economic activities from various sectors, reflecting five aspects in particular:

- the manufacturing activities of various business sectors in the economy, and the demands stemming from such manufacturing activities;

* This chapter was overseen by Zhaoyuan Xu from the Development Research Center of the State Council and Martin Haigh from Shell International, with contributions from Baosheng Zhang and Shouhai Chen from the China University of Petroleum, Lianzeng Zhao from the China Petroleum Planning Research Institute, Linji Qiao from ENN and Juan Han from Shell China. Other members of the research group participated in discussions and revisions.

Shell International and The Development Research Center (Eds.), *China's Gas Development Strategies*, Advances in Oil and Gas Exploration & Production, DOI 10.1007/978-3-319-59734-8_5

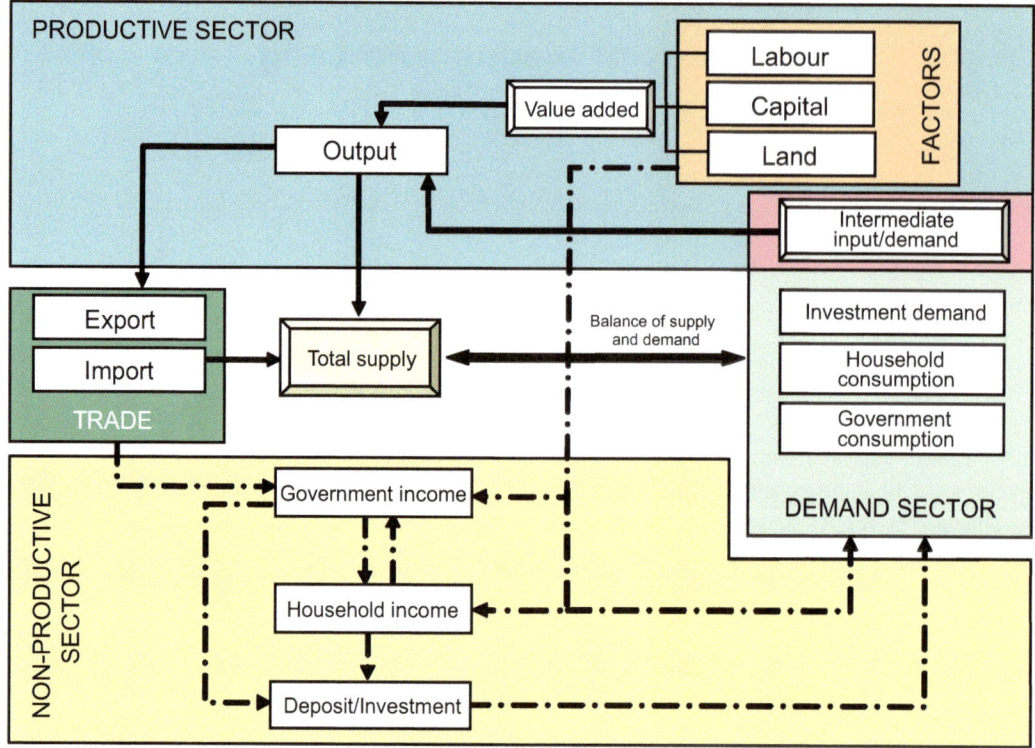

Fig. 5.1 Main structures of the CGE model used in this study

- the economy's demand for products: the total demand coming from the everyday activities of the citizens, the government and businesses, as well as from other countries (regions), including the three major demands of consumption, investment and exports;
- the income, consumption and reserves of each major economic body;
- foreign trade; and
- dynamic economic growth and development.

The relationships of these factors are shown in Fig. 5.1.

In this study, the model includes all of the 46 production industries (as seen in Table 5.1), specifically 1 from agriculture, 5 from mining, 22 from manufacturing, 3 from public works and construction, and 14 from the service sector. These industries illustrate the production activities of various business categories.

5.1.2 Main Intensifications and Adjustments Made Towards Studies of Natural Gas in This Model

Compared to typical CGE models, the one used in this study made the following major changes to reflect influences on the supply and demand equilibrium of natural gas.

1. **Detailed representation of primary energy production**

To better reflect the input-output and interchangeability of various energy sources, this study takes into account coal mining and processing, oil drilling, natural gas extraction, oil refining, coking, coal power, natural gas power, hydroelectricity, wind power, solar power,

Table 5.1 Subdivision of industries in the model of this study

Number	Industry	Number	Industry	Number	Industry
1	Agriculture, forestry and fishing industry	25	Non-metal mineral product industry	49	Heat production and supply industry
2	Coal mining and preparation industry	26	Ferrous metals smelting industry	50	Natural gas production and supply industry
3	Petroleum industry	27	Steel rolling industry	51	Water supply industry
4	Natural gas drilling industry	28	Non-ferrous metals smelting and rolling industry	52	Construction industry
5	Ferrous metals mining industry	29	Metal products industry	53	Logistics industry
6	Non-ferrous metals mining industry	30	Equipment manufacturing industry	54	Urban public transportation industry
7	Non-metal and other mining industry	31	Special equipment manufacturing industry	55	Other transportation and warehouse industry
8	Food and alcoholic beverage industry	32	Railway transportation and equipment manufacturing industry	56	Postal industry
9	Tobacco industry	33	Car manufacturing industry	57	Information transmission, computer and software industry
10	Textiles processing industry	34	Ship and flotation device manufacturing industry	58	Distribution and retail industry
11	Textiles, knitting and manufacturing industry	35	Other transportation equipment manufacturing industry	59	Hospitality industry
12	Clothing, shoe and hat manufacturing industry	36	Electrical appliances industry	60	Financial industry
13	Leather, fur, feathers (down) and related products industry	37	Electricity supply and distribution and control equipment manufacturing industry	61	Real estate industry
14	Carpentry and furniture manufacturing industry	38	Household electrical and non-electrical equipment manufacturing industry	62	Rental and commercial services industry
15	Paper making and printing industry	39	Other electrical machinery and equipment manufacturing industry	63	Research, testing and development industry
16	Education and physical education equipment manufacturing industry	40	Communication equipment and radar manufacturing industry	64	Integrated technical services industry
17	Oil refinery and nuclear fuel processing industry	41	Computer manufacturing industry	65	Water conservancy, environment and public facilities management industry
18	Coking industry	42	Electronic component manufacturing industry	66	Personal and other services industry
19	Chemical industry	43	Home audio-visual equipment manufacturing industry	67	Education
20	Fertilisers and agricultural chemicals	44	Other electronics manufacturing industry	68	Health, social security and benefits industry

(continued)

Table 5.1 (continued)

Number	Industry	Number	Industry	Number	Industry
21	Synthesised material manufacturing industry	45	Measuring instruments and devices manufacturing industry	69	Cultural, physical education and entertainment industry
22	Special chemical products manufacturing industry	46	Cultural, office machinery manufacturing industry	70	Public management and social organisation
23	Other chemical products	47	Handicrafts and other products (including waste products) industry		
24	Plastics and rubber products	48	Power generation and supply industry		

nuclear power, natural gas heating, coal heating and natural gas processing, giving a total of 14 types of energy source, thus better reflecting the interchangeability of energy sources, especially the potential of natural gas to replace other energy sources.

Based on the national energy policy, this model separately configured production functions and development scales of different energy sources, such as installed capacity of nuclear energy, hydroelectricity, solar power and others, in order to ensure that they are in accordance with national planned goals.

2. **Detailed representation of characteristics of the demands of the various sectors towards natural gas**

To examine natural gas demand in detail, this study subdivided industries, in particular industries with higher demand for natural gas. For example, the chemical industry was subdivided in order to reflect its demand for natural gas as a raw material and the transportation industry was subdivided in order to reflect its demand for natural gas.

5.2   Simulation Scenarios for Analysis Simulations of Natural Gas Demand

To carry out the simulated analysis of natural gas supply and demand, and to compare various policies, this study has devised two different scenarios: one is the standard scenario, where the current basic trends of economic growth are reflected, the other is the policy-driven scenario, which includes the effects of government policies favourable towards natural gas demand.

5.2.1   Key Assumptions of the Standard Scenario

In the standard scenario, backed by clearly designed policies, relatively assertive growth factors are observed in the Chinese economy. On the other hand, under the influence of ambitious government policies regarding environmentally friendly development and innovation, there are the following significant changes to the growth factors:

- Changes in total population and age structure reflecting the influence of the newest population policy adjustments (one-child policy, two-child policy) in terms of medium- to long-term total population and workforce. The population of China peaks at around 2032, when it will be around 1.463 billion, with the workforce peaking between 2017 and 2027, at around 1 billion.
- Growth in personal consumption expenditure, brought about by an increase in income levels. Particularly remarkable is the reduction in the percentage of expenditure taken up by food and other consumable goods, while travel, leisure, education and expenditure for other services continue to increase.

- Policies promoting energy conservation and emission reduction, influenced by the attempt to bring about a "new normal" and a variety of activities targeted at alleviating atmospheric pollution, where the government continues to push for energy conservation and consumption reduction, raising energy efficiency by between 3 and 2% points each year (a higher rate of increase at first, followed by a lower one).
- The new urbanisation policies progress smoothly, and more of the rural population are now living in cities. The urbanisation rate will hopefully reach 70% in 2030 and 75% in 2050.
- Personal savings remain at high levels, but this is expected to gradually reduce as personal income and level of social security provision improve. By 2030, the personal savings rate of urban populations should fall by 13%, from the current rate of 38% to around 25%.

In addition to the above factors, optimised policies include policies that concentrate in particular on transformation of the mode of economic development:

- A more significant ratio of renewable energy usage, mainly represented by rapid developments in non-fossil fuel energy sources such as nuclear power, wind power and solar power. For nuclear power, the installed capacity of 14.61 million MW in 2013 will quickly grow to 58 million MW by 2020, reaching 150 million MW by 2030 and 400 million MW by 2050. Installed capacity of hydroelectric power plants was 280 million MW in 2013, and will rise to 340 million MW in 2020, 400 million MW by 2030 and around 450 million MW by 2050. Other energy sources such as wind power and solar power will also grow at similar speeds.
- The government places great importance on innovation-driven development, enabling remarkable breakthroughs in improving business innovation, the results of which will contribute to an increased significance of improvement in core areas driving economic growth, which will become evident by the

13th and 14th Five-Year Plans, at which stage secondary industry in China will have a high sustainable all-factor growth rate of 4%.
- Mitigation of overcapacity is relatively smooth, allowing industry to grow continuously. This can be observed in the slowing growth in investment in heavy and chemicals industry while output growth stabilises, a trend that will benefit sustained development of the industry.
- The nurturing of tertiary industries, in particular the hastening of the development of productive service industries and tertiary industries, which will be beneficial in bringing in new investment to monopolistic industries previously difficult to access, such as finance, logistics, education and health.
- As a result of financial reforms, the proportion of the national economy accounted for by private income increases; this is advantageous because of raised income levels and the improved ratio of consumption and investment in the economy.

5.2.2 Key Assumptions of the Policy-Driven Scenario

In the policy-driven scenario, apart from making the same assumptions as the standard scenario, the government will also levy a carbon tax on coal, oil and natural gas—fossil fuels that result in greenhouse gas emissions. Pricing for this carbon tax collected through carbon trading will be implemented in 2015 for coal. The carbon price (or tax rate) for the first year will be 10% of price, rising to 20% in the second year, then remaining at 30% thereafter. The tax rate for petroleum (including both crude oils and refined oils) is the tax rate for coal multiplied by the difference in carbon emission rates, which works out at 0.64 times the tax rate for carbon price for coal, while the rate for natural gas is 0.2 times the rate for coal, converted based on a coal price of 500 CNY/ton once the rate reaches 30% (ad valorem). This is equivalent to a carbon emissions tax (or achieving a carbon price through carbon trading of an equivalent amount) of CNY 60 for every ton of carbon dioxide where coal is concerned.

5.3 Natural Gas Supply and Demand in the Standard Scenario

5.3.1 Speed of Economic Growth and International Comparison

Under the assumptions of the standard scenario, China will retain a relatively fast growth rate in the future. In the two years after the 12th Five-Year Plan, the average economic growth rate is expected to be around 7.3%, with GDP

growth during the 13th Five-Year Plan at around 6.66 and 5.56% from 2021 to 2025. In the period between 2026 and 2030, economic growth is expected to be around 4.64%, around 3.47% from 2031 to 2040 and around 2.73% from 2041 to 2050 (Table 5.2).

Based on simulation results, the per capita GDP of China will continue rising at a relatively fast rate. For example, by around 2020, per capita GDP in China is hoped to reach $10,000 (in 2013 figures, and taking into account an average annual 0.5% appreciation of the CNY), while by 2030 per capita GDP in China could reach $20,000, and $40,000 by 2050, thus achieving a similar level to that current in Japan. If the comparison is made using purchasing power parity, China's development level will be even higher (Fig. 5.2).

Table 5.2 Growth speed and growth momentum in the standard scenario

Year	2013	2014–2015	2016–2020	2021–2025	2026–2030	2031–2040	2041–2050
GDP	7.68	7.31	6.65	5.55	4.63	3.46	2.73
Speed of growth in workforce	0.3	0.1	−0.2	−0.3	−0.5	−0.7	−0.4
Speed of growth in capital deposit	11.6	10.9	9.3	7.7	6.5	4.8	3.2

Data source Based on calculation results

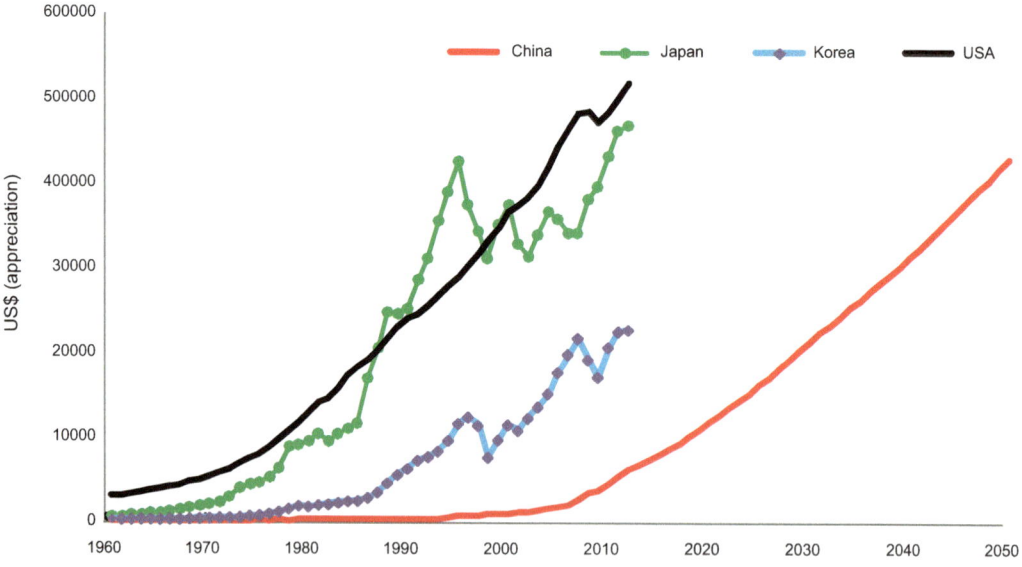

Fig. 5.2 Development levels in China and international comparisons in the standard scenario. *Data source* Figures for China after 2013 based on model calculation results, other figures from Wind data

In the standard scenario, the speed of growth of China's economy is still higher than the historical speed of other "catch-up economies", while also exhibiting the usual sustained growth characteristics encountered with "catch-up" economies. For example, before 2020, there is a possibility that per capita GDP growth in China can be kept above 6%, exceeding the historical figures for Japan, Taiwan and the United States, only being lower than historical figures for specific years in South Korea. The same goes for the period from 2020 to 2030, when growth speed in China will far exceed historical figures for other countries.

Similar results were obtained from long-term purchasing power parity calculations using the World Bank's International Comparison Program. Using this programme, per capita GDP in China will reach the level of South Korean per capita GDP by 2030 and by 2035 it will be on a par with Japanese per capita GDP, exceeding per capita GDP in the USA in 2050 (Figs. 5.3 and 5.4).

5.3.2 Mid- to Long-Term Changes in Industrial Structures

The simulation shows that the tertiary industry ratio will continue to rise. In 2013, the tertiary industry ratio was 46.1%, which is significantly lower than the level of most countries at a similar level of development. In 2015, it rises to around 48.1%, and from 2015 to 2020 it will rise by 4.9%, possibly exceeding 60% by 2030, reaching around 65% in 2050 (Table 5.3).

From experience gained with economic growth in other countries, as the level of development rises, the ratio of non-agricultural industries also goes up, especially the tertiary industry ratio, which seems to be a general norm. Major factors pushing the tertiary industry ratio higher are the changes in people's consumer styles, increased services export ratio and rising demand for services in various sectors, as well as government spending. Deceleration of export growth also holds sway over tertiary industrial structure, since the main exports are

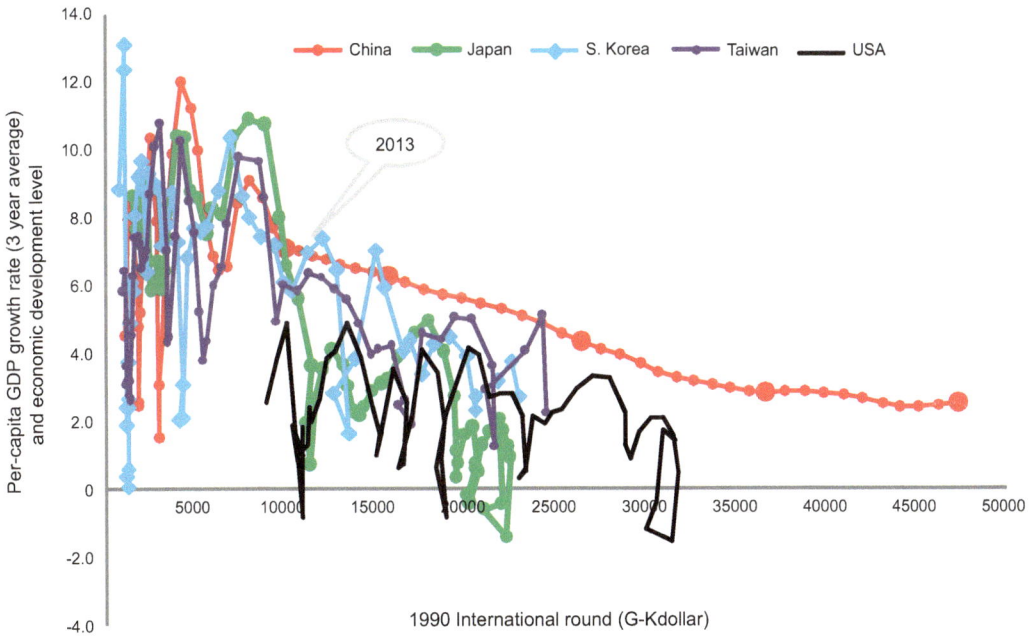

Fig. 5.3 Growth speed in China and international comparisons in the standard scenario. *Description* Data for Japan, South Korea, Taiwan and USA drawn from historical growth figures; horizontal axis is the international dollar calculation based on 1990 purchasing power parity calculated using Maddison's method. *Data source* Figures for China after 2013 based on model calculation results, other figures from Maddison

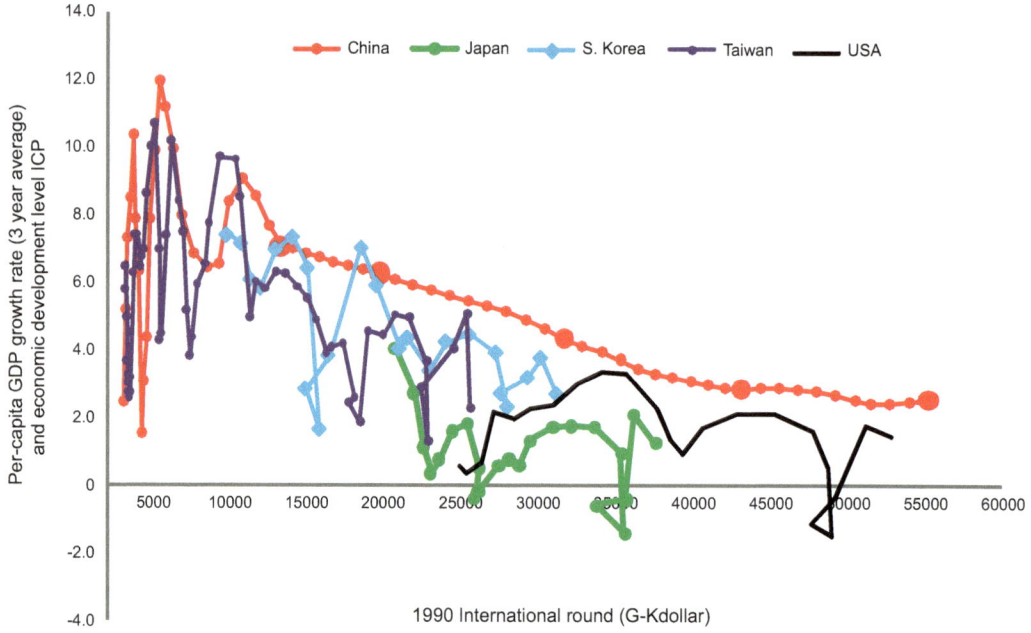

Fig. 5.4 Growth speed in China and international comparisons in the standard scenario. *Data source* Figures for China after 2013 based on model calculation results, other figures from Maddison

Table 5.3 Industry structures in the standard scenario

Year	2010	2015	2020	2025	2030	2035	2040	2045	2050
Primary industries	10.0	9.0	7.1	6.2	5.4	5.0	4.8	4.8	5.1
Secondary industries	48.2	43.0	39.8	35.7	33.4	32.1	30.8	30.0	29.3
Tertiary industries	41.8	48.0	53.1	58.1	61.2	62.9	64.4	65.2	65.6

Data source Results from model calculation

manufactured goods and therefore, when other conditions remain unchanged, a faster growth in exports will most likely lead to a larger corresponding secondary industry ratio (Fig. 5.5).

In the period from 2020 to 2025, China will essentially complete the industrialisation process, entering the post-industrialisation era. Looking at the progress of industrialisation in China, the demand for heavy industries such as steel and concrete is approaching its peak, while the eastern coastal regions are on their way into the post-industrialisation stage. This development trend conforms to the strategic plans of the 18th National Congress of the Communist Party of China. This is based on two indicators: per capita GDP (which will exceed $10,000 in 2020 (at

2013 prices) and $14,000 in 2025, achieving the development standard for post-industrialisation) and secondary industry's contribution to growth (which will be above 40% before 2020, but will go through a period of significant decrease after 2020). This is similar to the situation in South Korea in the 1990s (Table 5.4).

The proportion of high energy consumption industries will begin to fall. High energy consumption industries are a major factor affecting economic structures, and if the proportion of them is high, then economic growth requires more environmental resources, leading to more serious pollution. Under optimised policies, towards the end of the 12th Five-Year Plan, it is hoped that the ratio of high energy consumption

Fig. 5.5 Primary, secondary and tertiary industries in China and international comparisons in the standard scenario. *Data source* Results from model calculation

Table 5.4 Developed countries in the post-industrialisation era	Country	Industrialisation stage	Post-industrialisation stage
	United Kingdom	1760–1870	1950–
	United States of America	1790–1900	1950–
	Germany	1830–1913	1970–
	Japan	1885–1973	1973–
	South Korea	1960–1995	1995–

industries will begin a continual descent, from accounting for 32.4% of all industries to accounting for 28.5% by 2030 and 24.5% by 2050 (Fig. 5.6).

5.3.3   Energy Consumption and Structural Change

Even though China's national industries show significant optimisation in the standard scenario, industry still accounts for a considerable proportion of the economy, and along with the rise in personal income also comes higher personal energy use. China's national energy consumption is therefore expected to increase greatly in the future. It is estimated that by 2020 total national energy source consumption will reach 5 billion tons, an annual increase of 3.4% from 2010 to

2020. The rise should slow between 2020 and 2030 to a 1.3% annual increase, reaching around 5.68 billion tons, and between 2030 and 2040 the rise will be further reined into 0.8% per year, but by 2040 the total national energy consumption will still reach 6.06 billion tons. Around 2045, total national energy source consumption in China will peak and in 2050 it will stabilise at around 6.1 billion tons (Fig. 5.7).

Even though total energy consumption will continue to rise, energy use per unit GDP will decline considerably, falling from 0.89 tons of coal per CNY 10,000 GDP in 2010 to 0.62 tons of coal per CNY 10,000 GDP in 2020, and then to 0.43 and 0.25 tons of coal per CNY 10,000 GDP in 2030 and 2050, respectively (Figs. 5.8 and 5.9).

There will also be great structural changes in medium- to long-term energy source production.

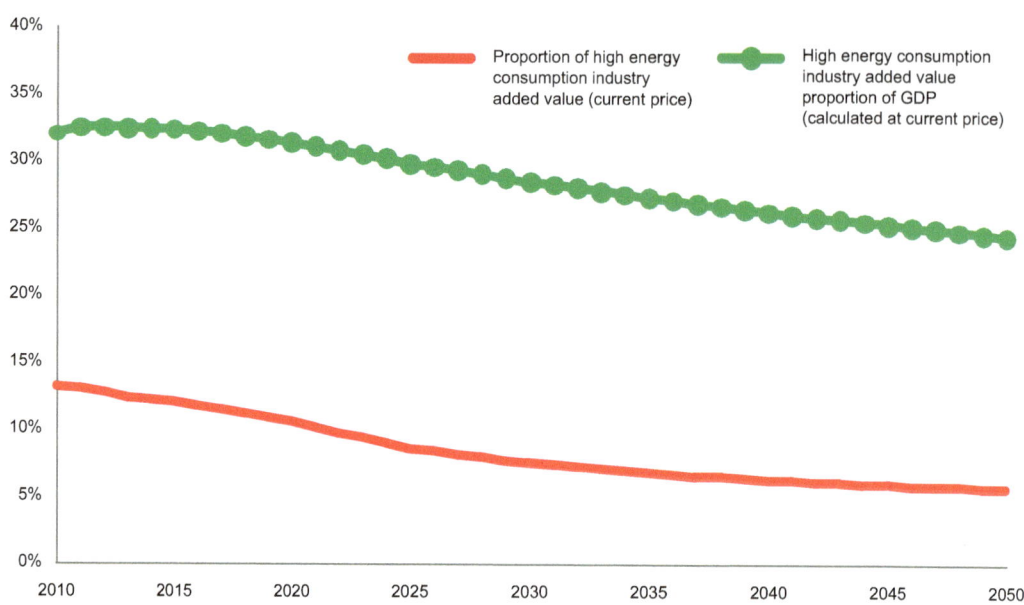

Fig. 5.6 Change in high energy consumption industry ratio under baseline scenario. *Data source* Results from model calculation (%)

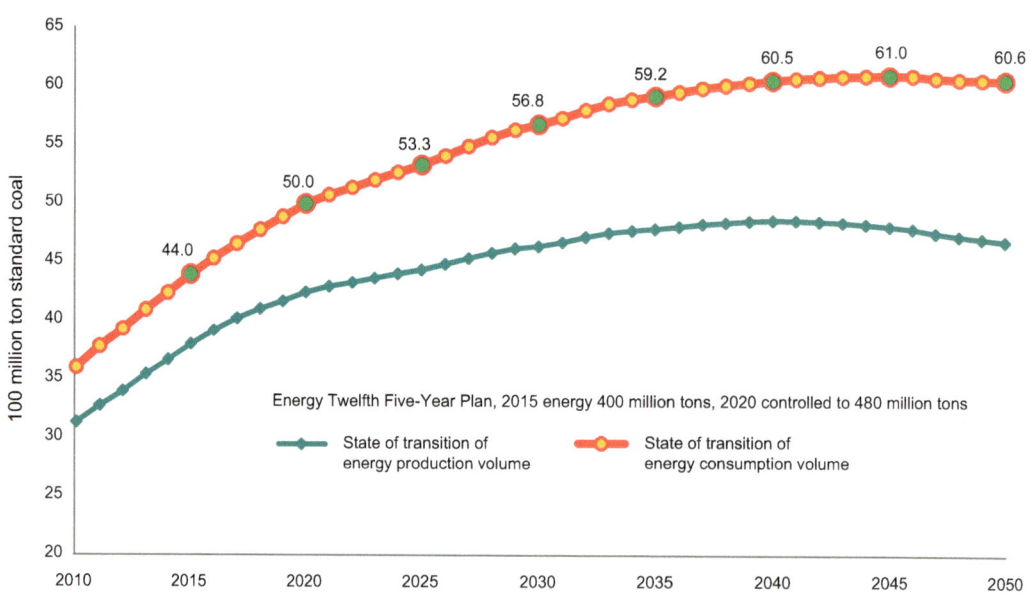

Fig. 5.7 Total energy consumption in standard scenario. *Data source* Results from model calculation

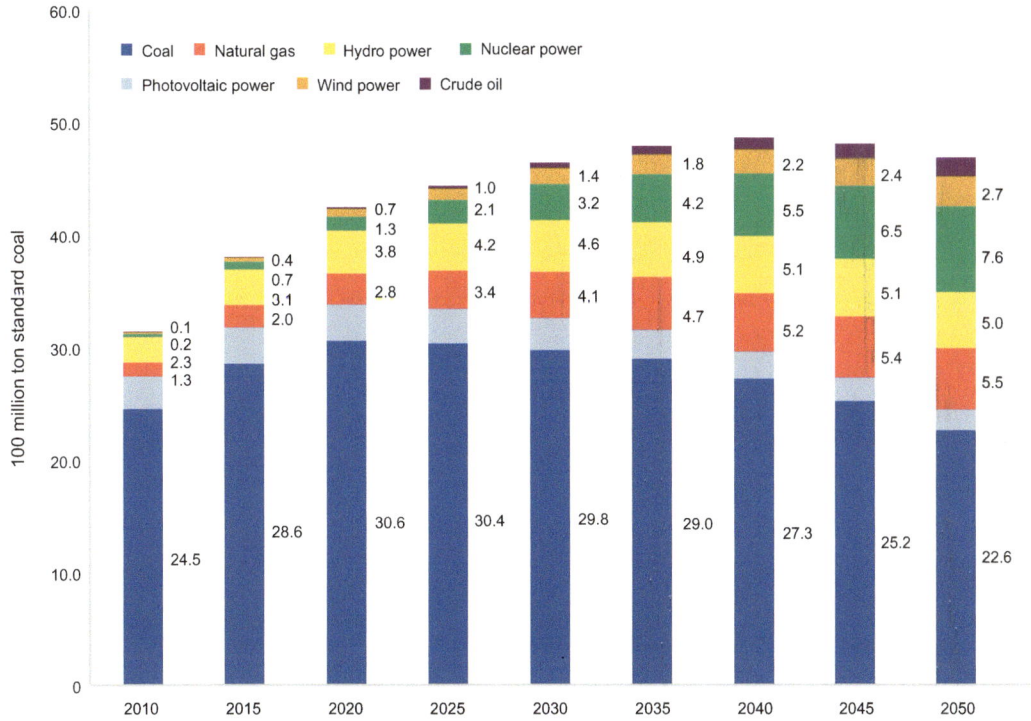

Fig. 5.8 Medium- to long-term energy production structure in China. *Data source* Results from model calculation

Production of the most important primary energy source, coal, will peak in around 2020 at 3.07 billion tons of standard coal, or a total of 4.3 billion tons of coal. From then onwards, coal production will fall annually, dropping to below 3 billion tons of standard coal by 2030, below 2.5 billion tons by 2040 and around 2.26 billion tons by 2050. As for clean energy, there will be greater development of nuclear power and hydroelectricity; hydroelectricity will grow from 230 to 380 million TSCE in 2020, and hopefully will have reached 460 million TSCE in 2030. From that point onwards, however, growth will slow to a crawl due to limitations in hydroelectric resources, reaching around 500 million TSCE by 2050.

Nuclear power appears to have better development potential. Based on current plans, if progress is smooth, nuclear power output in China will grow from 20 to 130 million TSCE by 2020, and then rise to 320, 550 and 760 million TSCE by 2030, 2040 and 2050, respectively.

Looking at the final energy production structure, electricity will see the biggest expansion in medium- to long-term energy consumption. In 2010 electricity consumption was 1.22 billion TSCE, which is anticipated to rise to 2.65 billion TSCE into 2030 and 2.9 billion TSCE in 2050. Consumption of refined oil will also see further increases, from 520 million TSCE in 2010 to 850 million TSCE in 2030 and 940 million TSCE in 2050.

5.3.4 Demand for Natural Gas and Main Increases in Consumption

In the standard scenario, demand for natural gas is expected to grow rapidly. It is expected to approach 200 billion m^3 in 2015, and exceed 300 billion m^3 by 2020. By 2030 it may exceed 450 billion m^3, and exceed 600 billion m^3 by 2050. Limited by domestic production growth, imports of natural gas (including liquid natural

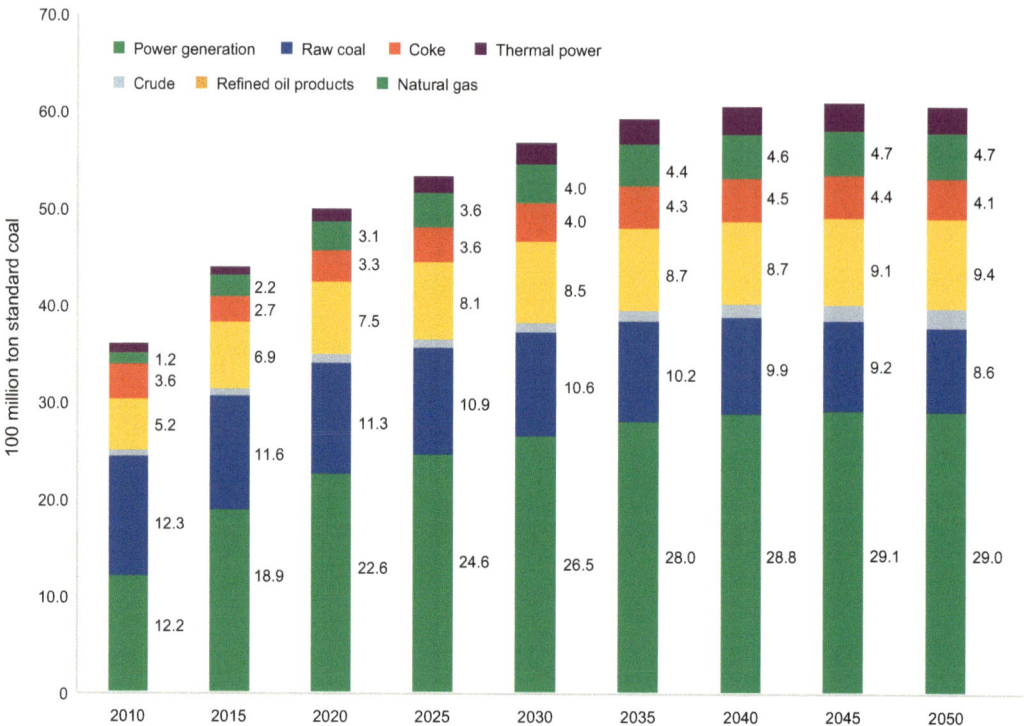

Fig. 5.9 Medium- to long-term energy consumption structure in China. *Data source* Results from model calculation. *Note* Raw coal refers to coal used for purposes other than generating electricity or coke refining

gas) are expected to see a major increase, from 50 billion m³ in 2015 to 100 billion m³ by 2020 and 180 billion m³ by 2030, stabilising by 2040 at around 200 billion m³ (Fig. 5.10).

Although fast growth is expected, growth in other energy sources, and in particular power generation, is rising, therefore the ratio of total energy sources accounted for by natural gas still lags quite a long way behind target levels. It is expected that natural gas will make up 6.0% of total energy sources by 2015, which is somewhat less than the goal of 7.15% set out in the 12th Five-Year Plan. In 2020 it will be 8.0%, which is still lower than the planned goal of 10% (Fig. 5.11).

In terms of the main areas where natural gas consumption is growing, the most important is gas use in power generation. If use of natural gas in power generation becomes economically competitive, allied with strengthened attempts to improve air quality, it is likely that natural gas use in power generation may rise from 25 billion m³

in 2015 to 50 billion m³ by 2020, and the figure should then double and exceed 100 billion m³ by 2030, finally stabilising after 2040 at around the 160 billion m³ mark. Other than natural gas use in power generation, there are three other major natural gas uses: central heating, transport and the chemical industry.

In the medium to long term, to achieve green, clean power generation it is essential that power source structures are optimised. One major aspect of the optimisation of generation would be a reduction in output of coal-based generation and a drop in the proportion of the total resources that it accounts for. Based on simulation analysis, the standard scenario is that coal power generation in China will peak at 5.1 trillion kWh in around 2020 and then decline from there on, to around 4.9 trillion kWh by 2030, going down further to 4.1 trillion kWh by 2040 and further still to 3.2 trillion kWh in 2050.

The reduction in coal power generation must be based on rapid developments in clean power

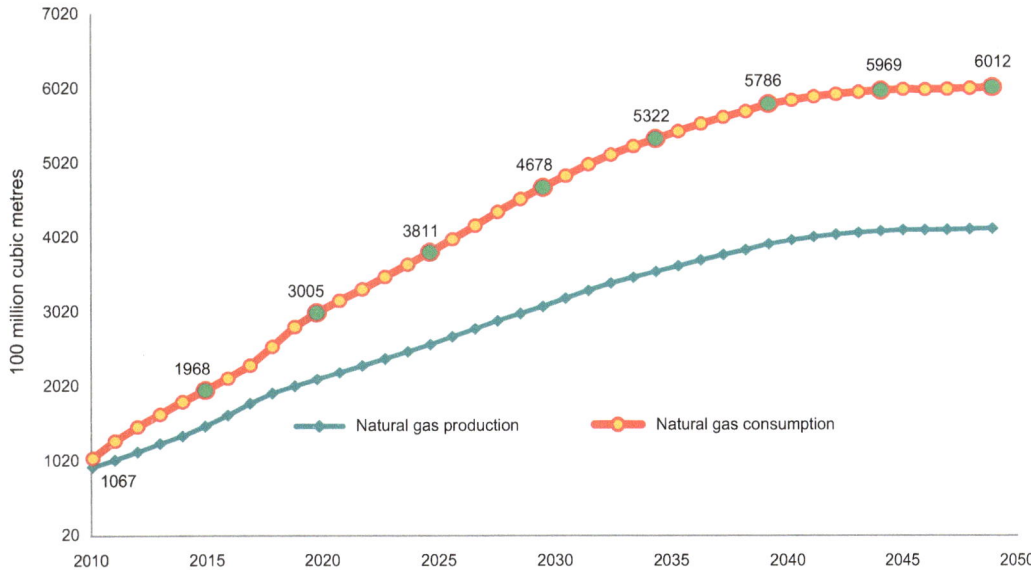

Fig. 5.10 Natural gas production and consumption in the standard scenario. *Data source* Results from model calculation

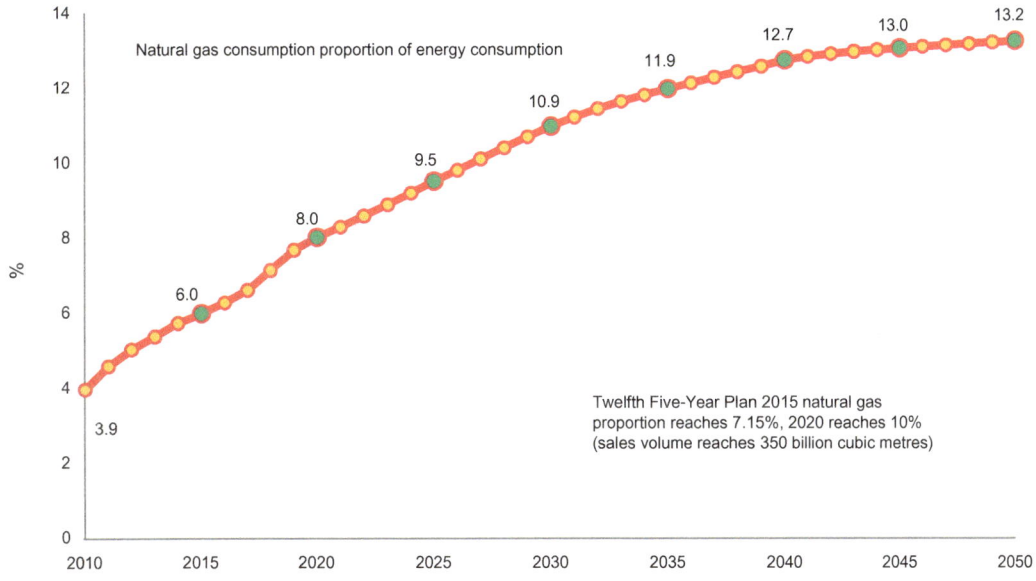

Fig. 5.11 Medium- to long-term changes in ratio of natural gas consumption in China. *Data source* Results from model calculation

generation. In the standard scenario, the biggest growth lies in nuclear power, while there is also much room for growth for natural gas power generation. Natural gas power generation will rise quickly from 77 billion kWh in 2010 to 280 billion kWh by 2020, 620 billion kWh by 2030 and around 920 billion kWh by 2050. Similarly, wind power, photovoltaic electricity, hydroelectricity and nuclear power will also undergo rapid development (Fig. 5.12).

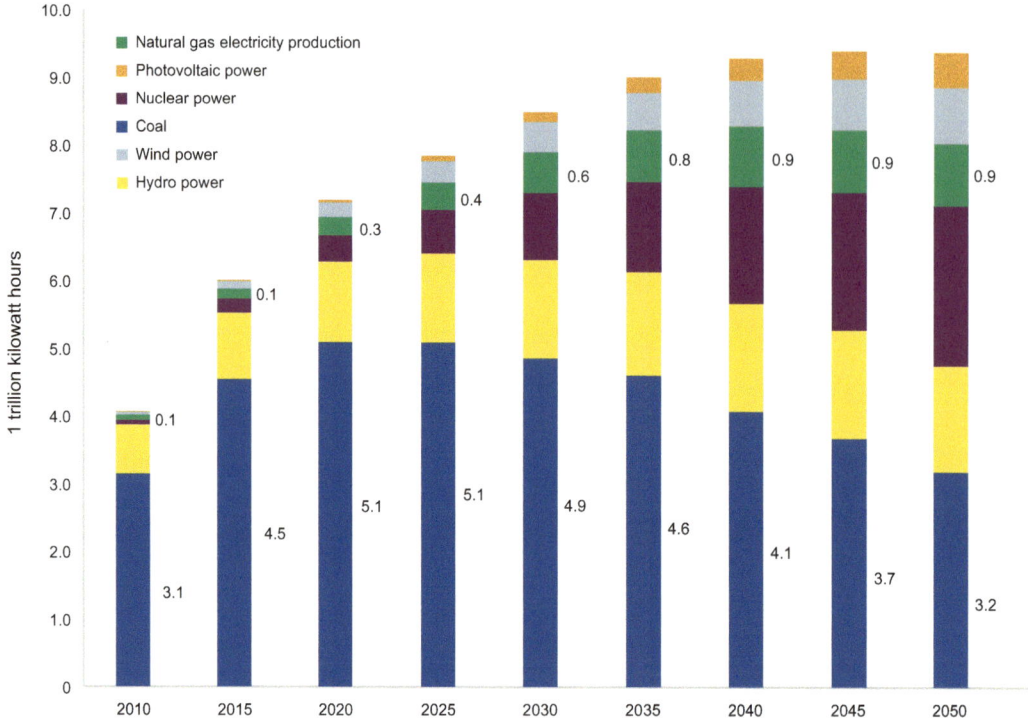

Fig. 5.12 Medium- to long-term power generation structural changes in China. *Data source* Results from model calculation

5.4 Natural Gas Supply and Demand in the Policy-Driven Scenario

In the standard scenario, even though natural gas consumption in China will rise to 300 billion m^3 by 2020, 450 billion m^3 by 2030 and 600 billion m^3 by 2050, it would still not reach the goals set out for natural gas development in the medium- to long-term energy development plan of the 12th Five-Year Plan. This indicates that, in order to achieve those goals, in addition to the current arrangements for energy sources, policies that are more effective must be put in place to boost natural gas demand.

5.4.1 Total Natural Gas Supply and Demand

Among the various possible economic measures, the most effective would be the much-vaunted

pricing policy based on carbon trading and other ways of realising a carbon price. Under the pressure of a global push for reduction of greenhouse gas emissions, the introduction of a carbon tax could help to encourage society to reduce its reliance on fossil fuels and reduce greenhouse gas emissions. The use of natural gas produces much less carbon than coal or oil, and thus carbon pricing will help to further promote the replacement of coal and oil with natural gas.

In the policy-driven scenario, the demand for natural gas will increase, due to changes in relative energy source prices that are advantageous to natural gas. Due to the sensitivity of natural gas to pricing depending upon different uses, the increase in demand encountered will vary. Some uses, such as household use of natural gas, already contribute a substantial proportion, and thus the increase would be limited in these areas. Other uses, such as power generation and heating, which are more sensitive to pricing, will see a greater relative increase in demand (Fig. 5.13).

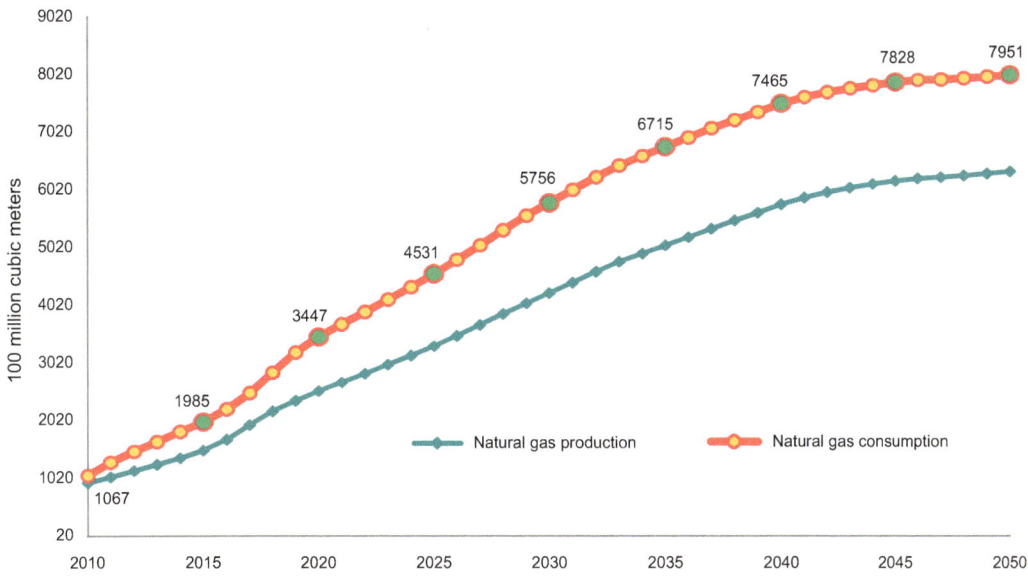

Fig. 5.13 Natural gas supply and demand in the policy-driven scenario. *Data source* Results from model calculation

With the right policies in place, natural gas demand in China could be expected to reach 344.7 billion m^3, achieving the planned goal of 350 billion m^3 stipulated in the 12th Five-Year Plan. By 2030, natural gas consumption may approach the level of 580 billion m^3, and be around 800 billion m^3 by 2050.

5.4.2   Effects on Pollutant Discharge of Natural Gas Consumer Demand Growth

Under the influence of policy, demand for natural gas will grow and will replace major coal applications, so the use of natural gas will be effective to some degree in restricting emissions of various pollutants (Fig. 5.14).

Based on emissions of 2.46 tons of CO_2, 8.5 kg of SO_2 and 7.4 kg of nitrogen oxides for every one ton of standard coal burned (when approaching zero emissions, each one ton of standard coal burned results in 0.14 kg of SO_2 and 1.12 kg of nitrogen oxides), under the policy-driven scenario, natural gas consumption will increase from 183 billion m^3 in 2014 to 575.6 billion m^3 by 2030, equivalent to at least 522 million tons of standard coal based on calorific value. Based on this, it would be possible to reduce emissions to 418 million tons of SO_2 and 4.354 million tons of CO_2 (replacing 20% of near-zero emission coal-fired power generation and 80% of dispersed coal use) by 2030. This is the equivalent of 21.3% of the national SO_2 emissions total of 20.439 million tons in 2013. There will also be a reduction in nitrogen oxides emissions of 2.339 million tons, which is the equivalent of 10.5% of the national nitrogen oxides emissions total of 22.274 million tons in 2013. Moreover, by 2020 there will be further reductions of CO_2 emissions by 172 million, of SO_2 emissions by 1.794 million tons and of nitrogen oxide emissions by 964,000 tons.

In a policy-driven environment, it is possible to markedly reduce greenhouse gas emissions. In a baseline scenario, greenhouse fossil fuel energy consumption in China is expected to continue to grow in terms of emissions in China each year from the 13th Five-Year Plan to 2030. In 2015, greenhouse gas emissions caused by fossil fuels are expected to reach 9.51 billion tons (due to carbon exchange and other factors, greenhouse gas emissions caused by fossil fuel consumption

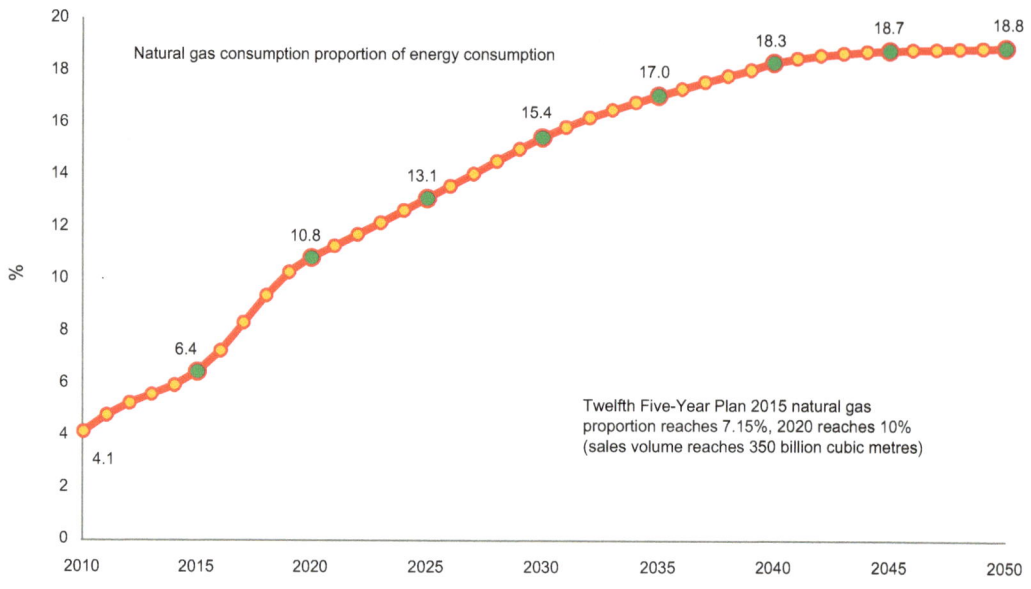

Fig. 5.14 Ratio of natural gas in total energy sources in the policy-driven scenario. *Data source* Results from model calculation

is not equivalent to the total national emissions total), and by 2020 will approach 10.5 billion tons, an annual increase of 1.93%. By 2030 it will approach 11 billion, growing annually by 0.95% from 2020 to 2030, peaking in 2033 at approximately 11.08 billion tons. Subsequently it will gradually decline. In a policy-driven environment, because overall fossil fuel energy consumption will see a major reduction, the resulting greenhouse gas emissions also markedly drop. In 2015, there is expected to be a reduction in emissions of approximately 400 million tons compared to the baseline scenario. CO_2 emissions reductions from 2020 to 2030 will be approximately 1.7 billion tons.

5.4.3 Main Areas of Natural Gas Demand

In the policy-driven scenario, areas where natural gas demand would grow quickly are still the ones more sensitive to costs and prices, namely natural gas power generation, transport and the chemicals industry, while changes in demand remain relatively small in other areas such as domestic

use. For example, with the policies in place, compared to the standard scenario, domestic use would only rise by 5 billion m^3 by 2050, while for natural gas power generation the consumption in 2030 would be 45 billion m^3 higher compared to the standard scenario. Natural gas heating possesses tremendous potential and is very sensitive to costs. With policies in place it could be hoped that consumption might rise by 30 billion m^3 by 2030.

Based on the simulation results, in the policy-driven scenario natural gas consumption in China reaches 600 billion m^3 by 2030. Of that total, urban use (including vehicular and waterborne transport) accounts for 35%, while 32% goes to power generation, 27% to industrial fuels and 6% is used in the chemicals industry. Urban use mainly consists of cooking, hot water for everyday use, gas use in public utilities (airport, government departments, staff canteens, kindergartens, schools, hostels, hotels, restaurants, malls, offices etc.), centralised heating, centralised air conditioning and being used by vehicles and ships. As for industrial use, natural gas is used as fuel for equipment used in various industries, including pottery, glass, steel, petrochemicals, textiles, aluminium oxide, titanium dioxide, fire-resistant materials, carbon resources and so on. In the chemicals industry, natural gas is mainly used as an ingredient for the synthesis of ammonia and methanol and in hydrogen production. In power generation, natural gas is used in peak regulation power stations, thermal power stations and by distributed energy source users. There is, therefore, massive potential natural gas demand (Table 5.5).

5.4.4 Demand Curve for Natural Gas in China in 2030

1. **Predicted replacement energy source price series**

With consideration given to the long-term trends in international oil prices, this analysis is based on an international oil price of $80 per barrel (Table 5.6).

2. **Demand curve for natural gas in 2030**

The demand for natural gas in China in 2030 is expected to be 576.6 billion m^3, subdivided between 31 provinces and 16 types of user. Assessment of terminal natural gas price tolerance for all users in all provinces was carried out based on the local price of replacement energy sources. Conversion was then carried out based on the Shanghai benchmark price conversion, with the natural gas price tolerance based on the Shanghai benchmark being ranked from high to

Table 5.5 Natural gas consumption based on primary usage (100 million m^3) according to use

Major fields of application		2013		2030		Growth (%)
		Consumption	Proportion (%)	Consumption	Proportion (%)	
Urban gas	Residential living	181	11.0	520	9.03	187
	Commercial services	104	6.3	300	5.21	188
	Centralised heating	97	5.9	360	6.25	271
Vehicular and waterborne transport	CNG taxis	59	3.6	160	2.78	171
	CNG buses	32	1.9	90	1.56	181
	LNG lorries	33	2.0	550	9.56	1567
	LNG vessels			45	0.78	
Natural gas electricity production	Peak regulation power stations	157	9.5	640	11.12	308
	Thermal power stations	126	7.6	1060	18.42	741
	Distributed energy sources	4	0.2	130	2.26	3150
Industrial fuel		606	36.7	1540	26.75	154
Natural gas chemical engineering	Ammonia	152	9.2	185	3.21	22
	Methanol	61	3.7	106	1.84	74
	Hydrogen production etc.	41	2.5	70	1.22	71
Total natural gas consumption		1653	100	5756	100	248

Data source Results from model calculation

Table 5.6 Natural gas energy source replacement price list by province (calculated at the international oil price of $80 per barrel)

No.	Region	Coal price	Fuel oil price	Naphtha price	93# petrol retail price	0# diesel retail price	Small bottle LPG price	Large bottle LPG price	Industrial electricity consumption price	Household electricity consumption price
		(CNY/t)			(CNY/L)		(CNY/kg)		(CNY/kWh)	
1	Xinjiang	316	2553	7096	5.86	5.74	7.19	6.83	0.36	0.53
2	Gansu	507	2553	7155	6.01	5.85	7.24	6.88	0.46	0.51
3	Qinghai	519	2553	7252	6.00	5.86	7.24	6.88	0.36	0.45
4	Tibet	507	2727	7083	6.00	5.86	8.20	7.79	0.36	0.45
5	Ningxia	478	2553	6959	6.03	5.83	5.76	5.48	0.43	0.45
6	Shaanxi	450	2553	7126	6.01	5.83	6.30	5.98	0.55	0.50
7	Shanxi	460	2595	6988	6.08	5.88	7.62	7.24	0.50	0.48
8	Inner Mongolia	444	2595	7080	6.04	5.83	7.62	7.24	0.48	0.47
9	Henan	565	2594	6990	6.04	5.85	6.86	6.52	0.60	0.56
10	Hubei	628	2594	6993	6.04	5.85	8.49	8.07	0.60	0.57
11	Hunan	606	2594	6997	6.08	5.90	8.03	7.63	0.65	0.59
12	Jiangxi	666	3722	6990	6.07	5.88	6.85	6.51	0.65	0.60
13	Anhui	674	3722	6985	6.06	5.87	6.93	6.58	0.65	0.57
14	Yunnan	539	2727	7018	6.18	5.98	7.62	7.24	0.48	0.48
15	Guizhou	506	2727	6991	6.16	5.95	8.20	7.79	0.51	0.46
16	Sichuan	561	2727	7028	6.19	6.02	7.28	6.91	0.58	0.47
17	Chongqing	534	2727	6974	6.19	6.01	6.98	6.64	0.63	0.52
18	Guangdong	675	3548	6992	6.09	5.89	8.23	7.82	0.69	0.61
19	Guangxi	677	3548	7000	6.14	5.94	8.44	8.02	0.59	0.53
20	Fujian	661	3722	6982	6.08	5.89	7.98	7.58	0.61	0.45
21	Hainan	678	3548	6960	6.14	5.94	8.19	7.78	0.66	0.61
22	Jiangsu	632	3722	6978	6.07	5.86	7.40	7.03	0.64	0.53
23	Zhejiang	701	3722	6978	6.07	5.88	8.14	7.73	0.65	0.54
24	Shanghai	615	3722	6946	6.35	6.24	7.20	6.84	0.69	0.62
25	Beijing	580	2595	6953	6.36	6.26	7.62	7.24	0.62	0.49
26	Tianjin	580	2595	6950	6.03	5.83	7.62	7.24	0.63	0.49
27	Hebei	596	2595	6993	6.03	5.83	7.02	6.67	0.56	0.52
28	Shandong	556	3722	6988	6.03	5.84	7.81	7.42	0.67	0.55
29	Liaoning	499	2891	7352	6.03	5.83	7.47	7.09	0.53	0.50
30	Jilin	645	2891	7320	6.03	5.83	6.51	6.18	0.58	0.53
31	Heilongjiang	623	2891	7394	6.03	5.83	6.33	6.01	0.58	0.51

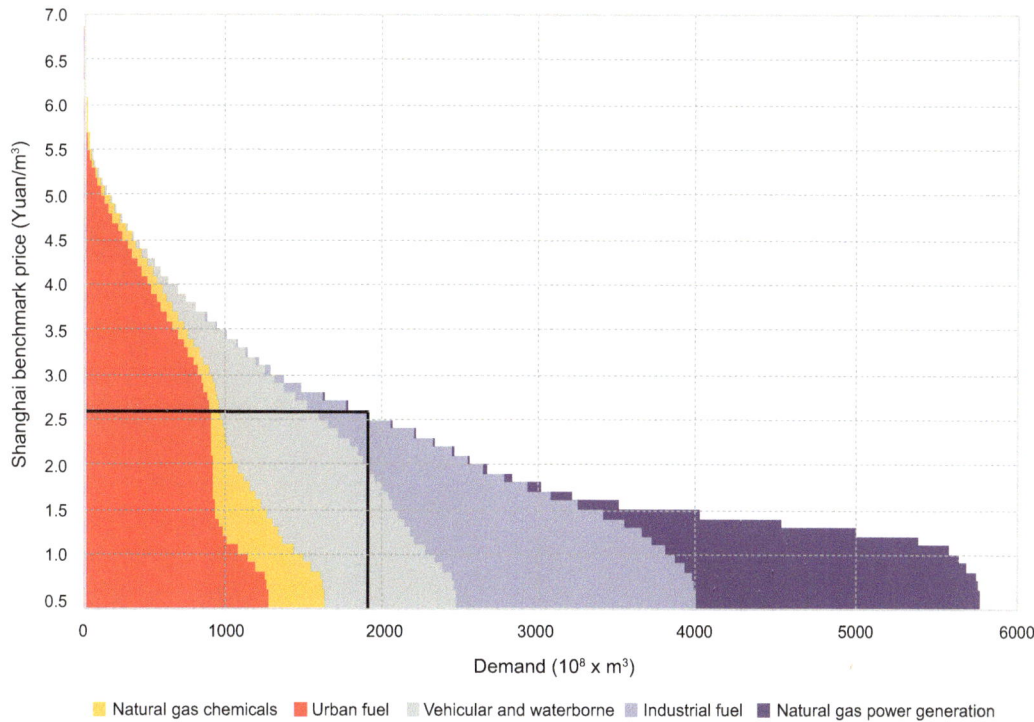

Fig. 5.15   Demand curve for natural gas in China in 2030

low, yielding the demand curve for natural gas in China in 2030 (Fig. 5.15).

3. **Various integrated policies would be required in order to achieve a 575.6 billion m³ natural gas consumption**

   If a pricing policy is not implemented, achieving a target of 575.6 billion m³ of natural gas consumption by 2030 would require a series of alternative policies to be introduced. Based on the national natural gas pricing reform plan of 2013, with an international oil price of $80 per barrel as the basis, it is calculated that the benchmark Shanghai natural gas price would be 2.6 CNY/m³. Looking at the effective demand curve for natural gas, the planned natural gas consumption in China by 2030 would be 575.6 billion m³, and with reference to the benchmark Shanghai natural gas price of 2.6 CNY/m³ and effective consumption of 206.7 billion m³, this would account for 32% of estimated demand. Urban natural gas centralised heating, power generation, synthesis of ammonia and methanol, and the majority of usage as fuel in other industries would not satisfy the conditions for generation of effective demand capacity based on the benchmark Shanghai natural gas price of 2.6 CNY/m³ (Table 5.7).

   At the current energy source pricing levels, market forces alone would not push demand for natural gas up to the goal of 575.6 billion m³. Systematic design of policy which reflects the environmental and social value of natural gas is necessary.

Table 5.7 Market demand for natural gas in China in 2030 (100 million cubic metres)

Major fields of application		2030		2030 market demand (no policy measures)		Actual degree (%)
		Consumption	Proportion (%)	Consumption	Proportion (%)	
Urban gas	Residential living	520	9.03	516	28.30	99
	Commercial services	300	5.21	292	16.02	97
	Centralised heating	360	6.25			
Vehicular and waterborne transport	CNG taxis	160	2.78	160	8.77	100
	CNG buses	90	1.56	90	4.93	100
	LNG lorries	550	9.56	366	20.09	67
	LNG vessels	45	0.78	10	0.57	23
Natural gas generation	Peak regulation power stations	640	11.12			
	Thermal power stations	1060	18.42			
	Distributed energy sources	130	2.26			
Industrial fuels		1540	26.75	315	17.29	20
Natural gas chemical engineering	Ammonia	185	3.21			
	Methanol	106	1.84	2	0.12	2
	Hydrogen production etc.	70	1.22	70	3.84	100
Total natural gas consumption		5756	5756	100%	1824	100

Data source Results from model calculation

Analysis of China's Natural Gas Use Policies and Suggested Reforms

6

6.1 Development of China's Natural Gas Use Policy

6.1.1 The Encouraging Consumption Stage

Before the Shaan-Jing natural gas pipeline began operation in 1997, China's only gas pipeline to reach a length of over 350 km was the Hong Kong offshore gas pipeline (Yacheng–Hong Kong, 778 km). Owing to the lack of cross-regional gas pipelines, natural gas in China was characterised by being produced in nearby fields. At that time, the two most important markets for natural gas were Sichuan and the North East. The Sichuan Basin has a long history of developing natural gas and already has a highly developed network of pipelines. The North East gas fields can produce high-quality oil-associated gas. The main uses of natural gas are primarily in the chemical industry, where it is used to produce methanol and fertilisers. Much associated gas is vented or consumed by oil companies during production. Urban gas and industrial fuel usage is limited to areas adjacent

* This chapter was overseen by Zhaoyuan Xu from the Development Research Center of the State Council and Martin Haigh from Shell International, with contributions from Baosheng Zhang and Shouhai Chen from the China University of Petroleum, Lianzeng Zhao from the China Petroleum Planning Research Institute, Linji Qiao from ENN and Juan Han from Shell China. Other members of the research group participated in discussions and revisions.

to sources, and only constitutes a fraction of total natural gas consumption (see Table 6.1).

China's natural gas exploration in the 1990s was extremely fruitful. In total, nine large gas fields and 32 medium-sized gas fields were discovered. Taken together, these make up the six major gas regions of Sichuan, Ordos, Tarim, Qaidam, the East China Sea and the Ying-Qiong area, of which the Jingbian gas field is the largest, with proven reserves of 290.09 billion m^3. The discovery of large natural gas reserves stimulates the construction of long-distance gas pipelines. These include the Shaanxi gas entry to Beijing, western gas being piped east, Sichuan gas being piped east and offshore gas terminals. In 2003 the IEA noted that the development of China's natural gas market had many obstacles to overcome, the first being increasing demand for natural gas. At that time, the main problems facing the natural gas industry were competing with cheap coal and creating demand for, rather than restricting use of, natural gas.

6.1.2 Restricted Usage Stage

With the completion of the western gas piped east project, a new era in the development of China's natural gas markets commenced. The markets changed from being regional to being national, freeing up supply and leading to a sharp increase in production. This brought about a shift in market trends. There had previously been worries that there were not enough consumers of natural gas, but suddenly there was not enough supply to meet the demand. In 2005, the National

© The Editor(s) and The Author(s) 2017
Shell International and The Development Research Center (Eds.), *China's Gas Development Strategies*,
Advances in Oil and Gas Exploration & Production, DOI 10.1007/978-3-319-59734-8_6

Table 6.1 China's natural gas consumption structure in 1998 (excluding Hong Kong)

Item	Gas used in the chemicals industry	Industrial fuels	Urban gas	Oil production	Miscellaneous	Total
Consumption ($\times 10^8$ m^3)	89.64	18.53	24.12	56.68	13.6	202.57
Percentage	44.25	9.1	11.9	28	6.7	100

Data source National Bureau of Statistics

Table 6.2 Natural gas prices in 2005 after price adjustments (CNY/1000 m^3)

		Sichuan-Chongqing gas fields	Changqing gas fields	Qinghai oil fields	Xinjiang oil fields	Other oil fields
First-grade gas	Gas for fertiliser	690	710	660	560	660
	Industrial gas	875	725	660	585	920
	Urban gas	920	770	660	560	830
Second-grade gas	980					

Development and Reform Commission tried to adjust prices in order to regulate the use of natural gas. This led to the decision to revise the manufacturer's natural gas pricing mechanism, raising the manufacturer's natural gas price by an appropriate amount, encouraging gas use efficiency, optimising gas use structures and promoting healthy and sustainable development of industrial gas use, all in order to ensure supplies to the Chinese natural gas market. It was made known that all oilfield gas not subject to planning would be treated as second-grade gas and would be priced based on the government guidelines at 980 CNY/1000 m^3. Price increases would be limited to 10%, but there would be no bottom limit. The price of first-grade gas was raised by an appropriate amount, the government being responsible for price regulation, with upper and lower limits of 10% (see Table 6.2).

The 2005 natural gas price adjustment still failed to effectively curb the rapid growth in demand for natural gas. This was largely due to the price of gas being relatively low, concentrated development of the gas-based chemicals industry in areas adjacent to gas production sources and a mad rush to develop low added value, short industrial chain methanol and fertiliser plants in

various areas. In August 2007, in order to resolve the conflicts arising in supply and demand, the National Development and Reform Commission issued a new natural gas use policy. The Commission divided natural gas use into four distinct categories (priority, permitted, restricted and prohibited) and took an interventionist policy toward its use. Such measures prioritised eco-friendly urban gas uses (see Table 6.3).

6.1.3 Flexible Restrictions Stage

The Chinese government has implemented many policies to meet the growing domestic demand for natural gas. First, it aimed to limit large-scale natural gas and chemical projects. Second, it raised gas prices to encourage natural gas production and imports. Between 2002 and 2012, China's natural gas production more than doubled, from 32.7 to 107.2 billion m^3. During this time, China was also actively importing overseas resources. On May 25, 2006 the first shipment of LNG entered Dapeng Bay in Shenzhen from Australia's North West Shelf. This marked the beginning of a new era for China's imports of natural gas. In 2010, the Central Asian gas

Table 6.3 2007 natural gas usage policy

Order of priority	Field of use	
Highest priority	Urban gas	1. Cities (especially medium-large cities), residential cooking, boiling water and other residential uses 2. Public service facilities (airports, government offices, canteens, nursery schools, schools, hotels, restaurants, shopping centres, office buildings, etc.) 3. Natural gas vehicles (especially dual-fuel vehicles) 4. Distributed co-generation and CCHP users
Permitted	Urban gas	1. Centralised heating (in city centres) 2. Individual household heating 3. Central air conditioning
	Industrial gas	4. Industrial fields in which natural gas may replace oil or LPG, such as for building materials, machinery, textiles, petrochemicals and metallurgy 5. Industrial fields in which natural gas is a more sustainable substitute than coal gas, or is economically beneficial. Examples include building materials, machinery, textiles, petrochemicals and metallurgy and other industrial fields where adoption of natural gas offers relatively high environmental and economic benefits 6. Building materials, machinery, textiles, petrochemicals, metallurgy
	Natural gas power generation	7. Construction of natural gas peak regulation infrastructure for main power loading centre(s) in regions with sufficient natural gas supplies
	Gas chemical industry	8. Economically beneficial hydrogen gas projects that consume gas on a small scale 9. Ammonia and fertiliser producers that would be incapable of moving or category 1 and 2 users to which the regulations do not apply
Restricted	Natural gas power generation	1. Construction of natural gas power stations in non-crucial power loading centres
	Gas chemical industry	2. Extension of pre-existing synthetic ammonia facilities relying on natural gas, conversion of synthetic ammonia facilities from coal to natural gas 3. Carbon 1 chemical engineering projects with primary methane-based products, including acetylene and chloromethane 4. New synthetic ammonia projects relying on natural gas, other than those included in item 9 of category 2
Prohibited	Natural gas power generation	1. Construction of base-load natural gas power stations in 13 large-scale coal mining regions including Shaanxi, Mongolia, Shanxi and Anhui
	Gas chemical industry	1. New construction or extension of methanol production facilities relying on natural gas 2. Projects for the replacement of coal with natural gas for the production of methanol

pipeline began delivering gas to China and in 2013 the China-Burma pipeline was brought online. Based on the Natural Gas Development Plan, China's annual imports of natural gas will reach 93.5 billion m^3 by 2015, which will account for one third of the entire gas supply. This increase in the supply of offshore gas resources, in conjunction with the rapid increase and growth of domestic natural gas production, means that the hitherto problematic situation of China's gas supply has been greatly eased.

In October 2012, the National Development and Reform Commission adapted its policy on natural gas use (see Table 6.4) and relaxed restrictions on a number of specific gas use fields:

- In terms of policy objectives, there was no longer an emphasis on relieving conflicts between supply and demand. Encouragement of a low-carbon economy was added, as was

Table 6.4 2012 natural gas usage policy

Order of priority	Field of use	
Highest priority	Urban gas	1. Urban areas (especially medium-large cities), residential cooking, boiling water and other residential usages 2. Gas used for public service facilities (airports, government offices, canteens, nursery schools, hospitals, hotels, bars, restaurants, department stores, office buildings, stations, welfare offices, old people's homes, harbours, pier terminals and bus stations) 3. Natural gas vehicles (especially dual-fuel and liquefied natural gas vehicles), including city buses, taxis, commercial vehicles, passenger vehicles, sanitation lorries and other transport vehicles that rely on natural gas as fuel 4. Centralised heating (in city centres) 5. Gas air conditioning
	Industrial gas	6. Building materials, machinery, textiles, petrochemicals, metallurgy 7. Interruptible hydrogen manufacturing projects relying on natural gas
	Other users	8. Natural gas distributed energy projects (comprehensive energy efficiency above 70%, including the use of renewable energy sources) 9. As fuel in sea-going, river-going and lake-going vessels (including gas-fuelled and dual-fuelled vessels) that rely on natural gas (especially LNG) 10. Towns that have an emergency and peak regulation natural gas storage facilities 11. Coalbed methane (marsh gas) power generation projects 12. Natural gas co-generation projects
Permitted	Urban gas	1. Individual household heating users
	Industrial gas	2. Projects in which natural gas is used as a replacement for oil or LPG, such as building materials, machinery, textiles, petrochemicals and metallurgy 3. New projects where natural gas is used as fuel, including building materials, machinery, textiles, petrochemicals and metallurgy 4. Projects where the adoption of natural gas as a replacement for coal offers environmental and economic benefits, including building materials, machinery, textiles, petrochemicals and metallurgy 5. Urban projects for conversion of industrial boilers to gas (especially in medium-large cities)
	Natural gas power generation	6. All natural gas power projects excluding item 12 in the first category and item 1 in the fourth category
	Gas chemical industry	7. Natural gas hydrogen production projects excluding item 7 in the first category 8. Small natural gas liquefaction facilities for peak regulation and storage
Restricted	Gas chemical industry	1. Extension of existing natural gas-based ammonia plants and conversion of ammonia plants from coal to natural gas 2. Carbon 1 chemical engineering projects with primary methane-based products, including acetylene and chloromethane 3. New urea fertiliser projects that use natural gas as raw material
Prohibited	Natural gas power generation	1. Base-load power generator construction projects in 13 large-scale coal mining regions including Shaanxi, Mongolia, Shanxi and Anhui (excluding coal bed methane (marsh gas) power generation projects)
	Gas chemical industry	2. New construction or extension of natural gas-based methanol plants and downstream methanol production projects 3. Projects for methanol manufacturing using natural gas as replacement of coal

increasing the proportion of primary energy sources accounted for by natural gas.

- Restrictions applicable to natural gas power were relaxed, but the prohibition on gas power construction projects (excluding coalbed methane (marsh gas) power generation projects) in 13 regions with large-scale coal bases continued, including Shaanxi, Inner Mongolia, Shanxi and Anhui. The categories of other natural gas-based power generation projects were changed from "permitted" or "prohibited" to "highest priority" and "permitted".
- There was relaxation of restrictions on use of natural gas as an industrial fuel, in centralised heating and by interruptible users in the fields of building materials, machinery and electronics, textiles, the petrochemicals industry, metallurgy and other industrial fields, who had their categories changed from being "permitted" to "highest priority".

6.2   "Optimisation" of Natural Gas Use Structures and Policies

China's policy on natural gas use clearly states that one of its main goals is to optimise the structure of natural gas use. So what exactly are "optimised" gas use structures, and how is their "optimisation" to be achieved?

6.2.1   China's Natural Gas Use Structure

The structure of natural gas usage refers to the proportion of natural gas used in different areas relative to overall consumption. Before 2000, due to a lack of inter-regional pipelines, oil field production self-usage and chemical engineering adjacent to oil fields accounted for the vast majority of China's natural gas consumption. Urban gas and gas used for generating power accounted for a very small proportion (see Table 6.5). Once long-distance pipelines were brought online, the gas-using regions spread from being concentrated around oil and gas fields towards the more economically advanced central and eastern regions (see Table 6.5). In 2013, out of a total consumption of 166 billion m^3 of natural gas in China, urban gas (including gas used in vehicular and waterborne transport) and industrial fuel consumption together accounted for 67% of total consumption.

6.2.2   Worldwide There Is no Such Thing as "Optimised" Gas Use Structure

In 1998, the chemical industry accounted for 44.5% of Chinese natural gas consumption. By 2007, this proportion had dropped to 30%. In 2007, when natural gas use policy was introduced, the structure of China's natural gas use was far

Table 6.5 2003–2011 China natural gas use structure

	2003	2004	2005	2006	2007	2008	2009	2010	2011
Oil field internal use	79.44	72.77	78.88	77.44	85.59	104.41	117.40	129.55	126.04
Power generation and heating	7.54	12.74	18.78	29.49	70.74	73.92	127.91	180.80	215.90
Industrial fuels	45.71	70.22	89.69	99.62	116.29	153.24	155.75	183.29	264.5
Chemical engineering raw materials	128.13	122.90	141.44	177.42	207.05	200.03	176.84	187.28	233.48
Transport and logistics	18.82	26.16	38.01	47.24	46.88	71.55	91.07	106.70	138.35
Commercial use gas	6.85	9.18	10.79	13.16	17.11	17.75	23.96	27.24	33.64
Residential			79.43	102.63	143.39	170.10	177.70	226.90	264.38
Other gas use		14.14	9.12	12.77	16.09	20.92	23.64	26.00	27.14
Total	286.49	328.11	466.14	559.77	703.14	811.92	894.27	1067.76	1303.46

Unit 10^8 m^3

from ideal, mainly reflected by the chemical industry accounting for a disproportionately large proportion of total natural gas consumption. In light of this, the natural gas chemicals industry should have been classified as a restricted or prohibited gas use field. In 2011, gas use in the chemical industry dropped to 18%, then 15% in 2013. In 2012, when the National Development and Reform Commission issued the gas use policy, the main objectives were still optimising gas use structure and restricting and prohibiting use of natural gas in chemical engineering.

Global natural gas use structures can be divided into three types: a balanced structure type, a mainly power generation type and an urban gas use type. The United States is a typical balanced structure, with urban gas, power generation and industrial gas (including industrial fuels and gas for use in chemical engineering) each constituting roughly one third. Japan, South Korea and Russia use natural gas predominantly for power generation. Japan, for example, used 94.5 billion m^3 of natural gas in 2010, of which 60% was used to produce electricity. After the Fukushima nuclear accident in 2011, an even higher proportion of natural gas was used to generate electricity. Consumption of natural gas in The Netherlands and the United Kingdom, two Western European countries, is predominantly urban. In 2010, the Dutch consumed 43.6 billion m^3 of gas, of which urban gas accounted for 56%, industrial fuels for 33% and power generation usage for 11% (Canqi 2014: 1–2). Looking at the historical development of China's natural gas output in the light of the experiences of other countries, it is clear that there is no such thing as a standard optimised structure in terms of natural gas use.

Even though there is no standard optimised structure, it doesn't mean that standards don't exist for the optimisation of natural gas use. The National Development and Reform Commission in its policy on gas use proposed that the use of natural gas should be optimised according to three criteria: social benefit, environmental benefit and economic benefit, though it does not define what is meant by social, environmental and economic benefits. However, based on the order in which they apply to gas use, "social

benefit" refers to living standards and the guarantee of residential gas supplies, in order to avoid social unrest due to gas shortages. This is why the domestic uses of cooking and boiling have been listed as prioritised areas of gas use. Environmental benefits include the reduction of emissions. This is the reason for the inclusion of gas-powered vehicles, centralised heating, industrial fuels, distributed energy resources and co-generation in the "permitted" and "top priority" categories. Finally, economic benefits concern profits derived from market sales, and therefore it is the overruling factor in deciding the prioritisation of natural gas use. The "optimisation" of natural gas use structures should therefore seek to maximise social, environmental and economic benefits.

6.2.3 Market Pricing Adjustments Would Automatically Result in "Optimisation" of Natural Gas Use Structures

In order to "optimise" natural gas use structures, the National Development and Reform Commission developed a specific gas use policy. This policy also classifies the use of gas as being either highest priority, permitted, restricted or prohibited, and thus established the order of priority. Chinese economist Hua Ben analysed the capacity and affordability of five markets in the lower reaches of the Yangtze in terms of power generation, urban gas, industrial and commercial fuels, transportation fuels and chemical raw materials. He then performed a feasibility study on natural gas, concluding that, given that natural gas costs 2.5 times more for the same calorific value when compared to coal and other fuels, a combined cycle power plant cannot compete with coal power generation in terms of basic load and can only be used for peak regulation. Urban use of gas fuel is less price-sensitive, but the volume used is limited. Industrial and commercial gas use is the largest end user market for natural gas. CCHP systems improve natural gas pricing tolerance and are currently key to the efficient use of natural gas in China. LNG has advantages as a vehicle gas fuel. Compared to coal, natural gas is not economically viable for use as a raw material in the chemical engineering industry. Based on Hua Ben's theoretical results, there is absolutely no need for the government to implement gas use policies, as price adjustments will automatically result in "optimisation" of gas use structures and natural gas is not economically viable for use in power generation or as a raw material in chemical engineering.

Regarding the problem of the use of natural gas in chemical engineering, however, it is likely that natural gas producers and local governments in gas-producing areas may have applied different methods of calculation. Where natural gas producers are concerned, due to the natural gas price being controlled by the government, revenues from external sales of natural gas are limited, while they are required to provide peak regulation and ensure supplies. However, development of industrial natural gas use near oil and gas fields is stable, since natural gas costs less when the price doesn't include a transport element. In addition, there is no regulation of the price of chemical products, allowing higher levels of income to be achieved. From the perspective of local governments in natural gas-producing regions, external sale of locally produced natural gas results in meagre earnings from resource tax (5% of the selling price). Therefore, local development of a natural gas chemical industry will not only help increase local tax revenue and provide employment, but can also promote local economic development. When one does the math, this is an additional impetus for natural gas producers and governments in gas-producing regions to develop the natural gas chemicals industry.

A basic approach to "optimising" gas use structure involves natural gas tax reform and environmentally-friendly policy, accompanied by reorganisation of the natural gas pricing mechanisms, all which should work together so that natural gas projects that bring greater social and environmental benefits have access to greater economic benefits as a result. It is preferable to

reduce the effects of government intervention in specific markets to a minimum. Experience of government interference in natural gas markets between the mid-1950s and the 1980s in the United States has already taught us the lesson that, whether pricing controls or usage controls are employed, neither are helpful in the development of a healthy natural gas industry.

6.3 Natural Gas Use and Natural Gas Price Controls

6.3.1 Natural Gas Price Control Policies in China

Due to the existence of a monopoly in the natural gas market, it has been difficult for market pricing mechanisms to emerge, and thus natural gas prices have always been controlled by the government in China. The prices at which upstream corporations sell natural gas are determined by the National Development and Reform Commission, and the prices at which the downstream corporations sell natural gas are determined by pricing departments within local government. Before the natural gas price reforms of June 2013, the National Development and Reform Commission set manufacturers' natural gas prices according to the price of alternative energy sources (limiting increases to 10%, with no bottom limit), with gas being priced differently for different uses. In the natural gas price reforms of 2013, the National Development and Reform Commission decided to use alternative energy sources as a pricing reference, and, based on considerations of the main directional flow of natural gas market resources and pipeline costs, established gate prices by the netback market value method. The gate prices are set temporarily and adjusted annually, and no further distinction is made regarding different uses, while the goal is to merge stored gas and incremental gas prices within three years. In the second half of 2014, international oil prices fell drastically, based on the price adjustment formula established in 2013, the National Development and Reform Commission then issued the Notice on Reorganising

Table 6.6 Maximum natural gas gate prices by province (region, city) after 2015 price adjustments

Province	Maximum gate price	Province	Maximum gate price
Beijing	2700	Hubei	2660
Tianjin	2700	Hunan	2660
Hebei	2680	Guangdong	2880
Shanxi	2610	Guangxi	2710
Inner Mongolia	2040	Hainan	2340
Liaoning	2680	Chongqing	2340
Jilin	2460	Sichuan	2350
Heilongjiang	2460	Guizhou	2410
Shanghai	2880	Yunnan	2410
Jiangsu	2860	Shaanxi	2040
Zhejiang	2870	Gansu	2130
Anhui	2790	Ningxia	2210
Jiangxi	2660	Qinghai	1970
Shandong	2680	Tibet	1850
Henan	2710		

Unit CNY/10^3 m^3 (including value added tax)
Data source Pricing variation order (2015)351: Notice on reorganising the price of non-residential natural gas, issued by the National Development and Reform Commission

the Price of Non-Residential Natural Gas on February 28, 2015, merging the prices of stored gas and incremental gas, as well as partially lifting the control over gate prices for commercial users. After adjustments, the maximum natural gas gate price for each province was as shown in Table 6.6.

6.3.2 The Effect of Pricing Mechanisms in Keeping Natural Gas Prices Low

Prior to June 2013, the National Development and Reform Commission set different prices for various uses of natural gas (see Table 6.7), but to a certain degree this pricing scheme conflicted with the natural gas use polices at the time. For example, gas use for chemicals was one of the areas prohibited by the policies, but in order to keep the price of fertilisers low, a relatively low price was set for natural gas used in fertilisers, which then acted as a signal that such usage was encouraged. Urban fuel gas use (non-industrial) was one of the natural gas use areas given

priority by gas use policy, resulting in a lower price for supplies. However, even though the lower prices were beneficial for increasing market demand, it did not encourage greater supply. Due to there being a shortfall between supply and demand in the natural gas market, and for financial reasons, upstream enterprises prioritised supplies to higher priced users such as stable industrial users, who could support a higher price, which did not actually benefit the development of gas as an urban fuel.

Under strict government control, natural gas prices in China remained at a relatively low level for a long time. in 2012, the land-based natural gas price (excluding tax) for domestic manufacturers was 1.06 CNY/m^3 (1.2 CNY/m^3 with tax added), which was equivalent to only a quarter of the WTI oil price at that time, a quarter of the CIF value of imported LPG, a third of the CIF value of imported fuel oil, half of the CIF value of imported Central-Asian natural gas, or a quarter of the CIF value of imported Qatar LNG. In 2012, the average price of residential gas in 36 major cities was 2.43 CNY/m^3, while LPG was priced at 7.65 CNY/kg and residential electricity

Table 6.7 Baseline manufacturer's prices (first-gate station) for domestic land-based natural gas after May 2010 price adjustments

Name of oil/gas field	Type of user	Manufacturer's baseline price (CNY/10^3 m^3)
Chuan-yu oil-gas field	Fertilisers	920
	Direct supply to industry	1505
	Urban use (industrial)	1550
	Urban use (non-industrial)	1150
Changqing oil field	Fertilisers	940
	Direct supply to industry	1355
	Urban use (industrial)	1400
	Urban use (non-industrial)	1000
Qinghai gas field	Fertilisers	890
	Direct supply to industry	1290
	Urban use (industrial)	1290
	Urban use (non-industrial)	890
Various oil fields in Xinjiang	Fertilisers	790
	Direct supply to industry	1215
	Urban use (industrial)	1190
	Urban use (non-industrial)	790
Various oil fields in Tibet	Fertilisers	790
	Direct supply to industry	1215
	Urban use (industrial)	1190
	Urban use (non-industrial)	790
Various oil fields in Dagang, Niaohe, Zhongyuan	Fertilisers	940
	Direct supply to industry	1570
	Urban use (industrial)	1570
	Urban use (non-industrial)	1170
Other oil fields	Fertilisers	1210
	Direct supply to industry	1610
	Urban use (industrial)	1610
	Urban use (non-industrial)	1210

(continued)

Table 6.7 (continued)

Name of oil/gas field	Type of user	Manufacturer's baseline price (CNY/10^3 m^3)
West-East Gas Pipeline	Fertilisers	790
	Direct supply to industry	1190
	Urban use (industrial)	1190
	Urban use (non-industrial)	790
Zhong-Wu Line	Fertilisers	1141
	Direct supply to industry	1541
	Urban use (industrial)	1541
	Urban use (non-industrial)	1141
Shaanxi Beijing Line	Fertilisers	1060
	Direct supply to industry	1460
	Urban use (industrial)	1460
	Urban use (non-industrial)	1060
Sichuan going east pipeline		1510

Data source Circular (2010)211 of the NDRC "Increasing the base price of domestic on-shore natural gas"

was priced at 0.53 CNY/kWh. The calorific value of natural gas is 8000 kcal/m^3, for LPG it is 12,000 kcal/m^3 and for electricity it is 860 kcal/kWh. In other words, for the same calorific value the price of natural gas was only 48% of the price of LPG, and 49% of the electricity price. Although a lower price for natural gas is beneficial in the short term for promoting gas use, it does not reflect the market value of clean and efficient energy sources, while also contributing to the conflicts between supply and demand, thus affecting both the supply of resources and sustainable healthy natural gas market development.

6.3.3 The Influence of a High Fixed Price for Natural Gas

In the 2013 natural gas price reform, the price restrictions on high-cost gas sources, such as LNG and unconventional natural gas, which make up a sixth of the entire market, were lifted, allowing the market to determine prices instead. However, price controls were kept in place for domestically produced and pipeline gases, which account for the remaining five-sixths of the market supply. Based on the 2013 natural gas price reforms, incremental gas was adjusted to 85% of the price of alternative energy in one move, and classifications according to usage were no longer applied. The price of stored gas was raised by 400 CNY/1000 m^3, while the price of domestic gas was not adjusted (see Table 6.8). Against the backdrop of insufficient overall natural gas market supply, removal of pricing restrictions on high-cost gas sources and adjusting the pricing mechanism for natural gas by a reasonable amount should have encouraged supply and alleviated conflicts between supply and demand. However, this was not a successful move as far as encouraging natural gas consumption, and resulted in a lack of market demand for natural gas.

In the 2013 price adjustments, the volume of stored gas was 112.0 billion m^3, but for

Table 6.8 Gate natural gas prices by province after the 2013 price reform

Province	Stored gas	Incremental gas	Province	Stored gas	Incremental gas
Beijing	2260	3140	Hubei	2220	3100
Tianjin	2260	3140	Hunan	2220	3100
Hebei	2240	3120	Guangdong	2740	3320
Shanxi	2170	3050	Guangxi	2570	3150
Inner Mongolia	1600	2480	Hainan	1920	2780
Liaoning	2240	3120	Chongqing	1920	2780
Jilin	2020	2900	Sichuan	1930	2790
Heilongjiang	2020	2900	Guizhou	1970	2850
Shanghai	2440	3320	Yunnan	1970	2850
Jiangsu	2420	3300	Shaanxi	1600	2480
Zhejiang	2430	3310	Gansu	1690	2570
Anhui	2350	3230	Ningxia	1770	2650
Jiangxi	2220	3100	Qinghai	1530	2410
Shandong	2240	3120	Xinjiang	1410	2290
Henan	2270	3150			

Unit (CNY/thousand m^3)
Data source Pricing Notice (2013)1246 of the National Development and Reform Commission concerning adjustment of the natural gas price

incremental gas it was only 11.0 billion m^3, accounting for 91 and 9% respectively. Although incremental gas only accounts for a small share, its price rose more, directly affecting the growth of market demand for natural gas. Taking the Beijing gate price of 3.14 CNY/m^3, for example, for the same calorific value it would cost 3.5 times more than coal (calorific value of 5500 kJ/kg, 620 CNY/ton), 0.87 times the price of oil fuel (calorific value of 10,000 kJ/kg, 4520 CNY/ton), and 0.70 times the price of LPG (calorific value of 12,000 kJ/kg, 6700 CNY/ton). Given the relatively high distribution costs, at this price natural gas is less competitive. In 2014, economic development in China entered the "new norm" era, and economic growth began to slow, while dropping coal prices and abundant hydroelectricity resulted in reduced demand for natural gas for power generation. The growth in natural gas consumption has been significantly lower than that predicted, with growth for last year of only 8.9%, which was considerably lower than the average of 17.4% over the last 10 years.

6.3.4   Removal of Natural Gas Price Restrictions Would See the Disappearance of Natural Gas Use Policies

Government natural gas price regulation, whether it is set high or low, is detrimental to creating equilibrium between supply and demand in the market and encouraging sustainable development of the natural gas industry. The experiences of the United States and China in managing the natural gas industry reveal that government restrictions on natural gas prices tend to cause a serious imbalance of market supply and demand, resulting in the need for further direct government intervention. Moreover, experience in the United States also indicates that with the removal of price restrictions the market will actively reallocate resources, and this automatically increases social benefits and optimises the economic gains associated with natural gas to their maximum potential. The best method to ensure the environmental benefits of natural gas is to increase the cost of pollution abatement associated with highly polluting fuels.

Among the different uses of natural gas, household use has the highest tolerance towards rising prices, not being sensitive to changes in gas price. Based on statistics for 2012, the national household use of natural gas was only 17.85 m^3/month per household, while in Beijing households it was 18.5 m^3/month per household. Given the gate price of 3.14 CNY/m^3, even after adding the distribution and peak regulation costs, the price of natural gas for household use would not exceed 5 CNY/m^3. In terms of competition between natural gas and electricity, with the price for household electricity use being 0.54 CNY/kWh, if the price of natural gas for household use exceeds 5 CNY/m^3, household users will stop using natural gas. Even if calculated at 5 CNY/m^3, the average natural gas cost for each Beijing household is only CNY 90/month. As the average annual disposable income for Beijing citizens was CNY 40,321 in 2013, natural gas consumption accounted for less than 1% of household income.

When compared to coal, natural gas has advantages as an industrial fuel. For example, it has a better calorific value, is capable of raising temperatures more quickly, is clean and non-polluting, and it can be easily switched on and off. It is also effective in tackling problems such as sulphur dioxide and phenol emissions, tarring and particulate pollution, thus improving quality and economic effectiveness in industries such as pottery, glass and non-ferrous metals, which also increases its competitiveness. When considered from the point of view of China's overall energy needs, China has only limited natural gas supplies, therefore natural gas could never completely replace coal. From the point of view of the atmosphere as a whole, "coal conversion to gas" does not completely overcome this problem, the more important issue being to increase clean coal use. The main uses for natural gas in the chemical industry are the production of ammonia and methanol, and these are the most sensitive to price changes in natural gas. Regardless of whether ammonia or methanol

production is concerned, coal is a serious competitor with natural gas for use as raw material in the chemicals industry. At the same time as the ratio of coal use in energy source structure is being reduced, coal prices are falling, thus offering greater advantages to the coal chemicals industry than to the gas chemicals industry. If natural gas pricing mechanisms were liberalised, this would result in a decline in natural gas's share of use in the chemicals industry, without the need for government intervention.

6.4 The Relevance of the 1978 U.S. Natural Gas Policy Act for China

In 1978, during the international oil crisis, in the face of panic caused by lowering domestic production and natural gas reserves on the brink of exhaustion, the United States Congress passed the Fuel Use Act (FUA), restricting oil and natural gas in power generation and for industrial purposes. The legislative background and the contents of the Fuel Use Act are similar to some extent to the "Natural Gas Use Policy" adopted by the National Development and Reform Commission. The way in which the FUA affected development of the natural gas industry in the United States provides a valuable reference for policies affecting China's natural gas industry.

6.4.1 Legislative Background

In order to protect the interests of independent producers and the end user, and prevent natural gas manufacturers and pipeline operators from colluding in monopolistic profiteering, in 1954 the Supreme Court of the United States gave its decision against Phillips Petroleum Co., authorising the Federal Power Commission (FPC) to regulate interstate natural gas wellhead prices. The FPC stipulated that wellhead prices must be determined "based on historical costs", regardless of stored amount, alternative energy source prices or changes in consumption. The main effect of this was to protect the short-term interests of consumers. However, it interfered with the investment necessary for manufacturers to increase future supply. In 1973, the production of natural gas in the United States reached a peak, at an annual volume of 615.4 billion m^3, and then began to fall. In 1978 production was at 555.8 billion m^3, and in 1986 it was as low as 454.7 billion m^3, dropping by a quarter from peak levels (see Fig. 6.1).

In October 1973 the first oil crisis broke out, causing massive oil price inflation, and yet the natural gas wellhead prices remained under regulation, worsening the natural gas supply situation. The price regulation caused an increase in demand and insufficient supply, and when combined with the effects of the oil crisis, this led to natural gas supply shortages in the United States in the 1970s. Between 1976 and the winter of 1977, at the worst point of the natural gas shortage, more than 9000 factories were forced to suspend production, 750,000 people lost their jobs and hundreds of schools in the Midwest and Northeast had to be closed, causing several billion dollars of losses to the US manufacturing industry. That period in the development of the US natural gas industry is now referred to as the "era of failure".

The catalyst for the Fuel Use Act was the peak gas theory, which predicts that natural gas will eventually be depleted after peak production is reached. In 1976, Exxon Oil published a report said to be the result of research conducted by over a hundred top geologists and geophysicists. The report stated that the total remaining natural gas resources in the United States were only 287×10^{12} cf (around 8×10^{12} m^3). In fact, from 1976 to 2012, the actual natural gas production in the United States was close to 20×10^{12} m^3, and at the end of 2012 there were still discovered resources of 8.5×10^{12} m^3.

6.4.2   Contents of the Legislation

Since the natural gas wellhead price restrictions only applied to interstate natural gas trade, pipeline companies buying natural gas within a state were not affected by the FPC price controls. As a result, there were no supply shortages in the major natural gas exporting states. When winter

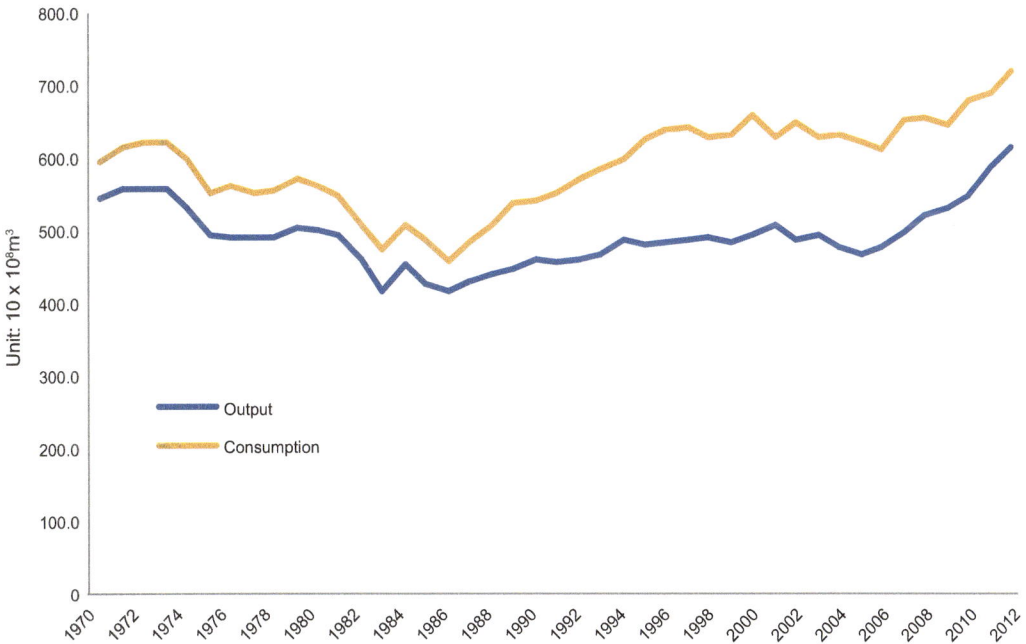

Fig. 6.1 Changes in production and consumption of natural gas in the United States from 1970 to 2012

approached, while the northern schools and commercial establishments had to close because of the lack of natural gas supplies, Texas was still using natural gas as a replacement for oil and coal in power generation, rather than shipping it up north. There were suggestions that the price restrictions should be lifted to let natural gas be traded at market prices. However, the public utilities representatives of both the major parties decided to trust the "experts" and their theory of resource depletion. They believed that the signs of depletion were already recognisable, and they felt that deregulation of natural gas prices was a selfish, short-sighted and idiotic solution. In their opinion it was necessary to continue with resource protection subsidies, terminal usage controls and pricing control, which would raise efficiency and ensure fairness all round.

In May 1978, the United States Congress passed two acts: the Power Plant and Industrial Fuel Use Act (FUA) and the Natural Gas Policy Act (NGPA). The NGPA was mainly a redesign of the regulatory regime and categorisation of natural gas well types, raising the price of natural gas from "old wells", and partially removing price restrictions on "new wells". The FUA is a supplementary act to the NGPA, authorising the Department of Energy (DOE) to issue orders restricting the use of natural gas to specific industries, in an attempt to directly control the use of natural gas. Apart from forbidding the building of new natural gas power plants and industrial boilers, the FUA further demanded that factories should abandon fuel switching capabilities, while encouraging factories using natural gas to switch to using other fuels. After 1990, the use of natural gas was banned in power plants and industrial boilers. In the transitional period, use of natural gas in public works was restricted to consumption volumes equivalent to those encountered during the energy crisis from 1974 to 1976.

6.4.3 Implementation

In the process of setting out the FUA, the United States Congress firmly believed that natural gas

resources were in the process of being depleted, and could not possibly have predicted that an excess of natural gas would arise. Yet, while the ink was still wet on the FUA, Secretary of Energy James Schlesinger announced that there were predicted natural gas reserves in excess of the 1×10^{12} ft^3. He did state, however, that this "bubble" was only temporary. While the "bubble" lasted, the Department of Energy supported a delay in switching to coal as demanded by the FUA, and encouraged industrial fuel users capable of switching energy sources to use "surplus" natural gas instead of oil.

The loosening of restrictions on wellhead prices increased the supply of natural gas, with high fixed prices limiting demand, and "take-or-pay" contracts locking up a large volume of higher-priced natural gas, which caused a "bubble" to occur in the natural gas market, with supply exceeding demand. As this bubble economy persisted, the FUA's restrictions on natural gas use became a historically significant mistake and resulted in the market imbalance worsening. In 1981, Congress overruled the FUA's power plant requirements and then in 1987 Congress repealed the FUA and withdrew all the regulations responsible for restricting natural gas consumption.

6.4.4 Lessons for Natural Gas Usage Policy in China

The Natural Gas Use Policy, as adopted by the National Development and Reform Commission, is not only similar in content to the 1978 Fuel Use Act of the United States, they also share a similar background, market environment and policy climate.

- The natural gas market in China has been growing rapidly since 2000; demand for natural gas has been rising, but the growth in domestic production remains slow, leading to increasing reliance on imported natural gas. This is similar to the uncertainties regarding future energy source supplies in the United States during the 1970s.

- The United States had yet to carry out natural gas industry reforms in the 1970s, and pipeline transportation and sales of natural gas were bundled together. The situation in which natural gas wellhead price restrictions were put into place was therefore similar to that when the Natural Gas Use Policy was enacted in China.
- Natural gas prices in China have been frequently adjusted since 2000, and from 2013 onwards natural gas prices were pegged to oil prices and LPG prices. Lifting regulation of unconventional natural gas and imported LNG prices would have the same effect as when the United States removed "new wells" pricing regulation in the FUA of 1978.
- Due to the higher value of natural gas, there has been a great deal of activity in development of unconventional natural gas in China, while a large number of natural gas import contracts have been signed with offshore partners. This is also reminiscent of how, after "new well" price regulation was lifted in the United States in 1978, there was extensive development of extraction of the more expensive "deep-level gas".
- After suffering more than a decade of shortages, United States pipeline companies eagerly signed long-term "take-or-pay" contracts with gas suppliers at high prices, in order to ensure capacity to satisfy the demands of users lower down on the supply chain, thus creating the conditions necessary for the "bubble" to occur. Contracts in the natural gas market in China are also in the form of "take-or-pay", and when supplies come under pressure, local governments opt for a "give me gas, disregard the price" approach out of fear.
- Due to the effects of the long-term shortage in the United States, in 1980, industrial and public power generation natural gas consumption was 20% lower than in the 1970s, with many industries having learned how to switch fuels and save energy; nearly all new power plants were coal- or nuclear-powered. As gas prices continue to rise, it will become difficult to continue using natural gas in power generation in China, and enthusiasm for natural gas in industrial uses will fall, with coal-fired power stations preferring to adopt a clean coal use approach rather switching to gas.

In a 2013 forecast, the gap between natural gas supply and demand in China would be as big as 22.0 billion m^3. In fact, due to the National Development and Reform Commission's large-scale adjustment of pipeline gas prices in June, stored gas price went up by 400 CNY/10^3 m^3, while the incremental gas price is now pegged against alternative energy sources, pushing it upwards to 1380 CNY/10^3 m^3. This has resulted in many gas power stations adjusting their power generation plans and has resulted in a decline in industrial gas use. In addition, with relatively warm winters, China National Petroleum Corporation was actually able to put gas into storage during the winter, all of which are signs of an excess in supply. What we can learn from the FUA in the United States is that it is necessary to make timely adjustments, and to repeal natural gas use regulations when it becomes necessary.

6.5 Policy Recommendations for Encouraging Natural Gas Use

The pressures of environmental protection and emission reduction mean that China needs to be more active in developing natural gas and encouraging its use. The nature of current natural gas use policies in China is such that, when demand exceeds supply and pricing mechanisms are no longer effective, the government interferes with supply and demand directly, mainly by applying restrictions to gas use in specific sectors. In the past decade, with the widespread development of natural gas sources, China has reached a stage where a range of locally produced gas, gas synthesised from coal, imported pipeline gas and imported LNG is available, greatly boosting the available quantity and stability of natural gas supplies. The "shale natural gas revolution" in the United States teaches us that as technology for natural gas extraction develops, the extent of natural gas reserves available for extraction will increase. As natural gas prices in China are adjusted to more acceptable levels, the scale of supply has increased, and there has been a reversal in the previous supply-demand scenario, currently manifesting as a lack of demand. China must therefore alter its current policy of restricting natural gas use, and instead promote and encourage a wider range of uses for natural gas.

6.5.1 Improve Environmental Supervision, Replacing Dispersed Coal Use with Natural Gas

The energy supply in China is said to be "coal-heavy, oil-deprived and short of gas". Even though various sources of natural gas have been introduced in the last decade, vastly enhancing the natural gas supply capacity, in the face of the tremendous demand for energy in China, the supply of natural gas available to China is still limited. Natural gas is not likely to completely replace coal, although it could be prioritised to replace dispersed coal use. In OECD countries, the ratio of coal used for power generation is 78%, while 90% of coal consumed in the United States is for power generation. Dispersed coal use is relatively concentrated in the steel and concrete industries, but very dispersed coal use is rare. In terms of total coal consumption in China, only half of it is used for power generation, the other half being used in dispersed coal burning. Of dispersed coal burning uses, half is used in metal smelting and cement sintering and other relatively concentrated uses, for which coal is essential, the other half being used in heating, textiles, paper manufacturing and other very dispersed fields. When coal use is too dispersed, energy efficiency is low and large volumes of pollutants result, resulting in increased pollution control costs. Natural gas substitution would therefore apply mainly to the replacement of very dispersed coal use.

Since coal is relatively cheap, natural gas does not have a competitive advantage in terms of price, therefore businesses prefer coal based purely on economic considerations. Encouragement of the replacement of coal by natural gas mainly comes from environmental protection policies.

- More rigid emission standards should be stipulated and, whether for concentrated or dispersed coal uses, all users should stay within emissions targets. By internalising what was originally an external cost of using coal, the competitiveness of natural gas would improve.
- A new supervisory structure should be devised, so that concentrated coal use can be supervised in real time and online, in order to prevent companies from installing but not using emission reduction and pollution control facilities, or only using them from time to time.
- In areas subject to urban planning or areas affected by acid rain, coal burning should be restricted or prohibited.
- Where gas supplies allow, in facilities where the cost of restricting emissions is too high or where it would be impractical to introduce emissions restrictions, conversion to gas could be obligatory. For those in regions where conditions do not allow for gas supplies, construction of new installations which require dispersed coal use which are not absolutely necessary in terms of everyday life should not be permitted. At the same time, more efforts should be put into building gas infrastructure, so that more regions could opt to use natural gas.

6.5.2 Optimise the Electricity Pricing Scheme, and Increase the Economical Natural Gas Power Generation

The key to increasing the ratio of natural gas in total energy source consumption structures is the development of natural gas power generation. Natural gas power generation is clean, occupies less space and provides effective peak regulation. It is well suited for provision of supplementary power to large cities or megacities. In developed countries, natural gas power generation usually takes up around 40–60% of total natural gas consumption. In China, on the other hand, only 15% of total natural gas used is for power generation. While there is much space for development, the biggest hurdle for expansion of natural gas power generation is the relatively high cost of natural gas power compared to coal power. Based on calculations, the cost of coal power generation is only 0.34 CNY/kWh, whereas natural gas power generation costs as much as 0.65 CNY/kWh (annual power generation of 35,000 h), or 0.615 CNY/kWh (annual power generation of 5000 h). Under the current subsidy regime, natural gas-generated power is more expensive and there is no incentive for power grid companies to buy natural gas-generated power. With a high natural gas price, there is nothing to be gained from natural gas power generation, and power generating companies have shown little interest in it.

There are two ways to promote natural gas power generation, both of which aim to improve the economic benefits of natural gas power generation. The first method involves introducing local production of core technology, especially the manufacturing of gas turbines, so that unit investment costs and long-term maintenance and inspection costs can be lowered. The second method is to increase power generation operational cycles, lowering the fixed cost of each kWh of electricity produced. However, the amount of leeway gained in terms of lowering costs is very limited. Based on calculations, using locally manufactured core technology would only bring the cost of power generation down by 0.024 CNY/kWh. Increasing the length of the power generation operational cycles would move natural gas power generation towards playing a base-loading role, lessening the gains it can achieve as a supplementary power source in peak generation. Currently there are no auxiliary service markets or peak/trough electricity pricing

mechanisms in China. The main policy adopted has been subsidising natural gas power generation by setting a higher electricity price. In large cities such as Shanghai, Beijing and Shenzhen, a two-part tariff is charged for electricity, the capacity price being 40–46 CNY/kW month, basically covering the fixed costs of natural gas power generation. On top of the two-part tariff, auxiliary service markets should also be developed, switching from government regulation of prices to market-regulated prices to provide such subsidies.

6.5.3 Strengthen Planned Guidance to Promote Natural Gas in Transportation

For transportation, the gases used are mainly CNG and LNG, which replaces petrol, diesel and ship fuel. Based on data provided by the Environmental Protection Department, around a quarter of all atmospheric pollution in China comes from exhaust gases. When natural gas is used as a replacement transportation fuel, vehicles emit less carbon dioxide compared to oil-burning vehicles, and they hardly emit any other pollutants such as nitrogen oxide and sulphides. Furthermore, the relatively large gap between natural gas and oil prices, and the absence of goods and services tax, means that natural gas vehicles are more economical than oil-burning vehicles. For example, in the case of a taxi that runs on CNG, there was a period of time when the price of CNG was half that of petrol, and hence the investment of several thousand CNY required for conversion could be recovered in a matter of months. Backed by atmospheric pollution control measures and its economical characteristics, use of natural gas in transportation in China has been growing quickly, from 97,000 vehicles in 2005 to 2.3 million natural gas-powered vehicles in 2014; the associated annual gas consumption reached 22.5 billion m^3, making up 12.3% of total natural gas consumption. LNG filling stations increased from 241 stations in 2011 to 2000 stations in 2014.

As natural gas transportation technology matures, it brings more economic benefits, but there are specific hindrances to the development of natural gas use in the transportation sector. Unlike household, industrial, power generation or other uses where the users stay in a fixed location, natural gas used in transportation is supplied to moving users and relies heavily on supply networks. Natural gas fuel, filling stations and transportation form the core of the natural gas transport use value chain, each segment affecting the decision of a user on whether to choose natural gas over all other alternatives; this in turn may slow down or hamper the development of natural gas use in the transportation sector. As the natural gas supply situation in China improves, the supply of natural gas fuel is assured for the foreseeable future. However, the government still needs to strengthen planning and guidance and refine related policies and standards, in order to increase the provision of infrastructure such as natural gas filling stations and standardised filling services, thus ensuring that the natural gas filling station network better fulfils the mobile needs of transportation and logistics users, thereby attracting users to adopt natural gas, accelerating growth in the natural gas transport sector.

6.5.4 Reduce Cross-Subsidisation Between Different Users, and Encourage Industrial and Commercial Natural Gas Use

In China, urban natural gas use is considered to be one of the benefits of city life. Since residential gas is associated with very low profit margins, with supplies sometimes being provided at a loss, the financial shortfall due to cheap household natural gas is made up by charging a higher natural gas price to the industrial and commercial sectors. For example, after the upstream natural gas prices were adjusted in 2013, the Beijing gate price for stored gas was

2.26 CNY/m^3 and for incremental gas was 3.14 CNY/m^3, but government pricing departments set the price of residential use gas at 2.28 CNY/m^3. It is clear that sales to household users were made at a loss, compensated for by industrial and commercial use. The subsidisation of household natural gas use by the industrial and commercial sectors is in fact passed on to society in its use of products and services. The higher household income is, the more hot water and heating a household uses, and the more it benefits from the natural gas subsidies. In contrast, households with lower incomes, or households that are not natural gas users or that only use a small amount of natural gas, would not benefit from, or only benefit slightly from, these subsidies. That households with higher incomes should benefit more from such subsidies cannot be seen as fulfilling the original objectives of government policy and is hardly fair.

Not only is an all-encompassing subsidy for household natural gas use unfair, it is also unsustainable and unnecessary. On the one hand, there is much room for growth in household natural gas use, and with volumes expected to increase from the 2012 figure of 28.8 billion m^3 to 54.8 billion m^3 by 2020, the continuance of such subsidies would be enormously difficult. On the other hand, household natural gas use is less price-sensitive, and hence there is no real need for subsidies. From the point of view of economic competitiveness, as long as the price is lower than 4 CNY/m^3, natural gas will still be competitive against oil fuel, LPG and electricity. City dwellers use an average of around 60 m^3, and even if the price of each cubic metre goes up by CNY 1, the annual increase is still only CNY 60. In 2013, low-income urban households (the lowest 20%) had a disposable income of CNY 11,434, and an increase in natural gas expenditure of CNY 60 is equivalent to only 5% of disposable income. Therefore, the price of natural gas for household users could definitely be set above the cost of supplying the gas, cancelling the cross-subsidising between household use and industrial and commercial use, reducing the cost of using natural gas to the industrial and commercial sectors, and thus encouraging the growth of industrial and commercial natural gas use. As for the specifics of the price adjustments, local governments should decide the timing and magnitude of them.

6.5.5 Extend Carbon Emission Trading Rights, and Actualise the Environmental Value of Natural Gas

The key to encouraging the use of natural gas in power generation as well as in the industrial and commercial sectors is to improve the competitiveness of natural gas against coal. If considering the price alone, natural gas cannot compete against coal, as the fuel cost per kWh of coal power generation is a mere CNY 0.21, while that of natural gas is CNY 0.50. As a clean energy source, natural gas has an environmental value, and when competing against coal this environmental value could be considered as the external environmental cost of coal. The external cost of coal burning could be broken down into specific pollutant emissions costs, such as dust, nitrogen oxides and sulphur dioxide, in addition to a carbon emissions cost. With increased investment, the pollutants released by coal burning can be lowered to a level close to that encountered where gas is used as the fuel. Newly constructed coal-burning power stations can basically achieve the same level of atmospheric pollutant emissions as natural gas power stations (when baseline oxygen content is 6%, dust, sulphur dioxide, nitrogen oxides concentrations are respectively above 10, 35 and 50 mg/m^3), while the increase in investment cost for each kWh

is only CNY 0.01–0.02. Therefore, simply strengthening enforcement of environmental regulations and restricting release of air pollutants would not effectively increase the competitiveness of natural gas against coal in the power generation sector.

Another method for boosting the competitiveness of natural gas is the levying of a carbon tax, but while a carbon tax can narrow the gap between the costs of coal power generation and natural gas power generation, thereby making natural gas comparatively more competitive, it would nonetheless increase the absolute cost of both forms of power generation. A better way of making natural gas more competitive is to set a carbon allowance for each kWh, and fully utilise trading of carbon emission rights. For example, if coal power generation has a carbon emission of 0.8 kg/kWh while gas power generation's carbon emission is 0.37 kg/kWh and the government sets the carbon allowance at 0.7 kg/kWh, then coal power stations have to purchase 0.1 kg/kWh of emission rights from the market, while natural gas power stations have 0.33 kg/kWh of emission rights for sale. If there are only coal and natural gas power generation to choose from, then the electricity generated by coal cannot exceed 3.3 times the amount generated by natural gas; in other words, the proportion of the quantity of natural gas power generation would be 23% of total power generated, otherwise there would be insufficient emissions rights available for purchase by coal power stations. This would allow the government greater flexibility to adjust the natural gas power generation ratio, getting gas power generators to directly subsidise natural gas power generation, and forcing coal power generation to provide a direct subsidy to gas power generation, thereby taking full advantage of the environmental value of natural gas.

1. China's natural gas resource volume is continually growing. According to the latest survey results (2013), China's conventional and low-permeability natural gas geological resource volume is 52 trillion cubic metres, with technically recoverable resource volume of 32 trillion cubic metres. In 2014, China's conventional natural gas production volume was 128 billion cubic metres, with ground coalbed methane production reaching 3.6 billion cubic metres and shale natural gas production reaching 1.3 billion cubic metres. It is expected that by 2020, China's natural gas production volume will be approximately 260–330 billion cubic metres, and by 2030 that number will reach approximately 430–560 billion cubic metres. In order to accelerate China's natural gas development, the development path should be made clearer, by adjusting the minimum exploration investment, maintaining reasonable prices, establishing trading mechanisms and trading platforms for proven natural gas reserves and accelerating unconventional natural gas development. Small- and medium-sized enterprises should be given special encouragement to join the unconventional natural gas development industry. In addition, based on international experience, system reforms should be extended to enhance the momentum of natural gas development and encourage innovation.

2. In 1993, China changed to become an importer of crude oil. In 2007, China became a natural-gas-importing nation. In just 7 or 8 years, by 2014, China's natural gas import dependency rate had exploded to 32%. This is a significant development. In future, the import status of natural gas will have a marked influence on supply and demand in China. Looking at natural gas supply and demand from a global perspective, natural gas resources are relatively plentiful, and natural gas will account for a large proportion of global energy consumption in future, playing a larger role than in the past. However, the drop in international oil prices has brought enormous uncertainty to the future development of the international natural gas market. It is expected that in the coming years imports of natural gas into China will rapidly increase, growing from 53 billion cubic metres in 2013 to 167 billion cubic metres in 2020, including 70 billion cubic metres of LNG and 97 billion cubic metres of pipeline natural gas. By 2030, imports of natural gas will have grown further, to approximately 210–240 billion cubic metres, with approximately 75 billion cubic metres of LNG and 135–165 billion cubic metres of pipeline gas. If domestic natural gas, especially shale natural gas, production volume achieves a market increase as a result of breakthroughs in technology and policy, then by 2030 China's natural gas import dependency can be kept to below 40%.

3. Since the West–East Gas Pipeline commenced operation in 2004, China's natural gas infrastructure has rapidly developed. As of the end of 2014, a pipeline distribution deployment of nearby suppliers has been formed from west to

east and from offshore to onshore, providing coverage to all provinces except Tibet. The two major import resource deployments of pipeline gas and LNG have also been completed, opening up access to Sino-Asian pipeline gas, Sino-Myanmar pipeline gas and LNG import channels. There are also underground storage and LNG receiving stations to act as major peak shaving options, covering coastal regions, gas production regions and the area around the Bohai Gulf. However, there are also shortages in pipeline construction and in underground pipeline reserve peak shaving facilities. In order to guarantee continued stable operation of China's natural gas infrastructure design, it is recommended that a safe and stable national natural gas infrastructure operating system be built. To strengthen infrastructure construction capabilities, there should be unified infrastructure planning, and construction investment should be diversified. To increase infrastructure usage efficiency, a fair third-party access (TPA) operations and management platform should be created.

4. Constructing a robust natural gas reserve peak shaving system is an effective approach for dealing with mid-term natural gas supply interruptions and ensuring stable operation of the natural gas industry as well as stability within the economy and society. Currently, China's natural gas reserve construction is in its infancy, and the peak shaving demands being faced are significant. The existing reserve peak adjustment capabilities are insufficient, and price and operations mechanisms have obvious issues. The peak shaving and emergency response assurance mechanisms are both insufficiently robust and present problems and challenges. The secure supply of natural gas should be made an important component of national security, accelerating the construction of reasonable reserve systems suited to the scale of market demand while giving greater attention to the formulation of natural gas peak shaving emergency response reserve planning, focusing on reserve facility law and regulatory construction, and clarifying natural gas business reserve obligations so as to promote natural gas peak shaving emergency response reserve facility construction. Active tax and price policies should be released to reform reserve gas management regimes and construct prompt and flexible warning systems for emergency response, continually improving China's natural gas supply assurance capabilities as well as its risk mitigation capabilities.

China's Natural Gas Resource Potential and Production Trends

7.1 Natural Gas Resource Potential

7.1.1 Resource Potential for Chinese Conventional and Unconventional Natural Gas

According to estimates of China's natural gas resources, China's conventional and low-permeability natural gas geological resource volume amounts to 68 trillion m^3, of which 40 trillion m^3 is technically recoverable. Onshore conventional and low-permeability natural gas resources are primarily distributed in the three large basins of Sichuan, Erdos and Tarim. Marine natural gas resources are primarily distributed in the Pearl River Estuary, the South-East Qiong and the East Sea Basin.

From a systematic appraisal of 121 regions in 42 coal basins (groups), China's onshore coalbed methane resources located less than 2000 m underground amount to 36 trillion m^3.

* This chapter was overseen by Zhonghong Wang from the Development Research Center of the State Council and Shangyou Nie from Shell International Exploration and Production. It was jointly completed by Xiaowei Xuan, Yingxie Yang, Yongwei Zhang, Jiaofeng Guo, Weiming Li, Beiqing Yao, Tao Hong and Yuxi Li from the State Resource Department Oil and Gas Centre, Xiaoli Liu from the National Development and Reform Commission's Energy Office, Guang Yang and Jianhong Yang from the China Energy Research Institute's Natural Gas Centre, Xiaobo Ju and Beijing Yao from Shell China. Other members of the topic group participated in discussions and revisions.

Technically recoverable resources less than 1500 m underground total 11 trillion m^3. These are primarily distributed in China's northern Qinshui Basin, Erdos Basin, the Yunnan-Guizhou-Guangxi area and the Eastern Yunnan Western Guizhou Area. In terms of geologic sequence, they are primarily from the Carboniferous, Permian and Jurassic. According to national shale natural gas resource potential survey appraisals and the identification of the most promising regions, China's onshore shale natural gas geological resources located less than 4500 m underground total 134 trillion m^3, with 25 trillion m^3 of recoverable resources, primarily distributed in the Sichuan Basin and surrounding areas (Table 7.1).

China's gas hydrate resources are also very abundant, and are primarily distributed in the South China Sea and East China Sea regions, as well as in the permafrost of the Tibetan Plateau. Geological resources are approximately 102 trillion m^3, and this is currently one of China's most abundant sources of clean energy, with the potential for large-scale development and the basic conditions in place to become a mainstream clean energy source (Table 7.2).

7.1.2 Changes in Natural Gas Resource Assessments

China has conducted five rounds of systematic national natural gas resource assessments. In 1986 the first round found that China's natural gas geological resource volume was

Shell International and The Development Research Center (Eds.), *China's Gas Development Strategies*, Advances in Oil and Gas Exploration & Production, DOI 10.1007/978-3-319-59734-8_7

Table 7.1 Resource potential and state of development of various natural gas resources in China

Distributed layer		Source rock		Migration layer		Trap layer		Middle-shallow layer
Resource type		Coalbed methane	Shale natural gas	Water soluble gas	Tight gas	Low-permeability gas	Conventional	Biogas
Recoverable resource potential	International	11×10^{12} m^3	25×10^{12} m^3	Not known	Disputed	32×10^{12} m^3		Not known
State of development	Domestic	Developed	Developed	Developed	Developed	Developed		Developed
				No development				

Material source Ministry of Land and Resources, revised

Table 7.2 China's natural gas resource volume

Type	Geological resource volume (trillion cubic metres)	Recoverable resource volume (trillion cubic metres)	Remarks
Conventional natural gas	68	40	
Shale natural gas	134	25	In cases where development from exploration is possible, relatively reliable resource volume is 12.85 trillion m^3
Coalbed methane	37	11	1500 m shallow coalbed methane recoverable resources
Gas hydrate	102		No likely commercial development before 2030
Total	341	76	

Source Ministry of Land and Resources

approximately 34 trillion m^3. In 1994, the second round found that natural gas resources were 38 m^3. In 2007, the third round found that, spread over 115 land and near-shore regions containing oil and gas basins, China's conventional natural gas resources were 35 trillion m^3, with recoverable resources of 22 trillion m^3. In 2012, the fourth round found that natural gas geological resources were 52 trillion m^3, with recoverable resource volume of 32 trillion m^3. The fifth round, in 2013, found that natural gas geological resources were further increased to 68 trillion m^3, with recoverable resources rising to 40 trillion m^3. In short, China's natural gas resource volume is continually increasing. This is especially clear from the 2013 results, which showed a marked increase over 2012 (Table 7.3).

Looking back at the history of exploration and development, it is clear that natural gas resources continually change along with the development of exploration. As demand continues to grow, natural gas exploration and development expands to satisfy it, with ever-increasing resource types

and resource volumes. Progress in the development of oil and gas geological theory and technology is continually expanding the types of natural gas and their scope, generally pushing natural gas toward a trend of ever-expanding development. Also, the influence of price changes will cause short-term fluctuations in the amount of natural gas that is judged economically recoverable.

7.2   Proved Natural Gas Reserves

7.2.1   Conventional Natural Gas

China's natural gas exploration and development lagged behind oil, with generally low levels of exploration. According to BP's annual statistical review, since 2006, proved natural gas reserves have consistently maintained high rates of growth, with annual proved natural gas original reserve volumes of over 500 billion m^3. Since 2010, annual proved conventional natural gas geological

Table 7.3 China's historical natural gas resource assessment results

Assessment time (years)	1986	1994	2007	2012
Geological resources (trillion cubic metres)	33.6	38.04	35	52
Recoverable resources (trillion cubic metres)			22	32

Source Ministry of Land and Resources, Crude Oil Department and Oil and Natural Gas Headquarters materials

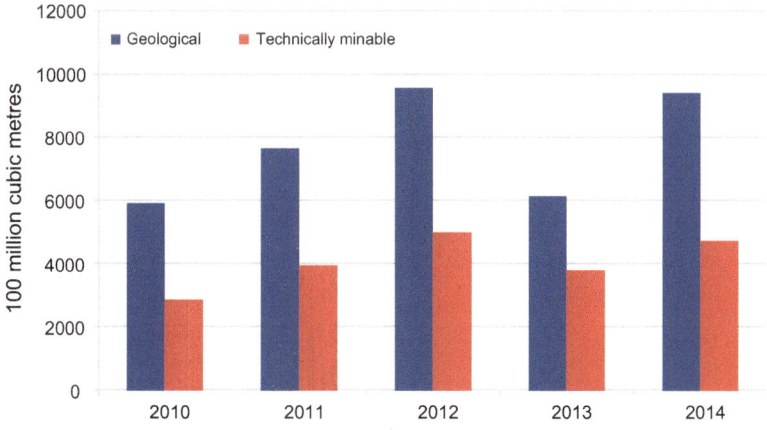

Fig. 7.1 China's annual increase in proved conventional natural gas reserves. *Source* Ministry of Land and Resources annual press conference materials

reserves have exceeded 600 billion m^3. Indeed, in 2012 and 2014, proved geological reserves were each nearly 1 trillion m^3 (Fig. 7.1). Conventional proved natural gas reserves are primarily in the Tarim Basin, Sichuan Basin and Erdos Basin. In addition, deep within the Songliao Basin there has also been rapid growth in proved reserves. As of 2013, China's proved natural gas geological reserves were 9.66 trillion m^3, with technically recoverable reserves of 5.54 trillion m^3.

The basins in Sichuan, Erdos, Tarim, Songliao, Chaidamu and Zhungeer are now the six major mainland natural gas exploration areas. Ying-Qiong and the East China Sea have become the two major near-shore exploration areas, forming a basic layout of eight large natural gas exploration areas. Erdos, Sichuan and Tarim are the three major gas regions with reserves over 1 trillion m^3.

China's proved natural gas reserve regions are relatively concentrated, and are primarily distributed in the western region, which accounts for 83% of resources, while the east region and offshore areas account for the other 17%.

From 1986 to 1990, proved natural gas source reserve growth was relatively small and less than 40 billion m^3. In 1991–1995, growth was close to 100 billion m^3, and in 1996–2000 growth was 150 billion m^3. In 2001–2005, growth rose considerably to approximately 300 billion m^3, while during 2006–2010, the growth was over 320 billion m^3. In 2011–2013, conventional proved natural gas resources growth exceeded 600 billion m^3 for three years in a row, with an additional 616.4 billion m^3 added in proved conventional natural gas reserves in 2013. In 2014, newly added proved conventional geological reserves reached 943.7 billion m^3, marking an increase of 53% year on year. Conventional proved natural gas volume increased consistently, setting a foundation for large natural gas production volumes.

7.2.2 Coalbed Methane

In 2012, national proved coalbed methane reserves added 134.4 billion m^3, rising year on year by 2.5%. In 2013, national proved coalbed methane reserves added 23.6 billion m^3 (Table 7.4). Proved coalbed methane geological reserves are 575.4 billion m^3, with technically recoverable reserves of 285.0 billion m^3, showing rapid growth in proved coalbed methane reserves. Three-quarters of coalbed methane reserves originate from Qinshui Basin, while one quarter is from the Erdos Basin, with proved reserves of coalbed methane relatively sparse in other areas.

In 2006, proved coalbed methane wells and developed wells numbered 1373, a figure which then increased rapidly, so that by 2014 there were 18,000 coalbed methane well drillings.

Over many years of exploration, the Qinshui Basin has become an especially large coalbed methane region, with 400 billion m^3 of geological reserves. The eastern edge of the Erdos Basin has become a target gas region, with over 100 billion m^3 of geological reserves, and the coalbed methane industry development has good foundations.

7.2.3 Shale Natural Gas

As of the end of 2013, China had not submitted proved reserves of shale natural gas. In June 2014, China's first proved shale natural gas reserves passed review and inspection by the relevant authorities. The proved reserve is located in Sinopec's Fuling Gas Field Jiaoshiba Region Coke-1 Well 3, over an area of 106.45 km^2, and the layers incorporating shale natural gas are a Wufeng group within Longmaxi. Experts believe that the Lingpei Shale Gas Field is a typical shale natural gas field. The gas field reservoir layer is a marine deep-shelf quality shale with significant

Time (year)	2008	2009	2010	2011	2012	2013
Proved geological reserves (100 million m^3)	225	244	1299	1311	1344	236

Table 7.4 Coalbed methane newly added proved reserve data table

Source Ministry of Land and Resources annual press conference materials

thickness and very high abundance. Distribution is stable, depth is suitable, and there are no intervening layers in between, markedly different from conventional gas deposits, and having the characteristics of typical shale natural gas, being similar to North American typical marine-shelf shale natural gas indicators. The Peling Shale Gas Field is a high-quality marine-shelf shale natural gas field that has excellent quality and quantity of resources. There is a high formational pressure, the natural gas components are good, gas well production volumes are high, and there have been good results seen from mining attempts. There has been a high production volume from pilot wells, with long periods of stable production. An expert appraisal panel has determined proved shale natural gas volume to be 106.5 billion m^3.

As of June 30 2014, the Jiaoshiba region's 29 wells with pilot mining recorded total daily gas production of 3.2 million m^3, with accumulated gas production of 611 million m^3. As part of this, the first exploration well Coke IHF has recorded stable production for a year and a half at 60,000 m^3 per day, with an accumulated gas production of 37.69 million m^3.

7.3   Growth of Natural Gas Production

7.3.1   Conventional Natural Gas

Prior to 1995, China's natural gas production growth was slow, rising from 14.3 billion m^3 in 1980 to 17.4 billion m^3 in 1995. From 1995 to 2005, natural gas production accelerated, growing from 17.4 billion m^3 to nearly 50 billion m^3, an average annual growth of 3.2 billion m^3. Since production reached 50 billion m^3, annual average growth has been approximately 8 billion m^3. By 2012 it had reached 107.2 billion m^3, 117.0 billion m^3 in 2013 and 128 billion m^3 in 2014, with net annual growth of 9.4% (Fig. 7.2). Since 2000, China's natural gas production volume growth has been approximately 9.6%, and annual average growth for natural gas over the next 10 years or so is expected to be roughly 9–10%.

Since 2007, conventional natural gas consumption volume has exceeded natural gas production volume, and natural gas imports have continually increased. By 2013, China's conventional natural gas consumption volume had reached 166 billion m^3 (see Fig. 7.3). In recent years, natural gas consumption volume has grown by 12%, markedly higher than natural gas production growth. Domestic natural gas production volume is still unable to meet the needs of domestic natural gas demand, with imports of natural gas continually increasing.

7.3.2   Coalbed Methane

China's coalbed methane production volume includes both extraction and coal mine well drainage. This research only deals with

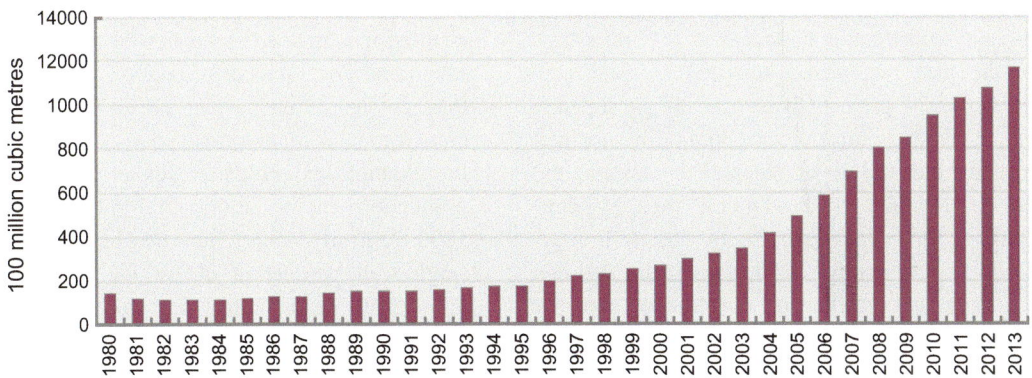

Fig. 7.2 Natural gas production volume in China. *Source* BP, 2014

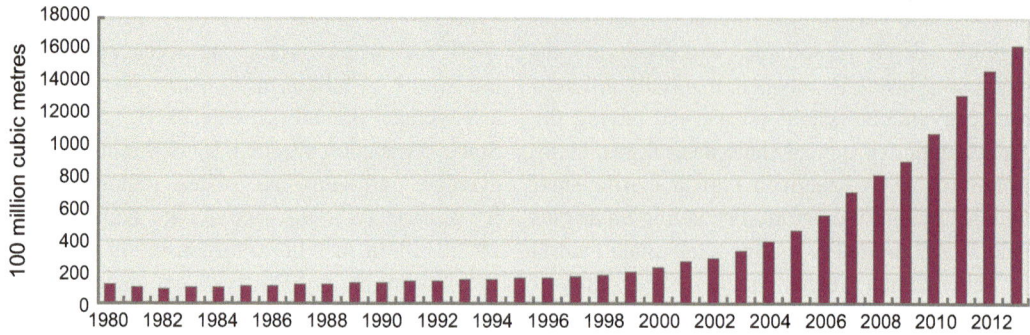

Fig. 7.3 Conventional natural gas consumption volume in China. *Source* BP, 2014

Table 7.5 China's coalbed methane production volume in recent years

Time (year)	2011	2012	2013	2014
Production volume (100 million m³)	23	25.7	29.3	36

Data source Ministry of Land and Resources annual press conference materials

extraction-based coalbed methane, and does not consider well drainage coalbed methane. As of 2011, China's ground extracted coalbed methane was 2.3 billion m³, with 2012 ground extracted coalbed methane production volume of 2.57 billion m³, a growth of 13% year on year. Coal mine region coalbed methane production grew by 9.9% year on year, with the rate of growth dropping during the same period. By 2013, ground extracted coalbed methane production volume was 2.93 billion m³, growing by 13.7%. In 2014, ground extracted coalbed methane production volume was 3.6 billion, marking a large jump upward (Table 7.5).

As of 2012, there were seven constructed coalbed methane development projects, with eight development projects in progress. Ground development coalbed methane production capacity construction had reached 4.9 billion m³. In the past two years, new regions and new development projects have continually opened up, providing huge momentum for coalbed methane industry development. For example, the Baode 1 billion m³ production capacity construction, the Yanchuan South 500 million m³ production capacity construction, the Mi'nan Region East 1.5 billion m³ construction capacity construction and the Nanzhuang South 1 billion m³ construction plans are all under way.

Moreover, total production capacity of approximately 1.5 billion m³ will be developed through projects including Shouyang, Liulin, Changzi, Mabi and Shunan.

Coalbed methane exploration and development exhibits a marked trend toward new regions, new layers and new sectors:

- In recent years, coalbed methane exploration has continually seen breakthroughs at depths lower than 800–1000 m, with coalbed methane seeking relatively good single well production volumes. For example, in the Yanchuan Region at depths of 1497–1503 m, daily gas production has been relatively high at 3600 m³, while the Qinshui Basin in Zhengzhuang at a depth of 1337 m has achieved daily gas production of 2336 m³. Deep coalbed methane breakthroughs have resulted in a marked expansion in the exploration and development of coalbed methane.

- In recent years there have also been significant breakthroughs in low-rank coal exploration, for example, in Huainan single-well daily production volumes at 660–880 m have been 1000–2100 m³, Hui Chun has seen stable single-well production of 1500–2200 m³ at 450–550 m and Yilan has seen stable single-well production of 1000–1200 m³ at 700 m in depth. In addition, in Huolin River,

Binchang and other places, low-rank coal regions have achieved production volumes of over 1000 m^3. Breakthroughs in low-rank coal have opened up the door to development of coalbed methane in northern regions, where low-rank coal is distributed.

- Taking the Erdos Basin Hedong region as a representative of the North China Palaeozoic coal stratum, there are typical coalbed methane, tight natural gas and shale natural gas characteristics between the layers. However, comprehensive development of multiple types of natural gas resources has not yet seen a breakthrough. In recent years, the Hedong region has increased exploration efforts within the coal stratum seeking tight sandstone gas, with marked breakthroughs already achieved and 6000–50,000 m^3 of production volume of tight natural gas already achieved, and individual well production volumes also higher.

7.3.3 Shale Natural Gas

In 2012 China released the Shale Gas Development Plan (2011–2015), in which there were plans that by 2015 proved shale natural gas reserve volumes would be 600 billion m^3, with 200 billion m^3 recoverable. In 2015, shale natural gas production was expected to be 6.5 billion m^3. Based on calculation of 115 billion m^3 of natural gas as 100 million tons of oil, 6.5 billion m^3 of shale natural gas is equivalent to 5.65 million tons of oil.

According to the latest information released by the Ministry of Land and Resources, in 2014 China's shale natural gas production volume was 1.3 billion m^3. As part of this, Jiaoshiba shale natural gas production volume approached 1.08 billion m^3, and shale natural gas production in Chuannan and other regions was 200 million m^3. The main natural gas production layers were in section 1 of the Wufeng-Longmaxi Formation.

From the perspective of exploration progress, China has achieved shale natural gas flow objectives in formations including the Qiongzhusi Formation, Wufeng-Longmaxi Formation, Longtan-Dalong Formation, Yanchang Formation, Xujiahe Formation, Ziliujing Formation and others. Currently, only the Wufeng-Longmaxi Formation has achieved successful large-scale development, with other target formations still needing continual exploration.

7.4 Conventional and Unconventional Natural Gas Production Forecasts

According to current natural gas increase trends, forecasts for conventional and unconventional natural gas production are as follows.

7.4.1 2020 Production Volume Forecast

1. **Forecasts based on the existing system**

Based on the annual average growth of natural gas production volume in the past several years of 8–10 billion m^3, conventional natural gas annual average production volume is forecasted to be 8 billion m^3. By 2020, conventional natural gas production volume will reach 180 billion m^3. Based on target forecasts for shale natural gas production volumes from the Chongqing and Sichuan regions, by 2020, shale natural gas production volume will be 40 billion m^3. Coalbed methane production volume data primarily shows a trend for growth in the extraction of coalbed methane, and by 2020 this will reach around 10 billion m^3. Production volume total for natural gas will be 230 billion m^3 (not including coalbed methane) (Fig. 7.4).

2. **Forecast in a partially changed existing system**

With the system unchanged on the whole, and only conducting new system explorations for shale natural gas in Chuannan and Chuandong, achieving shale natural gas development in

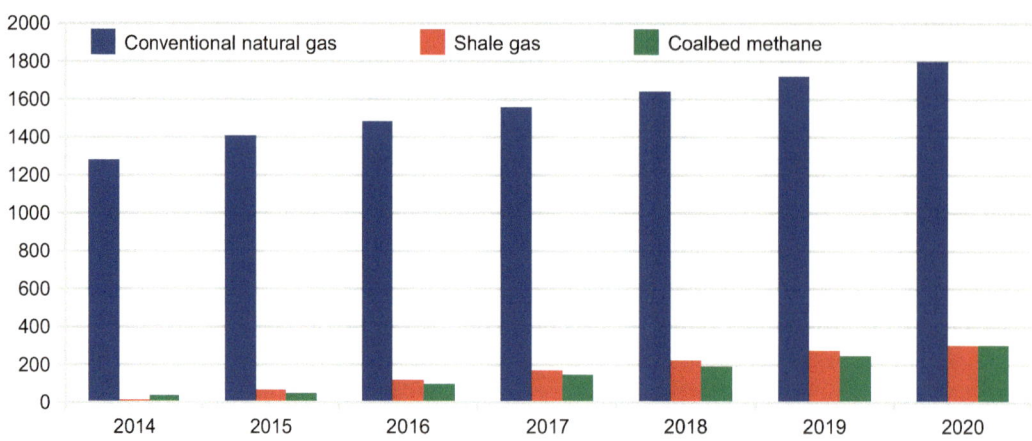

Fig. 7.4 Existing system natural gas production volume forecasts (100 million m^3)

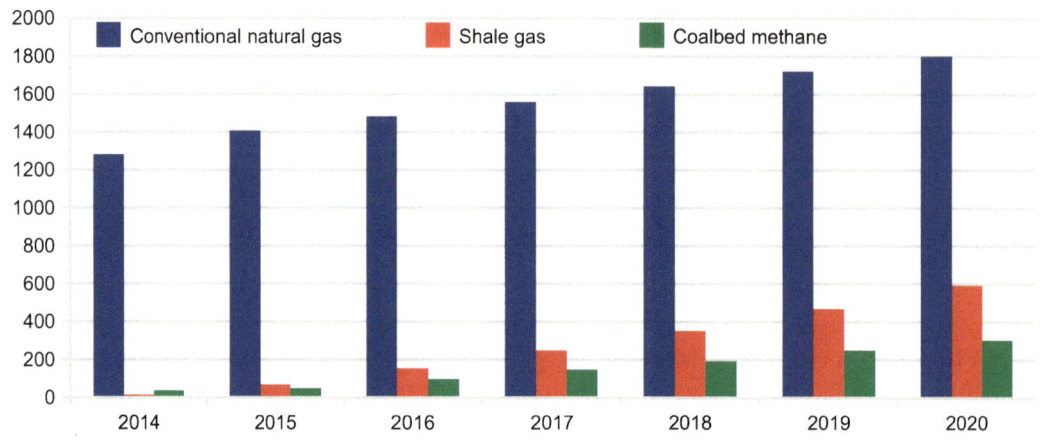

Fig. 7.5 Natural gas production forecasts in a partially adjusted system environment (100 million m^3)

general pilot regions for construction forecasts, conventional natural gas annual average growth will remain unchanged at 8 billion m^3, and by 2020 will reach 180 billion m^3. In Chuannan and Chuandong, if new system trials for shale natural gas development are conducted, by 2020 shale natural gas production volumes will be 60 billion m^3. Coal formation coalbed methane and tight natural gas general development efforts will be strengthened, with forecasts that by 2020, coalbed methane production volume will be 15 billion m^3 and tight gas and shale gas production volume within coalbed methane areas will be 15 billion m^3, for a total of 270 billion m^3 (not including coalbed methane) (Fig. 7.5).

7.4.2 2030 Annual Production Forecast

1. Forecast in existing system

From 2020 to 2030, conventional natural gas production annual growth rate will remain unchanged and maintain at 8 billion m^3. By 2030, conventional natural gas production volume will be 260 billion m^3. In Chuannan and Chuandong Longmaxi Formation, shale natural gas production volume will rise, with other shale natural gas formation production volumes also seeing breakthroughs. Shale natural gas production will rise from 40 billion m^3 in 2020 to 80 billion m^3.   Ground-extracted   coalbed methane production volume will reach 20 billion m^3, while coal formation tight natural gas and shale natural gas production volume will reach 20 billion m^3. Large-scale statistical coalbed methane production will be 40 billion m^3, for a total of 280 billion m^3 (not including coalbed methane).

2. **Forecast in a partially changed existing system**

If oil prices remain at the current normal level, enterprise natural gas profitability will rise, and enterprises will be more motivated to produce natural gas. If conventional proved natural gas reserves remain at a high level from 2010 to 2020, this will cause 2020–2030 annual production volume acceleration to rise slightly to 10 billion m^3, and reach 280 billion m^3 by 2030. If Chuannan and Chuandong shale natural gas new system pilots are particularly effective, and if they are expanded to achieve commercial breakthroughs in other shale natural gas systems, then by 2030 shale natural gas production volume will reach 150 billion m^3. Coalbed methane production volume will remain at 40 billion m^3. Total natural gas production volume will be 470 billion m^3 (not including coalbed methane).

China's natural gas production volume growth is controlled by myriad factors, including geological factors that can be resolved through technological progress. However, system and mechanism factors are also ubiquitous, and closely influence gains, and are difficult to resolve quickly. Cost factors in recent years have become an important factor affecting natural gas industry development, and since 2003 exploration and development costs have continually risen. If these cannot be effectively controlled, it will cause a marked influence on China's natural gas industry development and production volume goals. As a result, China's natural gas production volume increases could slow, and may be hard-pressed to exhibit explosive growth.

7.5 Accelerating the Development of China's Domestic Natural Gas Resources

7.5.1 Clarify Development Paths

Different paths of development require different policy measures. Given a clear development path, specific policy measures can be formulated to promote natural gas development.

1. **No change to the existing systems and mechanisms**

This path of development is an extension of the existing development paths. Based on State Council rules regarding oil and natural gas licensing, natural gas exploration, development and pipeline shipping will all be operated by existing state-owned oil companies. This business approach has the advantage of achieving intensive development of natural gas, and facilitates large-scale development. However, costs are also very high, and overall prices are somewhat excessive. This is especially true for prices in natural gas production regions, and supply markedly lags behind demand. For example, Sichuan, Chongqing, and Xinjiang natural gas prices and supply volumes have no marked advantage. Such a system causes cost increases for enterprises using natural gas as an energy source and source material.

2. **Using change to strengthen market guidance and diversify development**

Another path of development is to change the existing licensed operation system, and to use enhancements to market mechanisms to introduce multiple ownership system enterprises to develop natural gas and to carry out additional exploration, thus realising diversified development. Through market competition, the strongest and most competitive survive, carrying out new natural gas explorations, development projects and supply investments. This development approach requires that many issues are addressed, including high-quality preparation and fair allocation of mineral rights, the market liberalisation of oil and gas project services, the independent operation of natural gas pipelines and fair access to them, marketing and trading of proved reserves, and strong and effective oversight. Otherwise, major problems will arise that will destroy the market reforms.

It is suggested that existing systems and mechanisms are suitably adjusted, appropriately opening up conventional oil and gas operating rights by first opening up oil and gas mineral rights to large state-owned energy enterprises, such as state-owned energy enterprises that already hold shale natural gas mining rights, or opening up conventional oil and gas mining rights to energy enterprises that have already entered the oil and gas sector overseas, guiding these enterprises in their entry into domestic conventional and unconventional oil and gas exploration and development sectors.

7.5.2 Adjust Lowest Exploration Commitments

In 1996, China's Mine Law and subsequent accompanying legislation prescribed that oil and gas mine exploration rights holders had to invest at least 5000 CNY/km^2 in the first year, 7000 CNY/km^2 in the second year and 10,000 CNY/km^2 in the third year. In the past, the low levels of exploration spending commitment did not take inflation into account. Because oil prices have gradually risen since 2003, oil company exploration and development costs have also seen constant increase. Thus 10 years ago one exploration well only required CNY 3000–5000, whereas now CNY 10,000–20,000 might be needed to drill that same well, with some well fees being even higher.

Because of the steep increase in costs, maintaining such low exploration spending commitments has essentially led to annual declines in investment because of the changed

price calculations. Previously, an CNY 100 million exploration commitment could drill three exploratory wells, whereas now only one can be completed for the same outlay. Physical exploration commitments by area are only one third to one fifth of their level 15 years ago. A large area abundant in natural gas may thus be seriously lacking in survey work, severely impacting natural gas discovery and reserve production growth.

Shale natural gas tender region bidding standards and management requirements have raised the conventional oil and gas region lowest exploration commitment by a factor of 3–5 from current levels, with each square kilometre having a lowest exploration commitment of CNY 30,000–50,000. An increase in the lowest exploration commitment will increase survey efficiency and promote prompt exits from regions, thus reducing the occupied area of various sectors and changing the current situation in which areas are marked out but not explored. At the same time, the current system of authorisation through advance application could be changed, following the example of the shale natural gas region transfer approach, which uses a competitive mode to transfer existing conventional oil and gas blocks. International conventions provide examples of how to formulate effective block exit mechanisms. A regulatory team should be set up to oversee implementation of legislation and systems.

7.5.3 Maintain Reasonable Prices

Currently, calls for natural gas price increases are very loud, demanding that natural gas be priced at 70% of oil prices according to heating value. As a basic energy and a source material for economic and social functioning, a natural gas price that is too high or too low is not beneficial to overall development of the economy. When the price is too low, producer investment momentum is insufficient and natural gas development is restricted. When natural gas prices are too high, this will keep a large quantity of potential consumers out of the market, which is not beneficial to the economic development of the nation as a whole. In recent years, low natural gas prices in the United States have played an important role in economic recovery, and this is worth taking into consideration.

In the past 10 years, China's natural gas reserve growth has been rapid, but production volume annual growth has been less than 10 billion m^3. The reason for this is the gap between oil and natural gas heating value prices, especially with the existing special licensed operation system and high oil prices, which have made enterprises more willing to invest in oil production. Recently, oil price drops have caused oil and natural gas heating value pricing to become equivalent, and given China's relatively high natural gas price, oil companies have become more enthusiastic about the production of natural gas. In order to reach a natural gas price that both providers and buyers can accept in the market, the natural gas exploration and development market should be appropriately opened, and prices can then be formed through competitive mechanisms, thus establishing a relatively reasonable mechanism to set natural gas prices.

Different development mechanisms will result in different price-forming mechanisms. Therefore, natural gas price problems must be resolved after clarifying natural gas development mechanisms.

7.5.4 Establish Trade Mechanisms and Trade Platforms for Proved Natural Gas Reserves

In recent years, China's annual average proved natural gas geological reserves have consistently exceeded 500 billion m^3, and it is expected that this will continue to grow fast for the next few years. However, proved reserves have not been promptly developed, and some have even remained undeveloped for long periods of time. The reason for this is partly due to that portion of proved reserves lacking effective economic momentum.

It is recommended that a trade mechanism be established for proved and difficult-to-utilise reserves, permitting this portion of reserves to enter the market and be transferred. This will allow them to reach the hands of enterprises that can economically effectively develop such resources and thus achieve prompt development and exploration. In such a system, enterprises can transfer proved reserves to recover their exploration costs, while development enterprises can purchase reserves to obtain development opportunities. In order to guarantee the prompt development of these reserves, it is important to avoid simple circulation and non-development of reserves.

7.5.5 Accelerate Development of Unconventional Natural Gas

China has abundant unconventional natural gas resources, of which shale natural gas and coalbed methane are independent mine types, and are the focus of unconventional natural gas development. Shale natural gas can be developed rapidly, and success has already been seen in Chongqing Fuling Jiaoshiba, with a breakthrough also seen in Chuannan region, where success was realised within a single year.

1. **Pilot zones for shale natural gas development in Chuannan and Chuandong**

Since 2009, when CNPC deployed and implemented its Wei 201 Well and formally began shale natural gas exploration, shale natural gas production volume reached 1.3 billion m^3 by 2014. This demonstrates that shale natural gas is quick to enter production, quick to achieve efficiency and quick to develop.

The shale natural gas formation in the southern and eastern portions of the Sichuan Basin is the Longmaxi Formation. Based on CNPC and Sinopec shale natural gas resource assessment results, this shale natural gas formation has shale natural gas geological resource volume of 1.9 billion m^3 in Chuannan and Chuandong, with recoverable resources of 4.5 trillion m^3. Shale natural gas presents relatively low development risks, and has resource foundations and technical foundations to carry out scaled development.

Chuannan and Chuandong Longmaxi Formation are important areas when it comes to allowing shale natural gas to break through to large-scale production. Based on annual production of 60 billion m^3 by 2020, and calculating for stable production for 30 years, 1.8 trillion m^3 of shale natural gas recoverable resources are needed. It would only take a proportion of the 4.5 trillion m^3 of recoverable shale natural gas resources in Chuannan and Chuandong to achieve that goal. This means that by developing the pilot construction in Chuannan and Chuandong, it will be possible to effectively mobilise the Longmaxi Formation to provide sufficient shale natural gas to meet the long-term goal of 60–100 billion m^3 of shale natural gas per annum by 2020. At the same time, it is also possible to explore the direction which oil and gas system reform should take in the future using pilot regions here.

2. **Additional support to coalbed methane development**

Additional support for coalbed methane development could be provided with policies that encourage profitability in the sector, with the aim of reaping benefits for energy security and the environment.

The first goal of enterprises when they develop coalbed methane is to obtain coalbed methane commodities and to make profit. There is another major objective for surface and underground extraction of coalbed methane in coal mining regions though: to ensure coal mine safety and to reduce gas emissions. In many respects, obtaining the coalbed methane

commodities is actually the secondary aim. These two different mindsets require correspondingly different policy measures.

- Coalbed methane development in coal mining areas is motivated by safety and environmental protection considerations. Working closely with coal mining and developers, plans need to be advanced by five years in order to deploy surface extraction suitably far ahead of well extraction, by deploying Laotang drainage and other coalbed methane projects.
- Policies that encourage and facilitate coalbed methane can harness independent enterprises to carry out coalbed methane extraction at scale across multiple mine sites. Policies, especially subsidies that promote coalbed methane development, are primarily directed at enterprises that directly use this portion of coalbed methane (gas).
- When not involved with coal mining, coalbed methane development is a commercial activity, and can be promoted by incentive-focused industrial policy. Some coalbed methane mining regions are not within coal planning regions and coal mining regions in the next 10–20 years. The coal mining in these regions is commercial coalbed methane development, and coal mining safety and environmental problems do not arise. Such activity should not enjoy the preferential policies applied to coalbed methane development in coal mining areas. It is sufficient simply to use the current preferential development policies in such cases.

It is worth noting that subsidy policies that lack a time limit will cause enterprises to develop a reliance on national subsidies, resulting in the coalbed methane industry remaining in an immature state of development for longer than would otherwise be the case, rather than developing healthily and robustly.

Case 1: Establishing a 100 billion m^3 Shale Natural Gas General Pilot Region in the Sichuan Basin

The current international oil price is low, and US shale oil and gas companies have suffered the impact of this and even been forced to apply for bankruptcy protection. Asian LNG spot prices are also trending lower. However, looking to the long term, supply shortages and high import prices will be the major problems faced by China's natural gas development. Shale natural gas development is still a strategic choice for China's natural gas development. The Sichuan Basin's abundant shale natural gas resources and favourable development conditions make it possible to carry out pilots to test general reforms strategies. By using optimised systems and mechanisms, a preliminary shale natural gas development can be scaled up and commercialised to achieve annual production of 100 billion m^3 by 2025 in this large shale natural gas region. This is not only a strategic choice to increase secure domestic natural gas supply; it also represents an unusually attractive investment opportunity.

1. **Sichuan Basin has the basic conditions in place**

During multiple surveys of Sichuan and Chongqing, and after meetings with multiple investors and engineering technicians, and collecting opinions from the relevant experts regarding shale natural gas development resource conditions, technology, equipment, engineering, investments, applications, environmental loads and other factors, it is our opinion that with the appropriate system and policies, the Sichuan Basin will become a leading region for shale natural gas development in China. By 2025, this region could feasibly achieve 100 billion m^3 of shale natural gas

production volume. The reasons are outlined below.

(I) **Resource conditions are guaranteed**

The Sichuan Basin's shale natural gas resources are abundant, and there are wide distributions of two marine shelves of substantially thick black shale. According to Sinopec data, total shale natural gas resources in the entire Longmaxi Formation and Qiongzhusi Formation amount to 40 trillion m^3. Of this, the Longmaxi Formation marine shelf shale natural gas has a promising area of 75,000 km^2 that has already seen development breakthroughs, with a shale natural gas resource volume of 25 trillion m^3 and recoverable resources of 3.7 trillion m^3. As part of this 75,000 km^2, there are 35,000 km^2 of shale natural gas resources whose conditions are superior and geological resources of nearly 14 trillion m^3, with recoverable resources of 28 trillion m^3. If development is only pursued in the high-quality shale natural gas area of 20,000 km^2 of the Longmaxi Formation, beginning in 2015, the exploration and well layout planning could be completed and approximately 10,000 wells could be deployed within 2015–2025. By 2025, annual production of 100 billion m^3 could be achieved. Moving forward, active wells would number around 800, achieving stable production for 20 years. It could be said that as far as resource volume is concerned, Sichuan and Chongqing entirely satisfies the requirements to establish a region with annual production of 100 billion m^3.

(II) **Development technology and equipment is assured**

Currently, China has preliminarily come to terms with shale natural gas geophysics, well drilling, well completion, pressure fracturing and other technologies, giving it the ability to drill shallow water wells of 3500 m as well as staged fracturing. Non-seismic geophysical prospecting identification and forecasting technology, wholly-mobile rail drills, large-scale fracturing vehicles (3000 model, 3500 model), construction environment protection technology and other factors are now at similar levels of advancement to those seen internationally, and breakthroughs have also been achieved in domestic bridge plug manufacture. Compared to internationally mature technologies, China now only has shortcomings in drill geological positioning, on-drill measurements and micrometre-nanometre structure and composition analysis. However, the majority of well-known oil services companies from abroad are now participating in China's shale natural gas development, and can provide vital technologies. Sinopec's Chongqing Fuling (Fuling) Shale Gas Field has entered commercial development, attesting to the fact that China's shale natural gas development technology is now up to par. In addition, China's shale natural gas core equipment, such as sets of drills, fracturing vehicle groups, well tools and accompanying project services, all have powerful capabilities, with existing domestic drill quantities and drill production capacities entirely meeting the requirements placed on well drilling quantities during construction and production stages.

(III) Good mining efficiency, guaranteed economic benefit

As of November 30, 2014, Sinopec (Fuling) Region had completed fracturing test gas sites at 69 wells and obtained medium to high production of shale natural gas flows, with an average well test production volume of 320,000 m^3 per day. Based on unhindered flow volume of 1/3, 1/4 and 1/5 accompanying set production, average single well annual production volume is approximately 21.60 million m^3 or higher, with average single well costs of CNY 80 million (including exploration, mining and shipping). Adding simple processing such as water extraction to this, well costs are less than 1.5 CNY/m^3, whereas current shale natural gas well prices are 2.78 CNY/m^3. If a national financial subsidy of 0.4 CNY/m^3 is added in, the actual price is 3.18 CNY/m^3, and single well annual revenue once production stabilises could reach CNY 68.688 million. In addition to Fuling Region's shale natural gas mining achieving commercial viability, in the Chuanning, Changning and Weiyuan region the shale natural gas is buried relatively deep, but shale natural gas and conventional gas overlap in distribution and both possess general economic value. The eastern Chongqing region and Western Hubei region shale natural gas areas also feature light oil, and the economic value of mining this is also relatively high. The economic success of the Sichuan Basin's shale natural gas development is guaranteed.

(IV) Environmental controls can draw on experience elsewhere, and mining water is assured

Currently, the United States has over 10,000 shale natural gas wells, and with the participation and oversight of regulators and the public, there have been no environmental incidents affecting society. The majority of the shale natural gas-producing layers in the Sichuan Basin are at over 2000 m in depth, and so long as casing work is handled appropriately or assurance measures are added (such as installing an encircling casing), fracturing water has no chance of permeating into the water table system. In terms of water usage, water usage volume for 100 billion m^3 of gas is approximately 100 million L^3 of water, which is 1.6% of the water usage of Sichuan Province. Since the Sichuan Basin has abundant water resources, these water resources are entirely assured. However, recently in the United States, including Texas, there have been reports of small earthquakes and other events caused by development of shale oil/gas that are worthy of public attention.

2. **A shale natural gas development pilot area should be established quickly**

Basing calculations on annual production of 100 billion m^3, an investment of CNY 800 billion would be required during the construction and production period (2015–2025 well completion for around 10,000 wells). Based on the current system (relying only on CNPC's south-west branch companies and Sinopec's south-west branch companies to invest), capabilities are extremely limited. This is especially true for the Longwangmiao large gas field discovered by CNPC's south-west branch company, with reserves of over 400 billion m^3, located in central Sichuan. Conventional gas field construction likewise requires a large quantity of investment into exploration and development. Therefore, there is a need for new ideas for shale natural gas development, and it is recommended that a general pilot zone for shale natural gas development be constructed in the Sichuan Basin and surrounding areas. With a controlled and safe environment, preliminary trials can be attempted, restructuring shale natural gas development using a new model, with new mechanisms and new rules. This will promote shale natural gas development and also offer new

points of reference for China's oil and gas reforms so that successful experiences can be replicated and promoted. Important measures include:

(I) **Innovative mining rights management and market access management**

The first issue is mining rights management. Resources that the three major oil companies have not begun to produce, explore or are in the process of developing and exploring should be the subject of tenders. The second issue is market access. Not only should other state-owned enterprises be introduced, but also private enterprises and foreign investment enterprises. In terms of foreign investment enterprise entry, not only should large transnational corporations be introduced, but small and medium-sized oil and gas enterprises with abundant experience and innovative capabilities should also be enabled to enter. These enterprises have played a critical role in the shale oil and gas revolution in the United States. In terms of foreign investment access, so long as technical and environmental standards are met, and national security reviews are passed, foreign investment enterprises could enter in an independent fashion.

(II) **Make use of the economic advantages offered by mixed ownership**

CNPC and Sinopec have taken steps towards establishing new ownership paradigms, for example CNPC has established the Sichuan Changning Natural Gas Development Co., Ltd in Changning and Sinopec has established Chongqing Shale Gas Exploration and Development Co., Ltd in Chongqing, each of which is a mixed ownership company. Mixed ownership offers state assets, private assets and foreign assets the opportunity to enter the shale natural gas development sector. It is also beneficial to central government enterprises, enabling greater use of societal capital and more rapid development. With local state-owned asset participation, there is greater breadth for enabling resettling, road repairs and natural gas nearby utilisation. With private enterprise and foreign enterprise participation, company government can be improved, thus raising the level of mining efficiency and service. The next step in shale natural gas development mixed ownership reforms requires efforts in terms of corporate governance, strategic co-ordination and the division of labour and partnerships. In order to improve overall activity, it is not enough simply to overlook mechanism transitions to obtain monetary funds. Likewise, the sole focus should not be on bringing in mixed capital, while essentially refusing to allow outside investors to participate in the corresponding management and operational activity.

(III) Exploring effective shale natural gas regulatory models

It is possible to aim to create a pilot zone for policy systems, management standards, oversight rules and regulatory systems relating to shale natural gas exploration, development, storage, and shipping usage within 2–3 years, including information search and sharing platforms. In terms of environmental oversight, central government standards and policies should be established, with local organisations implementing the environmental regulation. The Ministry of Environment is in the process of researching the establishing of shale natural gas development assessment guidelines. These could be tested in this region, with a preliminary launch and implementation of national norms and standards for such aspects as land usage, vegetation restoration, water resource usage, waste water processing, waste

processing, gas emissions, drilling and well completion. In terms of geological material and information sharing, comparison can be made to the United States' administrative legislation approach to the mandatory collection of shale natural gas geological information, thus establishing a geological engineering information management platform with government-controlled resource information. This will achieve information sharing and promote highly efficient development.

Case 2: Chinese Coalbed Methane Production Volume Forecasts

1. Coalbed methane resources

China has abundant coalbed methane resources, which are typically found in multiple coal basins with multiple coal formations and a wide variety of coal types and coalbed methane deposits. Due to the wide variance in the coal-containing basins in China, subsequent construction and restructuring is intense, resulting in a very complex geological environment for coalbed methane resource deposits in China. Coalbed methane resources are primarily distributed in China's north, north-west, south and north-east regions.

China's proved coalbed methane volumes have grown rapidly. In 2012, national proved coalbed methane reserves grew by 134.4 billion m^3, rising year on year by 32.2%. In 2013, national coalbed methane exploration proved new reserves of 23.577 billion m^3. Proved coalbed methane geological reserves now sit at 575.4 billion m^3, with technically recoverable reserves of 285 billion m^3.

The regions that are abundant in coalbed methane are primarily the Qinshui Basin, the Erdos Basin and the Yunnan Guizhou North Depression. In 2012, new proved geological reserves in the Qinshui Basin were 90.242 billion m^3, with the Erdos Basin newly adding proved coalbed methane reserves of 37.11 billion m^3. The Yunnan Guizhou North Depression in 2012 deployed and implemented 11 assessment wells and three well groups, for a total of 18 wells as production capacity experiments. A total of 17 wells have achieved coalbed methane flows, with single well daily production capacity of 500 m^3.

2. **Coalbed methane reserve forecasts**

(I) **Characteristics of coalbed methane-intensive regions in China**

Coal layer gas factors are generally assessed by usage of gas per ton of coal, absorption saturation, concentration of methane and resource abundance. Gas per ton of coal is gas volume, and is a determining factor in coalbed methane resource quantity. Resource abundance is a general reflection on the amount of gas in the coal and the thickness of the coal layer. Absorption saturation is a gas volume factor related to gas mining feasibility. Methane concentration is a major standard in assessing gas quality.

Regarding regional distribution, average coalbed methane gas volume is highest in southern China, followed by the north-east regions and northern China. The north-west region has the lowest amounts. The methane concentrations of each region are similar, and from highest to lowest are: southern China, northern China, the north-west and the north-east. The average resource abundance in each region varies significantly, with the north-west region markedly higher than other regions, followed by the north-east. Northern China is slightly lower than eastern China, with southern China the lowest. Average absorption saturation is led by north-east China, followed by southern China, northern China and the north-west.

From the perspective of coalbed methane layers, the Permo-Carboniferous, upper Permian series and Lower and Middle Jurassic series are representative of good gas contents. Other eras are relatively poor. From a general analysis, the northern region's major basins have good gas contents in the Permo-Carboniferous, and this is the primary gas layer for important basins.

From the perspective of coal rank distribution, the resource volume of high coal rank lean anthracite III (Ro = 1.9–2.5%) is 7.8 trillion m^3, accounting for 21.1%. The resource volume of medium-rank gas/lean coal (Ro = 0.7–1.9%) is 14.3 trillion m^3, accounting for 38.9%. Low coal rank brown—long flame coal (Ro < 0.7%) resource volume is 14.7 trillion m^3, and accounts for 40%.

(II) **Key target areas for coalbed methane exploration**

Based on summarising coalbed methane abundance patterns and gas characteristics, there are 18 target regions that have been selected nationwide as preferential locations, including Pucheng, Ji County—Hancheng, Shenmu, Hengshanbao, Ningwunan, Panguan, Gemudi, Pingle, Changji, Dajing, Wuermu, Wushenqi, Sanjiang-Mulinghe, Huolinhe and Yimin. Of these, six are Class I target regions, five are Class II target regions and seven are Class III target regions, giving a total area of 54,700 km^2. The target Class I coalbed methane resource volume for the region is 3.63 trillion m^3, primarily distributed in the Qinshui, Erdos, Zhunge'er and Ningwu basins. Coalbed methane geological conditions are good, representing coalbed methane near- and mid-term reserve and production volume realisation target areas. Class II and Class III areas have a total of 2.87 trillion m^3 of resource volume, and are distributed across Tuha, Santanghu, Sanlian Basin, eastern Yunnan and Western Guizhou, and the Pingle Basin, and have potential for mid- to long-term reserve growth and subsequent replacement blocks. As part of this, the Qinshui Basin south portion Jincheng region, Erdos Basin south-east region Ji County—Hancheng region, and the Zhunge'er Basin south-east Changji-Dajing are promising coalbed methane regions with reserves of 100 billion m^3 (Table 7.6).

This research is based on national coalbed methane reserve historical data, in conjunction with coalbed methane geological characteristics and resource distribution traits, making use of the Weng model method and Gompertz method to forecast proved coalbed methane reserves and production volume. It is expected that by the end of the 12th Five-Year Plan, 10 advantageous target regions will see newly added proved geological reserves of 700 billion m^3. During the period of the 13th Five-Year Plan, the scope of explorations will be further expanded to south

China, and the north-east and north-west regions, with newly added proved reserves of 350 billion m^3 in the Qinshui Basin, Erdos Basin, Zhunge'er Basin, Ningwu Basin and Erlian Basin. From 2013 to 2030, annual proved reserves will be at approximately 100 billion m^3. After 2030, as coalbed methane exploration and development technology continues to progress, coalbed methane at 2000–4000 m will also be possible to confirm and mine, with large increases in proved reserves expected in coalbed methane (Table 7.7).

(III) **Coalbed methane production volume forecasts**

As coalbed methane exploration and development technology matures and mining costs drop, coalbed methane development will become larger-scale and will develop toward greater industrialisation, gradually forming 10–15 coalbed methane production bases. Using the Weng

Table 7.6 National coalbed methane results of selection of favourable target regions

Class	Basin	Region	Major depth (m)	Major coal thickness (m/layer)	Gas contents (m^3/t)	Area (km^2)	Resource volume (10^8 m^3)
I	Qinshui	Qinshui north	200–1200	8–17/2	10–32	5150	11,864
	Erdos	Erdos east	300–1500	7–22/2–3	9–20	7430	11,805
	Zhunge'er	Changji-Fukang	300–1000	25–32/3	5–15	7010	7460
	Qinshui	Yangquan-Heshun	150–1300	9–12/2–3	8–35	1200	2500
	Ningwu	Ningwu south	800–1500	11–14/1	11–21	534	1665
	Sanlian	Huolinhe	300–900	7–34/5	5–8	380	1025

Data source Ministry of Land and Resources and other material analysis and preparation

Table 7.7 National proved coalbed methane reserve forecast (100 million m^3)

	2015	2020	2025	2030
New proved geological reserves	7000	3500	4800	7200
Accumulated proved reserves	8619	12,119	16,919	24,119

Data source Ministry of Land and Resources and other material analysis and preparation

Table 7.8 National coalbed methane production growth forecasts (100 million m^3)

	2015	2020	2030
Production volume	45	100–300	400

Data source Ministry of Land and Resources and other material analysis and preparation

model and the Gompertz method to analyse China's future coalbed methane production volume growth, it is expected that 2015 coalbed methane production volume will be 4.6 billion m^3, rising to 10–30 billion m^3 by 2020. From 2021 to 2030, 20–30 coalbed methane production bases will be established, with 40 billion m^3 of production volume by 2030 (Table 7.8).

Case 3: Forecast of China's Coal Methane Production Potential

1. **Coal methane development**

With the implementation of a sustainable development strategy and the strengthening of environmental and other policies, domestic natural gas consumption markets are continually expanding, and there are more channels and more methods of expanding the natural gas resource supply, and optimising the gas source structure has

become an important strategic part of optimising the national energy structure. Converting coal region resources into natural gas can to a certain degree supplement China's shortage of conventional natural gas supply.

China has approved its first two coal methane projects: Keqi, with coal methane production capacity of 4 billion m^3/year, divided into three series, with each series having production capacity of 1.33 billion m^3; and Qinghua, with 5.5 billion m^3/year, divided into four stages of construction. The first stage has production capacity of 1.375 billion m^3/year. In 2013 the two coal methane projects in total had gas volume of 31 million m^3.

Approved and in-progress coal methane projects include the Fuxin and Huineng coal methane projects, with total production capacity in the region of 5.6 billion m^3/year, and expectations to begin gas supply in 2015. In-progress and operating coal methane projects are listed in Table 7.9.

Table 7.9 Operating and in-progress coal methane projects

Type	Project name	Date of operation	Scale	Total production capacity (100 million m^3)
In operation	Datang Keqi Coal Methane Project	2013	13.3	40
	Xinjiang Qinghua Coal Methane Project	2013	13.75	55
Under construction	Huineng Coal Methane Project	2015	–	16
	Liaoning Datang Fuxin Coal Methane Project	2015	–	40
	Inner Mongolia Shenhua Erdos Coal Methane Project	2015	–	20
Total		27.05	171	

Data source Public materials analysis and organisation

2. **Progress in coal methane project planning and construction**

In recent years, China has seen an acceleration in approvals of coal methane projects, primarily concentrated in Inner Mongolia and Xinjiang. As of the end of June 2014, the NDRC had approved 18 coal methane projects, with a combined production capacity of 84.2 billion m^3/year. Provincial planning and filing showed 37 projects with production capacity scale of 129.6 billion m^3/year. Operating projects, or those with road permits and

approvals, as well as those under construction, had a total production capacity of 230.9 billion m^3/year. Currently, only a small number of these projects have begun construction, and considering that current natural gas prices are a problem, as well as the current policy environment, not all projects will necessarily be finished. It is also not a given that the completed projects will be able to satisfy the load of production, but production capacity will still offer basic assurances for China's natural gas supply security (Table 7.10).

Table 7.10 China's current coal methane projects with road permits

No.	Province	Project name	Production capacity (100 million m^3/year)
1	Inner Mongolia	Xinmeng Energy Investment Co., Ltd Coal Methane Project	40
2	Inner Mongolia	Erdoa Coal Methane Industrial Park and 12 billion m^3 Coal Methane Project	120
3	Inner Mongolia	State Grid Inner Mongolia Guxinganmeng Coal Methane Project	40
4	Inner Mongolia	Huaneng Yimin Coal Methane Project	40
5	Inner Mongolia	Inner Mongolia China Star Energy Limited Coal Methane Project	40
6	Xinjiang	Huaneng Xinjiang Coal Methane Project	40
7	Xinjiang	China Power Investment Corporation Xinjiang Huocheng Coal Methane Project (three phases)	60
8	Xinjiang	Xinjiang Fuyun Guanghui Coal Methane Project	40
9	Xinjiang	China Coal Energy Xinjiang Coal Methane Project	40
10	Xinjiang	State Grid Pingmei Coal Methane Project	40
11	Xinjiang	Xinjiang Longyu Coal Methane Project	40
12	Xinjiang	Huadian Xinjiang Coal Methane Project (Xiheishan coal methane)	40
13	Xinjiang	Sinopec Great Wall Energy Coal Methane Project	80
14	Xinjiang	Xinjiang Yili Xintian Coal Methane Project (Xinwen phase one)	20
15	Xinjiang	Suxin Energy Xinjiang Coal Methane Project	40
16	Xinjiang	CPCEC Yinan Coal Methane Project (three phases)	60
17	Shanxi	CNOOC Shanxi Datong Coal Methane Project	40
18	Anhui	Anhui Huai'nan Coal Methane Project	22
Total			842

Data source Public materials analysis and organisation

3. **Coal methane production potential analysis**

According to the construction progress of coal methane projects, and in conjunction with project construction cycles and production capacity plans, an arrangement has been made of all coal methane project production capacity.

Projects in 2020 with full road permits will have total production capacity of 84.2 billion m^3/year, and, considering they are based on three phases, with basic commencement post-2018, by 2020 it will be possible to achieve two-phase scaled capacity of 60 billion m^3/year, reaching 95 billion m^3/year in 2025 and 102 billion m^3/year in 2030.

However, what is worthy of note is that, moving forward, coal methane development still faces many uncertainties. One factor is the currently relatively low international oil and natural gas prices, which have greatly inhibited the profit space for coal methane, especially given significant difficulties with project cost difficulties. A second factor is water consumption and environmental pollution, which causes significant controversy. Therefore, while coal methane production capacity will be built in the future, specific production capacity is nonetheless still highly uncertain.

Case 4: Rural Organic Waste Natural Gas Effects and Policies

It has been estimated that nearly 76% of organic waste in China comes from agriculture, with crop stalks and livestock manure being the two major sources. China's villages have 600 million tons of crop stalks dispersed among fields, and farmers generally deal with these by burning them, which is inefficient and not environmentally friendly. As livestock intensity becomes increasingly concentrated, concentrated processing of manure is a problem that not only hinders enterprise development but also endangers the rural environment. Resolving rural organic waste transitioning has always been one of the major issues facing agriculture. Beijing Deqingyuan has explored a series of technical models that transform rural organic waste into clean energy, and their experience is worthy of attention and wider promotion.

1. **Improving rural environments and changing rural energy usage structures**

Beijing Deqingyuan is an egg producer in Beijing that has won industry fame and market share by giving its eggs an "identity document" (using edible ink to print the brand name, production date and a fraud prevention code on each egg's shell). However, after they expanded their production scale, they discovered there were also attendant environmental problems, with 2 million laying hens producing the same amount of manure each day as 300,000 people. If this waste is not processed, it causes contamination to the groundwater and soil. To this end, Beijing Deqingyuan specially established a clean energy company to resolve the problem of transforming and using chicken manure. After several years of study, a breakthrough took place in the three technical areas of sand removal, high-concentration purity extraction and biological desulphurisation, achieving power generation from chicken manure methane in 2007. Following on from this, they next developed a mixed-source material production of methane gas from crop stalks and chicken manure, making natural gas after purification compression. The break throughs in these technologies have yielded feasible solutions to bring about rural organic waste transformation and to allow farming using renewable energy.

The first point to note is the improvements to organic waste conversion and to the rural environment. Deqingyuan can process 100,000 tons of manure per year, equivalent to reducing CO_2 emissions by 71,000 tons, and 35 tons of nitrites. A total of 25,000 tons of crop stalks are processed annually, equivalent to a CO_2 reduction of 40,800 tons and 25 tons of nitrites. Not only has the manure generated by chickens been used effectively, the approach has also provided a solution for the problem of stalk burning that farmers have always faced.

Also notable is the formation of a full industry chain. Deqingyuan's methane power generation plant achieved connectivity in 2009 with North China Power Grid, and now produces upward of 14 MW of power annually, along with 160,000 tons of organic fertiliser. In 2013, biogas sales revenue was CNY 600,000. In 2014, this reached CNY 2.7 million. The remaining liquid methane biogas residue is processed into solid and liquid organic fertiliser, with annual production of 160,000 tons of organic fertiliser sold to local farms in Yanqing and prompting the development of 10,000 Chinese mu of organic apples and 120,000 acres of environmentally friendly corn crops. Through ordered procurement, 60% of Yanqing's corn is purchased, which is then used to process into chicken feed to be fed to the 3 million layer hens at the farm. Thus a complete industry chain has been created: ecological cultivation, food processing, clean energy, equipment manufacturing, organic fertilisers, agriculture and organic planting.

Finally, there is the improvement to rural production and living conditions that has occurred. Deqingyuan transports the natural gas generated from purification and compression by tanker truck to various village facilities for tank storage. It is then sent via pipeline to countless homes. In 2013, biogas was supplied to 2000 homes, which had risen to 5000 homes by July 2014, with a further rise to 7000 homes expected. Deqingyuan and Yanqing County are co-operating in new biomass energy projects, with plans within the next three years to enable the 10,100 homes in 39 villages within Yanqing County to use biogas. During an on-site survey of the area, we found that farmers used liquid methane and remnant methane to carry out organic planting, selling the organic apples they produced to consumers, while the environmentally friendly corn was sold to Deqingyuan to raise chickens. The farmers no longer cooked, heated their homes and bathed in hot water that was reliant on crop stalks and coal, ending their high expenditure on coal. This has resulted in an enormous change to their living and working conditions.

2. **Potential for larger-scale rural organic waste energy conversion**
The first factor behind this scheme's success is the mature technology:

- Cow manure and chicken manure contain approximately 10–15% sand. In the past, due to technical restrictions on sand removal, pure chicken manure could never be used as a source for fermentation of methane gas. After Deqingyuan invented organic separation technology, achieving separation of chicken manure sand and organic materials, it was possible to extract over 80% of non-soluble solids from chicken manure.
- The methane gas produced after fermentation of livestock manure contains 2000–5000 ppm of hydrogen sulphide. Traditional methods were never able to

solve the problem of desulphurisation. However, Deqingyuan invented bio-desulphurisation of methane gas, using sulphur-consuming bacteria to turn sulphurised hydrocarbon into elemental sulphur. It is a low-cost approach and does not generate secondary pollution.

- Conventional methane concentrations are generally 2–5%. After Deqingyuan invented high-concentration methane gas fermentation technology, chicken manure fermentation and crop stalk fermentation concentration rose to 10–15%, which reduced water consumption and fermentation system area by over 50%.

- Deqingyuan invented a mixed-source material fermentation technology that used crop stalks as a source of carbon, liquid methane as a source of nitrogen and fermenting microbes to achieve joint fermentation. This resolved various problems during fermentation, including saturation of stalks, floating of stalks, rapid hydrolysis of zymophytes by the stalks, as well as the need for large swathes of land. The company's membrane purification equipment uses only one-tenth of the area of similar projects, and has reduced energy consumption by over 50%.

The second factor behind the scheme's success is a more reasonable economic approach:

- Investment costs are low. Deqingyuan's newly constructed daily production 8000 m^2 biogas project can satisfy the residential gas use requirements of 10,000 households with an investment cost of approximately CNY 30 million, CNY 16 million lower than industry competitors.

- Gas production costs are lower. Deqingyuan, because it can collect and use the various source materials it requires from the area 20 km around the methane gas station, has an average gas manufacture cost of 2 CNY/m^3, which is markedly lower than the gas price currently provided in the city of Beijing (Beijing's current natural gas station pricing is 3.28 CNY/m^3).

- There has been an increase in economic value: 1 m^3 of methane gas can conventionally be generated into power at a value of CNY 1.19, whereas Deqingyuan's membrane purification method enables conversion of 1 m^3 of methane into CNY 2.4 worth of natural gas, thus achieving added value of 1.21 CNY/m^3 of methane gas.

The third success factor behind the scheme's success is the greater economic utility to farmers. Deqingyuan sells the natural gas to farmers at a price of 2.5 CNY/m^3, while also accepting the crop stalks provided by farmers, converting these source materials into natural gas and costing the farmers that provide crop stalks very little actual expense. Taking a household of 3–4 people as an example, in the past, costs would be CNY 100–120 per month using tank storage liquefied gas, whereas this is now only CNY 30–40 per month using biogas.

3. **Policy recommendations**

The first recommendation is that support for rural and small town biogas and pipeline network construction should be increased. As the pace of urbanisation and rising living standards quickens, China's natural gas consumption is entering a period of rapid growth, and natural gas supply and demand conflicts are being highlighted. If all of China's 600 million tons of crop stalks were fully utilised, annual production of methane gas could reach 120 billion m^3, providing 71.6% of 2013 total natural gas consumption. Livestock manure can also be converted into production capacity of 122 million tons of standard coal. These organic wastes are significantly concentrated, and crop stalks are concentrated in grain-producing regions while manure is concentrated in livestock-cultivating areas. Forestry leftover materials are concentrated in forest regions, while industrial waste is concentrated in agricultural product processing plans, with the characters of the major portions being concentrated, thus facilitating promotion of rural and small and medium-sized town energy construction. It is recommended that a perspective be borrowed from energy supply and demand and from rural development, starting analysis of the issue at the level of overall national strategy and deployment in order to provide the best policy support for rural organic waste resource energy conversion.

The second recommendation is to resolve the problems faced by enterprises in energy conversion innovation. During our Deqingyuan survey, enterprises reported that some current policies do not support enterprises in innovating. For instance, there are grid connection issues. After ensuring sufficient rural gas usage, the remaining biogas has no channel through which it can be sold, and it is recommended that biogas be permitted to merge with urban natural gas pipeline networks to allow access to natural gas companies involved with urban sales.

The third recommendation relates to pricing problems. Currently, Deqingyuan's natural gas sales price is 2.5 CNY/m^3, while Beijing's industrial user gas price is 3.65 CNY/m^3 and transportation and shipping gas prices are even higher. It is recommended that biogas sales prices be freed up to open market competition.

The fourth recommendation relates to tax waivers. Currently, China does not provide price subsidies for biogas, and also collects 17% VAT. Considering biogas's low carbon impact on the environment, and given that a certain amount of policy support is given to development for solar power, wind power, and biofuel energy, it is recommended that biogas be exempt from VAT.

Case 5: Upstream Policy for Unconventional Oil and Natural Gas

Unconventional oil and natural gas development and production is a development trend in the majority of regions worldwide. Currently, the United States and Canada are leaders, followed by Argentina and Australia. Russia, the United Kingdom and other nations have also attempted to ensure that their shale natural gas/oil resources have potential. This paper summarises financial policies in the United States (onshore), the United Kingdom, Canada, Russia, Argentina and Malaysia, focusing on unconventional resource development incentive measures and making recommendations for China.

1. Argentina[1]

With the exception of China, Argentina has the world's largest shale natural gas resources while at the same time also being one of the nations with the most abundant shale oil resources (after Russia, the United States and China) (Nulle 2014: 60).

Currently, Argentina is implementing a special licensing rights regime. Since 1991, all licences have been managed under a special licensing fee/tax collection system, including special rights usage fees, income

---

[1]Argentina summary materials are sourced from HIS PEPS, WoodMac and Apache Enterprise website.

taxes, provincial sales taxes, other signing deposits/lease amounts and oil/gas export tariffs. This system was implemented during the period of economic crisis in 2002 and has continued on to the present day.

In 2014, Argentina passed the new Oil and Natural Gas Law, prescribing special licensing usage fees and permit issuance systems controlled centrally, and also giving management rights to various provincial regulatory institutions. Prior to the launch of this approach, oil and natural gas licensing and operational jurisdiction belonged to the provincial governments. Thus provincial governments could benefit by issuing their own oil and gas rights licences.

ENARSA plays a purely commercial role, owning all unlicensed near-sea exploration on federal land within 12 nautical miles of the shore. All activities in these regions must be conducted in co-operation with ENARSA.

To describe Argentina's upstream industries as "uncertain" would be very apt. The RSC system was changed to the current licensing rights usage fee/tax regime, and the existing laws thus entirely covers the "2006 Short Law", giving oil and gas jurisdiction to provincial governments (with the exception of federal marine regions).

In order to satisfy national demand, Argentina has consistently increased use of the oil and gas industries. In recent years, energy demand has continually grown, and the government has admitted that it must become more involved, and thus it has implemented multiple investment incentive measures in the hope that the newly passed Oil and Natural Gas Law (October 2014) will help attract investment in unconventional industries. Some particular features are:

- The bidding procedure and licensing rights usage fee system is controlled by the central government, and provincial governments are only responsible for handling oversight.
- Provincial oil and gas company preferential equity holdings in the exploration stage have been cancelled.
- Conventional and unconventional assets are differentiated, with unconventional asset development terms extended to 35 years (as opposed to 25 years) and exploration terms limited to 13 years.
- The minimum investment required to secure international prices for 20% of the crude produced was reduced to US$250 million from US$1 billion (for 3 years). This provides a window of opportunity for small exploration and production companies. However, analysis by Woodmac has shown that 20% of export volume over 3 years has essentially no effect on improving project IRR. Removing the 3-year restriction, on the other hand, would increase IRR dramatically.
- The Gas Plus project, which raises the price of natural gas produced using unconventional gas deposits.
- Establishment of an unconventional production permit, allowing 5 years of pilot production to establish commercial viability, and the splitting of adjacent geological concessions into units.

- Once the pilot project stage is complete, unconventional project licensing usage fees are reduced by 25%.
- Unconventional industry assessments/appraisals periods are extended (1–5 years).
- Provide 20% of export supply.

However, there are still many problems that have yet to be addressed, and resolving these could achieve a lot of what is hoped to be realised:

- Pre-emptive measures must be taken to address problems of a lack of trained labour and labour allocation. One example of such measures is the province of Neuquén, which improved shale natural gas technical capabilities by establishing an Unconventional Oil and Gas Field Technology Centre.
- Lack of an environmental regime: Because existing oil and gas fields are far from communities, there is no problem at present, but as the industry develops, problems could potentially arise.

For many reasons, it is difficult to say whether Argentina has actually succeeded in increasing unconventional sector activity, but it has certainly encouraged unconventional resource development recently, and appears to be on the right course in its efforts. However, the government must avoid the trap of intervening in current success, or it will damage the confidence of the international oil companies.

Licensing Authority

- ENARSA areas—ENARSA
- Provincial areas—Provincial authorities

Regulatory Authority

- Federal level: Energy Secretariat—Has overall responsibility for the hydrocarbons sector. Responsible for policy formulation and regulation

(continued)

- Provincial level: Provincial authority—In general, regulatory duties fall on the same person responsible for licensing

NOC

- ENARSA—Created in 2004 by Law No. 25.943. The company is owned by the government (53%) and the provincial governments (12%), with the remaining 35% listed in the Buenos Aires Stock Exchange
- Most provinces have their own Provincial Companies (e.g. G&P Neuquén)

2. Canada[2]

Canada (one of the countries with the most abundant shale resources in the world) implements a tax/licensing rights system through provincial governments. Multiple factors determine basic system adjustments, such as contract terms, production circumstances, oil prices and so on. Even though there is no obligation to supply domestically, operators must still obtain government approval to export resources.

Canada's various provinces play a greater role by being in charge of implementing and overseeing operators. The majority of provincial governments have their own regulatory institution to carry out licensing and oversight of oil and gas operations: for example, in Saskatchewan the Oil and Natural Gas Office of its Department of Energy and Resources, and in British Columbia the Ministry of Energy and Mines and the Oil and Natural Gas Division.

Generally, with the exception of Newfoundland, no province has a NOC. In 2007, the Energy Act prescribed that Canada should have an energy company to hold and manage oil and natural gas interests. Nalcor Energy seems to have played this role, and this company has also

---

[2]In addition to HIS PEPS and WoodMac, materials relating to Canada are sourced from Manitoba and Alberta government websites as well as oil and natural gas PWC tax collection reports.

co-operated in participating with Newfoundland's three marine development projects.

Essentially all governments have used licensing to capture economic lease funds —when oil prices rise, the government revenues increase, and when prices are low, the limits are adjusted. In order to avoid continual changes, a price connection mechanism was established, with the majority of governments already linking licensing rights usage fees with oil prices, production and contract years (Fig. 7.6).

According to some provincial finance systems, the licensing fee rate taxes prior to recovery of investment costs are different, allowing contractors to recover costs with relatively low risk. There are also some provinces that implement multi-tiered fee

rates based on rates of return, thus forming a tiered mechanism.

Canada's Provincial Institutions

• British Columbia Ministry of Energy, Mines and Petroleum Resources—The Oil and Gas Commission
• Saskatchewan Ministry of Energy and Resources—Petroleum and Natural Gas Division
• Minister of Energy—The Alberta Energy Regulator
• Department of Aboriginal Affairs and Northern Development Canada (AANDC)—Northern Oil and Gas Branch of the AANDC
• Canada Nova Scotia Offshore Petroleum Board
• Canada—Newfoundland and Labrador Offshore Petroleum Board
• The Ministry of Energy in Alberta
• Department of Energy, Mines and Resources of Yukon
• Ministry of Natural Resources in Québec

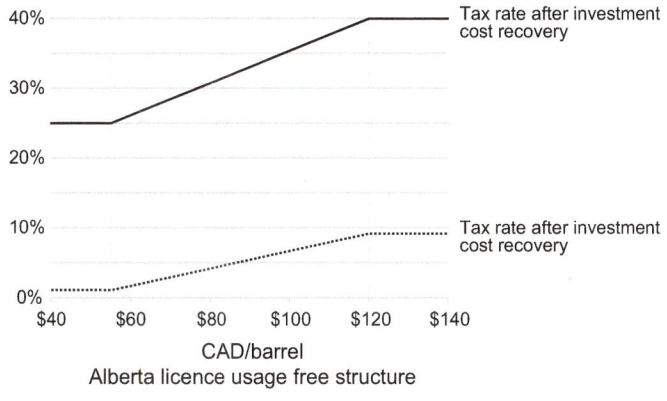

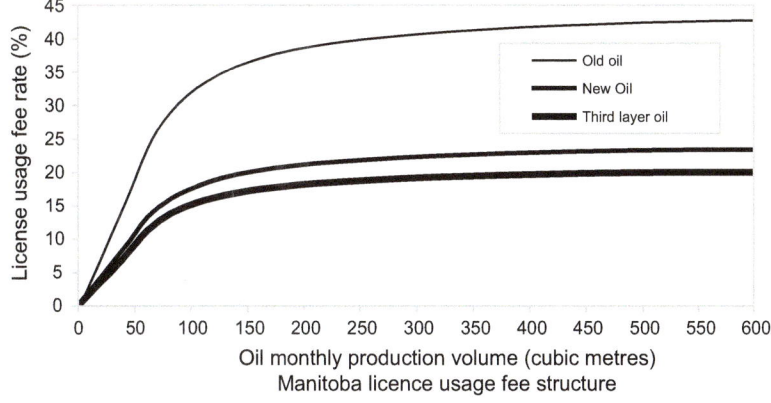

Fig. 7.6 Licensing rights usage fee structure

Canada's three tiers of taxes:

- **First tier**: Federal income tax collected for "taxable income" from oil and natural gas projects.
- **Second tier**: Provincial income tax (differentiated from federal taxes by the different tax-exempt amounts for each province) collected for "taxable income" from oil and natural gas projects.
- **Third tier**: A provincial and national resource tax collected for ownership rights of Canada's resource assets. In order to allocate some tax amounts to the provincial governments, the federal tax has a 10% exemption. These exemptions are only applicable to income generated within Canada.

Like many countries, Canada has taken a very active role in terms of policy, and is very willing to use tax exemptions/reductions/offsets to provide correct incentives. Currently, overall measures taken are as follows:

- Early-stage 10% Investment Tax Credits (ITCs) for oil and natural gas projects.
- Provide other ITCs for scientific research and experimental development (SRED) activities to incentivise industry innovation.
- Unused tax offsets can be carried forward over 20 taxation periods.
- Unconventional projects enjoy higher capital cost subsidy rates, directly affecting the taxes that contractors pay.
- Implemented programmes include encouragements for tight natural gas and shale natural gas development net profit licensing rights, low production capabilities usage fee reductions, oil sand usage fee programmes, marginal oil field usage fee programmes and so on. These are aimed at satisfying

special needs related to resources and/or geographical restrictions, and many other programmes are available.

In order to encourage innovation, the Canadian government connects oil prices to production and usage fee structures, along with various tax exemptions, which play an important role in upstream industry development. For example, the development of steam-assisted gravity drainage technology was highly beneficial to oil sand projects.

3. **Malaysia**

Malaysia has a series of highly optimised and stable systems that implement various financial terms based on production volume share contract (PSC) terms—the most recent revenue-over-cost PSC contract conducting licensed business (Fig. 7.7).

In conjunction with IOCs, Malaysia's Petronas uses an exploration and development subsidiary (Petronas Carigali) as a joint venture enterprise investor (sometimes as operator) to participate in upstream industry activities. In addition, it also uses the subsidiary company MPM to act in a regulatory capacity, overseeing all of the nation's upstream industry activities. According to PSC requirements, IOCs and operators (including Petronas Carigali) pay usage fees and taxes to MPM and the local government. In the majority of contracts, Petronas Carigali holds incidental interest rights to all exploration regions, generally between 15 and 25% (negotiable).

Recently, Malaysia has introduced a Risk Service Contract (RSC) to increase activity both in mature fields and in stranded fields.

Recently, Malaysia has begun implementing risk service contracts (RSCs) to strengthen the development of oil and gas fields and latent oil and gas fields. This includes terms that not only protect

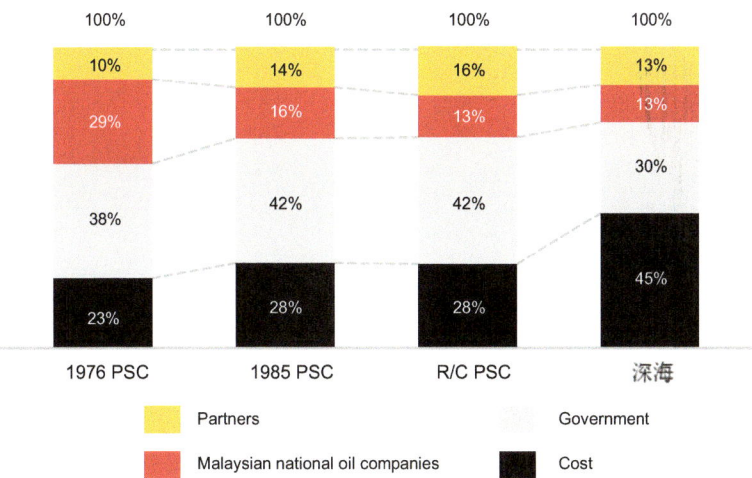

Fig. 7.7 Malaysia production share contracts. *Note* In 1976 and 1985, R/C production share contracts were based on 40 mmboe reserve volume. Moreover, deep-sea production share contracts assume oil volume of over 1 bnboe

contractors from downside risk but also limit the upside. (It should be noted that Petronas is not permitted to sign this type of contract. The contract's main purpose is to encourage and cultivate local Malaysian enterprises to participate in upstream oil and natural gas projects).

The potential for Malaysia in unconventional resources has not yet been recognised, and the country is in the process of learning from the experience of other nations (Australia and Canada). In fact, recent industry reports state that Petronas could be able to partner with US company Hess to assess and learn how to develop China's unconventional resources/shale oil or shale natural gas.

As cost structures change, or when new cost structures arise (as was the case with marine oil), Malaysia's government adjusts Petronas's and its own proportions of ownership to achieve a balance. At different periods, the government has used a variety of mechanisms, including reductions to taxes and usage fees, acceleration of depreciation, reduction of export tariffs

and so on, to attract more investment. In 1985s revised version, there were many adjustments that improved the rate of cost recovery, adding to the profits of contractors while encouraging exploration and development of small gas and oil fields found through earthquake research. At the same time, Petronas's participation became mandatory, allowing it to play a larger role. In 1991, the government launched marine oil incentive measures and subsequently signed revisions of contracts with more attractive terms for two oil fields with Mobil Oil, exhibiting the pragmatic style of the government and regulatory institutions. The increase in contractor profits and the proportion of cost recovery, as well as reductions in taxes and export tariffs, have raised revenue proportions for these contracts, prompting development of marine oil.

In terms of adjustment to financial terms in the macro environment, Malaysia has always been very pragmatic and flexible. When appropriate, it will take suitable measures to attract more investment and turn standard PSC development into R/C PSC, increasing average purchase promotions of contracts and enabling one to see Malaysia's time arrangements during a transitionary period. The pragmatism of Malaysia is also seen at appropriate times when the country is expanding state projects and encouraging commercial structures, and the country does not use a general structure (for example, an enhanced oil recovery project increased production of mine oil and used relatively relaxed rules).

4. Russia[3]

Russia currently has many types of contract. They are briefly summarised below:

- **Government reserve**: A financial system based on export taxes and mineral extraction taxes—usage fees (also called "licensing rights").
- **Government control**: Terms of 5–7–25 years (according to "Underground Law") for underground exploration and/or exploration or production licences that are distributed through auction or tender.
- **Product share agreements (PSAs)**: Even though such agreements do still exist in Russia, they should no longer be part of any new project contract templates, now that the PSA law has been revised.
- **Strategic Industry Law and "Underground Law" amendments regarding "strategic oil and gas fields"**: Russia's institution in charge of underground management is Rosnedra, which comes under the country's Ministry of Natural Resources. Rosnedra is responsible for

managing all issues related to the issuing and revocation of permit, as well as licensing.
- **NOCs**: NOCs (Gazprom, Rosneft and Zarubezhneft) have special status. Gazprom monopolises pipeline exports and must hold at least 50% of shares in offshore projects. In fact, infrastructure is monopolised by Gazprom and Rosneft.

Whether or not a project is approved is determined by project type—foreign company direct licensing and foreign company shares in Russian companies, as well as joint ventures between foreign and Russian companies, are each restricted by a specific category of laws.

In order to encourage the commercialisation of unconventional resources, Russia has implemented several financial and regulatory incentive measures to attract investment in the sector:

- Starting with a resource consumption deduction of 1%, a 15-year zero mine tax and reduced remote-region export tax, as well as an extension of exploration permit periods from 5 to 7 years, have been implemented.
- Profit-based taxation for some projects and a complete revision of the licensing system for unconventionals is currently under way, as well as the introduction of finance terms to attract investment into Russian unconventional resources. Russia has among the most abundant reserves of unconventional resources of any country, and this approach could be slowed due to recent sanctions placed on Russia by the West.

---

[3]In addition to HIS PEPS and WoodMac, reference was also made to Shell reports on Russia.

- In order to support conventional resource development, VNIGUI, VNIGNI and the Shpilman Research Institute are in the process of studying the legal status of unconventional resource projects in Russia—definitions, resource types, reserve reports, permit issuances, commitments and regulation.

Some traditional problems, such as local authorities presenting complex processes and bureaucracy, continue to cause foreign investment to be hesitant to enter Russia. Based on macro environment changes, Russia's systems are also continually changing. One example is that after a new law was passed in December 2013, private companies obtained the right to export LNG.

5. **United Kingdom**[4]

As a country whose oil sector is relatively mature in its development, the United Kingdom is continuing to implement simple licensing usage fee/tax regimes. Currently, special licensing only exists in Northern Ireland, with all other regions already essentially being purely tax regions. Unconventional resources follow similar systems to conventional resources, but there are also special exemptions aimed at encouraging their development.

Through the newly established Energy Development Unit, the United Kingdom's DECC issues production, exploration, and development licences, which provide operators with well drilling/development permissions and construction permits for exploration and production activities. Well drilling/development and construction permits are issued by local authorities, and require interaction between various local government institutions.

Unconventional activities are relatively new (compared to the United States) and management methods await further improvement. Mineral resources are all state assets, but an industry group, UKOOG, wishes to launch a new method of compensating the communities affected by the shale gas operations, involving fixed fees and a linkage to total revenue. This model has been welcomed by the UK government.

In 2003, all oil and gas field usage fees were discarded, with post-2003 approved oil and gas revenue taxes being annulled. Oil and gas income taxes approved after 2003 were waived, and in 2006, as the oil price rose, surcharges were introduced. In addition, since the price environment was not sufficiently attractive, various types of tax exemption were introduced. All of this shows that the UK government took active and timely measures to intervene. Furthermore, based on different oil and gas field history, the United Kingdom has different tax collection structures, thereby achieving lease fee risk sharing through a positive balance between government and operators.

With regard to unconventional resources, the following incentive measures have already been announced:

- for surrounding supplementary expenditure applicable to scope expansion for shale and unconventional oil and gas resources—an extension of the loss structure carry forward period to 10 accounting periods (originally 6); and
- "pad" subsidy for shale projects (the definition of a "pad" is a well and mining area; a portion of generated revenue is exempted from the surcharge, reducing the tax rate from 62 to 30%) (Fig. 7.8).

---

[4]In addition to HIS PEPS and WoodMac, material sources also include the DECC and UKOOG websites.

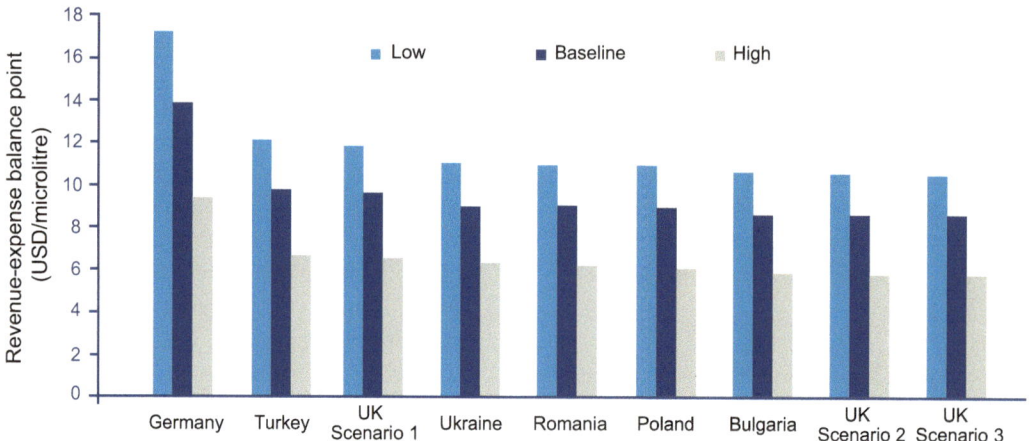

Fig. 7.8 UK shale finance system competitiveness (Woodmac study)

Wood Mackenzie used the Namurian shale region cost hypothesis (as well as three cases of oil well production capacity) to carry out an analysis of the new conditions and calculate product break-even price (the pad has five producing oil wells, and another five will begin production in five years). The scenario's configuration is essentially based on three different tax circumstances:

- the current land permit regime (not including small oil field subsidies);
- a new pad allowance based on 50% of all capital expenditure—this expenditure can be used to offset payable SCT from profits over the following five years, and all subsidies can be carried forward;
- a new pad allowance based on 75% of all capital expenditure—this expenditure can be used to offset payable SCT

from profits over the following year, and all subsidies can be carried forward.

However, negative appraisals still exist, and the government is currently solving these. It is now permitted to transfer cross-pad exemptions from failed pads to successful pads (though this must be done within three years), and it is also permitted to replace operating asset development discounts (in the early stages companies could not implement oil/natural gas sales revenue development periods unless there was reinvestment into the British mainland shelf). Subsidies can be used to pay capacity fees and fair trade expenses.

Financial incentive plans have improved the competitiveness of British shale natural gas in Europe. However, oil well efficiency and local approval process simplification are keys to the success of British shale natural gas.

UK Regulatory Framework

- Minister of Energy and Climate Change: Oil permit licences are issued following a decision by the minister
- DECC: Responsible for the issuance of oil licences, with specifics overseen by the EDU
- UOGO: UOGO is DECC's internal unconventional oil and gas activity direct point of liaison

6. United States—Mainland[5]

Unlike other countries, land owner mineral asset resources ownership rights are the foundation of the US upstream industry investment regime. In the majority of cases, this means that mainland resource ownership rights belong to the individual, and thus there are no licence-issuing institutions and only private leasing. For federal and state land, the government will issue lease agreements based on direct tax collection and usage fee systems.

The United States works on a federal system, and thus the majority of laws and regulations governing the oil and natural gas industry are determined and implemented by the states. The federal government is responsible for federal lands (namely federal water regions such as the Gulf of Mexico (GOM)) and interstate trade, including interstate pipelines. The US government's regulatory systems (state and federal) are relatively optimised and efficient.

Historically, US regulatory systems have created an overall environment and then a series of conditions that support private investment development. In order to support mining industry development, the United States has consistently avoided

nationalisation. The current shale revolution is also a result of this overall environment.

After eliminating the price controls on natural gas and its shipping (deregulation), the United States began to rely on the market to satisfy demand (increased supply). Shipping capability rights and pipeline ownership rights are separated, allowing producers to compete with pipeline capacity.

Interestingly, the US government offers direct support for public research and development expenses. For example, the primary goal of the Eastern Shale Gas Project (1975–1992) was the development of shale natural gas production technology. The United States also offered direct funding support for several industry pilot programmes such as the first multi-break horizontal well in 1986. It also gave its assistance to the 1991 hydraulic fracturing law in conjunction with water level wells company Mitchel Energy. After several decades of stable industry development, US technology and professional assistance is far ahead of the rest of the world's oil and natural gas industries (the United States has the world's largest quantity of well exploration equipment).

Based on the positive experiences of the United States, some recommendations are (based on Nulle 2014):

- flexible land transfer rules and flexible commercialised statement requirements;
- advance provision of reliable pipeline and shipping infrastructure usage rights;
- implement free prices and guarantee prices;

---

[5]In addition to HIS PEPS and WoodMac, reference was also made to Nulle (2014), the Eenergy website news section and Daniel Johnston's work on financial terms.

- import expert labour and equipment capabilities;
- encourage high-cost/risk unconventional development tax regimes;
- provide direct R&D support, and improve geological and technological knowledge and skills;
- enable free return of profits to investors; and
- stability in terms of laws and policies.

The US government's measures have seen broad success, but it is worth noting that this success was obtained only after a long period of time, and was the result of multiple policies (financial incentives, R&D funding support, deregulation) and structural elements related to the US economic and regulatory system (capital availability, mine asset private ownership (land owners also benefit), stability, high degree of professional technology and experienced human capital, existing pipeline capacity tenders and so on).

Effects of Financial Incentives

The federal government implements specific policies for unconventional resources, and these policies were launched in 1980 and expired in 2002. The Crude Oil Windfall Profit Act played an important role, for example in offering subsidies to unconventional resources developed from shale natural gas and in the formulation of the "Intangible Development Cost Expenditures" provisions, which permitted producers to write off a large portion of development costs as expenses. The government's share of revenue derived from primary producers on the Outer Continental Shelf of the Gulf of Mexico is very low. Indeed, because of fee usage refunds and tax offsets, there is an effective 0% royalty rate

7. **Recommendations for China**

1. **Encourage provincial/community management of shale oil/gas development and production**: In the United States, Canada and Argentina, local authorities/governments play a critical role and for a period of time there are direct regulatory institutions. Sometimes, provinces (such as in Argentina) will invest in partnerships and form joint venture enterprises (rather than being an "incidental" partner).

2. **Implement low licensing usage fees/tax rates for unconventional natural gas to provide tax reductions and exemptions**: In some countries (such as Canada), tax rates are linked to actual oil prices, and this allows investors to know what the tax burden is at different oil price levels without having to adjust strategy based on changes in oil price. For example, Russia is currently considering a cancellation of unconventional oil/natural gas mineral extraction tax, and mineral extraction tax is a significant tax burden for Russian companies. Other tax leverages include tax carry forward (such as in the UK).

3. **Extend unconventional oil and gas field small-scale trials, exploration and production terms**: Formulate transition timetables. In the majority of countries, shale oil/gas field licences are extended for 5–10 years.

4. **Deregulate, and introduce competition mechanisms**: Provide reliable pipelines and shipping infrastructure usage rights in advance and cancel price controls, which is believed to be the key to the US's success. In particular, separate shipping infrastructure usage rights from pipeline ownership rights, allowing producers to compete over pipeline capacity.

5. **Support development of dedicated operation companies**: The government

plays a unique role in development R&D and in supporting dedicated operation enterprises. These companies can provide effective assistance to accelerate the development and production of China's shale natural gas/oil. To this end, the government should support the development of Chinese companies or foreign professional technology companies, with particular attention given to the growth and development of small-scale enterprise, which plays a vital role.

Case 6: Mexico Oil and Natural Gas Industry Upstream Opening Policies

In 2013, Mexico revised its national constitution in an attempt to revitalise its oil and natural gas industry. Mexico's government took specific and transparent measures to promote Bid Round Zero. During this period, Pemex submitted an application and obtained 83% of proved oil field/natural gas field mining rights and 21% of unproved oil/gas area exploration rights. This was viewed as an effective and practical method to balance the nation's overall interests and the interests of the country's oil companies. At the same time, it also offered sufficiently attractive opportunities for foreign business and private investors.

It is recommended that China use the Round Zero procedure to enable existing licensed state-owned oil companies to name portions they wish to hold and the portions that can be provided to new investors, thereby prompting investment and in the mid- to short-term accelerating

change in unconventional energy shale natural gas production volume. Using this approach, China could improve non-renewable energy structure usage efficiency for natural gas while at the same time controlling the degree of reliance on foreign imports—which has already grown from 0% in 2005 to 32% in 2013.

1. **Mexico prior to 2013: awaiting change**

Mexico's oil industry history is very long, with oil discovered for the first time back in 1904. Production volume had reached 150,000 barrels/day by 1917. The IOC likewise rapidly joined, helping to raise oil production volume. By 1921, Mexico's oil daily production had exceeded 530,000 barrels, accounting for one-fourth of the world's total production volume. However, fierce disputes between unions and the IOC ultimately led to Mexico nationalising all assets of the IOC in 1938, and through this establishing a NOC. Since then, PetroleosMexicanos (Pemex) has always been the sole rights holder of Mexico's oil and natural gas resources.

For more than 70 years, Pemex had operated its own oil and natural gas business under the ISC. Such a monopolising national oil company has gradually become sluggish, partly because budget and financial management has been tightly controlled by Mexico's Ministry of Finance and Public Credit. As seen in the diagram, Pemex's production volume reached its peak in 2004–2005, with daily oil output of approximately 3.6 million barrels, and daily natural gas production of approximately 3 billion ft^3. Since 2004, production volume continuously dropped, and the prospects appeared bleak. Wood-Mac made forecasts of rapid decline in production volume over the following several years. Mexico was then awaiting change.

In order to relieve production volume declines, Mexico launched an energy policy reform in 2008, allowing Pemex to outsource some mature marginal oil, gas fields and service contracts. The IOC was not particularly enthused. In 2011, Pemex licensed out production volume based on oil and gas fields in two service contracts to French company Schlumberger and UK company Petrofac. In 2012 and 2013, Pemex again held two service contract tenders and still only attracted international service companies (Fig. 7.9).

Deep Water Bay, Mexico has consistently been a major success story for the US oil and natural gas industry. Deep Water Bay, Mexico is on par with West

Africa and Brazil and is known as the world's golden triangle for exploration and production. However, despite having the majority of jurisdiction over the GOM, Mexico is essentially behind all other industry participants, and has conducted no production of deep water oil/natural gas.

Recently, the US shale natural gas/oil revolution has been another jolt to Mexico, since a portion of the shale oil/gas distributed in Texas should, from a geological perspective, extend into Mexico.

The Mexican government and Pemex have realised that radical change is needed and that the time has come to start a new chapter in the history of oil and gas in Mexico.

2. **Energy reforms in 2013 with the rise of a new president**
On December 1, 2012, Enrique Peña Nieto became president of Mexico. His party PRI beat the PAN to win the presidency.

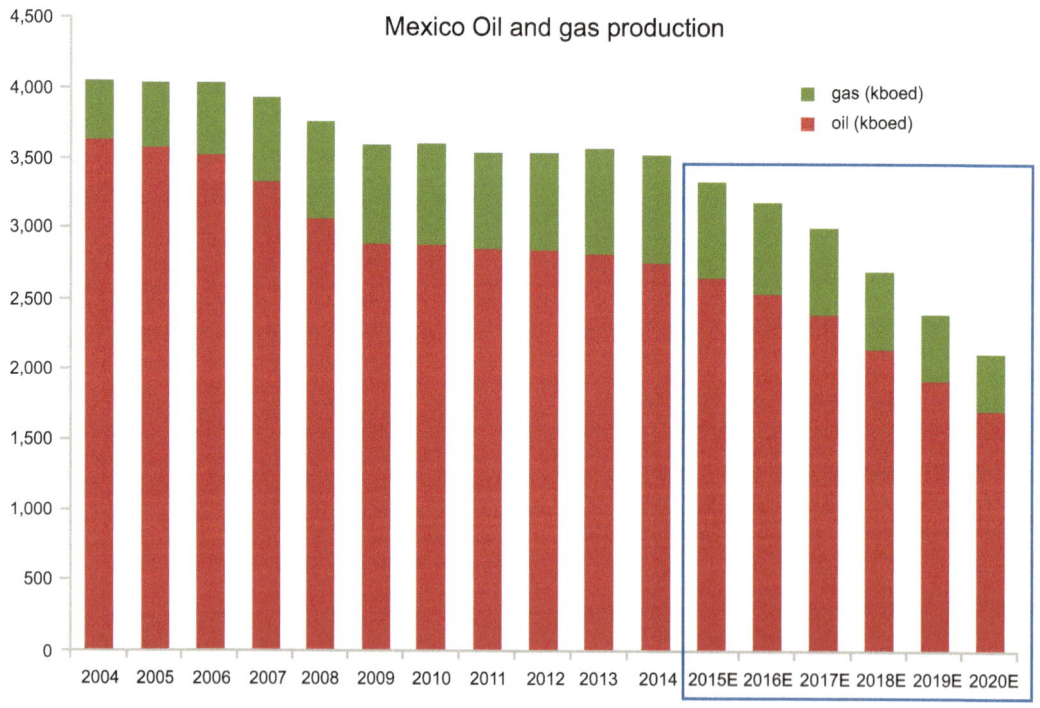

Fig. 7.9 Mexico oil and gas production

The PAN had been the ruling party from 2000 to 2011. The PRI had controlled Mexico since 1928, ruling for 71 years.

In August 2013 President Nieto submitted an energy reform bill with a total of 27 specific laws to the Mexican congress. The congress had to pass the bill and execute reforms. By December 2013, Mexican law prescribed that once again it was permitted for international oil companies and private investors to invest in oil and natural gas through three types of contracts: service contracts; production volume or profit-sharing contracts; and licence contracts (Fig. 7.10).

3. **Round Zero: Balance between protecting national oil company interests and attracting foreign investment**

Pemex, in terms of the overall oil industry, is far behind China. China has multiple national oil companies, including the top three, CNPC, Sinopec and CNOOC. These three companies are all listed on the NYSE.

In order to use Pemex equity to protect overall national interests, in 2014 the Mexican government took the unique approach known as "Round Zero". Pemex had to submit its choices for retaining and abandoning oil and gas fields prior to March 2014. The government's regulatory institution energy minister and CNG had six months to decide which oil fields/Pods would be retained by Pemex. In order to retain any production assets, Pemex had to explain its capabilities in the application regarding technology, assets and operations. In terms of exploration area, Pemex likewise had to show that it had explored wells and/or was currently exploring underground areas.

Based on estimates by the CNH, the result of Round Zero was that Pemex retained approximately 83% of proved reserves (2P reserves), as well as 21% of

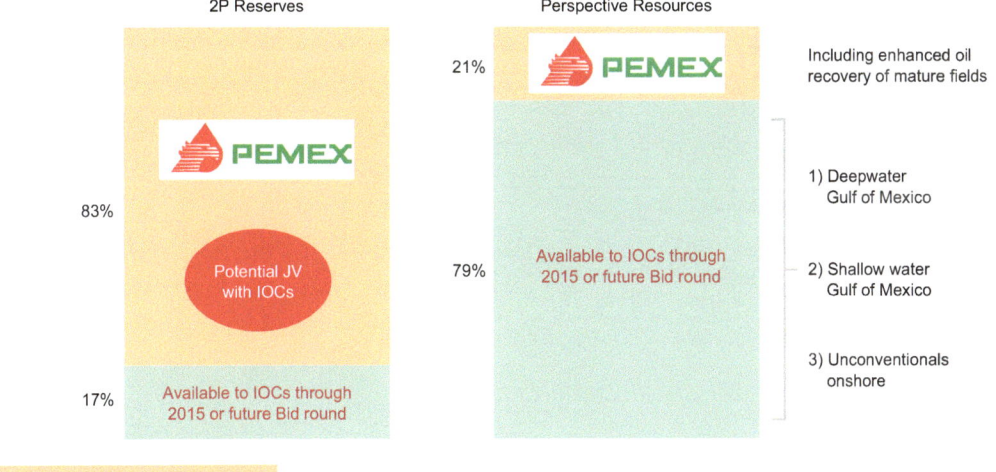

Fig. 7.10 Mexican NOC prior to reforms

unproved reserves. For the held assets of Pemex, this still allowed it to invite foreign investment and transnational companies to co-operate in exploration/development/ production.

4. Mexico's Round 1 Bid Round: What to expect in 2015?

Since December 2014, the CNH has begun to carry out the first round of bidding in stages, including for shallow water exploration, shallow water development, land and major production field Chicontepec capital injections, unconventional energy and deep water. The CNH has continued to invite potential international investors to participate in the Round 1 Bid Round.

Since June 2014, the drops in oil prices have forced the Mexican government to reconsider whether or not to make 2015 the year to attempt attracting international oil company investment. The current decision appears to be to carry out all bidding rounds according to plan, including the most attention-grabbing deep water portion. The date for completion of bidding is reportedly mid-July 2015. Leading international oil companies, including Chinese national oil companies (such as CNOOC) and other foreign national oil companies, are expressing great interest in this first round of bidding in Mexico. There is still some concern within the industry about whether Mexico will continue to fully control and occupy the oil/natural gas assets, whether there will be major changes to Pemex and whether there will be attractive financial terms for major deep water regions so as to accelerate exploration and mining of deep water oil and gas in Mexico.

Of course, the industry has also consistently recognised the speed and determination of the Mexican government in its energy reforms, as well as its detailed and transparent approach. This includes the Round Zero opening up of over 20% of proved reserves and 80% of unproved reserves to international oil companies.

It is worth considering whether these methods are suitable to China, especially given its attempts to accelerate unconventional energy and shale natural gas exploration and development and to speed up progress attracting suitable international oil companies and private funding.

Another issue the oil and natural gas industry should consider is whether Mexico can maintain its open policies if there is a leadership change after Nieto's six-year term (see Fig. 7.11).

Mexico Round Zero bidding reform prospects					
	Shallow water bay: exploration	Shallow water bay: development	Land	Chicontepec region and unconventional	Deep water bay
Project date of announcement	December 2014	January 2015	February 2015	March 2015	April 2015
Data centre date of establishment	January 2015	January 2015	March 2015	April 2015	May 2015

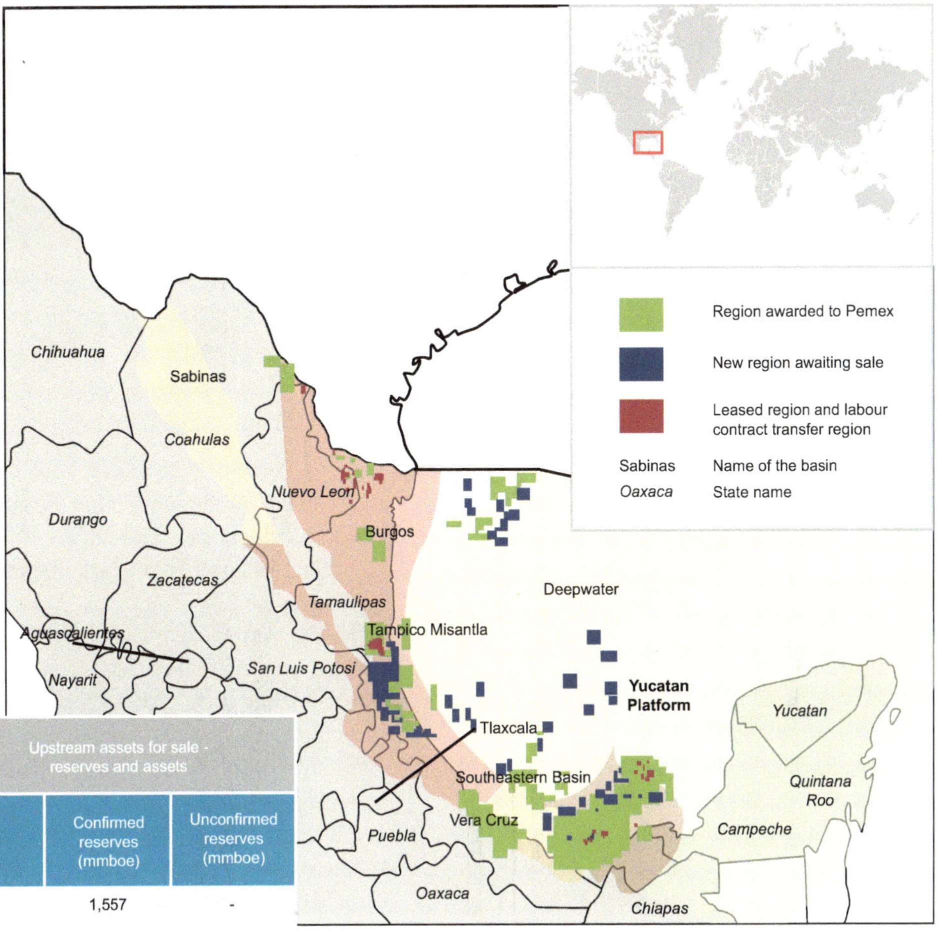

Upstream assets for sale - reserves and assets		
	Confirmed reserves (mmboe)	Unconfirmed reserves (mmboe)
Leased	1,557	-
New regions	3,782	14,606
Total	5,339	14,606

Fig. 7.11 Mexico's Round Zero reform prospects

Case 7: Analysis of the Approach to Opening Up Upstream Industries in Oil and Natural Gas in Various Countries

Generally speaking, the opening up of a country's oil and natural gas industry is closely linked to its natural resources and its development strategy. For example, in an oil and natural gas-exporting nation with abundant resources and good mining conditions, the terms on potential investor contracts will be stricter. Thus in Saudi Arabia, oil exploration and development are essentially not open to outside interests. Instead, Saudi Arabia hires international service companies to provide their oil industry with technical services. In importing nations with few resources and difficulties with mining, the degree of openness is greater. For example, in the United States, Canada, the United

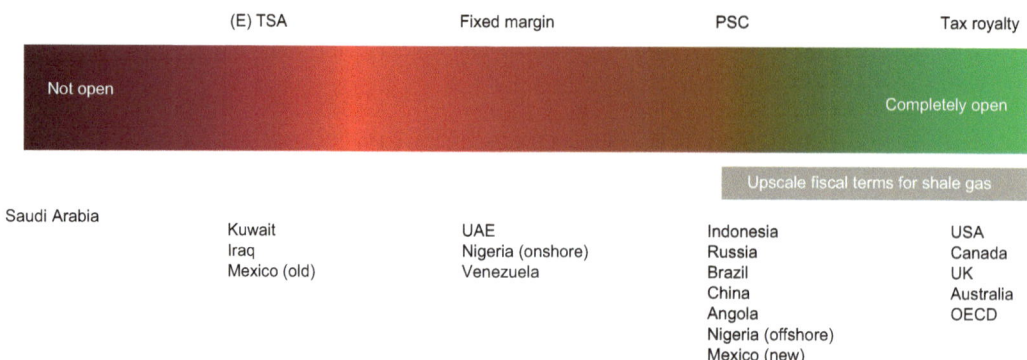

Fig. 7.12 World natural gas upstream openness

Kingdom, Australia and other OECD nations, oil and natural gas upstream industries are entirely open. The tax and royalty system between the nation and states/provinces ensures that their interests are protected. In terms of shale natural gas, US private land owners can also discuss royalties from the engaged company (Fig. 7.12).

Upstream contract terms are of four types, from stringent to loose:

- **Service contracts (or Technical Service Agreements (TSA), Enhanced TSA)**: Countries employing this approach are primarily countries with very abundant oil and gas resources, such as Qatar, Iraq and Mexico. International oil companies and international service companies are each able to apply for these terms. However, Qatar and Iraq only employ international oil companies, using their comprehensive management for oil and gas development. It is worth noting that in recent years Mexico only used service contracts, but has been unable to change its precipitous slide in oil and gas production. They are therefore in the process of using new and more attractive

contracts to attract foreign investment and technology, especially in deep water oil and gas exploration and development.

- **Fixed margin contracts**: Some countries, such as the UAE and Nigeria (mainland), have used a special contract term approach. For each barrel of oil produced, the foreign party obtained a fixed return, for example $1 per barrel. This term's downside for foreign parties is a low rate of return, but the benefit is that there is very little investment risk and a high degree of certainty. For the country in question, this term's greatest weakness is that it encourages cost controls by investors.

- **Production sharing contract (PSC)**: In 1972, Indonesia led the way in proposing PSCs. This is an international approach, and in particular a popular approach in third-world countries, that encourages foreign investment. It caters for the interests of both the country in question and the foreign parties. Investors generally assume all exploration risks and expenses. Prior to the discovery of commercial opportunities, they receive their reward by recovering investments and production volume sharing. It is worth noting that investors see large differences in shared proportions, and the recovery factor (also known as the R-factor) determines the differences.

- **Tax and royalty**: In Western countries such as the United States, the United Kingdom, Canada, Australia and others, domestic oil companies and foreign oil companies are essentially competitors on the same playing field. On average, tax systems offer investors rates of return that are slightly higher. However, this also depends on the degree of taxation. Some countries, such as Canada, use a tax rate that is linked to international oil prices, which enables investors to have a measure of certainty regarding investment risk. However, tax systems are not equivalent to a loss of control by the country in question. Using a tax system still allows the nation to continue to play an important role, especially in signing off on oil and gas exploration and development licences, approval environments and safe production plans. The country can also use increases or decreases in tax rates to encourage or inhibit a given direction of development.

In summary, the more oil and gas resources there are, the more that nations tending to export will prefer to use TSAs. Moreover, in oil and natural gas-consuming nations, countries with open markets tend toward using tax and royalty methods. Some countries such as Nigeria offer fixed-return contracts where land development is not particularly difficult and risk is low, and offer production volume sharing contracts for marine oil mining where there is higher difficulty and greater risk.

In China, conventional oil and gas sectors use relatively general production sharing contracts. Shale natural gas is buried deeper, there are greater difficulties and costs in its development, and it is sold domestically. If it is hoped to attract more domestic and foreign investors to join in shale natural gas exploration development ventures, then a reasonable improvement would be to offer PSC or tax and royalty systems that are more attractive than China's existing production volume sharing contracts.

International Natural Gas Supply and Quantities Available to China

8

8.1 Preface

This section analyses potential future changes in international natural gas market supply and demand and the available natural gas import volume for China, as well as what reasonable measures China should take, including suitable policy adjustments, to promote its own natural gas supply and demand security.

Currently, China's natural gas market is in a key stage of development, with demand, supply and system mechanisms each undergoing intense change. This state of flux is having a major effect on the evolution of the natural gas market. The NDRC has already issued goals and policy measures[1] that differentiate base supply gas and

incremental increase volume, and that merge the price structure of the two different types of gas by the end of 2015, setting a general road map for natural gas market reforms.

However, as the price of natural gas continues to rise, there will inevitably be impacts on downstream demand, and backward energy replacement by some users (using coal instead of natural gas) and other phenomena could occur. There are also some who believe that the continued increase in natural gas prices will lead to natural gas supply in China exceeding demand in 2017 (Jiaofeng et al. 2014). In addition, the large fluctuations in international oil market prices seen recently will inevitably have a far-reaching influence on the evolution of China's natural gas market. How to determine the future supply and demand circumstances for the natural gas market is thus of critical importance to the smooth execution of China's natural gas pricing reforms.

In 2013, China imported approximately 53 billion m^3 of natural gas, which already accounted for 32% of total consumption volume. How natural gas future imports change will undoubtedly have a significant impact on the structure of China's supply and demand. Because natural gas imports are long-term (the contract length is long) and stable (take-or-pay), it is still possible to rely on current circumstances to make some reasonable predictions about imports and the

---

* This chapter was overseen by Zhonghong Wang from the Development Research Center of the State Council and Shangyou Nie from Shell International Exploration and Production. It was jointly completed by Xiaowei Xuan, Yingxie Yang, Yongwei Zhang, Jiaofeng Guo, Weiming Li, Beiqing Yao, Tao Hong and Yuxi Li from the State Resource Department Oil and Gas Centre, Xiaoli Liu from the National Development and Reform Commission's Energy Office, Guang Yang and Jianhong Yang from the China Energy Research Institute's Natural Gas Centre, Xiaobo Ju and Beijing Yao from Shell China. Other members of the topic group participated in discussions and revisions.

---

[1]See National Reform and Development Commission Notice Regarding Conducting of Natural Gas Price Formation Mechanism Pilot Sites in Guangdong and Guangxi Autonomous Region (FGJG [2011] No. 3033); National Reform and Development Commission Notice Regarding Adjustment to Natural Gas Price (FGJG [2013]

(Footnote 1 continued)

No. 1246); National Reform and Development Commission Notice Regarding Adjustment to Non-Residential Reserve Natural Gas Price (FGJG [2014] No. 1835).

© The Editor(s) and The Author(s) 2017
Shell International and The Development Research Center (Eds.), *China's Gas Development Strategies*, Advances in Oil and Gas Exploration & Production, DOI 10.1007/978-3-319-59734-8_8

near future, even though there may still be some uncertainty regarding a contract's final process of execution. Such predictions could be highly significant in helping to understand the changing trends in China's natural gas market and could thus contribute to better natural gas price reforms.

This chapter is broken down into three parts. There is first a discussion of current international natural gas supply and potential future changes. This is followed by an analysis of China's current natural gas imports and future trends. Finally, recommendations are made regarding future natural gas trade policy for China.

8.2 Current and Future Sources of Global Natural Gas Supply

8.2.1 Current and Projected Global Natural Gas Resources

According to IEC determinations, current global natural gas resources are approximately 784 trillion m^3 (28,000 trillion feet^3). Based on current consumption levels, this could last sustainably for approximately 200 years or more. Based on proved resources that are recoverable

with today's economics and technology, the global natural gas resource reserve to production ratio is 54.8 (BP 2014: 20). Therefore, on the whole, the future supply of global natural gas resources is relatively ample.

In the global energy layout, natural gas is playing an increasingly important role. In the past 10 years (2005–2014), global natural gas consumption growth has been around 2.7%, making it the fastest-growing energy type (BP 2014). The IEA estimates that natural gas consumption will continue to maintain relatively strong growth up until 2030, when growth rates will be at approximately 2% (IEA Current Policies Outlook) (see Fig. 8.1). According to forecasts by Shell, natural gas supply and demand will rise from 3.1 trillion m^3 in 2010 (3100 billion m^3) to 5 trillion m^3 (5000 billion m^3) in 2030.

In addition, according to an analysis by ExxonMobil (2014), global energy consumption growth will be approximately 1% from 2010 to 2040, whereas natural gas growth will reach 1.7%, markedly higher than overall energy consumption growth. It is clear that natural gas will play a more significant role in global energy structures moving forward.

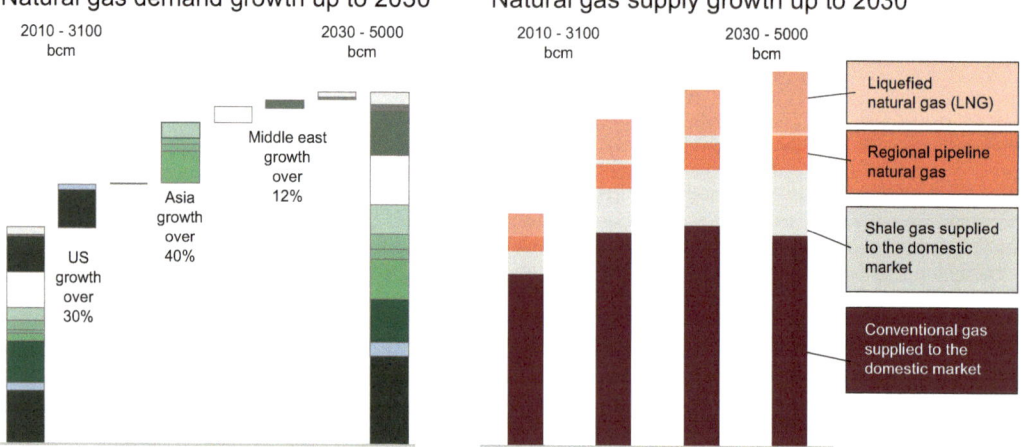

Data source: Shell Analysis

Note: all the tight gas / shale gas is calculated to standard value LPG

Fig. 8.1 Growth in global natural gas supply and demand totals

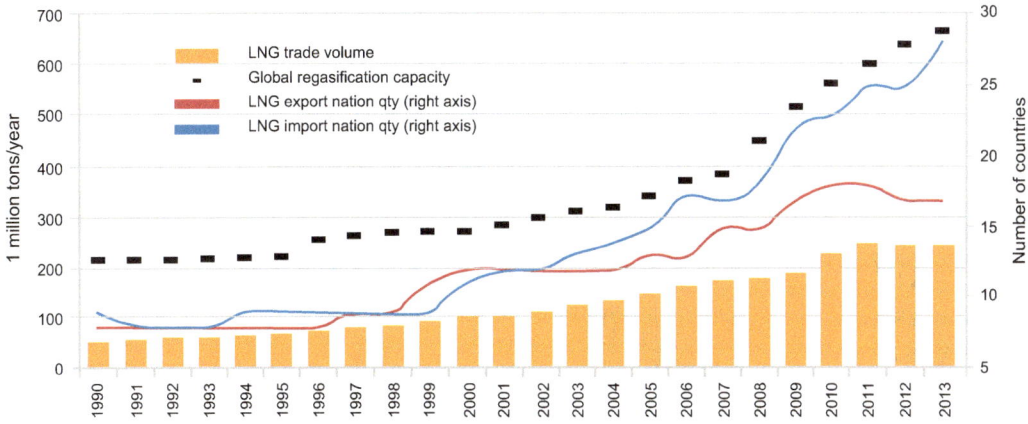

Fig. 8.2 International LNG trade volume (1990–2013). *Source* IGU (2014)

8.2.2   Global LNG Trade Development

For long periods, global natural gas trade has focused on pipeline imports as the major shipping mode, but as liquefied natural gas (LNG) technology develops, LNG is accounting for a larger proportion of trade. Beginning in 1959 with the world's first shipment of LNG across the ocean, by 2013, 320 billion m^3 of the total international natural gas trade of 1 trillion m^3 is LNG, 32% of total trade volume (IGU 2014) (Fig. 8.2).

LNG trade volume growth has broken through the shipping restrictions of pipeline natural gas, helping Japan, Korea and other countries to import natural gas. In addition, for countries facing geographical restrictions, LNG has also expanded potential import sources, helping to lower import risks. Moreover, LNG is more flexible than pipeline natural gas, and can change its export destination according to market changes. Thus LNG trade circulation has had a major influence on global natural gas trade layouts and supply and demand relationships as well as price trends.

8.2.3   Trends in Global LNG Trade

Since 2000, the global LNG trade volume annual average growth rate has reached approximately 5%. This trend will continue for the foreseeable future. At the same time, global LNG trade structure is becoming more complex. In 1990, there were eight LNG exporting nations and nine

importing nations; those numbers are now up to 27 and 31, respectively. It is accepted that in the next 10 years, the number of LNG exporting nations will rise to 50 and the number of importing nations to 25. As changes to trade networks add to uncertainties over market supply and demand relationships, the future structure of global natural gas trade is going to become more complex.

Figure 8.3 shows the 2012 production capacities of the major LNG exporting regions and their respective proportions. In 2012, global LNG annual production capacity was nearly 390 billion m^3. Qatar was the largest exporting nation, with production capacity of 77 billion m^3, accounting for 27% of global capacity. Together with other Middle East nations such as Oman, Yemen and UAE, and North African countries such as Egypt and Algeria, 47% of global LNG production capacity was controlled by the Middle Eastern and North African regions, followed by south-east Asian regions including Malaysia and Indonesia, which accounted for 23% of production capacity. This is then followed by Australia and other African countries, which both account for approximately 9%, and then South America with 7% and Europe, including Russia, with 5%.

8.2.4   Developments in the Global LNG Export Market

In recent years, global LNG supply deployments have seen relatively major changes. First is that

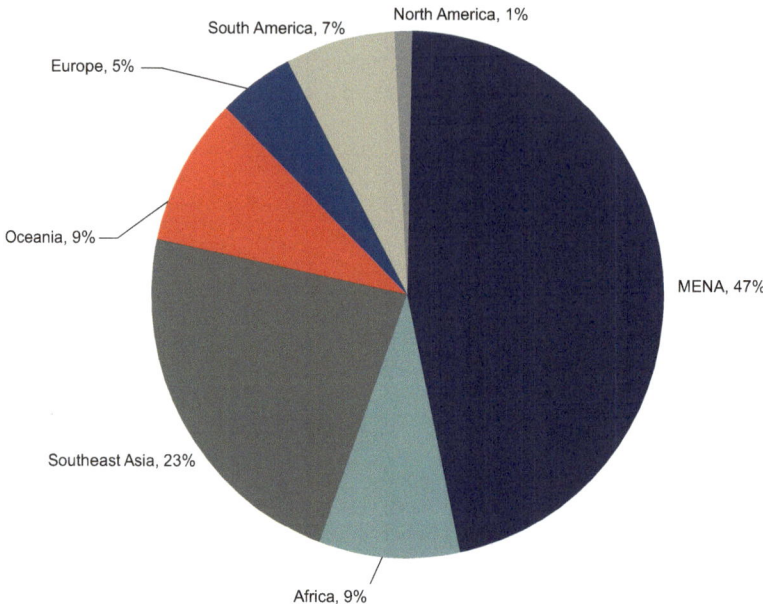

Fig. 8.3 Global natural gas regional production capacity (2012). *Source* PFC global LNG supply and demand report (2013) and IGU global LNG report (2013)

Australia's production capacity is in the process of rapid expansion, and is expected to exceed that of Qatar, making it the world's largest LNG producer. It is in the process of investing in the construction of seven LNG projects, with production capacity totalling approximately 85 billion m^3. Adding the in-progress projects of its neighbouring country Papua New Guinea brings the total to 95 billion m^3. Following this is the United States, with its successful development of shale natural gas, which has potentially transformed it from a natural gas-importing nation to a natural gas-exporting nation, with in-progress projects alone reaching 25 billion m^3. Figure 8.4 shows global regional in-progress projects, with a total of approximately 130 billion m^3.

If all the in-progress projects are completed as expected, global LNG production capacity will grow from around 390 billion m^3 in 2012 to approximately 520 billion m^3 in 2017. The proportion accounted for by Australia and Papua New Guinea will rise from 9% in 2012 to 25%, with the proportions of the Middle East and North Africa dropping from 47 to 36%. The United States' proportion will rise from essentially zero to about 5%. This means that the

Middle East and North Africa in natural gas supply markets will see their leading positions weakened, and market competition will be further strengthened (Fig. 8.5).

In addition to in-progress projects, there are also many LNG projects in planning or being proposed. There is capacity of about 30 billion m^3 in projects that have already submitted a final investment decision, and 370 billion m^3 in projects that are in some stage of Front-End Engineering Design. There is another total of 500 billion m^3 (IGU 2014) in projects that have submitted a motion but which are still in preliminary stages. Of course, not all projects will be completed, especially those in the early stages, and there is a variety of factors and risks that could end in their cancellation. However, this does show that current international LNG supply markets are in a relatively active state. In addition to Australia, the United States and Canada are two countries that will join the list of LNG-exporting nations and which have major potential for exports: East Africa's Mozambique and Tanzania, which have both discovered major deep sea natural gas fields. The exports from these countries will have a major impact on future global natural gas trade

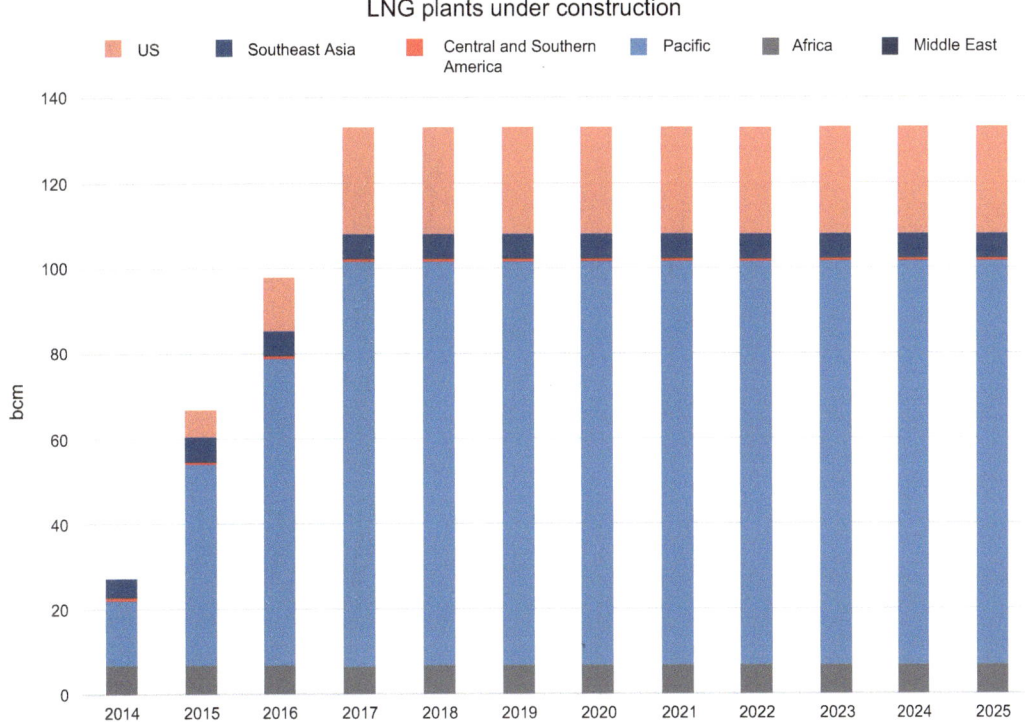

Fig. 8.4 Global in-progress LNG plant regional production capacity. *Source* Collected and assembled from PFC Global LNG Supply and Demand Report (2013), company reports and authors

layouts, and will affect the interests and risks for China's natural gas imports.[2]

8.2.5 Growing Natural Gas Demand in China, India and Other Emerging Markets

In the long term, natural gas demand growth will come primarily from China, India and other emerging economic entities. In Europe, North America, Japan and other developed nations, due to weakness in economic recovery and replacement with cheap coal, as well as subsidies given toward renewable energy development among other factors, natural gas consumption growth

will be limited. Developed nations are seeing sustained increases in natural gas demand in the transportation sector, especially with increased controls and higher emissions standards, as well as growth in natural gas consumption in marine shipping fuels. However, on the whole, developed nations will see relatively stable natural gas demand growth moving forward (Fig. 8.6).

By comparison, non-OECD emerging nations will have the strongest demand for natural gas. As economic development and income levels rise, emerging economic entities will see growing demand for natural gas. At the same time, the majority of emerging economic entities still have relatively low levels of expansion of natural gas usage, leaving major room for growth in the future. Moreover, due to strengthened atmospheric pollution governance and the need to deal with climate change, emerging nations will turn more toward natural gas for new momentum.

---

[2]For the influence of LNG development in Australia, the US, and Canada on Chinese imports, further detailed analysis will be provided in the section dealing with China's Future LNG Imports.

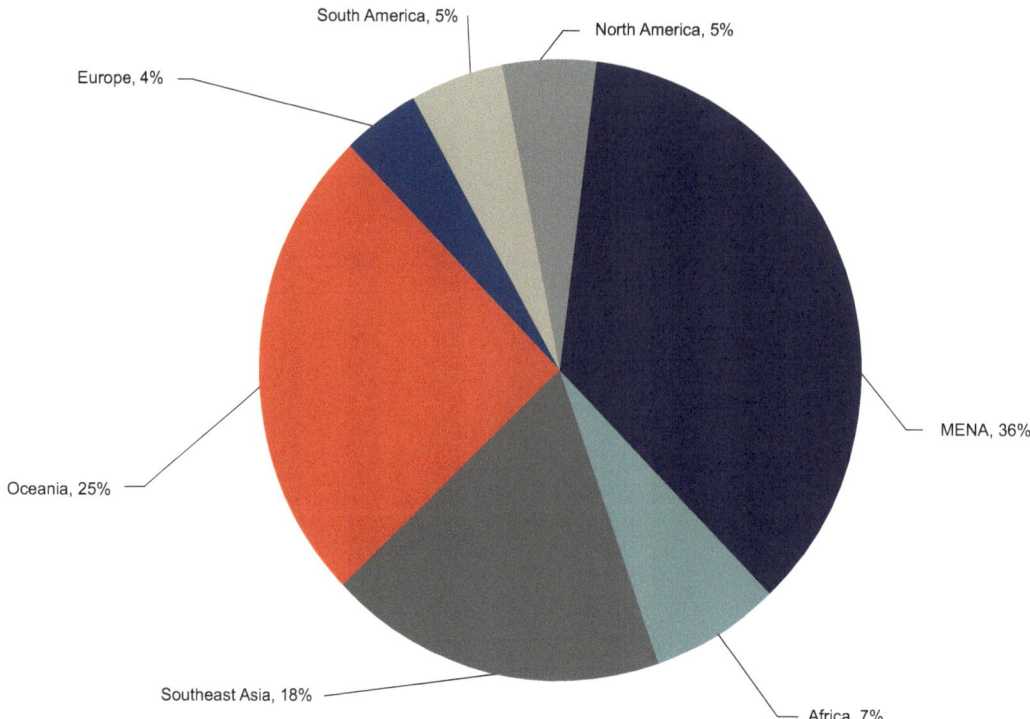

Fig. 8.5 Global LNG regional production capacity (2017 existing and in-progress capacity). *Source* Collected and assembled from PFC Global LNG Supply and Demand Report (2013), company reports and authors

8.2.6 The Influence of Oil Price Declines Future Natural Gas Trade

International crude oil prices dropped from over $100 per barrel in June 2014 to a low of CNY 43 in January 2015, driven by the US unconventional oil and gas revolution, which has seen major increases in US crude oil production volumes, falling international oil market demand and various other factors. By May 2015, the price was at approximately $60 per barrel. Because many natural gas contract prices are linked to the price of oil, international crude oil price fluctuations also immediately set off major changes in the international natural gas market, and will have far-reaching impacts on future natural gas trade layouts (Fig. 8.7).

In the short to medium term, in terms of demand, the JKP price has dropped from $20/MMBtu in 2014 to $7. This means a reduced

cost for natural gas-importing nations such as China, especially for spot and short- and mid-term contracts. For suppliers, the largest impact is felt by existing projects and in-progress projects, such as the projects currently under construction in Australia, where some project investments will face sunk cost. Because many projects are established based on the economic expectation of high oil prices, a major drop in oil prices will lower the expected gains from the projects, and some could even face losses.

In the long term, major oil price drops affect natural gas projects that are in planning. These projects can have their progress halted as a result, for example the Browse project in Australia and the Pacific Northwest project in Canada, which have currently delayed the time for making their final investment decision.

A drop in international oil prices also causes Asian natural gas exports to lack their previous appeal. This is because US natural gas pricing

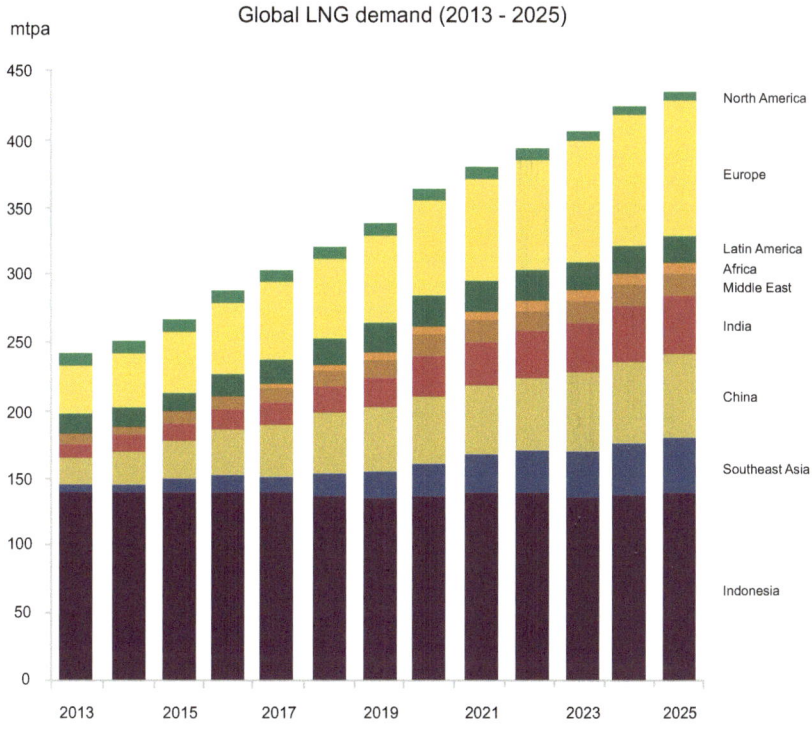

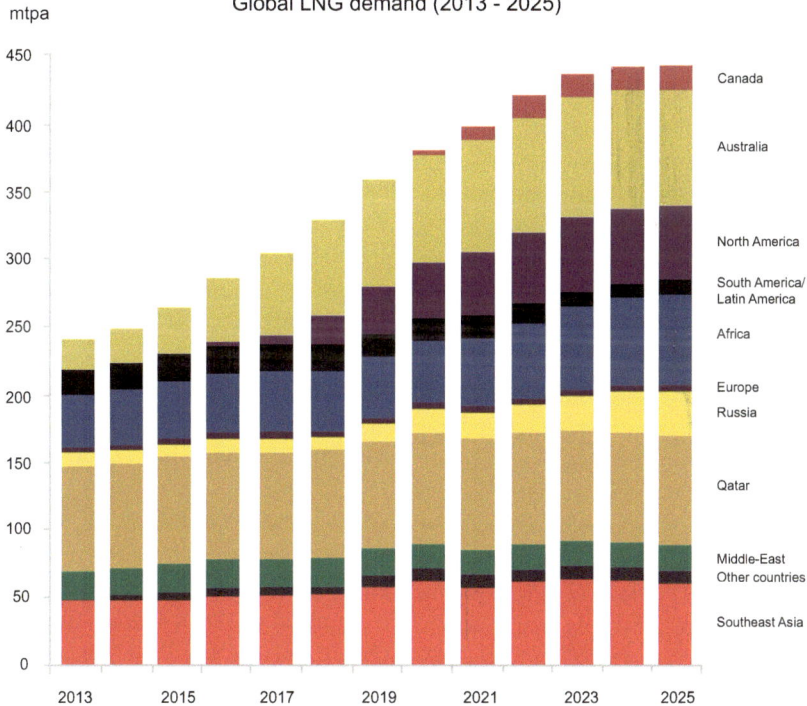

Fig. 8.6 Global LNG future supply and demand layout

Fig. 8.7 WTI oil price trends (2004–2015). *Source* Energy Information Administration, US

mechanisms are different from Asia, and are not linked to oil prices but are directly determined by supply and demand of natural gas itself. When oil prices plunge, US natural gas price advantages are markedly reduced, and Asian market demand is thus reduced, so a portion of projects could turn toward Europe or Latin America, while another portion could be cancelled. It could be said that international oil price drops have the effect of reshuffling the deck for natural gas trade, causing markets to enter a new phase of balancing out.

If Japan's Fukushima nuclear reactor disaster of 2011 is said to have ignited a new wave of natural gas export, international oil price drops can be viewed as having poured cold water on that momentum and resulted in a more conservative approach. As can be expected, sustained low oil prices will cause natural gas international supply reductions, and this to a certain degree will help natural gas prices to avoid becoming excessively low. However, in the long term, what will allow natural gas to find a new stable price will rely upon many factors, and there are many major uncertainties. Another trend is that LNG industry participants will work hard to co-operate to reduce LNG project expenses while increasing their projects' economic competitiveness.

8.3 China's Current Natural Gas Imports and Future Trends

8.3.1 China's Current Natural Gas Imports

In 2013, global natural gas consumption was approximately 3.5 trillion m^3, with natural gas trade totals of 1 trillion m^3. LNG accounted for about one third of this, while pipeline natural gas accounted for two thirds. China's 2013 natural gas imports were 53.4 billion m^3, at an average price of 10.4 \$/MMBtu (about 2.6 CNY/m^3) (Chen 2014). As part of this, LNG imports were approximately 25 billion m^3, at an average price of 10.5 \$/MMBtu. Pipeline natural gas imports were approximately 28 billion m^3, at an average price of 10.4 \$/MMBtu. China's natural gas consumption volume, imports of natural gas, LNG imports and pipeline imports correspond to 4.8, 5.3, 7.6 and 4.2% of global amounts, respectively (Table 8.1).

In terms of LNG, China's major import sources are Qatar, Australia, Malaysia and Indonesia. Import volumes are 9.2, 4.8, 3.6 and 3.3 billion m^3, respectively, at prices of 17.9, 3.5, 8.1 and 3.9 \$/MMBtu. In terms of pipeline gas, major importing nations are Turkmenistan, Uzbekistan,

Table 8.1 China's natural gas imports, by source

Natural gas imports	LNG and pipeline imports		Major importing nation	Imported amount (100 million m³)	Import price ($/MMBtu)	Proportion of China's total imports (%)
Total: 534,100 million m³ Price: 10.4 $/MMBtu Import dependency: 31.6%	LNG	Total: 250,100 million m³ Price: 10.5 $/MMBtu Proportion: 14.9%	Qatar	92	17.9	17.7
			Australia	48	3.5	9.3
			Malaysia	36	8.1	7.0
			Indonesia	33	3.9	6.4
	Pipeline	Total: 284,100 million m³ Price: 10.4 $/MMBtu Proportion: 16.7%	Turkmenistan	244	9.6	47.0
			Uzbekistan	29	9	5.6
			Kazakhstan	1	3.5	0.1
			Myanmar	10	11.5	1.9

Source "BP Statistical Review of World Energy (2014)" and author calculations

Myanmar and Kazakhstan, with import volumes of 24.4, 2.9, 1.0 and 0.10 billion m³, respectively, at prices of 9.6, 9, 3.5 and 11.5 $/MMBtu.

Looking at the numbers, Turkmenistan is currently China's largest importer, accounting for 47.1% of all natural gas imports, followed by Qatar, Australia, Malaysia and Indonesia, accounting for 17.7, 9.3, 7.0, and 6.4% of import volumes, respectively. These five countries account for 87.4% of China's import volume. From the perspective of price, Qatar's LNG price is highest, at 17.9 $/MMBtu (approximately 4.48 CNY/m³), and the Australian LNG price is the lowest, at 3.5 $/MMBtu (approximately 0.88 CNY/m³) (Fig. 8.8).

Compared to other international natural gas prices, China's current imported natural gas average price of 10.4 $/MMBtu is in the mid-range, and much higher than the price on the Henry Hub of 4 $/MMBtu. However, it is markedly lower than Japan's imported LNG price (approximately 16 $/MMBtu) and roughly equal to the imported natural gas prices of the United Kingdom NBP[3] (about 10.4 $/MMBtu) and Germany (about 11.3 $/MMBtu) (Fig. 8.9).

Overall, since China began importing natural gas in 2006, import volumes have risen rapidly and import prices have also markedly risen. As Australia and other nations that signed early

import contracts with China have had their import contracts come to term, China is now seeing a new wave of natural gas contract signings, which are having a decisive influence on China's future natural gas imports.

8.3.2 Potential Source Nations for China's Future LNG Imports

1. **Australia**

Australia currently has 22 proposed LNG projects, with annual production capacity of a total of 200 billion m³. These projects are at different stages of development. Of these, seven projects are under construction and represent 85 billion m³ of annual production capacity, with 50 billion m³ of signed sales contracts, primarily with China, Japan and South Korea. As part of this, China and Australia have signed contract agreements for approximately 20 billion m³: 5 billion m³ between CNOOC and Queensland Curtis, 4.7 billion m³ between CNPC and Gorgon and 10.5 billion m³ between Sinopec and Australia Pacific. This results in China importing 25 billion m³ of natural gas from Australia, accounting for over 35% of signed LNG contract agreements.

From the perspective of project progress, Australia's project development is in a leading position, with seven projects planned to

---

[3]NBP, National Balancing Point. This is the UK's virtual gas trading hub price.

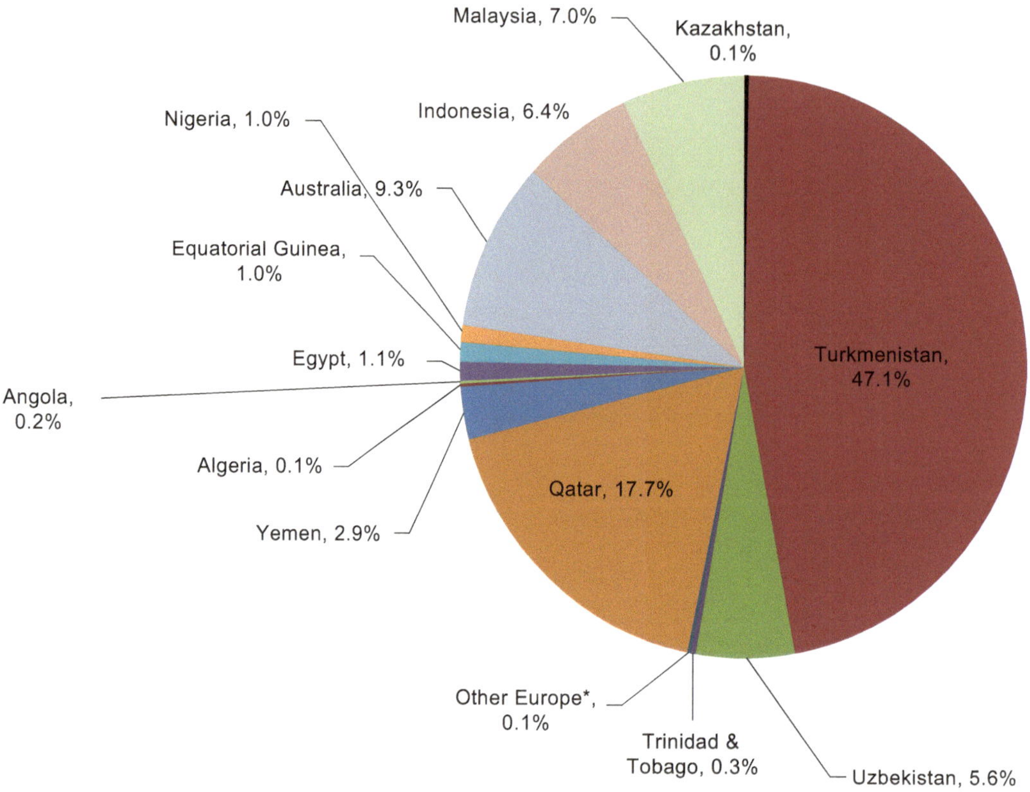

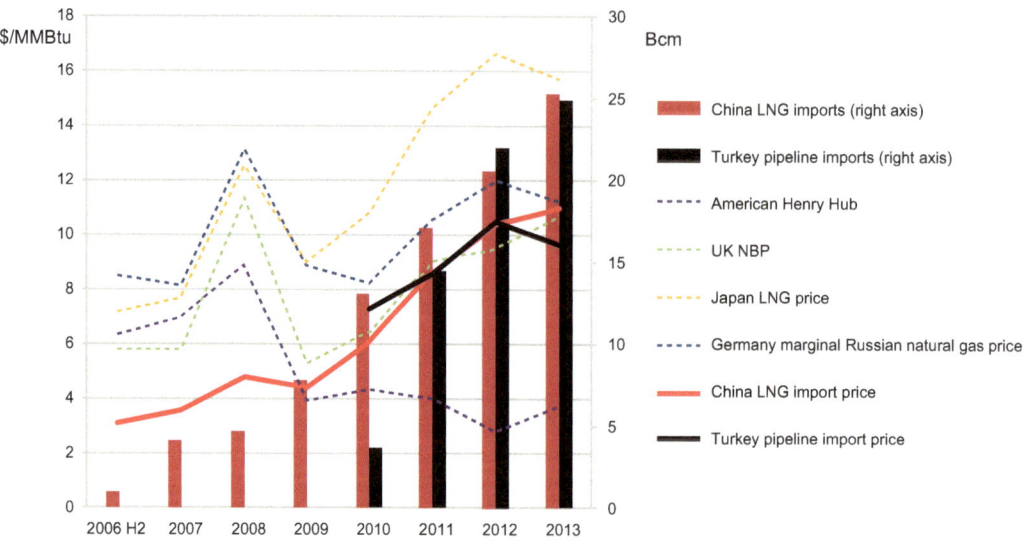

Fig. 8.8 Chinese natural gas import sources. *Source* BP Statistical Review of World Energy (2014)

Fig. 8.9 Comparison of China's natural gas import prices and other international natural gas prices. *Source* Michael Chen (2014)

commence prior to 2017. Moreover, because of advantages in geographical location, shipping fees to China are lower compared to the United States and Canada. However, these projects in Australia are all scheduled to develop simultaneously, causing source material and labour market shortages and price increases that have caused project cost rises. In some projects, for example Gorgon, costs have even risen by as much as 40% (Forster 2013). These cost increases could affect natural gas pricing. In addition, even if cost increases are only temporary, subsequent projects could face similar problems, and Australia could face issues of insufficient natural gas supply. This problem could then further affect subsequent projects. These risks must be considered as China further signs LNG purchase contracts, especially since so many LNG contracts have already been signed with Australia.

2. **United States**

The US shale natural gas revolution is currently changing American and even global energy market layouts. According to materials from the US Energy Information Administration, in the past several years, American shale natural gas production volume increases have seen an over 20-fold increase. Today, the United States has surpassed Russia to become the world's largest natural gas producer. Research by international consulting company ICF has shown that the United States can use the natural gas reserves held as of 2011 for 130 years of consumption. Shale natural gas development has drastically reversed the United States' reliance on imports. In 2007, the United States was still planning to invest enormously in the construction of natural gas import ports, whereas as of July 2014, there were 29 LNG export projects currently seeking approval from the Department of Energy, for a total of 430 billion m^3.

American natural gas export targets include the FTA and non-FTA nations. According to the United States Natural Gas Act, revised in 1992, natural gas approvals for export to FTA nations can be obtained automatically, but to export to non-FTA nations requires authorisation from the Department of Energy, and approval is only given if the proposal is in the interests of the American public. China is a non-FTA nation.

The United States has seen fierce debate over the last two years about whether it should export natural gas. Opposition to export includes US manufacturers, the chemical industry and other natural gas downstream enterprises, who believe that exporting natural gas will cause domestic natural gas prices to rise, as well as American environmentalists, who believe that shale natural gas will harm water resources as well as add to greenhouse gas emission increases. Supporters of exporting natural gas include natural gas upstream development enterprises and free market supporters, who believe that export of natural gas will be beneficial in stimulating the US economy and creating jobs. There are also some international energy political considerations that relate to whether American exports of natural gas will help put an end to halts placed on European nations by Russia. Finally, based on its analysis and research, the US Department of Energy believes that export is on the whole beneficial to American economic interests, and thus approved a series of natural gas export projects. By July 2014, there were seven projects that had received approval, with a total of 100 billion m^3 of capacity exporting to non-FTA nations, including Sabine Pass (2.30 billion m^3), Freeport (20 billion m^3), Lake Charles (20 billion m^3), Cameron (18.0 billion m^3), Cove Point (8.0 billion m^3), Jordan Cove (8.0 billion m^3) and Oregon LNG (13.0 billion m^3).

The major reason for American natural gas attracting such excitement from natural gas buyers is that US LNG imports improve the diversification of import sources in those nations. This diversification is in two respects: one is in terms of geographical location, since prior to this natural gas exports primarily came from the Middle East and North Africa. The other is from the perspective of pricing mechanisms. For a long period, international natural gas pricing was essentially linked to oil prices, but due to its own natural gas market, and because natural gas markets are somewhat independent from oil markets, the United States prices domestic natural gas using the natural gas price on the Henry Hub as a baseline. Thus the natural gas exported from the United States might not be entirely linked to oil prices, and provides a hedge against oil risk in natural gas import channels. From the perspective of actual price, the US Henry Hub price is currently between $3 and $5. In the long term, it could rise, especially if international oil prices plummet. There is also the consideration that the majority of US export ports are in the GOM, and require the Panama Canal in order to reach Asia, and shipping fees are going to be more expensive than Australia.

From the perspective of project progress, the United States currently only has one LNG project under construction, Sabine Pass, which is expected to be completed in 2015/2016. However, American LNG has been the focus of significant attention, especially among Asian buyers such as in Japan, South Korea, India and China, with approximately 120 billion m^3 of natural gas already signed under purchase contract: 65 billion m^3 signed with portfolio players, 23 billion m^3 with Japan, 7 billion m^3 with Spain, 9 billion m^3 with India and 7 billion m^3 with South Korea. Currently, China has not signed any natural gas purchase agreements or contracts.

3. Canada

Canada's LNG projects got under way later than those in Australia and the United States. Currently, there are proposals for a total of 19 LNG projects for a total of 355 billion m^3. No project has as yet entered construction. Approximately 20 billion m^3 have already been signed under some kind of purchase agreement and contract, including 5.8 billion m^3 between CNPC and Pacific Northwest, 830 million m^3 between Huadian and Pacific Northwest, and 1.38 billion m^3 under a memorandum between Guangzhou Natural Gas Group and Woodfire.

As far as distance is concerned, Canada is closer to China than the United States, and thus shipping fees will be slightly lower. Moreover, political barriers could be less of a problem. Project development has lagged behind Australia and the United States, and thus there is significant

project uncertainty. At the same time, in some communities, including the native residents, opposition is still an important factor holding back project progress. In addition, severe weather and the need to construct infrastructure could increase the project development costs and thus affect future natural gas prices. Regardless, as a long-term potential natural gas-exporting nation, Canada's natural gas export dynamics and development are worthy of attention and study.

8.3.3 Trends in China's Future Natural Gas Imports

1. Trends in LNG imports to China's

Since China's natural gas consumption has increased rapidly in recent years, the three major oil corporations are actively seeking out opportunities to purchase natural gas from international natural gas markets to satisfy future demand. Table 8.2 shows long-term signed LNG contracts that China currently holds, including sales and purchase agreements, heads of agreement and memorandums of understanding. It is estimated that in the next few years China will see rapid growth in LNG imports. New LNG imports from Australia will be largest part of this increase, rising from approximately 5 billion m^3 currently to 25 billion m^3 by 2015. Other LNG import increases currently planned will be from Russia (approximately 4 billion m^3), Papua New Guinea (approximately 3 billion m^3) and from third parties (approximately 9 billion m^3). In addition, China is currently in talks with Canada to sign a purchase of LNG contract letter of intent or memorandum of understanding for approximately 8 billion m^3.

However, these projects are still uncertain. First, contract letters of intent or understanding could potentially end without the final transition to a purchase contract taking place. In addition, while these projects plan to begin imports to China in 2017–2019, they have still not entered construction, and the repercussions of the international oil price drops could introduce uncertainty into the expected timeframe. For example, the Pacific

Table 8.2 China's LNG long-term contracts

Type	Exporting nation	Name of plant	Buyer	Volume (bcm)	Start year	End year
Spot	Australia	Withnell Bay	CNOOC	4.55	2006	2030
Spot	Indonesia	Tangguh	CNOOC	3.59	2009	2033
Spot	Malaysia	Malaysia LNG Tiga	CNOOC	4.14	2009	2029
Spot	Qatar	Qatargas	CNOOC	2.76	2009	2034
Spot	Portfolio seller	Portfolio	CNOOC	1.38	2010	2024
Mid-/long-term	Portfolio seller	Portfolio	CNOOC	6.90	2015	2035
Contract letter of intent	Portfolio seller	Portfolio	CNOOC	2.07	2019	2039
Purchase contract	Australia	Queensland Curtis	CNOOC	4.97	2014	2034
Spot	Qatar	Qatargas	CNPC	4.14	2011	2036
Purchase contract	Russia	Yamal	CNPC	4.14	2018	2038
Purchase contract	Australia	Gorgon	CNPC	3.10	2014	2033
Purchase contract	Australia	Gorgon	CNPC	2.70	2014	2033
Purchase contract	Australia	Australia Pacific	Sinopec	10.49	2015	2035
Purchase contract	Papua New Guinea	PNG LNG	Sinopec	2.76	2014	2034
Contract letter of intent	Canada	Pacific Northwest	Sinopec	4.14	2019	2039
Contract agreement	Canada	Pacific Northwest	Sinopec	1.66	2019	2039
Contract agreement	Canada	Pacific Northwest	Huadian	0.83	2019	2039
Memorandum of understanding	Canada	Woodfibre	Guangzhou Natural Gas Group	1.38	2017	2042

Source International Group of Liquid Natural Gas Importer (GIIGNL), LNG Industry 2006–2014, and author compiled and organised

Northwest project recently announced a delay in the project's final investment decision due to international oil price drops. Therefore, even if in the long term international oil prices support the continued progress of these projects, they are still very likely not to be executed until after 2020. Taking this into consideration, China is likely to see a rise in LNG imports from the current 25 billion m^3 to approximately 57 billion m^3 in 2020. By 2030, if the Canadian contract is realised, LNG imports will rise to approximately 65 billion m^3. Two east African countries are also likely to become LNG-exporting nations in the latter half of the 2020s.

Looking at importers (see Table 8.2), CNOOC accounts for a major share of the current LNG long-term contracts, and has signed contracts for approximately 30 billion m^3. Sinopec and CNPC currently have approximately 14 billion m^3 and 20 billion m^3 of LNG contracted, while Huadian Group and Guangzhou Natural Gas Group have signed for 0.83 billion m^3 and 1.28 billion m^3 in LNG contracts respectively (Fig. 8.10).

In addition to long-term contracts, spot and short-term contracts are also used to import natural gas. In 2013, Chinese spot imports of LNG were approximately 5.4 billion m^3, accounting

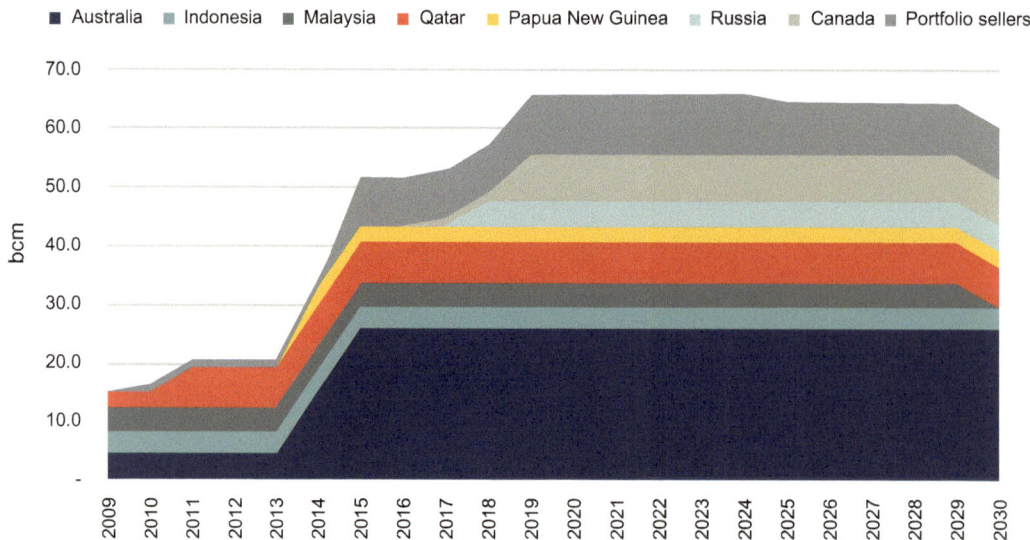

Fig. 8.10 China LNG imports. *Source* International Group of Liquid Natural Gas Importer (GIIGNL), LNG Industry 2006–2014, and author compiled and organised

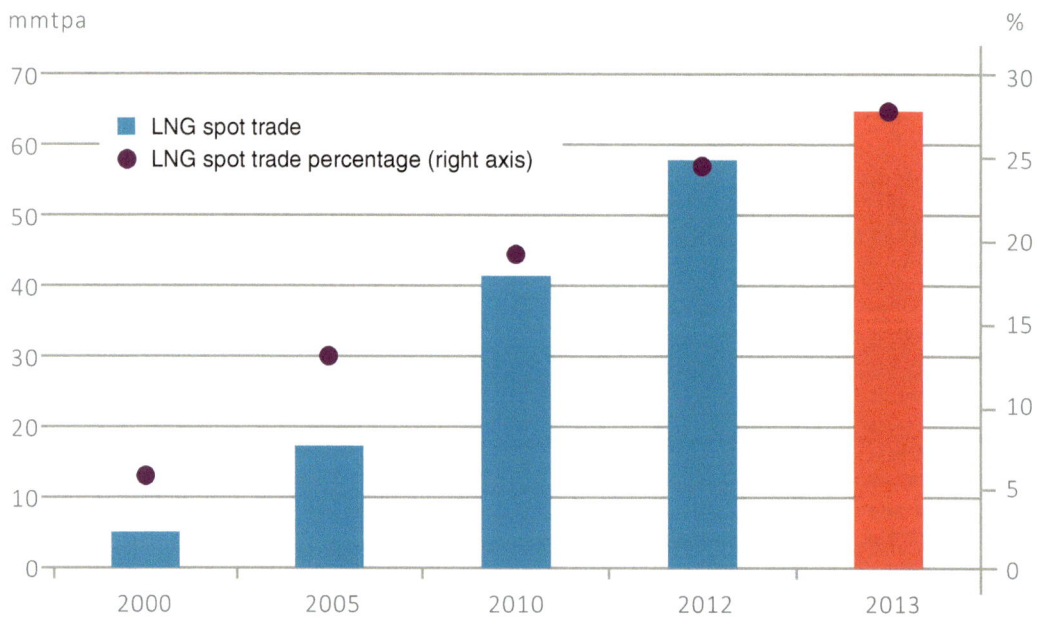

Fig. 8.11 Global spot trade market change trends this year. *Source* GIIGNL 2013. *Data source* International Group of Liquid Natural Gas Importer (GIIGNL), LNG Industry in 2013

for around 22% of the total LNG import volume, slightly less than the current global LNG spot transaction volume percentage of all trades of 27%. As shown in Fig. 8.11, global spot transactions have seen a rising trend in recent years, primarily due to their flexibility, which allows them to top up supply shortages from long-term contracts and enables arbitrage of price

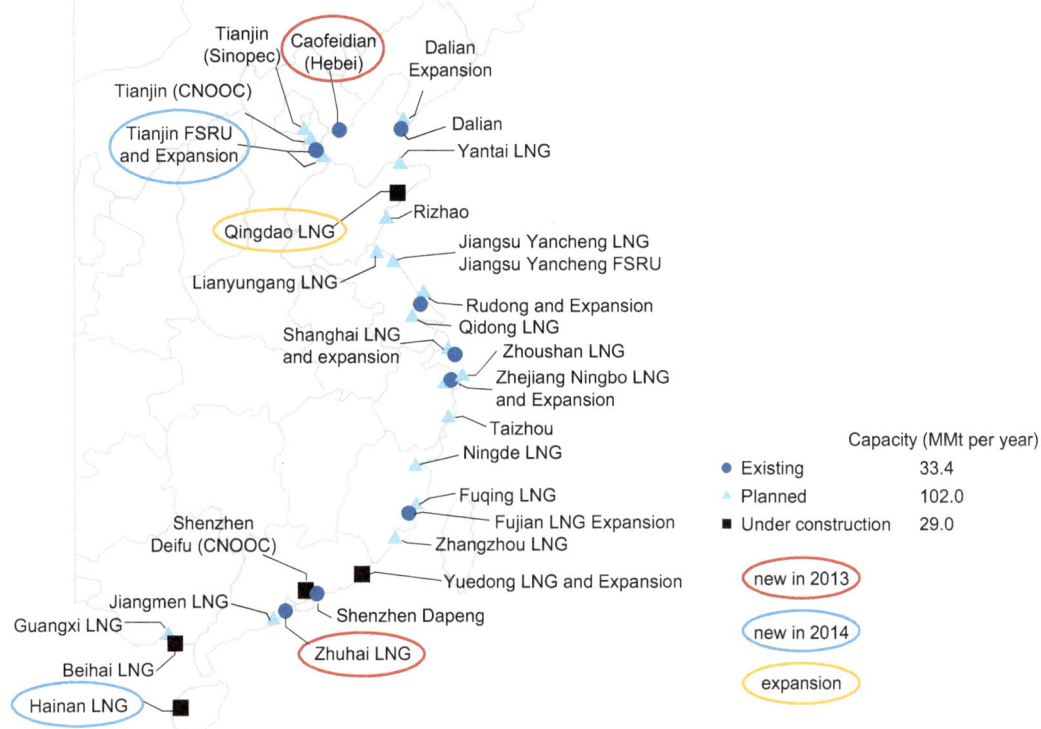

Fig. 8.12 China natural gas receiving station construction

between different LNG markets. As China's natural gas trade market develops, awareness of risk aversion will strengthen and the construction of a natural gas financial market is sure to result in spot and short-term markets also seeing further growth through natural gas import increases. Taking 20–25% as an estimate of spot LNG trade's percentage of total LNG trade volume, by 2020 spot LNG trade volume could reach approximately 10 billion m^3, and approximately 15 billion m^3 by 2030.

2. **China's LNG receiving station construction trends**

Natural gas receiving stations are a critical part of the infrastructure for the LNG trade chain, and can also restrict the ability to import natural gas. China's first LNG receiving station, the Shenzhen Dapeng Receiving Station, started operations in 2006. China currently has LNG receiving station capacity of 45.4 billion m^3/year (33.4 MMt/y), with projects under construction of 40 billion m^3/year (29 MMt/y). There is also planned construction of 140 billion m^3/year (102 MMt/y).

Based on these figures, by 2020 China's LNG receiving station capabilities could reach 85–225 billion m^3/year, which would enormously surpass the possible scope of future LNG imports. These planned projects could be increased or reduced along with China's natural gas demand and international natural gas trade market changes. However, considering current signed natural gas long-term contracts and expressions of intents on the part of China, at first sight it seems clear that the overall capabilities of China's LNG receiving stations will not be a bottleneck for the import of natural gas (Fig. 8.12).

Table 8.3 China's future pipeline natural gas import capacities

	Current (100 million m^3)	Future imports (100 million m^3)	Remarks
Central Asia pipeline	550	850	
A line and B line	300	300	Gas sourced from Turkmenistan, having entered operation in December 2009 and October 2010 respectively
C line	250	250	Gas source: Turkmenistan (10 billion m^3), Uzbekistan (10 billion m^3) and Kazakhstan (5 billion m^3), began operation in June 2014
D line		300	Expected to begin operation in 2016
Myanmar	120	120	Began operation in August 2013, and as of August 2014 had received 4 billion m^3
Russia		380–680	
East line		380	Began construction in September 2014, expected to open in 2018
West line		3000–6000	Came to understanding memorandum in November 2014, expected to open in 2019. Signed intention is for 60 billion m^3, but 30 billion m^3 is a more realistic expected goal
Pipeline natural gas	670	1650–1950	

Data source Company websites, news, and author collected and arranged information

3. **China's future pipeline natural gas imports**

China's existing pipeline natural gas imports primarily come through central Asia and Myanmar, with the future Russian pipeline set to become a major import source as well (Table 8.3).

The Central Asia line is divided into the four lines: A, B, C and D. The A and B lines are sourced from Turkmenistan and currently have capacity of 30 billion m^3/year. The C line is 25 billion m^3/year (10 billion m^3 from Turkmenistan, 10 billion m^3 from Uzbekistan and 5 billion m^3 from Kazakhstan, began operation in June 2014). The D line shipping capability is 30 billion m^3/year and is expected to open in 2016. The Myanmar line began operation in August 2013 and by August 2014 had imported 4 billion m^3, with capacity for 12 billion m^3/year.

With regard to Russia, currently Russia and China have signed an East Line contract as of May 2014, with shipment capabilities to reach 38 billion m^3/year. The Western Line has had a memorandum of understanding signed as of November 2014, and could include scope of 30 billion m^3/year. Of course, until a formal contract is signed, nothing is certain. Although both projects are slated to commence prior to 2020 (the Eastern Line in 2018, and the Western Line in 2019), the actual start time could yet be after 2020. First, since 2014, Russia's project progress and development has slowed, with approvals not yet obtained from the Russian government. In addition, Russia's geographical location and severe weather, among other factors, will make it more difficult for the project to make progress. Moreover, the Eastern Line and Western Line both require large quantities of funds, labour and resources to be invested, which is a challenge for Gasprom, especially as international oil prices have dropped.

In summary, it is expected that in 2020 China's pipeline imports will be in the region of 97 billion m^3. In 2030, China's pipeline import capacity will reach approximately 135–165 billion m^3, with the main uncertainty in this figure

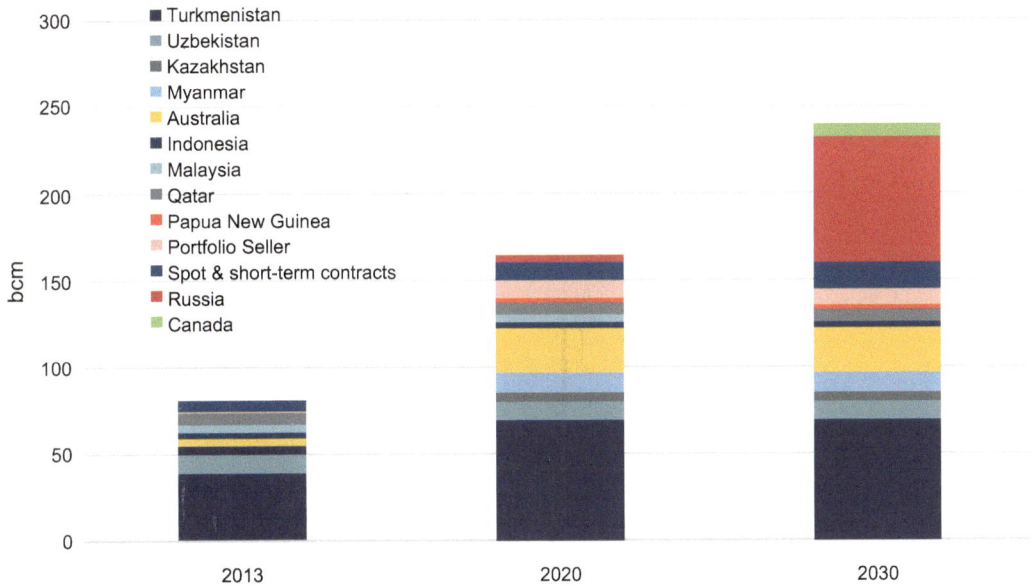

Fig. 8.13 China's future natural gas imports. *Note* Long-term contracts include purchase contracts, co-operation intention agreements and memorandums of understanding

coming from the progress of Russia's Western Line.

4. **China's future natural gas import trend outlook**

In summary, given the long-term LNG and pipeline contracts that China has already signed, China's natural gas imports will grow rapidly in the next few years, from 53 billion m^3 in 2013 to 167 billion m^3 in 2020, including 70 billion m^3 from LNG and 97 billion m^3 from pipeline natural gas. By 2030, imported natural gas will further expand to approximately 210 billion m^3, including 75 billion m^3 in LNG and 135 billion m^3 in pipeline supply. If Russia's Western Line is able to implement its 30 billion m^3, then imported natural gas will grow further to around 240 billion m^3, including 75 billion m^3 in LNG and 165 billion m^3 in pipeline supply.

Figure 8.13 shows China's 2030 natural gas import source country proportions based on long-term contract calculations. Currently signed contracts and intentions clearly indicate that Turkmenistan and Russia will become the major natural gas importers to China, each accounting for approximately 30%. Following this will be Australia, whose natural gas imports will account for 12%. Another approximately 30% will be sourced from eight nations, including Myanmar, Qatar, Uzbekistan and Papua New Guinea, as well as from third-party natural gas suppliers. This is a major change from the figures in 2013: the proportion of China's imports that are sourced from the Middle East will have been reduced the most, from 17.7% in 2013 to just 3%. Turkmenistan's proportion will move from nearly 47 to 30%, while Russia will move from 0 to 32%. In addition, Australia will see a slight increase. It is clear that from 2020–2030, China's will increase the diversity of its natural gas import sources, primarily by achieving historical breakthroughs with Russia in natural gas contract negotiations, thus enabling China to reduce its reliance on Central Asia's natural gas (Fig. 8.14).

What should be noted is that the predictions of China's future natural gas import trends outlined

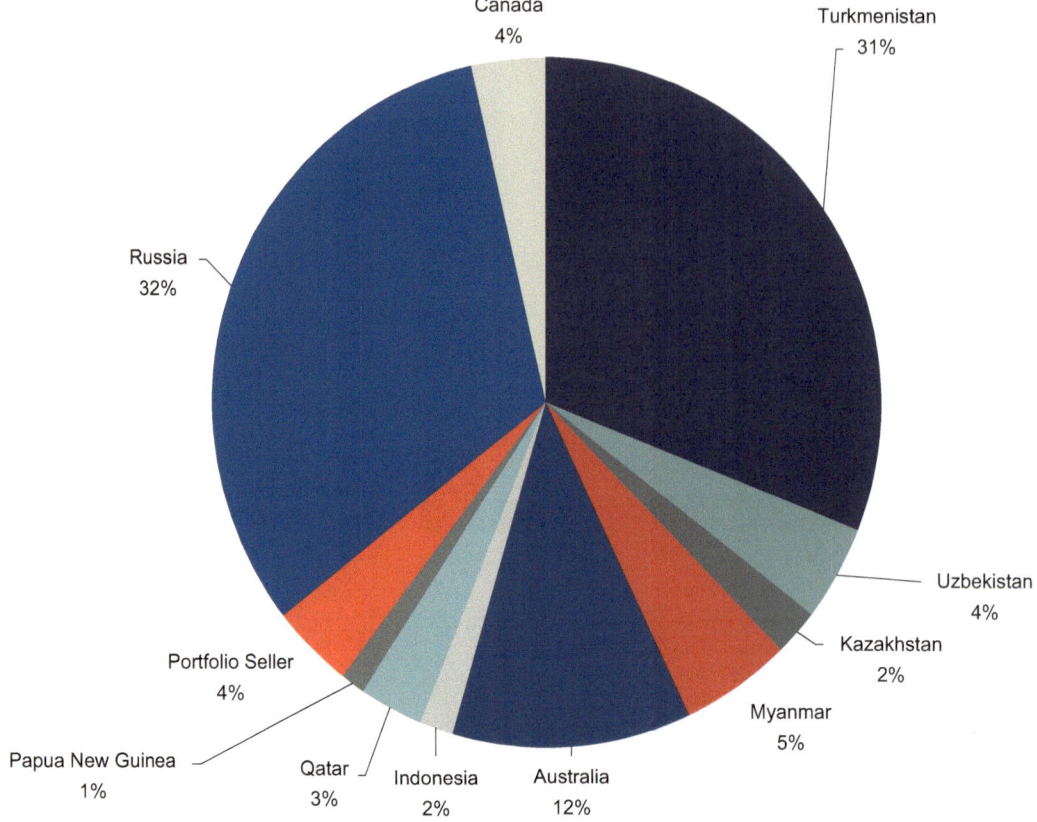

Fig. 8.14 Future Chinese natural gas import source nation proportions. *Source* GIGNL and author collection and arrangement

above are based on natural gas import capabilities and quantity of contracts. However, import capabilities are not entirely equivalent to the quantity of actual imports. For example, after a pipeline is constructed, the volume of actual shipments can be less than its maximum shipping capability. Moreover, even though natural gas long-term contracts are relatively stable, they are still not completely fixed, and both parties are still able to negotiate and make adjustments according to actual changes in circumstances. Thus China's actual natural gas import volume in the future will be determined not only by import capabilities and contracts, but also by domestic demand.

At the same time, the risks and uncertainties facing China's future natural gas imports must be acknowledged:

- First, the global natural gas market structure still presents several variable factors. Many Front-End Engineering Design projects are in planning stages and may not be ultimately completed, especially those in early stages, which could be influenced by various factors and risks and cancelled. Currently there is significant export potential in Australia, the United States and Canada, and various project planning capabilities are very promising, but how these projects will ultimately affect the global natural gas trade is hard to say.
- Second, international oil price changes present a major variable to the international trade in natural gas. A drop in international oil prices represents a reshuffling of the cards, and currently markets have begun to enter a

new process of balancing out, and long-term natural gas prices will have their new points of balance determined by multiple factors. There are significant uncertainties, and this will affect China's interests and risks as it imports natural gas.

- Furthermore, along with cost increases, political risk and various other factors are also having significant influences on future Chinese natural gas imports. Currently, Australia's in-progress projects are seeing significant cost increases, while if China wishes to achieve imports from the United States and Canada, there are significant political barriers that must be overcome, presenting major political risks. Therefore, even if China insists in future upon a strategy of natural gas import diversification, the success of this strategy will still depend on the circumstances of the international natural gas market and the political environments within individual nations.

- Finally, there is a certain amount of uncertainty over China's future imports of pipeline natural gas. Even based on current circumstances, pipeline natural gas imports will play a determining role in future natural gas imports for China, but the final realised pipeline natural gas import volume is as still a significant variable. For example, in the Russian pipeline natural gas project, actual project progress since 2014 has not been ideal, and the project has encountered difficulties as a result of Russia's geographical location and severe weather. Russia still faces a major challenge to satisfy the enormous requirements for funding, labour and resources that are necessary to complete the project.

In brief, looking at the current LNG contracts and pipeline natural gas agreements, Chinese natural gas imports are set to expand. However, close attention must be paid to risk and uncertainty, as well as to the rapid changes in international circumstances and specific conditions in nations which can lead to fluctuations in international natural gas markets. China must prepare for a rainy day and make suitable preparations to ensure its own security of supply as it pursues a strategy to diversify natural gas imports.

8.4 China's Natural Gas Trade Policies and Recommendations for Reform

8.4.1 China's Current Natural Gas Trade Policies

China's natural gas trade policies can be grouped into the following major sectors, based on the stages of the natural gas import process:

- **Commodity trading policy**: Treating natural gas as a trade commodity while managing imports. This includes answering questions such as: Do domestic entities have the right to import natural gas? Do approvals from relevant authorities need to be obtained? Are approvals from authorities required for a domestic entity to sign a contract with a foreign company?

- **Facility construction and operation policy**: After natural gas is imported into China, managing facilities (receiving stations, gas reserve pipeline networks etc.), construction and operation. This includes answering questions such as: Which entities can construct receiving stations and other infrastructure? Are approvals required, and how are they handled? What policies are there for infrastructure operation? How are policies implemented?

- **Price and taxation policy**: Policies relating to imported natural gas pricing and taxes. This includes answering questions such as: Is there regulatory interference in imported natural gas pricing? For imported natural gas, what tax policies apply, for example subsidies and tax rebates?

1. Commodity trading policy

In terms of the foreign trade sector, natural gas does not constitute a state-run commodity, and is viewed as a "general commodity".[4] At customs, conducting natural gas trade does not require prior approval, with customs currently overseeing statistics for natural gas commodity trading. Requirements for corporate reporting and filing have also been cancelled, making it no different than other general commodities.

Therefore, in theory, natural gas imports are open to all qualifying enterprises, and there is no substantial barrier arising from commodity trading policies. So long as the business scope includes "natural gas trade" (when handling administration for industry and commerce registration) and all Ministry of Commerce approvals have been obtained (routine procedures), the enterprise can carry out natural gas import business, and the signed trade contracts do not require approval and review from relevant authorities.

2. Facility construction and operation policy

The main facility construction and operation policy approvals are at the facility construction stage. According to national policy, a construction project beyond a certain size requires approval from the NDRC, and natural gas sector building construction is included within the scope of approvals. Currently, there are different policy directions for the different natural gas facilities. For LNG receiving station project construction, since receiving capabilities are far ahead of actual shipping volume, China is strictly controlling such projects. For pipeline networks and gas reserves and other facilities, China is implementing an attitude of encouragement, encouraging various ownership systems to participate in gas reserve facility investment, construction and operation.[5] Therefore in practice the true barrier to an enterprise conducting LNG trade is whether or not it has corresponding facilities to accept the imported natural gas.

Even if the nation is attempting to encourage more enterprises to participate in natural gas import trade, thereby expanding imports, enlivening markets and increasing the degree of competitiveness, thanks to restrictive factors such as the limit on receiving stations, there is still only a limited number of enterprises engaged in natural gas trade markets (the three major oil companies, Huadian, Jiufeng, ENN and a small number of other companies).

Regarding receiving station construction, third-party access is not yet in place (policies have just been released) and thus various entities still wish to construct their own receiving stations. As a result, current receiving station capabilities have become much greater than actual shipping volume, but even so, domestic parties are still actively seeking to build new receiving stations. At the same time, the NDRC is strictly controlling construction of new receiving stations. In practice, whether or not approval can be obtained depends on an interplay between the central government, local government and enterprises, and there is a lack of openness and transparency in the approval standards and procedures, resulting in significant uncertainties.

With natural gas trade being restricted by facilities such as receiving stations, and receiving station construction also being closely controlled by the central government, the core question affecting natural gas trade is therefore: Can current domestic receiving station facilities provide true third-party access?

There is already a clear provision that has been provided in the recently released Natural Gas Infrastructure Construction and Operation Management Measures (implemented April 1 2014):

---

[4]See Commerce Department Office Opinions Regarding Several Issues Involving LNG Commodities, June 5, 2007.

[5]See NDRC Guiding Opinions Regarding Acceleration of Reserve Gas Facilities, April 5, 2014; NDRC Opinions Regarding Establishment of Effective Mechanisms to Ensure Stable Supply of Natural Gas, May 5, 2014.

Natural gas infrastructure operators shall announce and provide information regarding service conditions, obtained service procedures, and remaining service capacity, fairly and justly providing all users with pipeline shipping, storage, gasification, liquefaction, and compression services.

Natural gas infrastructure operators may not use their control of infrastructure to exclude other natural gas operators. Where service capacity exists, users meeting requirements may not be refused provision of service, or be presented with unreasonable requirements. Existing users preferentially obtain natural gas infrastructure service.

However, the third-party access policy requirements for natural gas infrastructure described above are somewhat vague, lacking specific details and a relevant series of procedures, and thus lacking practicality. In reality, because of the imperfection of the above policy provisions, and also due to current market structure, monopolistic powers are still strong, and without proper regulation the current policies relating to third-party access are essentially a blank document. As far as international natural gas pipeline trade is concerned, the enormous investment involved and complex geopolitical factors, and the extreme requirements these place on the relevant entities, currently appear to have resulted in an exclusive monopolistic structure, and in the short term this is unlikely to see fundamental change.

3. Price and tax policy

There are no review and approval requirements for trade commodity natural gas prices. So, from a policy perspective, natural gas price decisions are open, and are decided by the buyer and seller involved in the trade. This is particularly true of LNG, though in pipeline natural gas pricing there is indirect reliance on national negotiating power. However, this reliance is indirect, and pipeline natural gas contracts are still ultimately determined by the two parties involved in the trade. In terms of tax policy, policies on VAT rebates or discounts on natural gas imports are determined and adjusted by the Ministry of Finance. As China's natural gas pricing becomes better adjusted in the future, this practice could be halted (Table 8.4).[6]

8.4.2 Recommendations for Adjustments to China's Natural Gas Trade Policy

1. Encouragement imports and focus on expanding domestic production

In terms of natural gas import policy focus, China needs to balance three main conflicting factors: dealing with environmental pollution; the continued growth of domestic natural gas consumption as economic growth and residential living standards rise; and ensuring natural gas supply security. The first two require more natural gas usage, and to a certain degree will result in the import of more natural gas. The third, however, requires restraining natural gas import dependency at a certain level, and thus there could be a need to take some sort of control of the rapid growth of natural gas imports.

In the next period of time (from the present to 2020, 2030), the direction that should be taken in terms of natural gas import policy is: prudent encouragement. This is because currently China's natural gas sector's most fundamental problem is how to rapidly increase domestic consumption of natural gas, and whether it be in terms of proportion of energy consumption (6.5%) or residential diffusion rate (16%), domestic natural gas consumption is still in need of major increase. Considering that as economic development increases and residential living standards rise there will be a greater requirement for energy consumption models, there is a need

---

[6]Relevant policy references: Finance Department General Administration of Customs National Tax Bureau Notice Regarding Issues Surrounding Ratio-Based Return Import Stage VAT for 2011–2020 Imported Natural Gas and Pre-2010 "Central Asia Pipeline" Project Imported Natural Gas (CGS [2011] No. 39); Finance Department General Administration of Customs National Tax Bureau Notice Regarding Adjustments to Imported Natural Gas Preferential Support Policies (CGS [2013] No. 74); and Finance Department General Administration of Customs National Tax Bureau Notice Regarding Adjustments to Preferential Tax Policy Natural Gas Import Projects (CGS [2014] No. 8), April 21, 2014.

Table 8.4 China's existing natural gas trade policies

	Commodity trade	Facility construction	Facility operation	Price	Taxes	Foreign investment
Policy	"Natural gas" is considered to be a general commodity. So long as the business scope includes "natural gas trade" (when dealing with Administration for Industry and Commerce registration) and all Ministry of Commerce approvals have been obtained (routine procedures), the enterprise can carry out natural gas import business, and the signed trade contracts do not require approval and review from relevant authorities	Pipeline networks, gas reserves and other facilities are encouraged in construction, with participation possible by a variety of ownership structures. LNG receiving station construction is strictly controlled and foreign investment is barred from participation	Requires existing natural gas infrastructure operators to provide their excess capacity information and offer third-party access. Natural gas operators cannot control their infrastructure to exclude other natural gas operators. With regard to service capacity, they may not refuse users who meet the conditions, nor may they make unreasonable demands on them	–	For natural gas import VAT proportional rebate discount policies, the target of rebates is determined and adjusted by the Ministry of Finance	Approval required for foreign investment over a certain amount
Reality	Market entities involved in domestic natural gas trade are still very limited (only the three major oil companies, Huadian, Jiufeng, ENN, and a small number of other companies)	Currently, receiving station capacity is far higher than actual shipping volume, but gas receiving and pipeline facilities are severely lacking	Because of the imperfection of the policy provisions and due to the current market structure, monopolistic powers are still strong, and without proper regulation the policies released relating to third-party access essentially amount to a blank document	Price is determined by the trade participants and there are no approval or review requirements	–	–
Problems	In practice, the true barrier to an enterprise conducting LNG trade is whether or not it has corresponding facilities to accept the imported natural gas	Whether or not approval can be obtained depends on an interplay between the central government, local government and enterprises, and there is a lack of openness and transparency in the approval standards and	Natural gas infrastructure third-party access policy requirements are somewhat vague, lacking specific details and relevant series of procedures and missing corresponding operability	–	Not all natural gas importers are treated equally	–

(continued)

Table 8.4 (continued)	Commodity trade	Facility construction	Facility operation	Price	Taxes	Foreign investment
		procedures, resulting in significant uncertainties				
Recommendations	–	Strengthen facility construction approval standards, openness and transparency	Consider international experience in releasing operable rules for third-party access and strengthen corresponding regulation	–	–	–

for energy consumption that is more efficient, safer and cleaner. From this perspective, domestic natural gas consumption has entered a stage of rapid growth, and there is already strong inherent growth momentum.

Given the serious need to control environmental pollution, especially in terms of limiting atmospheric smog, a major increase in domestic natural gas consumption has become a pressing measure that must be addressed. So, driven by the need to rapidly increase natural gas domestic consumption, the formulation and announcement of policies for this sector must address this major issue. Thus policy incentives to increase natural gas supply and promote natural gas domestic consumption should be enacted.

At the same time, given that China has abundant natural gas resources (including conventional and unconventional gas), and that current exploration and development levels are still relatively low, ensuring natural gas supply security requires controlling external reliance on natural gas to a certain level, and therefore increasing natural gas imports to satisfy growing domestic demand for natural gas is not a viable option. A more reasonable policy is, at the same time as expanding natural gas imports, to give greater attention to domestic natural gas production, implementing a principle of "the nation first". In this way, natural gas import growth can be encouraged to satisfy domestic demand while also preventing dependency on imports and as domestic natural gas production volumes quickly rise. Therefore, if import dependency is to be controlled to a reasonable level without inhibiting imports, then creative efforts should be made to increase the domestic production volume— and this will also solve the dilemma caused by domestic natural gas consumption being insufficient. In addition, even at the same level of imports, import methods (import source diversification, contract and pricing diversification) and domestic infrastructure conditions (whether gas reserve and peak shaving capabilities are sufficient, and whether pipeline network facilities are well established) also have an important influence on natural gas supply security.

In summary, China on the one hand needs to enact policies that are supportive and encouraging toward natural gas imports, effectively expanding them and striving to diversify the country's import sources and import modes. On the other hand, it must also focus on expanding domestic production and optimising relevant gas reserves, pipeline networks and other facilities. Therefore, China should take the following attitude toward natural gas import policies in the coming years: prudent encouragement.

2. Make LNG an integral part of China's natural gas supply

Given the unique properties of LNG, its flexibility and the natural gas consumption centres nearby China's eastern regions, LNG can be used in peak shaving and complementary approaches. In addition, international experience has shown that connecting Chinese LNG procurement with international markets can introduce and promote competition, which will in turn promote market liberalisation in the China's natural gas market.

In the European Union market, LNG both balances out EU natural gas supply and demand and also balances out global LNG markets. This is beneficial to EU natural gas consumers, and LNG in Chinese markets can play a similar role. LNG can promote market openness while also playing a role in optimising energy structures, giving consumers greater benefits and also prompting further progress in the market liberalisation of the China's natural gas market.

3. Diversify imported natural gas, encouraging multilateral participation

China should take measures on the international market for import of natural gas. First, it should include long-term, mid-term, short-term and spot portfolios to facilitate increased flexibility to deal with and control market changes. Second, integration should be achieved with oil indexes and reserve typical diversified pricing indexes. Furthermore, when assessing price competitiveness, full consideration must be given to project risk, commercial structure and other factors, and through rolling contracts suited to one's own supply/demand balance, procurement of LNG can be achieved to relieve risk and achieve a natural gas supply portfolio that is sourced from diverse geographical locations.

At the same time as this, it is important to encourage more large-scale end users to purchase natural gas from international LNG markets. Whether it is power companies or large urban fuel enterprises, all of these will bring benefits to the market. This is because this will help to reduce intermediary stages and improve natural gas value chain efficiency. In order to achieve this goal, it is necessary to have third-party access, which is a key to supporting liberalisation.

4. Strengthen regulation promoting third-party access

Policies in the Chinese natural gas trade sector tend, on the whole, to be open and encouraging, though some specific rules still need to become more practical. This openness is clear from policies such as: seeking to diversify the countries imported from; importing more natural gas to resolve the issue of "domestic natural gas consumption being insufficient"; and promoting increased market supply while raising market competitiveness. Therefore, whether it is treating natural gas imports as a commodity, open trade and encouraging third-party access to facilities or VAT rebate discount policies, on the whole, further natural gas imports are being encouraged.

However, in reality, the construction and operation of receiving stations and other facilities have become key influences on natural gas imports. Third-party access has become a core issue affecting Chinese natural gas trade. Based on the current policies, even though there are relevant provisions, there is still a lack of common-sense workability and there is a major doubt as to how effective they will be.

Thus, on the one hand it is important to achieve better standard processes and greater openness and transparency in the approval of building new receiving stations and other facilities. On the other hand, it is necessary to learn

Table 8.5 China's natural gas supply volume (100 billion m³)

	2015	2020	2030
Conventional gas	1400	1800	2600–2800
Shale natural gas	65	400–600	800–1500
Coalbed methane	45	100–300	400
Coal methane	0	300–600	500–900
Import pipeline	390–420	970	1350–1650
Imported LNG	390–410	700	750
Total	2290–2340	4270–4970	6400–8000

Data source Analysis based on Ministry of Land and Resources and State Council Research Center study results

from international experience in order to bring in rules for third-party access and effectively strengthen the relevant regulation.

8.5   Conclusions for Chinese Natural Gas Supply and Availability

In summarising domestic natural gas production volume and international natural gas available volume, we can come to the following preliminary conclusions:

- **China's natural gas available volume total will rise from 234–239 billion m³ in 2015 to 425–495 billion m³ in 2020. The 640–800 billion m³ in 2030 will be a relatively stable and ample supply**: Specifically, by 2015, the amount is expected to reach 234–239 billion m³. As part of this, conventional gas production volume will reach 140 billion m³, shale natural gas will reach 6.5 billion m³, coalbed methane will reach 4.5 billion m³, imported pipeline gas will reach 39–42 billion m³ and imported LNG will reach 39–41 billion m³.

  By 2020, the amount is expected to reach 425–495 billion m³. As part of this, conventional gas production volume will reach 180 billion m³, shale natural gas will reach 40–60 billion m³, coalbed methane will reach 10–30 billion m³, coal methane production will reach 30–60 billion m³, imported pipeline gas will reach 95 billion m³ and imported LNG will reach 70 billion m³.

  By 2030, the amount is expected to reach 640–800 billion m³. As part of this, conventional gas production volume will reach 260–280 billion m³, shale natural gas will reach 80–150 billion m³, coalbed methane will reach 40 billion m³, coal methane production will reach 50–90 billion m³, imported pipeline gas will reach 135–165 billion m³ and imported LNG will reach 75 billion m³.

- **From the perspective of natural gas availability structure, domestic conventional gas proportion will drop, and domestic unconventional gas proportion will rise, with a rise in import pipeline gas and a drop in LNG imports**: In 2015, conventional gas proportion of total production volume availability will reach 60%, while shale natural gas will account for approximately 2.7%, coalbed methane approximately 2%, import pipeline gas approximately 17–18% and imported LNG approximately 17%.

  In 2020, conventional gas proportion of total production volume availability will drop to 36–42%, while shale natural gas will account for approximately 8–13%, coalbed methane approximately 2–7%, coal methane will account for 7–13%, import pipeline gas approximately 19–22% and imported LNG approximately 14–16%.

  In 2030, conventional gas proportion of total production volume availability will be 33–42%, while shale natural gas will account for approximately 11–21%, coalbed methane approximately 4–6%, coal methane will account for 6–13%, import pipeline gas

Table 8.6 2015 China natural gas supply volume and structure

Gas source	Domestic gas (100 million m³)				Imported gas (100 million m³)		Remarks
	Conventional natural gas	Shale natural gas	Coal methane	Coalbed methane	LNG	Pipeline gas	
Production volume or import volume	1400	65	45	50	390–410	390–420	
Total	1560				780–830		
	2340–2390						

Data source Integration, analysis, and compilation of study results from the Ministry of Land and Resources and other institutions

Table 8.7 2020 China natural gas supply volume and structure

Gas source	Domestic gas (100 million m³)				Imported gas (100 million m³)		Remarks
	Conventional natural gas	Shale natural gas	Coalbed methane	LNG	Pipeline gas	Conventional natural gas	
Production volume or import volume	1800	400–600	100–300	300–600	700	950	
Total	2600–3300				1650		
	4250–4950						

Data source Based on the above analysis and compilation

Table 8.8 2030 China natural gas supply volume and structure

Gas source	Domestic gas (100 million m³)				Imported gas (100 million m³)		Remarks
	Conventional natural gas	Shale natural gas	Coalbed methane	LNG	Pipeline gas	Conventional natural gas	
Production volume or import volume	2600–2800	800–1500	400	500–900	750	1350–1650	
Total	4300–5600				2100–2400		
	6400–8000						

approximately 18–25% and imported LNG approximately 9–12%.

- **From the perspective of foreign reliance, China's reliance on foreign natural gas will continue to rise significantly and be controlled to within 36% by 2030**: In 2015, China's reliance on foreign natural gas was approximately 33–35%, while in 2020 that number will reach approximately 33–39%. If domestic natural gas, especially shale natural gas, production volume achieves major increases through breakthroughs in technology or policy, then in 2030 China's foreign reliance in natural gas will be at approximately 27–36%. Namely, by 2030, China's reliance on foreign natural gas can be controlled within 36%. However, the prerequisite is that domestic natural gas production, especially unconventional natural gas, sees a breakthrough increase (Tables 8.5, 8.6, 8.7 and 8.8).

Case 9: Global natural gas/LNG market outlook

1. Global energy and natural gas market development

Global energy demand is continuing to rapidly grow, especially in nations and regions outside of OECD countries. As populations quickly grow and economies flourish, people have more numerous channels through which to obtain reliable power supply, and this will cause demand to grow for energy across the world by 50% from 2010 to 2050 (Shell 2014).

In the past several decades, natural gas has consistently been one of the fastest growing energies. From 2005 to 2014, average annual growth maintained at approximately 2.7% (BP 2014). Currently it is commonly recognised that up to 2030, natural gas demand will maintain annual average growth rates of 2% (IEA, Current Policies Outlook). Global natural gas resource reserves are massive and broadly distributed, and at current production speeds can be used for more than 200 years.

With the exception of the United States, growth in unconventional resources in other countries is very uncertain, because these countries have very little or no momentum toward production. Argentina (23 trillion m^3; source: US Energy Information Administration) could have the best development conditions, but faces severe financial and non-technical risk limitations.

Even though, according to the US EIA, China has approximately 32 trillion m^3 of unconventional resources, the Chinese government recently adjusted its recent development expectations for shale natural gas downward by half, and thus China's energy growth could be reduced. (China's natural gas is distributed across 500 basins, but because there are currently only a few companies, and there is a lack of international company involvement in this sector to carry out development, complex terrain, costs, infrastructure construction, waste water treatment and technical innovation insufficiencies and other problems are faced).

The national authorities and relevant policies will affect long-term LNG demand growth, and it is not a case of simple determination of project cycle costs, especially in Europe and Asia. Policymakers incorporate four aspects into energy policy portfolios, namely energy security, cost and competitiveness, environment and health, and energy availability. The choices they make will determine the energy structure of nations and regions.

This means that even though broad areas have become homogenous in terms of energy growth, nonetheless local and short-term dynamic factors have a high degree of uncertainty and changes are occurring rapidly.

2. Global LNG market outlook

Since 2000, global LNG demand has grown at a rate of approximately 5%, and this trend will continue for the foreseeable future (BP 2014).

In the past 10 years, there have been 31 nations engaged in LNG imports, and 27 nations engaged in export. Therefore, LNG demand and supply has enormously diversified, and estimates are that in the next 10 years, LNG demand nations and supply nations will grow respectively to 50 and 25.

Looking to the long term, LNG growth in demand will primarily come from Asian emerging markets (in particular China, India, and south-east Asia) and Europe. Even though LNG is only in early stages of development in the transportation sector, nonetheless LNG has major potential to become an important niche market.

Japan, South Korea and Taiwan currently account for approximately 60% of the global LNG market, and will continue to be a major demand centre in a moving world. Even though as nuclear power returns and renewable energy develops (especially solar power) Japan's demand for LNG will be limited, LNG demand will still continue to remain at a high level. Another uncertainty is future degree of market liberalisation.

In south-east Asia and India, because domestic gas reserves are declining, and because electricity and other industries are increasing demand for LNG, south-east Asia and India LNG demand will increase further. The major uncertainty in this region is price tolerance and the infrastructure construction needed for import and shipping of LNG.

China's policy to increase natural gas market share is critical. Given China's released policies for pollution control and GDP growth, China has already and will continue to become a market with rapid growth in natural gas. China has the potential to influence global market balance and price levels. Uncertainties are potential slowdowns in GDP growth, as well as degree of implementation of ambitious policies.

With regard to Europe, in addition to local production and pipeline import energy, LNG offers flexible supply. Europe has ample regasification capabilities (in 2013 for each 140 mtpa, 40 mtpa could be produced), and the region could become a flexible market for LNG. Europe's LNG used in transportation sectors has potential for growth, especially in marine shipping, because strict air quality laws will drive a fuel switch.

From 2010 to 2013, Europe's economic growth slowed, and with cheap American coal and significant subsidies in Europe for renewable energy development, natural gas demand in Europe was pushed lower. Even though to a certain degree the factors causing this reduction will exist in the long

term, European market natural gas demand could still rebound, with the precondition being reduced import of coal and reductions in subsidies supporting renewable energy growth.

3. Global LNG supply

By 2030, the United States, Australia, and Eastern Africa LNG supply will account for over 80% of total supply (source: 2015 Shell analysis).

- North America has already proposed utilising its over 6 million tons/year production capacity for LNG export (shared equally between the United States and Canada).
- A total of over 44 million tons/year of natural gas has already received a final investment decision and is under construction (locations all in the United States). North American natural gas supply quantities rely upon permits, financing, construction costs and speed.

Russia and Australia have a potential new project list. Cost increases and non-technical risks have slowed the progress of many projects. In addition, sanctions could delay or destroy the Russia's LNG plans.

Eastern Africa (primarily Mozambique and Nigeria) could in the next 10 years become new gas supply sources. Challenges faced include the fact that Eastern Africa has large swathes of undeveloped regions, and the majority of companies are not LNG companies with stable market presence.

- Nigeria's non-technical risks (such as political, regulatory and financial uncertainties, damage risks and domestic power demand) continue to represent the major obstacles to LNG supply growth.
- Due to domestic market demand, Egypt has seen a reduction to zero LNG

exports, and will even need to import LNG. Egypt is currently actively seeking LNG exporters (and has already signed contracts with Algeria's Sonatrach and Russia's Gazprom; additional natural gas supply could come from Israel or international enterprises).

- Qatar has not yet clarified its direction post-2016 for its northern oil fields, and thus there exists some supply uncertainty. Despite this, Qatar has 77 mtpa of LNG supply capability, and because it is equidistant from Europe and Asia, it will continue to hold an important price-determining position.

- For Middle Eastern regions, the removal of sanctions on Iran could bring about major effects, and natural gas supply is able to catch up to domestic demand growth. Therefore, LNG imports will be incorporated into planning.

4. **A focus on North American markets**

The United States was once one of Shell's largest markets, and now is preparing to export LNG. Potential export projects have not been announced for 300 mtpa in the United States and over 130 mtpa in Canada.

Initial project liquefaction costs will be relatively "low" (approximately $3.5/MMBtu) (addition of LNG import end), but subsequent costs will successively rise (we have already seen the initial price of <$3/MMBtu rise to the current quote of >$4/MMBtu) (Fig 8.15).

Such projects are equivalent to traditional Asia Pacific project levels, and are even at HH$4/MMBtu.

Other barriers are: approval, construction speed, construction costs (labour) and users seeking resource portfolios (if major Asia Pacific users restrict North American project development risks to 20–30%, then this will turn into 100 mtpa supply).

However, the former view of American LNG being "cheap" has changed, as oil prices have significantly dropped. Add to this the market uncertainty (for example Japan's deregulation of electricity), and this has caused users to limit LNG investment portfolio Henry Hub quoted supply volumes.

Therefore, when users are unwilling to sign large volumes of Henry Hub quoted long-term agreements, project final investment decisions will face major challenges, because final project investment decisions require the support of long-term agreements.

5. **Global LNG market is more active**

We have seen continuous increases in the market share held by natural gas in fundamental energy. Other energy industries are continually growing, but natural gas growth rates have been faster. We expect that in the next 20 years, global natural gas demand will grow at an annual rate of 2–3% (Fig. 8.16).

As natural gas usage increases, cross-border natural gas trade using pipelines and LNG will continue to increase.

The growth in natural gas trade requires that natural gas import and export infrastructure continue to receive investment.

Large new pipeline projects are continually achieving progress. For example, last year an early announcement was made of the Sino-Russian natural gas pipeline export project having early stage shipping volume that should reach 38 billion m^3/year, with the possibility of further increases. LNG has also become a major part of natural gas trade. Currently, LNG accounted for 10% of natural gas trade, and this will rise to 15% by 2025.

In recent years, LNG supply and demand has markedly diversified. There are now 30 LNG importing nations and 20 exporting nations, with this expected to change in the next decade to 50 LNG

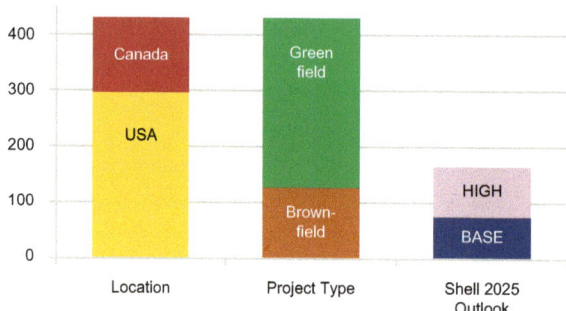

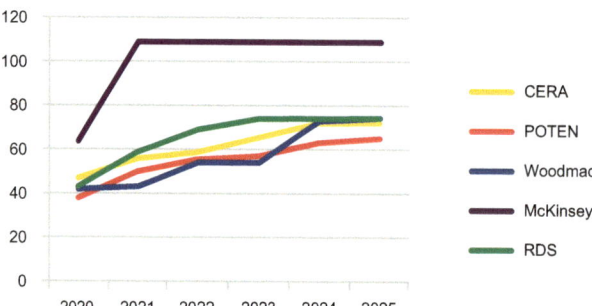

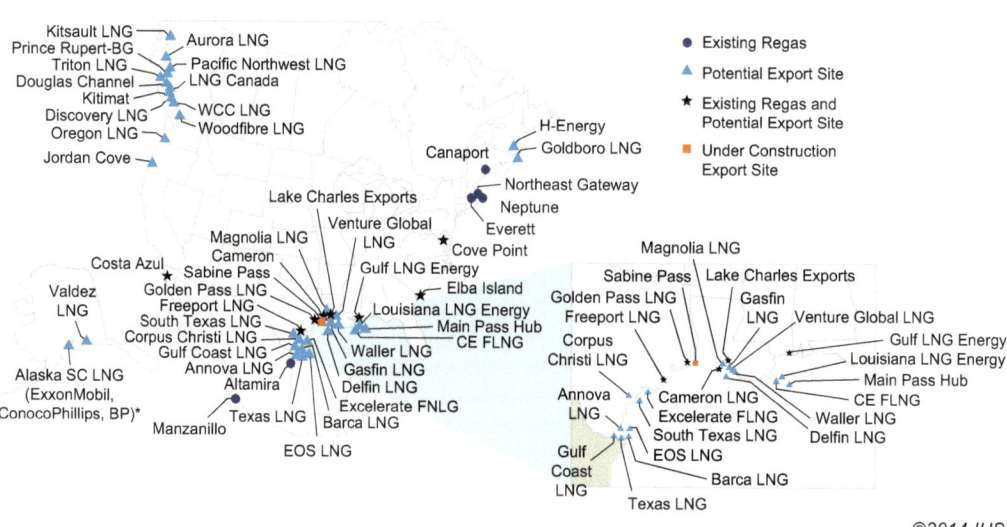

©2014 IHS

Fig. 8.15 North American LNG supply. *Source* IHS Cera. * Potentially located near Valdez

exporting nations and 25 importing nations. This is different from the situation in the 1990s when the global LNG market was only 50 million tons per year, and there were only eight exporting nations and nine importing nations.

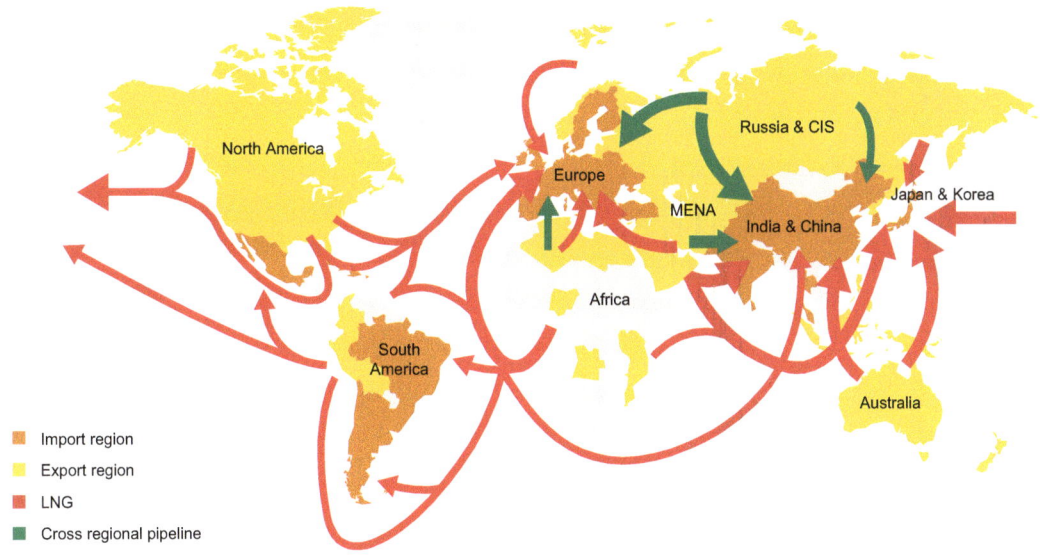

North America

Europe

Russia & CIS

MENA

India & China

Japan & Korea

Africa

South America

Australia

■ Import region
■ Export region
■ LNG
■ Cross regional pipeline

Fig. 8.16 Global natural gas import and export

From now until 2025, a major change will occur as North America rises as a substantive natural gas exporting region.

6. **Recommendations for China**

(I) **Multiple supply channels to ensure energy supply security**

In the past decade, China's natural gas consumption volume growth has been fierce, and in 2014 reached 183 billion m^3. In addition, natural gas supply has comprised domestic supply, pipeline natural gas imports and liquefied oil and gas imports. This shows that diversified supplies are an important guarantee for energy security. In the next decade, the government has designated the natural gas portion of energy structure to grow from 4% to over 10%. This will require multiple supply channels to satisfy market demand. Furthermore, considering the market scale at that time, the wide diversity of supply channels will play an important role in guaranteeing energy security.

In the north-east Asian regions of Japan and South Korea, reliance is on single natural gas supply sources, and this is markedly different from China. Precisely because of this diversity in supply portfolios, China enjoys benefits when procuring natural gas from international markets, ultimately helping Chinese consumers.

(II) **LNG's unique role makes it an integral part of China's natural gas supply portfolio**

Considering that China has constructed import channels to import natural gas via pipeline corridors (whether it be from Western Central Asia, Eastern and Western Russia or south-western Myanmar), and these are maturing by the day, this case study focuses on the possible choice of LNG as a supply source on the international market.

Given that LNG has the unique properties of being flexible and near to the eastern consumption centres, LNG can play a unique role in peak shavings and as a complement to pipeline gas.

In addition, prior experience has shown that because Chinese LNG procurement is directly linked to international markets, this can introduce and promote competition, through which Chinese natural gas liberalisation can be catalysed.

Previously we have discussed the role of LNG in EU markets. On the one hand it balances out EU natural gas supply, while on the other it balances out the global LNG market, benefiting natural gas consumers in the European Union. We have reason to believe that LNG can also play a similar role in the Chinese market, but it is important to realise that the reason that LNG could play such a role in the EU was because of the historical progress of market liberalisation. Therefore, we would say that LNG on the one hand can promote market opening, bringing benefits to consumers, while on the other hand it can also assist as China completes the liberalisation of its natural gas market.

(III) **LNG procurement considerations**

By observing similar markets, it is possible to obtain some lessons:

1. Experienced LNG buyers on international markets often have a diversified portfolio, including long-term, medium-term, short-term and spot portfolios to facilitate improvements to flexibility and to meet market changes. They also have a connection to oil indexes or a hub as a platform of diversified pricing mechanisms, because when the market changes, for example when oil prices change, the price competitiveness of each sales and purchase agreement will likewise change. In addition, when assessing price competitiveness, it is necessary to wholly consider project risk, commercial structure and other factors, thus avoiding only considering price labels, otherwise it is not a valid comparison. LNG supply portfolios must have a diversity of geographical sources. Another commonly made error is to

believe one is smarter than the market and able to manipulate the market. Therefore, a commonly seen method of risk avoidance is to procure LNG through rolling windows to satisfy one's own supply and demand balances.

2. If more numerous large end users can purchase natural gas from international LNG markets, whether it be power companies or large urban fuel enterprises, this would bring benefits to the market. This is because such actions help to reduce intermediary stages and make the natural gas value chain more efficient. To achieve this, though, third-party access is needed, and this is a key factor to supporting progress in liberalisation. We are very happy to see companies such as Huadian becoming active in the realm of liberalisation.

3. However, this is not to say that every enterprise should be encouraged to establish its own receiving terminal and carry out importing, since this would cause a large quantity of infrastructure to be less than amply utilised, resulting in price wars and ultimately harming the sustainable and healthy development of natural gas markets. In this regard, we recommend that managers make adjustments through formulation of game rules rather than artificial controls.

4. As regards the game rules for the natural gas sector, including natural gas pipeline networks and receiving station infrastructure, management standards are extremely important. This is because as pipeline network standardisation becomes more mature and plays a greater role, it will increase natural gas value chain transparency, thus improving market efficiency. Taking the European Union as a typical mature market, they primarily use the following typical pipeline management rules

to manage major issues: (a) requiring non-discriminatory third-party access rule restrictions; (b) capacity auctions, balance, pricing, distribution and receipt issuance; (c) holding full negotiations; (d) natural gas delivery pipeline network capacity release and price schedules; (e) auction trades held in secondary markets; (f) separation of input/output, separation of capacity/commodity; (g) third-party access to receiving stations, gas storage, and pipeline networks, whether it be established through bilateral negotiations or based on TPAs.

5. Maintaining natural gas/fuel oil portfolios is also of assistance, because this will enable natural gas users to optimise gas portfolios through switching, thus avoiding gas users seeking to resolve all problems through LNG SPA, which pushes LNG purchase contract negotiations into a stalemate.

Case 10: China international pipeline natural gas imports

China has an advantageous geographical location for pipeline natural gas imports, with Russia to its north, central Asia (especially Turkmenistan) to its west and south-east Asia (especially Myanmar). Existing and planned import pipeline capacity is approaching 190–200 billion m^3/year, which is likely to be one quarter to one third of the natural gas consumption volume of China expected in 2030.

Pipeline gas imports are one of the three main supply sources for China, with the other two being domestic natural gas production and LNG imports (respectively having multiple supply channels). These three fundamentals ensure China's natural gas supply while also forming competition with regard

to baseline prices, thus satisfying the natural gas demand that continues to grow in China.

Regarding the greatest factors of uncertainty for these international natural gas pipeline imports, they are as follows:

1. Russia's natural gas supply base for the east-west line.
2. Whether Russian companies have the ability to develop the eastern Siberian natural gas field and create natural gas pipelines and deliver natural gas within the promised timeframe. Given that the United States and European Union are implementing sanctions on Russia, this is especially critical.
3. In order to enrich client groups, what decisions will Turkmenistan make? For example, participating in the Turkmenistan-Afghanistan-Pakistan-India (TAPI) natural gas pipeline and/or overcoming legal and political difficulties to co-operate with Azerbaijan and directly supply Turkmenistan natural gas to Europe through Azerbaijan and Turkey.

1. **Pipeline natural gas import sources**

As can be seen from Table 8.9, the world has 14 nations whose conventional natural gas reserves exceed 100 trillion m^3, and based on its advantageous geographical location, China can import natural gas from essentially any country.

Currently, Iran, which is subject to United Nations sanctions, has the largest proved natural gas reserves in the world, with proved conventional natural gas reserves of nearly 120 trillion m^3. If developed, Iran will very likely become like Qatar, another LNG and unconventional natural gas supplier to China.

Russia and Turkmenistan have natural gas resource reserves that respectively rank second and fourth globally. Russia

Table 8.9 Most proved conventional natural gas reserves by country

	Proved reserve quantity		2013 production volume	R/P
	Trillion m^3	Trillion feet^3	1 billion m^3/year	Year
Iran	33.8	1193	167	202
Russia	31.3	1104	605	52
Qatar	24.7	872	159	155
Turkmenistan	17.5	617	62	282
United States	9.3	330	688	14
Saudi Arabia	8.2	291	103	80
UAR	6.1	215	56	109
Venezuela	5.6	197	28	199
Nigeria	5.1	179	36	141
Algeria	4.5	159	79	57
Australia	3.7	130	43	86
Iraq	3.6	127	0.6	5980
China	3.3	116	117	28
Indonesia	2.9	103	70	42

Source BP Data Review (2014) (only conventional natural gas)

produces 605 billion m^3 of natural gas each year, to satisfy enormous domestic demand and to meet 30% of Europe's demand for natural gas. Their reserve mining term is 52 years, which is relatively short compared to other major natural gas resource reserve nations. Russia has also promised to supply China with nearly 90 billion m^3 of natural gas per year through the Eastern and Western lines. In doing so, unless new resources are discovered, Russia's reserve mining term will be markedly shortened. This could perhaps explain why Russia is so urgently wishing to develop unconventional natural gas, including its domestic coalbed methane and shale natural gas.

In comparison, the landlocked Turkmenistan has 617 trillion m^3 of proved natural gas and over 280 years of natural gas reserves to mine. It is also very likely to enrich its own natural gas demand market, supplying natural gas to China, Russia and Iran. Two considerations for the future are worth noting:

- TAPI can provide natural gas to two countries with major populations—India and Pakistan. According to previous reports, Chevron and other US companies are looking to join this plan, and recent reports show that French company Total might also participate in relevant discussions.

- The European Union urgently wishes to increase its supply diversity, and the pipeline through TAPI is one important potential source. The European Union, including Vice President of the European Commission Maroš Šefčovič, is actively encouraging Turkmenistan to plan out this pipeline (Financial Times 2015), despite the fact that in the short term there are many legal and political roadblocks to its achievement. If the TAPI pipeline is realised, Europe and China could face direct price competition/connectivity.

Kazakhstan and Uzbekistan and other central Asian nations are not major natural gas reserve nations, and in the foreseeable

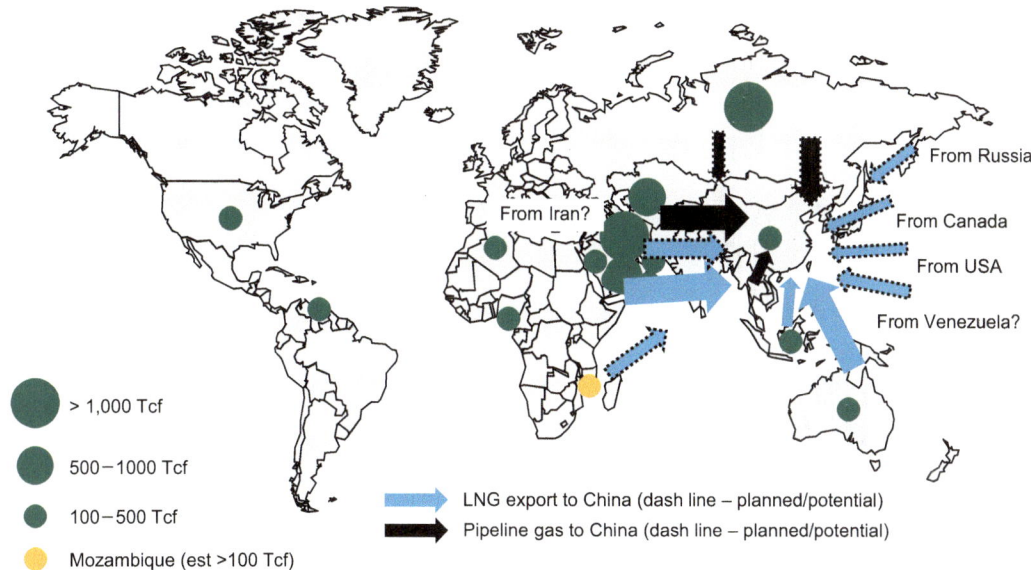

Fig. 8.17 Major natural gas resource reserve nations and China's potential natural gas export nations

Table 8.10 China's existing and planned natural gas import pipelines

	Capacity (1 billion m³)	Announced term	PFC expected time	Remarks
Imported natural gas pipelines (operation)				
Central Asia natural gas pipeline Line A	15	2009		
Central Asia natural gas pipeline Line B	15	2010		
Myanmar-China 1	12	2013		Not yet at full load operations (>50% idle capacity)
Sub-total	42			
Imported natural gas pipelines (planned)				
Central Asia natural gas pipeline Line C	25	2014		10 billion m³ comes from Turkmenistan, 10 billion comes from Uzbekistan, and 5 billion comes from Kazakhstan
Central Asia natural gas pipeline Line D	30	2016	2017	
Siberian pipeline	38	2018	2021	Signed natural gas export agreement. Under construction
Siberian pipeline extension	30			
Azerbaijan (West Line)	30			Signed memorandum
Kazakhstan–China	5			
Myanmar–China 2	7			
Sub-total	165			
Pipeline import capacity total	207			

Source WoodMac, HIS. Basic arrangement by study group

future there is still limited possibility for them to become natural gas pipeline supply nations to China.

In south-east Asia, Myanmar is currently supplying approximately 5 billion m^3 of natural gas per year to China through Yunnan. Likewise lacking in natural gas are south-east Asian nations, and Myanmar is currently supplying Thailand with pipeline gas. Recently, Myanmar successfully obtained a deep sea oil tender, attracting the attention of large transnational companies. According to geological analysis, these deep sea regions have the potential to have more natural gas than oil. In the next 5–10 years, could Myanmar find a large quantity (larger than 1 trillion m^3) of deep sea natural gas to become the Mozambique of Asia? If it does happen, China should consider making Myanmar's (deep sea) natural gas another source of pipeline delivery. It could necessitate a round of competition, with Myanmar perhaps hoping to develop other LNG markets in Asia to improve its own export capabilities in LNG, for example Japan and South Korea, so as to expand its buyer nation group.

2. **Existing and planned natural gas import pipelines**

Table 2.16 lists the import natural gas pipelines from Russia, Central Asia and Myanmar to China. Currently, China's pipeline import capacity is approximately 40 billion m^3/year, but this is not to say that each line is always at capacity. In 2013, China's pipeline import volume was approximately 25 billion m^3, accounting for approximately 15% of China's natural gas consumption of 166 billion m^3.

Once they are fully completed, these pipelines will have total capacity of nearly 200 billion m^3/year, and will account for one quarter to one third of all of China's estimated natural gas consumption by 2030. Therefore, the focus of the 13th Five-Year Plan may not be confirming added pipeline natural gas import volume, but joint efforts with Turkmenistan, Russia and Myanmar to ensure that these projects can be completed on time so that China can import pipeline natural gas at a competitive price (Fig. 8.17, Table 8.10).

Analysis of China's Natural Gas Infrastructure Development Strategy

9

9.1 Current Development of Natural Gas Infrastructure

Since the West-East Pipeline began operating in 2004, China's natural gas infrastructure has developed rapidly. As of the end of 2014, a pipeline network distribution layout has been formed with the West-East Pipeline, marine gas and near-region supply, providing coverage to all provinces with the exception of Tibet. There are also both pipeline gas and LNG import resource channels, opening up central Asia's pipeline gas, central Myanmar's pipeline gas and marine LNG import channels. There is also underground gas storage and LNG receiving stations as two major peak shaving methods, covering the seaboard region, gas-producing regions and regions around the Bohai Sea.

* This chapter was overseen by Zhonghong Wang from the Development Research Center of the State Council and Shangyou Nie from Shell International Exploration and Production. It was jointly completed by Xiaowei Xuan, Yingxie Yang, Yongwei Zhang, Jiaofeng Guo, Weiming Li, Beiqing Yao, Tao Hong and Yuxi Li from the State Resource Department Oil and Gas Centre, Xiaoli Liu from the National Development and Reform Commission's Energy Office, Guang Yang and Jianhong Yang from the China Energy Research Institute's Natural Gas Centre, Xiaobo Ju and Beijing Yao from Shell China. Other members of the topic group participated in discussions and revisions.

9.1.1 Current State of Infrastructure Development

As of the end of 2014, China had constructed 82,000 km of natural gas pipeline, of which 23,300 km are national trunk pipelines, 16,500 km are national branch pipelines and 13,800 km are provincial trunk pipelines, with major progress achieved in terms of urban gas pipeline networks. In addition, 11 LNG receiving stations have been built, with total receiving capacity of 39.40 million tons. Constructed underground gas storage bases number 19, with total warehousing capacity of 45.2 billion m^3 and designed working gas volume of 15.1 billion m^3, and effective working gas volume of 4.2 billion m^3.

1. **Natural gas pipeline networks**

Since the 1960s when China built its first natural gas pipeline, the "Bayu Line", over many decades of development, China's natural gas pipelines have undergone significant growth. This is especially true with the 2004 West-East Pipeline, which accelerated China's natural gas pipeline construction. As of the end of 2014, China had the West-East Pipeline, Shaanxi-Beijing Line, Sichuan to East Gas Pipeline Project, the Yuji Line and China-Myanmar natural gas pipeline as major national trunk pipelines, and the Jining Line, Zhongwu Line, Zhonggui Line and Huaiwu Line as major connection pipelines, with regional pipeline networks between the Yangtze River Delta, Sichuan and

Shell International and The Development Research Center (Eds.), *China's Gas Development Strategies*, Advances in Oil and Gas Exploration & Production, DOI 10.1007/978-3-319-59734-8_9

Chongqing, and North China. This has achieved a long-distance transport pipeline connectivity to major consumption markets, as well as to underground gas storage and LNG receiving stations, providing coverage to all provinces with the exception of Tibet and preliminarily forming a national natural gas grid.

2. LNG receiving stations

As of the end of 2014, China's operating, approved and under construction LNG receiving stations numbered 15, of which 11 were currently in operation, and four had been approved and were under construction, located in over 10 provinces and regions including Guangdong, Fujian, Shanghai, Zhejiang, Huainan, Jiangsu, Liaoning, Shandong, Tianjin, and Tangshan. Currently operating LNG receiving stations have a receiving and unloading capability of 39.40 million tons per year, which is equivalent

to 55 billion m^3 of natural gas. In 2014, annual imported LNG resources amounted to 18.79 million tons, equivalent to 26.3 billion m^3 of natural gas (Figs. 9.1 and 9.2).

3. Underground storage

As of 2014, China had constructed the following underground storage facilities: Lamadian North, Dazhangtuo, Ban 876, Banzhong North, Ban 808, Ban 828, Jinyun, Jing 58 Storage, Wen 96, Suqiao, Xiangguo Temple, Hutubi, Shuang 6 and Bannan. These were primarily distributed in eight provinces—Jiangsu, Tianjin, Hebei, Liaoning, Heilongjiang, Xinjiang, Chongqing and Henan—with total designed capacity of 45.2 billion m^3 and designed total working gas volume of 15.1 billion m^3 and effective working gas volume of 4.2 billion m^3 (Fig. 9.3).

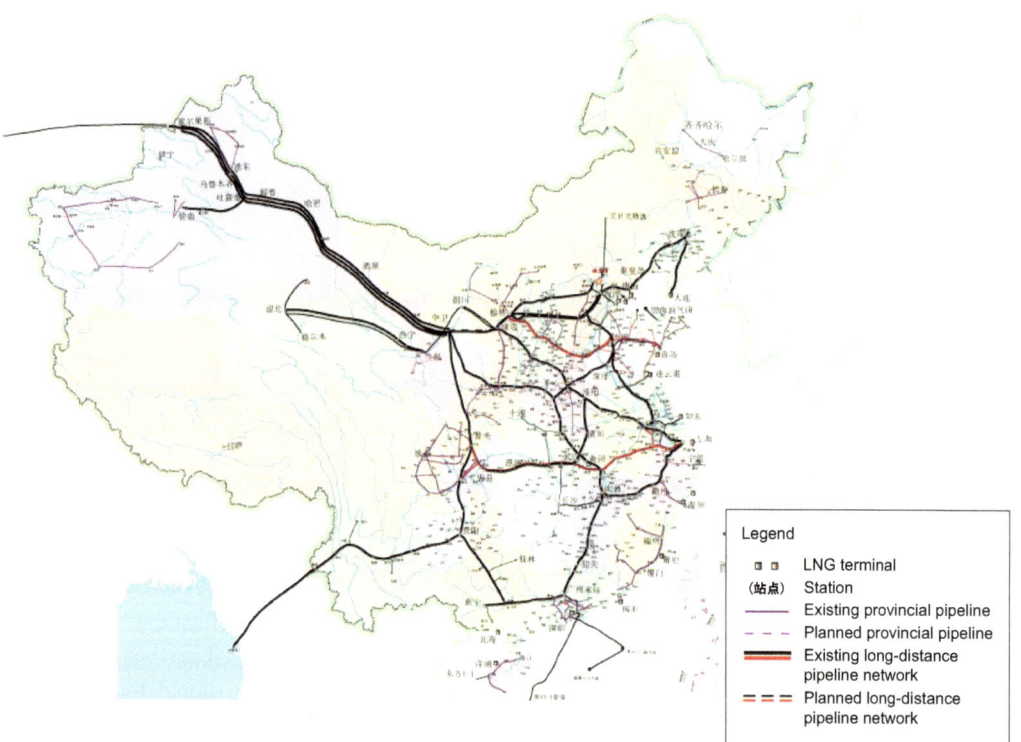

Fig. 9.1 China's natural gas pipeline network distribution

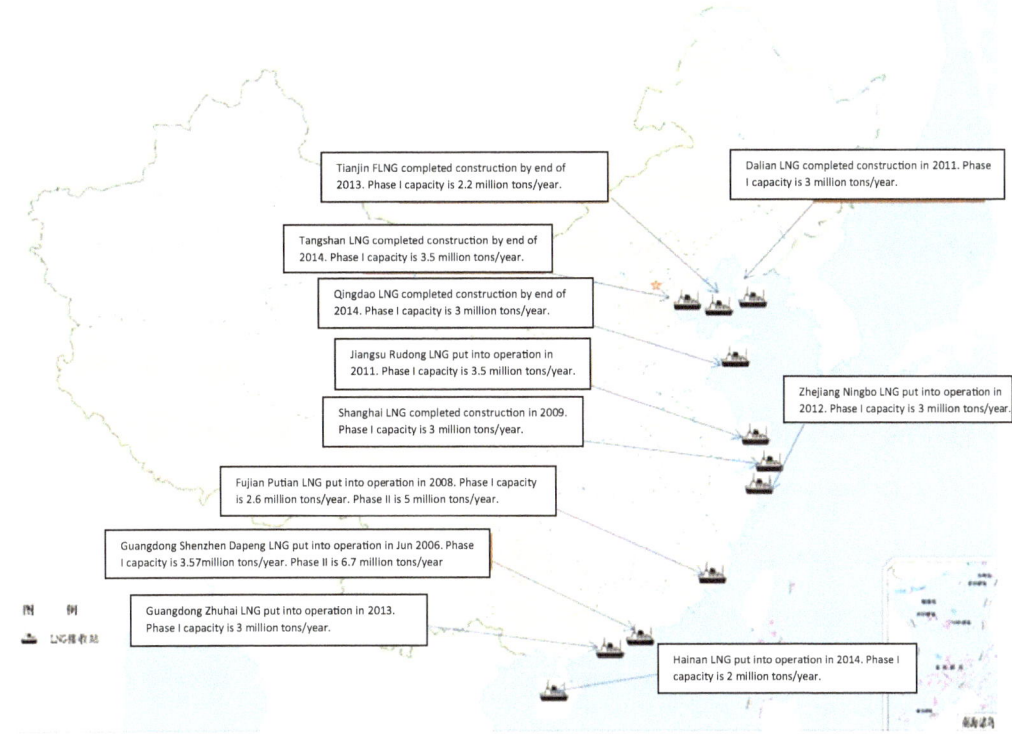

Fig. 9.2 LNG receiving station distribution map

9.1.2   Assessment of Development Levels and Existing Problems

1. Pipeline construction capacity shortcomings

With the completion and operation of the West-East pipeline in 2004, China's natural gas pipeline length reached 25,000 km. In 2010, after the West-East Pipeline 2 and the Sichuan to East Pipeline, China's natural gas pipeline length reached 56,000 km. As of the end of 2014, pipeline network length was 82,000 km, with natural gas consumption volume having reached 183 billion m³. According to the natural gas industry development experience of the United States, when consumption volume is 130 billion m³, pipeline length was 175,000 km.

Whether it is in terms of satisfying China's natural gas usage demands or when compared to the history of developing nations, China's natural gas pipeline construction is markedly underdeveloped.

As of the end of 2014, China's natural gas pipeline network distribution capabilities were 240 billion m³, with natural gas consumption exceeding 128 billion m³. Based on market consumption volume of 1.2 fold as a peak shaving demand, the required pipeline distribution capacity was 220 billion m³. From this it is clear that the current pipeline network distribution capabilities can only meet the market demand of the next several years, and if future natural gas market development is to be supported, accelerated promotion of pipeline network facility construction will be required (Fig. 9.4).

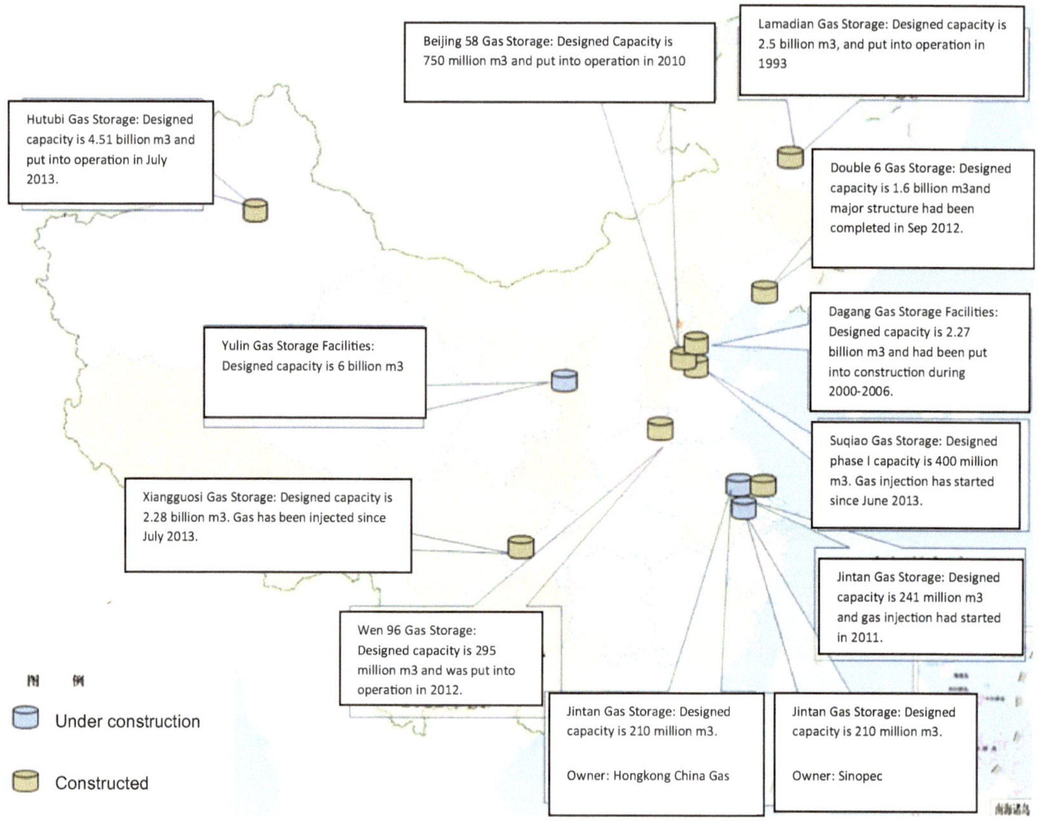

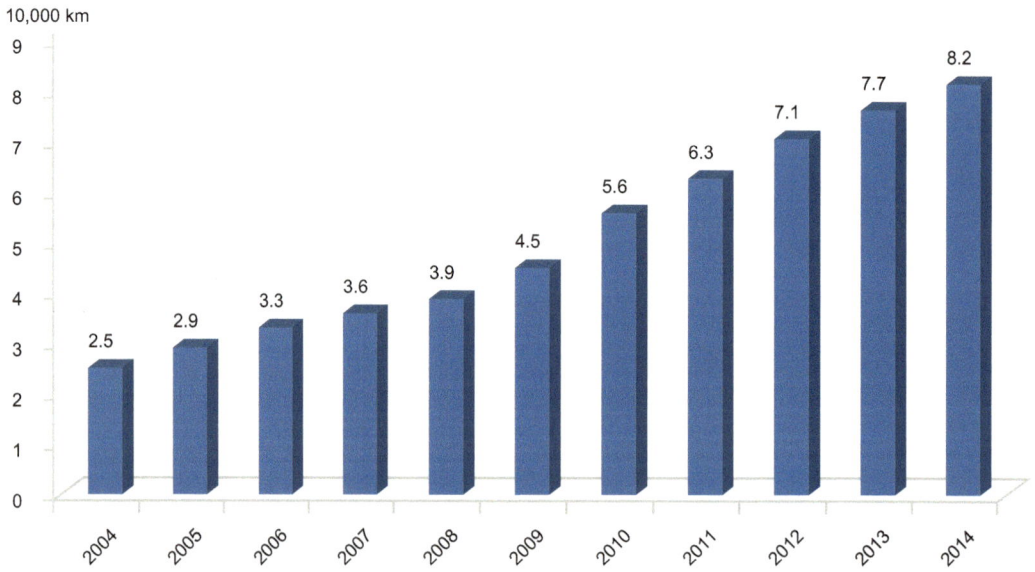

Beijing 58 Gas Storage: Designed Capacity is 750 million m3 and put into operation in 2010

Lamadian Gas Storage: Designed capacity is 2.5 billion m3, and put into operation in 1993

Hutubi Gas Storage: Designed capacity is 4.51 billion m3 and put into operation in July 2013.

Double 6 Gas Storage: Designed capacity is 1.6 billion m3and major structure had been completed in Sep 2012.

Yulin Gas Storage Facilities: Designed capacity is 6 billion m3

Dagang Gas Storage Facilities: Designed capacity is 2.27 billion m3 and had been put into construction during 2000-2006.

Suqiao Gas Storage: Designed phase I capacity is 400 million m3. Gas injection has started since June 2013.

Xiangguosi Gas Storage: Designed capacity is 2.28 billion m3. Gas has been injected since July 2013.

Jintan Gas Storage: Designed capacity is 241 million m3 and gas injection had started in 2011.

Wen 96 Gas Storage: Designed capacity is 295 million m3 and was put into operation in 2012.

Jintan Gas Storage: Designed capacity is 210 million m3.

Owner: Hongkong China Gas

Jintan Gas Storage: Designed capacity is 210 million m3.

Owner: Sinopec

Under construction

Constructed

Fig. 9.3 Underground gas storage distribution

10,000 km

Year	Value
2004	2.5
2005	2.9
2006	3.3
2007	3.6
2008	3.9
2009	4.5
2010	5.6
2011	6.3
2012	7.1
2013	7.7
2014	8.2

Fig. 9.4 China's natural gas pipeline construction

2. Development of LNG infrastructure

Since 2006 when the Dapeng LNG facility was completed, China's LNG receiving station construction speed has gradually accelerated. As of the end of 2014, constructed and operating LNG receiving stations numbered 11 in total, with receiving and offloading capacity of 39.40 million tons per year, equivalent to 55 billion m^3 of natural gas.

Because current LNG receiving stations primarily satisfy city demand in a radius around the receiving station itself, and each region's natural gas development has had a relatively long and staged development, LNG receiving stations are quick to enter production once they are completed. Also, because of regional preferential development of urban fuel, and limited increases in volume, in order to resolve the lack of market space for LNG receiving stations as a result of their rapid growth, accompanying fuel gas plants have been established to consume the exceed gas volume. As markets continue to develop, urban gas demands are gradually growing, and price tolerance is higher than power plants, making it easy for a power plant to have ample gas volume in early stages of development of a receiving station, but once the market develops to need LNG as a major gas source, the power plants could see supply shortages.

Therefore, China's current LNG receiving stations have played an important role in increasing supply capacity, and their strong abilities in terms of peak shaving offer critical effects in maintaining secure and stable operation of the market. However, their rapid growth can lead to uneven development of natural gas infrastructure.

3. Shortage of underground gas storage peak shaving capability

China's natural gas peak shaving method is primarily based on underground gas storage. Location is very particular for gas storage, with long construction cycles and a significant quantity of gas required to build the reserve. Underground gas storage has high requirements for geological conditions, and it is difficult to find

suitable locations. Salt cavern underground storage facility construction periods are long, for example the Jinyun underground storage location took over 10 years from commencement of construction to operation. In addition, underground gas storage requires significant input for the initial reserve, and with a designed gas volume of 200 million m^3, approximately 100 million m^3 are required as an initial reserve.

China's research and construction of underground storage began relatively late, and there is a relatively major gap as compared to developed nations. According to operational experience in foreign markets, if natural gas market operational risk remains at relatively low levels, then the storage dissipation ratio (referring to the ratio between natural gas peak shaving volume and natural gas demand volume) should reach 15%. In 2014, China's natural gas apparent consumption reaches 183 billion m^3, with underground storage peak shaving volume of 4.2 billion m^3, making peak shaving volume account for less than 3% of total natural gas consumption, thus meaning that peak shaving capabilities are severely lacking.

9.2   Opportunities and Challenges

9.2.1   The Energy Development Strategy Action Plan (2014–2020)

The Energy Development Strategy Action Plan proposed major development of natural gas, with non-renewable energy consumption totals to be controlled by 2020 to within 4.8 billion tons of standard coal, and natural gas to account for over 10% of non-renewable energy consumption. Based on principles of developing land and marine resources and conventional and unconventional sources, acceleration of conventional natural gas production would be sought, resolving development battlements as quickly as possible in unconventional natural gas, thus prompting natural gas reserve production volume growth. Based on the gas supply layout involving the West-East Pipeline, the North to South Gas Pipeline and the Sea to Land Pipeline, natural gas pipeline and reserve facilities were to be constructed more rapidly, thereby forming a national natural gas trunk pipeline network connecting production regions and consumptions regions. There was a clear proposal to reach over 120,000 km of natural gas trunk pipeline by 2020.

9.2.2 Policy Catalyses Rapid Development of the Natural Gas Industry

"Several Opinions Regarding Establishing an Effective Mechanism to Ensure Stable Natural Gas Supply" proposes acceleration of the development of natural gas supply, with 400 billion m^3 of natural gas supply capability by 2020 and striving to reach 420 billion m^3, supporting the progress of the "Coal to Gas" project. Further support is proposed for various market entities to participate equally in gas reserve facility investment, construction, and operation, with various regions to strengthen their gas reserve peak shaving facilities, LNG receiving stations, and storage facilities. "Guiding Opinions of the NDRC Regarding Acceleration of Progress in Gas Reserve Construction" proposes strengthening the sense of urgency regarding gas reserve facility construction, speeding up in-progress construction and encouraging various ownership models for construction and operation of gas reserves. Gas reserve facilities are the guarantee for safe and stable natural gas operation, and the nation supports various entities in participating in gas reserve facility construction, operation and management, promoting the construction of such facilities.

9.2.3 The Atmospheric Pollution Prevention Action Plan

In order to deal with atmospheric pollution and to improve air quality, the Atmospheric Pollution Prevention Action Plan was formulated, which proposes an acceleration of clean energy substitution, raising supplies of natural gas, coal natural gas and coalbed methane. By 2015,

newly added natural gas trunk line pipeline capabilities will reach over 150 billion m^3, offering coverage to Beijing-Tianjin-Hebei, the Yangtze River Delta and the Pearl River Delta. Optimisation of natural gas usage methods will be overseen, adding new natural gas that preferentially ensures residential usage or substitution of coal, thus encouraging development of natural gas distributed energy and other highly efficient utilisation projects while limiting the development of natural gas chemical engineering projects. This approach enables an orderly development of natural gas peak shaving power stations. Infrastructure for natural gas is an important assurance for market development, and the powerful implementation of "Coal to Gas" has accelerated efforts in infrastructure construction.

9.2.4 Existing Pipeline Network Capabilities Are Insufficient

According to resource support and market demand analysis, it is expected that by 2020, China's natural gas market demand for natural gas will be 400 billion m^3, and by 2030 it will be 600 billion m^3. If 1.1 times the demand volume can provide resource supply assurance, demand volume of 1.15 times shipping capabilities can assure market distribution. By 2020, pipeline network distribution capabilities should reach 260 billion m^3/year, and 690 billion m^3/year in 2030. As of the end of 2014, China's resource distribution pipeline network capabilities were 240 billion m^3, only satisfying 52% of distribution demand expected for 2020 and 35% that of 2030. There is a need to accelerate promotion of infrastructure construction to satisfy the needs of market demand.

9.2.5 Peak Shaving Capabilities Are Severely Lacking

Calculating based on 15% at the storage dissipation ratio, in 2020 China's natural gas market demand will be 400 billion m^3, and peak shaving demand will be 60 billion m^3. By 2030, peak shaving demand will be 90 billion m^3. Currently, underground storage peak shaving capabilities are only at 4.2 billion m^3, and LNG receiving stations have peak shaving capability of 2.6 billion m^3, calculated based on 0.2 times that steady distribution of gas at market peak shaving. From this it is clear that current peak shaving facility capabilities are far from adequate to meet future natural gas market peak shaving demand.

9.2.6 Increasing Pressure for Safe Operations

As natural gas markets continue to grow, infrastructure construction is likewise building, and there is increasing pressure for safe and steady operations. Currently, China's trunk pipelines continue to use a point-to-point resource supply, with different operators lacking interconnectivity between their pipelines. Provincial trunk line investment entities are more diversified, and pipeline supplies do not intersect. At peak gas periods, resource allocation is impossible, and when a pipeline network encounters malfunction there is no effective energy response approach. In future, China's natural gas trunk pipelines should establish interconnectivity and provincial trunk lines should be freely allocable.

9.3 Natural Gas Infrastructure Development Strategy

China's overall natural gas business and policy guidance has formulated natural gas development objectives, proposing a natural gas infrastructure development strategy to guide the rapid development of natural gas infrastructure.

9.3.1 Guiding Considerations

Taking the safe and stable operation of national pipeline networks as a target function, infrastructure construction should be planned to achieve diversified supply. The aim should be to construct flexible allocation infrastructure systems, strengthening gas pipeline hub construction, speeding up peak shaving facility construction,

establishing interconnectivity and securing gas supply systems that can be flexibly allocated. Modern natural gas industry systems need to be constructed, with stable supply, efficient operation and co-ordinated upstream, midstream and downstream development, and a national unified infrastructure development strategy needs to be established.

9.3.2 Development Objectives

Bearing in mind the guiding considerations for natural gas infrastructure development, and in the context of China's natural gas industry development, the natural gas infrastructure development goals for China by 2030 are as follows:

- Achieve interconnected gas supply systems for various operators at top-tier natural gas pipelines. Pipeline network length will reach 150,000 km by 2020, and gas delivery capability will exceed 460 billion m^3. 2030 will see 250,000 km of pipeline network and gas delivery capability of over 690 billion m^3.

- Achieve a four-pronged layout that includes West-East Pipeline, North to South Pipeline, Sea to Land Pipeline, and local supply. Optimise the natural gas supply and allocation stations in Ningxia Zhongwei, Hubei, Hebei Yongqing, Shanghai and Guangdong Wuqu. Construct six major underground reserve groups in the central plains region, northern China, the north-east, in Changqing, in the north-west and in the south-west. Aim to achieve 62 billion m^3 of peak shaving capability by 2020, with an effective underground reserve working volume of 44 billion m^3. Achieve 90 billion m^3 of peak shaving capability by 2030, and an effective underground reserve working volume of 77 billion m^3.

- Establish four major LNG import channels based on north-west China-Asia imports, south-west China-Myanmar imports, northern China-Russia imports and eastern imports. From 2020 to 2030, achieve 85 billion m^3 of distribution capability on the China-Asia

Lines A, B, C and D, with 12 billion m^3 of capability on the China-Myanmar line, 38–68 billion m^3 on the China-Russia line, and 100 million tons of receiving capability through LNG imports.

9.3.3 Development Strategy

In order to guarantee that China's natural gas infrastructure is designed for safe and stable operation, interconnected systems between the nation's various entities involved in natural gas infrastructure should be firmly established. To strengthen infrastructure construction capabilities, there should be overall planning for facility construction, building up diversity of investment and implementation. Finally, in order to improve infrastructure usage efficiency, fair third-party access operation and management platforms should be created.

As natural gas market demand further grows, China's natural gas infrastructure must be further optimised, and by 2020 important pipeline construction must be completed, including the China-Russia natural gas pipeline, the West-East Pipeline 3, West-East Pipeline 4, West-East Pipeline 5, Shaanxi-Beijing, Xinjiang coal methane pipeline, Erdos coal methane, coastal arteries and the Sichuan-Chongqing-Hubei natural gas pipeline. There is also a need to bring national pipelines to 150,000 km, with long-distance distribution pipelines to reach 44,000 km. By 2030, based on market demand, efforts will intensify in large pipeline network deployments, with the newly added West-East Pipeline 5, West-East Pipeline 6, the China-Russia east-west pipeline and construction of at least 250,000 km of national pipelines with long-distance distribution of up to 60,000 km and gas distribution capabilities to exceed 690 billion m^3.

The ultimate goal in achieving interconnectedness between different operating entity infrastructure is to achieve secure and stable operations, reasonable resource allocation and connection between resources and markets and between gas reserves and markets. The implications of interconnectivity between the infrastructure of various operating entities are: (1) national long-distance trunk pipelines are interconnected through hub stations and connection lines; (2) provincial trunk pipelines are freely allocated through split hubs; (3) underground gas reserves and LNG receiving stations are connected to national long-distance trunk pipelines; and (4) neighbouring provincial and municipal markets form a regional network.

1. **National long-distance trunk pipeline interconnectivity**

   China's long-distance trunk pipelines are primarily constructed by CNPC and Sinopec, with CNOOC primarily constructing marine gas shore connection and LNG receiving station external distribution pipelines. Long-distance trunk pipeline interconnectivity is not only to achieve connection with internal pipelines of oil companies, but also to achieve connection with pipelines between oil companies. National long-distance trunk pipeline interconnectivity relies on pipelines connected by central hubs and through connection lines.

   Forming central hubs: Based on China's current natural gas infrastructure development and future infrastructure development, central hubs are to form in Ningxia Zhongwei, Hebei Yongqing, Hubei, Shanghai and Guangdong. As part of this:

   - Ningxia Zhongwei is a point of coalescence between western resources and eastern distribution, sending western gas to eastern distribution systems, and connecting to the Xinjiang coal methane pipelines and Shaanxi Beijing line systems, the Zhongwei-Jingbian pipeline, the Zhonggui Line and other pipelines. This achieves connection between imported China-Asia gas, Xinjiang gas fields, Changqing gas fields and Xinjiang coal methane.
   - Hebei Yongqing is the hub for northern China, and connects the Shaanxi-Beijing line, the China-Russia natural gas pipeline, the north China gas reserve groups and the Tianjin-Hebei resource pipeline. In northern China it adjusts natural gas supply, optimising natural gas resource allocation and relieving demand for gas at peak periods.
   - Hubei has many national long-distance trunk pipeline channels, and is the point of

coalescence for the central and southern region hubs, forming a connection with Xinjiang conventional gas, Xinjiang coal methane, imports from the China-Asia line, imports from the China-Russia line, Sichuan-Chongqing conventional gas and Sichuan-Chongqing shale natural gas.

- Shanghai is China's natural gas market centre, and from the perspective of supply structure, Shanghai is the central hub for West-East Line 1 and 2, the Sichuan to East pipeline and other important natural gas lines while also receiving East Sea gas and importing a large volume of LNG. From its characteristics, Shanghai is a natural gas trading centre but also a national natural gas pricing baseline location, and in future will become an important natural gas hub for China.

- As an important resource hub, Guangdong is set to become a point of integration for multiple gas sources. From the perspective of supply structure, there is pipeline gas, marine gas and LNG, with pipeline supply including the West-East Line 2, the Xin'aozhe line, the West-East Line 3, marine gas and multiple LNG receiving stations. From the perspective of properties, supply resources include conventional natural gas, coal methane and LNG. With a variety of resources coming together, it is a complex province with national gas sources.

Construction of connection lines: Connection lines are important complementary resource bridges between national long-distance trunk pipelines, with constructed examples being the Cheng Zhong Gui Line, the Huai Wu Line and the Ji Ning Line. Of these, the Zhong Gui connects the West-East Pipeline, Changqing gas fields, Sichuan-Chongqing gas region and the China-Myanmar pipeline resources. The Huai Wu line connects the West 1 Line, West 2 Line and the Zhong Wu Line. The Ji Ning Line connects the Shaanxi-Beijing Line, the West 1 Line and the Rudong LNG resources.

China's coastal LNG receiving stations provide supply essentially point-to-point, with supply regions and peak shaving capabilities being limited. Moving forward, the focuses of construction for connection lines will be along coastal arteries, with pipelines to connect receiving stations from Liaoning Dalian LNG to Guangxi's North Sea LNG receiving stations, with intermediary LNG receiving stations, marine gas to shore pipelines and the transited national trunk pipelines (Fig. 9.5).

2. **Provincial trunk line free allocation**

China's various provincial natural gas infrastructure entities are diverse, with national enterprises and private corporations competing with one another, and without interconnected business communications, which causes overlapping construction of infrastructure, low usage efficiency and an inability to optimally allocate resources. Provincial natural gas infrastructure must form a unified management and operations platform, carrying out unified planning and construction for provincial infrastructure while unifying operations and management, resource co-ordination and distribution. Provincial trunk line interconnectivity is required to achieve flexible allocation of resources and prefecture-level dual gas source supply deployment.

Current provinces in China that have only one provincial natural gas company include: Beijing, Tianjin, Hebei, Jiangsu, Shanghai, Zhejiang, Guangdong, Shaanxi, Anhui, Inner Mongolia, Hubei and Chongqing. Provinces with two or more provincial natural gas companies include: Shanxi, Jiangxi and Hunan. Each provincial natural gas company's business model can be categorised as: unified purchase of resources, agent distribution, mixed operation, monopolistic construction of infrastructure and free construction. In order to realise the goals of provincial pipeline unification planning, provincial trunk line free allocation and provincial market balanced development, each province should actively establish provincial natural gas companies, with unified operation and management from the provincial natural gas companies, whose business models should be divided into two categories:

- The first category includes provincial trunk line unified planning, staged implementation and unified purchase of resources. In order to avoid overlapping infrastructure construction

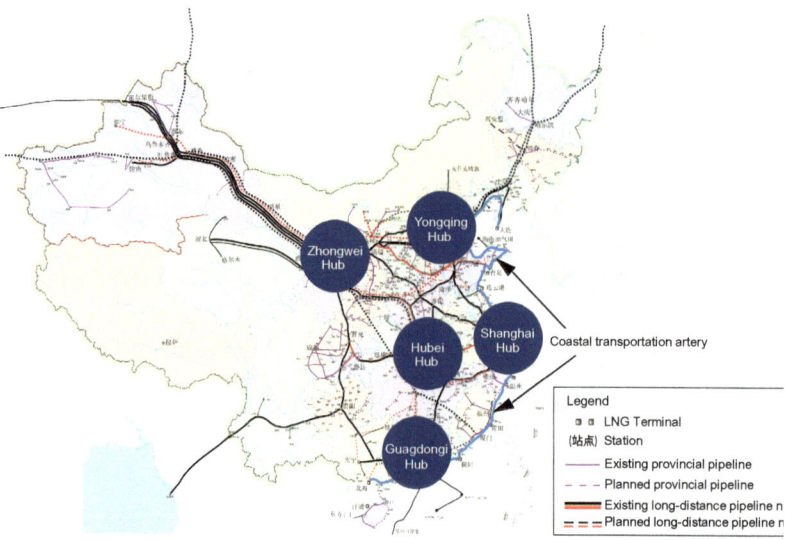

Fig. 9.5 Interconnectivity between various entity natural gas infrastructures across the country

and the resulting inefficiencies, there should be overall planning for provincial natural gas infrastructure. Based on the degree of market development, staged implementation should take place, with provincial trunk lines and peak shaving and allocation facilities constructed and operated by provincial natural gas companies. Resource procurement should be procured in unison by the provincial natural gas companies, with the natural gas resources that provincial users require negotiated with the provincial natural gas company and sold at a standard price that is not differentiated by distance.

- When infrastructure construction stages are in line with the above, but provincial companies do not carry out resource procurement and sales, the downstream users carry out negotiations directly with upstream gas suppliers, and gas suppliers then distribute the natural gas through the provincial natural gas companies. The provincial natural gas companies collect a management and distribution fee, and downstream users pay resource fees to the upstream gas supplier and a distribution fee to the provincial natural gas company. To safeguard the interests of users relatively distant from long-distance pipelines, provincial natural gas companies provide a uniform price to users within the province.

3. **Underground gas reserves and LNG receiving station peak shaving co-ordination**

Currently, China has two major large peak shaving facilities. One is underground gas reserves, and the other is LNG receiving stations. Of these, the underground gas reserves are distributed across the mainland and are distributed via pipeline. For these, the peak shaving principle is to adjust between long-distant and nearby pipelines. LNG receiving stations are distributed along the coastline, with their peak shaving principle being adjustment of nearby supplies and liquid adjustments. After 2020, underground reserve peak shaving will be the primary approach, with auxiliary peak shaving from LNG receiving stations.

Two types of peak shaving facilities: Currently, China's natural gas peak shaving facility capabilities are limited and are far from being able to satisfy future development demands, given that peak shaving capabilities should reach 62 billion m^3 by 2020 and effective underground gas reserve volume will need to be 44 billion m^3. By 2030, peak shaving capabilities must reach 90 billion m^3, with effective underground reserve volume of 77 billion m^3.

Based on China's current natural gas circumstances, in future the country's newly built underground gas reserves will primarily rely upon exhausted oil and gas deposits. By 2030 a group of six reserves will be formed in the central plains region, northern China, the north-east, in Changqing, in the north-west and in the south-west. As part of this, the central plains reserve group connects the Yu Ji Line, Xin'aozhe Yu Lu Sub-Trunk Line and the West 2 Line Tai'an Sub-Trunk Line. The northern China group connects the Shaanxi-Beijing Line and the China-Russia import pipelines. The north-east group connects the China-Russia import pipelines. The Changqing group connects the Shaanxi-Beijing Line and the West-East Pipeline. The north-west group connects to the West-East Pipeline. The south-west group connects the Zhong Gui Line and shale natural gas external distribution pipelines.

LNG receiving stations are connected through coastal arteries, and exchange resources with coastal region natural gas markets during peak shaving. Using tanker trucks for external distribution, peak shaving is provided for urban fuel and enterprise LNG reserve gas.

Regional peak shaving method distribution and options: Reserve facilities for natural gas are an important measure used to ensure the safe and stable supply of natural gas, and form an integral component part of natural gas delivery systems. Currently, the construction of gas reserve capability is lagging seriously behind, and according to national natural gas pipeline network layouts, the construction of such facilities should be accelerated, aiming to ensure that natural gas peak shaving emergency response requirements are met. In long-distance pipelines, details must be determined according to specific local needs, ensuring reasonable deployments, clear focal points and staged implementation of accessory construction of peak shaving facilities.

There was relatively early construction of gas reserve facilities in provinces and cities, including Beijing, Tianjin, Hebei, Shanxi, Liaoning, Jilin, Heilongjiang and Shandong, and these regions have good foundations. Moving forward, the focus will be on gradual optimisation of existing gas reserves and newly built underground reserves, complemented by small and medium LNG facilities and LNG receiving station storage. In conjunction with existing gas storage facilities, there will be optimised construction within exhausted oil deposits in Liaohe, Dagang, North China, Daqing and Shengli, including Liaohe Shuang 6, Qi 13, Shengliyong 21, Dagangban South, North China Su 1, Gong 20, Su 4, Su 49, Guxinzhuang, Wen 23, Saqing and Jilin oil fields.

Gas reserve facility conditions are relatively poor in Shanghai, Jiangsu and Zhejiang. Priority should be given to constructing LNG reservoirs, with underground storage and small and medium reserves as auxiliary peak shaving systems. Primary projects include the Jiangsu salt cavern gas reserve and the Jiangsu oil field exhausted oil deposit gas reserve. Prior to 2015, the focus had been on LNG gas reserves, relying on Jiangsu and Zhejiang existing LNG receiving station expansions to grow LNG reserves, thereby forming an LNG gas reserve system comprising Jiangsu and Zhejiang.

Fujian, Guangdong, Guangxi, Hainan and Yunnan have focused on LNG receiving station reserves, with small and medium storage and underground gas reserves, as well as small and medium liquefaction facilities providing auxiliary support. This region is working to construct an LNG reserve system prior to 2015 that is based on the LNG receiving station expansions in Fujian, Guangdong and Hainan, so as to satisfy regional peak shaving demand. Prior to 2030, once a reasonable network has been established, new LNG receiving stations will be constructed to increase gas reserve capabilities, while at the same time building a certain amount of underground gas reserve working gas volume, thereby establishing multiple interrelated peak shaving measures that satisfy local gas reserve system needs and the needs of the surrounding region.

Anhui, Hubei and Hunan have a certain amount of advantageous geography, and can mainly construct underground gas reserves, with auxiliary peak shaving systems of LNG small and medium reserves and small and medium liquefaction facilities. Major projects include Hubei Yingcheng, Yunying and Huangchang salt cavern gas reserves.

Shanxi, Henan, and Sichuan must use exhausted oil and gas deposits to build underground gas reserves while at the same time using upstream gas fields to resolve a portion of peak shaving problems. This is supplemented by peak shaving of users that can have supplies halted and small and medium liquefaction facility peak shaving. Major projects include Zhongyuanwen 23, Zhongyuanwen 96 and South-West Xiangguosi, among other exhausted oil and gas deposit gas reserves.

Shaanxi, Gansu, Qinghai, Ningxia, and Xinjiang are focused primarily on underground gas reserves for their gas reserve systems, building gas reserves from exhausted oil and gas deposits, including in Xinjiang Hutubi and Yulin.

4. **Regional networks between provinces**

Due to administrative zoning and approval procedures, regional pipeline networks are primarily concentrated within provinces (with the exception of long-distance pipelines and connection lines that run between provinces), resulting in relatively large differences in natural gas consumption between neighbouring provincial economies. In future, government guidance and support for construction of trunk line connections between provinces will allow resource supplementation, and thus boost the development of regional natural gas markets and optimise the regional pipeline network systems.

9.4 Standardising Infrastructure Planning and Diversifying Investment

9.4.1 Infrastructure Construction Planning and Project Progress Oversight

Infrastructure development involves foreign import pipelines, LNG receiving stations, national trunk long-distance pipelines, provincial pipelines and underground gas reserves, among other facilities. As part of this, pipeline network construction is critical to developing the natural gas market, but currently China's natural gas infrastructure is primarily constructed and operated by three large oil companies that have not established effective interconnectivity. As infrastructure continues to be optimised and interconnection of infrastructure deployments becomes critical to ensuring the secure and stable operation of the national natural gas network, effective infrastructure interconnection will become increasingly important, with the construction of Xin'aozhe, the West 4 Line, West 5 Line, China-Russia East-West Line, coastal receiving stations and underground gas reserves. Only with effective co-ordination in planning stages to form a complete industry chain can effective infrastructure connectivity be realised. Therefore, government intervention is required to create entities to oversee infrastructure construction and to strengthen oversight and controls of such development.

The major reason for a low degree of infrastructure planning is the lack of review mechanisms for planning construction projects. There is planning for projects, but no subsequent follow-up regarding progress in project implementation. A review needs to be carried out of enterprises involved in the planning of infrastructure projects at the national level, reviewing project implementation progress and timeline goal completion, taking measures to reward and penalise as required. A review should also be carried out for departments involved in provincial infrastructure, with consistent review criteria; provincial departments should actively implement relevant work involving the planning of provincial projects, working hard to co-operate and creating review systems for units under their supervision to promote active construction of infrastructure and optimise the outcomes of the practical work.

9.4.2 Promote Construction of Entities and Investment Diversification

The country encourages and supports various forms of capital participation and investment structure in the central planning of natural gas infrastructure. The three major oil companies should release investment and construction rights to national trunk pipelines, introducing private capital. Some of the provincial natural gas company monopolies on provincial infrastructure construction need to be broken. The introduction of private capital to infrastructure construction can increase participation in construction and improve the capacity for it.

The construction of CNPC West 3 Line brought together equity from the national social insurance fund council, an urban infrastructure industry investment fund and Baosteel Group, diverging from the previous model of investment contribution that was monopolised by CNPC's own investments and for the first time allowing private capital to be included in natural gas long-distance pipeline construction. The planned West 4 Line, West 5 Line and Xin'aozhe, among other national natural gas trunk pipelines, should carry out joint investment construction with private capital, based on the West 3 Line model. The uniformity seen in national long-distance pipeline construction entities has not yet been disrupted completely, though, and entities outside the three major oil companies should be introduced to diversify investment in the nation's long-distance pipelines.

Provinces with provincial natural gas companies should co-operate actively in investments with qualified private capital, accelerating the construction of provincial infrastructure. Province sub-trunk pipeline construction rights should be fully opened, and permission should be granted for qualified enterprises to construct and operate infrastructure within the ambit of national and local government planning.

9.5 Establishing Fair Third-Party Access to Infrastructure

9.5.1 Create the Conditions for Third-Party Access

As at the end of 2014, China had put into operation 16 natural gas national trunk pipelines, over 80 national sub-trunk pipelines, 11 LNG receiving stations and 19 underground gas reserves. There are no statistics available regarding the operating conditions of most infrastructure. By using open and fair operating systems, it would become possible to learn about infrastructure operating conditions, and excess capabilities could be put to use. When natural gas pipeline network facility operators have excess pipeline network facility capabilities, they can fairly open their pipeline network facilities to third parties, providing delivery, storage, gasification, liquefaction and compression services. In this mutually beneficial scenario, good use of can be made of facility capabilities while ensuring that existing users continue to have access to services, with a preferential order based on signed contracts with newly added users so as to realise fair, non-discriminatory open usage of natural gas pipeline network facilities.

Considering the fact that third-party access conditions are not yet mature, barriers between companies are difficult to eliminate, though some suggestions to catalyse third-party access are as follows:

- **Construct third-party review institutions to fairly and equitably verify excess infrastructure capabilities**: Considering the lack of transparency regarding infrastructure excess capabilities and the difficulties involved in downstream user verification, which result from asymmetry in information, third-party independent verification institutions should be established in order to fairly and equitably verify oil and gas pipeline infrastructure, storage, gasification, liquefaction and compression excess capability.
- **Pipeline network independence**: Currently, infrastructure enterprises provide transmission, storage, gasification, liquefaction and compression services while at the same time having businesses such as exploration and sales. This has resulted in local or partial monopolies within the natural gas industry at its various stages of development. Partition production shipping and sales business. At

the sales stage, break the monopolies to achieve highly efficient infrastructure usage. At the same time, due to the uniqueness of natural gas shipping and the heavy reliance on pipeline networks, separate national pipeline networks into several independent pipeline networks, with competition and co-operation between and among them to achieve highly efficient infrastructure utilisation.

- **Enterprise establishment and implementation measures**: On June 24, 2014, CNPC reviewed and passed in principle the CNPC Natural Gas Group Oil and Gas Pipeline Network Infrastructure Open Access Implementation Measures (Pilot), stipulating that oil and gas pipeline network facilities be fairly opened, as well as assigning management responsibilities to the relevant departments. Each enterprise should establish and implement measures based on its own infrastructure operating capabilities, operating costs and operating scope, including time used, scope and price.
- **Infrastructure interconnectivity, fee rates and liberalised market operation**: Through connectivity between pipelines and other infrastructure, break through natural gas transmission stage monopolies. Pipeline transmission fees should be liberalised in their operation, with the market to determine final resource allocation.

9.5.2 Establish Open and Transparent Oil and Gas Management and Operation Release Platforms

According to the Natural Gas Development Twelfth Five-Year Plan, natural gas infrastructure interconnectivity and services for third-party access should be implemented. At the end of the Twelfth Five-Year Plan, China will achieve partial connectivity of its pipelines, and have the conditions in place to achieve third-party access.

The utilisation of excess infrastructure capability cannot be entirely resolved simply by establishing operating entities, and each facility's operations can only be fully understood by the operating entity, with third parties unable to obtain information about them. Therefore, open and transparent oil and gas management and operation release platforms must be established, to cover all existing excess capacity in the infrastructure, with operators allowing third-party access to terms and fees, and operator-related contact information. Third parties should carry out specific negotiations with operators based on their needs. Because the status of infrastructure operators can be variable, they should provide the latest data relating to their infrastructure operations to the platforms, and the platform can then provide timely updates that are publicly available.

Analysis of China's Peak Shaving and Natural Gas Storage Systems

10

Natural gas security is one of the core components of energy security, and is an important component part of national security. Experience in many nations has shown that the establishment of a robust natural gas storage and peak shaving system is an effective means to address short-term and mid-term natural gas supply halts and to ensure natural gas industry stable operation as well as stability in the economy and society itself. As China's natural gas demand continues to grow, scaling up this infrastructure in various ways and reasonable deployments of effective emergency-response natural gas reserve systems is the direction and long-term goal of China.

10.1 Importance of Natural Gas Reserves for Peak Shaving

As China's economy continues to grow quickly and living standards are rising, as a clean, high-quality and efficient energy, natural gas's share of China's energy consumption is gradually

* This chapter was overseen by Zhonghong Wang from the Development Research Center of the State Council and Shangyou Nie from Shell International Exploration and Production. It was jointly completed by Xiaowei Xuan, Yingxie Yang, Yongwei Zhang, Jiaofeng Guo, Weiming Li, Beiqing Yao, Tao Hong and Yuxi Li from the State Resource Department Oil and Gas Centre, Xiaoli Liu from the National Development and Reform Commission's Energy Office, Guang Yang and Jianhong Yang from the China Energy Research Institute's Natural Gas Centre, Xiaobo Ju and Beijing Yao from Shell China. Other members of the topic group participated in discussions and revisions.

increasing. Since the official completion in 2004 of the West-East Pipeline project, natural gas consumption has rapidly expanded, and in 2014 it reached 183 billion m^3, more than a double the figure in 2004. At the same time, China's natural gas transmission pipeline network is also continually growing and being optimised, with natural gas now among main energies used in urban operations and development and involved in daily lives and urban economic functions. If natural gas supply encounters a shortage or halts, this significantly affects lives and the security of society.

In recent years, China has seen some major shortfalls in natural gas supply in some regions, with winter shortages even leading to a lack of supply for urban residential heating or for taxis wanting to refuel. As a result, ensuring natural gas supply security is of utmost political and economic significance for gas supply enterprises and the cities that use gas. Establishing a natural gas peak shaving reserve emergency response system that corresponds to the scale of city demand is an important undertaking for the sake of China's natural gas supply security.

Systems of natural gas involve upstream production, midstream transmission and downstream utilisation stages, and if any stage exhibits problems, this will affect the stable operation of the entire system. Because downstream users include residents, public services, commerce, industry, power plants, vehicles and air conditioning unit users, various types of user use gas in their own patterns, resulting in fluctuations in gas usage that can result in peaks and troughs in demand. Moreover, the room for production adjustments

Shell International and The Development Research Center (Eds.), *China's Gas Development Strategies*, Advances in Oil and Gas Exploration & Production, DOI 10.1007/978-3-319-59734-8_10

in upstream production is limited, and normally 20% additional gas volume can be provided at most, making it difficult to satisfy actual peak demand. Therefore, it is necessary to establish peak shaving emergency response facilities to satisfy the peak demand and to deal with sudden gas supply halts.

China's large-scale use of natural gas has only just begun, and natural gas reserve construction is still in its early stages. From the perspective of current development, as China's gas usage grows, the requirements for secure supply will grow. Therefore, it is necessary to integrate China's natural gas market development layout and reserve facility construction, and also to learn from overseas experience in constructing natural gas reserve systems.

10.2 Issues and Challenges

10.2.1 Rapid Increase in Natural Gas Consumption and Peak Period Demand

1. **China's natural gas demand continues to grow**

China's natural gas consumption growth has been rapid. As China's natural gas resource exploration and development continues to see breakthroughs, proved reserves and production volumes are rising, especially in the Shaanxi-Beijing line, West-East Pipeline and other project locations that are currently involved in commercial operation. China's natural gas industry has entered a stage of rapid growth, and natural gas consumption markets are expanding quickly, with natural gas accounting for an ever-rising proportion of energy consumption structure. In 2014, China's natural gas consumption volume reached 183 billion m^3, marking a six-fold increase over 2000, and annual average growth of 15%. The proportion of natural gas among non-renewable energy consumption structure has risen from 2.3 to 6%.

For a significant period into the future, as China's emerging industrialisation and urbanisation accelerates, there will be increasing pressure to improve regional environmental quality and to deal with climate change, actively adjusting energy structures to promote energy conservation and emissions reductions. This is especially true for atmospheric pollution prevention and other environmental protection policies that will be the major policy direction of China in the future, powerfully developing natural gas and other low-carbon energies. This is already a strategic task in the sustainable development of energy in China.

China's natural gas demand will continue to grow rapidly, leading to it becoming one of the major consuming nations of natural gas in the world. The research team believes that in a baseline scenario, natural gas demand in 2015 is likely to approach 200 billion m^3, surpass 300 billion m^3 in 2020, exceed 450 billion m^3 in 2030, and exceed 600 billion m^3 in 2050. In a policy environment of implementation of more active and effective environmental and energy transitioning (such as carrying out carbon trading or collecting a carbon tax), China's demand in 2020 could reach 344.7 billion m^3, essentially reaching the goal of 350 billion m^3 noted in the 12th Five-Year Plan. By 2030, consumption volume would be likely to reach 580 billion m^3, and approach 800 billion m^3 by 2050.

2. **Urban fuel grows fast, as does peak period demand**

Since 2000, China's natural gas consumption structure has tended toward diversification, with urban fuel becoming the fastest-growing sector of gas usage. From 2000 to 2013, China's urban fuel natural gas rose from 4.3 to 73.2 billion m^3, with the proportion of total consumption volume rising from 17.6 to 40%. Moving forward, with emerging urbanisation and industrialisation requiring more energy support, and at the same time as ecological cultural construction is strengthening efforts in lifestyle improvements and environmental protection, urban gas in China is set to be pushed toward continued rapid growth in four respects in particular:

- newly added natural gas users, accelerating the use of LPG and artificial coalbed methane as a replacement in residential gas usage;
- changes made by multiple provinces in relation to coal boiler standards for $PM_{2.5}$, with increased gas use for heating;
- gas used in transportation and other accompanying infrastructure, which will see rapid rises; and
- distributed energy projects that will successively enter production.

The rapid growth in urban fuel has greatly increased natural gas peak demand. In recent years, in order to improve atmospheric environment quality, northern Chinese regions such as Beijing and Tianjin have developed large quantities of natural gas heating users, including large gas power plants, gas boilers and residential wall-installed heaters. Winter season natural gas usage has skyrocketed. Taking Beijing as an example, its heating season (November–March each year) accounts for approximately 76% of annual gas usage, with winter and summer peak and trough having a ratio of 12.5:1. Currently, eastern Chinese region winter seasons have yet to implement residential heading, but they also exhibit seasonal characteristics, as peak values occur in the hot summer period and there is a peak to trough ratio of 1.6:1. CNPC's West-East Pipeline company has calculated winter peak demand based on northern and eastern Chinese region natural gas usage load curves, and the two regions' seasonal peak demands are respectively 31 and 4% of annual gas usage volume (Table 10.1).

In the future, as various regions increase usage of gas heating, seasonal peaks and troughs will continue to increase. According to forecast data by CNPC in 2013 for China's various regional natural gas demands, as well as comparative operation of peak shavings in northern and eastern regions in China, it is possible to calculate China's natural gas peak demand potential. By 2020, natural gas seasonal peal demand will reach 45 billion m^3, rising to 76 billion m^3 in 2030. In addition, as gas heating and electricity proportions increase, daily peaks and hourly peaks, among a wide range of problems, will also appear, and must be comprehensively resolved.

10.2.2 Gas Reserve Peak Shaving Capabilities Insufficient

1. **Great advances in gas reserve facility construction**

Reserve natural gas peak shaving facilities are an important component part of natural gas

Table 10.1 China's natural gas peak demand calculations (100 million m^3)

w	2020			2030		
	Natural gas demand volume	Peak factor (%)	Peak gas volume	Natural gas demand volume	Peak factor (%)	Peak gas volume
Circum-Bohai	630	31	195	1102	31	342
North-east	280	31	87	522	31	162
North-west	245	31	76	348	31	108
Yangtze River Delta	595	4	24	928	4	37
South-east coast	595	4	24	870	4	35
South China	385	4	15	870	4	35
South-west	455	4	18	638	4	26
Mid-west	315	4	13	522	4	21
Total	3500		452	5800		764

Data source 2020 data from CNPC Planning Institute, 2030 data from study group arrangements

transmission system projects. They not only ensure stable pipeline supply but are also integral to strategic reserves and commercial cycles. They are important measures to ensuring the secure and stable supply of natural gas.

China's large-scale gas storage construction began in 1999 with the commencement of the Shaanxi-Beijing Pipeline, in order to resolve imbalances in Beijing's seasonal gas usage and to relieve peak demand, with subsequent large power gas reserve groups and northern China gas reserve groups. Later, along with the construction of the West-East Pipeline, construction of the Jinyun and Liuzhuang storage facilities also began, so that West-East Pipeline market users could use gas as expected. From 2013, China's gas reserve construction saw major progress, and existing gas reserve working gas volume capability is 17.3 billion m^3, approaching 10% of natural gas consumption volume. As far as gas reserve capabilities are concerned, there is still a gap to the global average of 11.3%.

According to the national Natural Gas Development 12th Five-Year Plan, in the next 5–10 years China will continue to heavily promote underground gas reserve construction: "accompanying construction of gas reserve peak shaving facilities are required along long-distance pipelines according to local requirements, with reasonable deployments and clear points of focus, implementing staged construction". It is likely that by about 2020, China's gas reserve total working gas capabilities will exceed 30 billion m^3 (Table 10.2).

LNG commercial reserve capabilities have begun to take shape. As a large quantity of LNG receiving capability comes online, LNG commercial storage in the receiving station, transfer state commercial storage and urban peak shaving LNG storage is receiving increasing attention. Currently, China has yet to have a guiding strategy for LNG reserves, and the country's various large first-stage LNG receiving station designs likewise lack commercial storage beyond ensuring basic load supply. However, they can absorb spot LNG at low prices and release at high prices, while at the same time ensuring that supply security in commercial reserves remains a

major point of further development. This is especially true for small and medium-sized LNG receiving and transfer centres, which have relatively small annual transfer capabilities, and approval procedures that are relatively simple. If projects and existing large LNG receiving stations could interconnect, this would not involve LNG import approvals, and would be somewhat simpler, and thus it is currently a focus in the industry. Prior to 2013, China only had two projects in operation, Shanghai Wuhaogou and Dongguan Jiufeng, with annual receiving capabilities of 1.5 million tons in total. China now has more than 10 in-construction, reported or planned LNG receiving and transfer centre projects, and there has been active enthusiasm for investment and construction from local energy companies and private corporations.

In addition to LNG receiving station accompanying construction of reserve facilities, urban fuel is also gradually becoming a major force in LNG reserves. China currently has a certain amount of LNG volume in reserve stations in Beijing, Shanghai, Changsha, Wuhan, Xi'an and Chengdu, ranging from several hundred to over 100,000 m^3.

China's commercial reserves are in their infancy, but strategic gas reserves have just begun to build, with only Xinjiang Hutubi representing a functioning large-scale underground strategic gas reserve.

2. Gas usage load region peak shaving capabilities are insufficient

Even though there has been some progress in gas reserve facility construction, nonetheless current peak capabilities are still markedly insufficient. Because large-capacity reserves such as Hutubi, Xiangguosi, Suqiao and Jinyun have only been built in the past two years and are still in the gas injection stage, peak shaving capabilities are as yet insufficient, and designed working gas volume is hard-pressed to come into full play.

Currently, winter peak shaving primarily relies upon gas reserves, LNG, gas fields and pressure reduction market gas volumes among other joint peak shaving approaches. Gas field adjustments

Table 10.2 China's constructed underground gas reserves

Constructed sites	Name of reserve	Capacity/10^8 m^3	Working gas volume/10^8 m^3	Working pressure/Mpa	Number of wells	Daily gas injection capabilities/10^4 m^3	Daily gas collection capabilities/10^4 m^3
Dagang Reserve Group	Dazhangtuo	17.81	6	13.0–30.5	19	320	1,000
	Ban 876	4.66	2.17	13.0–26.5	7	100	300
	Zhongzhong North	24.48	10.97	13.0–30.5	15	300	900
	Banzhong South	9.71	4.7	13.0–30.5	10	225	600
	Ban 808	8.24	4.17	13.0–30.5	8	360	600
	Ban 828	4.69	2.57	15.0–37.0	6	360	600
Jing 58 Reserve Group	Jing 58	8.1	3.9	11.0–20.6	13	210	350
	Jing 51	1.27	1.2	8.6–16.5	4	210	350
	Yong 22	7.4	3	7.0–31.4	5	190	250
Jiangsu	Liuzhuang	4.55	2.45	7.0–12.0	10	150	204
Central plains	Wen 96	–	2.95	–	–	–	–
Shuang 6	Liaohe	–	16	–	–	1200	1500
Hutubi	Xinjiang	–	45	–	–	1393	2855
Xianguosi	Chongqing	–	23	–	–	1393	2855
Suqiao Reserve Group	North China	–	23	–	–	–	–
Ban South	Dagang	7.8	5	–	–	1550	2800
Jiangsu	Jinyun	26.4	17.1	–	–	–	–
Total			173.18				

and pressure reduction markets are still the two major modes of peak shaving. For example, in 2012 with its total peak shaving volume of 12.8 billion m^3, gas reserve contributions accounted for 16%, LNG for 24%, gas field peak shavings 28% and pressure reduction markets 31%.

Regions with the largest variance between gas usage peaks and troughs have the most marked shortfalls in peak shaving. The region surrounding the Bohai Sea is the region with China's largest gap between peak and trough, and peak shaving demand is significant. According to forecasts, even if currently constructed and under-construction gas reserves were all successfully established, in 2015 there would still be a shortfall in peak demand of 3.1 billion m^3, which will grow to 4.3 billion m^3 by 2020. There are relatively many oil and gas fields around the Bohai region, and there are good conditions for selection of reserve sites. However, there are not many high-quality gas reserve site locations, and the decision can be difficult and the investment requirements high. If the issue of gas peak shaving cannot be resolved, winter peak shaving gas collection capabilities will fall short and major problems could arise.

3. Pronounced problems with gas reserve prices and operation mechanisms

Gas reserve facilities are unable to unilaterally create prices, and newly constructed gas reserve facilities are restricted in their development. Currently, China's gas reserves have been supplemented by pipeline facilities and lack independent pricing. Gas reserve stage investments and costs are all linked to pipeline economic efficiency calculations, and corresponding transfer fees are incorporated into pipeline transmission fees, collected together with pipeline transmission fees, and there is no natural gas pricing system to independently establish an item for "gas reserve fees". For example, in 2003 the NDRC announced that the average shipping price (including gas reserve fees) for West-East Pipeline line-wide pipelines was 0.79 CNY/m^3.

Prior to the 12th Five-Year Plan, underground gas reserve investments were generally

incorporated and collected through pipeline investments as pipeline transmission fees, and gas reserve construction and investment did not have accompanying peak shaving gas prices and transfer fee policies. The nation invested in natural gas commercial reserve facilities, with construction investments (including baseline gas) and 30% working gas procurement expenses in the form of a rebated income tax that was paid for by the nation, with assets belonging to the group or company and gas reserve and shipping fees assumed by the enterprise. In its guiding opinions regarding acceleration of gas reserve facility construction, the NDRC noted that a reserve price must be established for independently operated gas reserve facilities, but the policy was not implemented.

Gas reserve facility and long-distance pipeline bundled operation affect long-term development. China's natural gas industry is still in the operational mode of having a unified upstream, midstream and downstream. Production, shipping, storage and sale of natural gas is primarily operated and managed by CNPC, Sinopec and other large oil companies. Gas reserves are similar to the early stage of operations in the United States and Europe, and with pipeline auxiliary facilities bundled along with pipelines, there is no independent stage for the natural gas industry chain development. Even though after 2010 there were national investments into construction of gas reserves, nonetheless gas reserve operational models have not undergone fundamental changes. Currently, China's gas reserves are primarily used to co-ordinate supply and demand and peak shavings, as well as for optimisation of production and pipeline network operations, including emergency response and strategic reserves.

4. Peak shaving and emergency response assurance mechanisms await improvements

Responsibilities for gas reserve peak shaving are unclear. Currently, China's accountability for natural gas industry chain peak shaving at each stage is unclear, and this causes a divesting of responsibility in emergency natural gas peak shaving within the natural gas industry chain.

Currently, upstream enterprise peak shaving tasks are excessively challenging, as they not only assume responsibility for regional and seasonal adjustments to gas supply, but also for daily adjustments in key regions. At the same time, China's natural gas emergency response reserve construction investment mechanisms need to be optimised, as they lack the necessary pricing mechanisms and incentive policies. This has led to lack of enthusiasm in the natural gas industry chain upstream, midstream and downstream, causing severe limitations to winter season natural gas peak shaving abilities.

An effective natural gas warning mechanism and emergency response mechanism have yet to be established. Warning forecasts and emergency response mechanisms are important measures for short-term halts of energy supply. Currently, China has yet to optimise mature operating mechanisms in this regard.

In addition, China also lacked progress in matters such as participation in international energy security co-operative frameworks. China has co-operative relationships with essentially all global and regional energy organisations, but not much substantive co-operation, which is not helpful when it comes to using international efforts to co-ordinate dealing with energy security risks. Overseas oil and gas resource co-operations lack effective emergency response mechanisms for sudden incidents, and robust political, foreign relation, economic and even military means are lacking to tackle anti-terrorist emergency response systems and many other plans. As soon as an incident occurs, the nation easily becomes a passive player in events.

10.3 Key Objectives and Considerations Going Forward

10.3.1 Basic Considerations

Comprehensively implement overall national security requirements, making natural gas supply security an important component part of national security while accelerating deployment of robust and reasonable reserve systems appropriate to the size of the market. As soon as possible, build prompt and flexible warning systems for emergency response, constantly improving China's natural gas supply assurance capabilities, risk avoidance and handling abilities.

10.3.2 Major Objectives

Based on the scale of China's future natural gas consumption, current natural gas reserve peak shaving capabilities, underground gas reserve resource conditions and progress in the construction of LNG receiving stations, and with reference to overseas major natural gas-consuming nation experience, this research proposes some 2020 and 2030 objectives for the natural gas commercial and strategic reserve sector in China, as well as emergency response system construction objectives.

In 2020, establish underground gas reserves and LNG reserve integrated peak shaving systems with underground reserves of approximately 50 billion, and LNG reserves of 5–10 billion for a total gas facility working volume of 35–40 billion. This will account for up to 10–11% of total natural gas consumption volume, changing the current peak shaving method, which is primarily focused on pressure to force reduction of user demand. At the same time, establish an emergency response mechanism for China's natural gas as soon as possible, optimising special project emergency response preparation plans, establishing warning systems for emergency response and ensuring key region gas supply security.

In 2030, further expand the scale and capabilities of the reserve peak shaving system, with gas facility working volume to reach 65 billion and account for 12% of natural gas consumption volume at that time, reaching global levels. With regard to 2030 natural gas import scale, strategic reserves should reach approximately 5% of import volume. In keeping with large-scale and full coverage for national natural gas market demands, form effective emergency response reserve systems for fluctuations in natural gas supply and demand.

10.4 Recommendations for Developing Natural Reserves for Peak Shaving

10.4.1 Accelerate the Formulation of Natural Gas Peak Shaving Emergency Response Plans

In recent years, some regions in China have encountered frequent gas shortages during the winter season, clearly exposing China's lack of development in natural gas peak shaving reserve facilities, compared to the rapid development of the natural gas market as a whole. In order to guarantee natural gas supply security and the secure operation of the economy, accelerated construction should be pursued by the nation and provincial and municipal governments as well as enterprises to jointly participate in the construction of a top-down natural gas peak shaving emergency reserve system and management system and mechanisms.

It is recommended that national natural gas authorities accelerate the drafting of natural gas peak shaving emergency response reserve plans, making good use of local and enterprise enthusiasm for involvement, ensuring the scale of reserve objectives, the number of emergency reserve days, and deployment and location selection while also incorporating the mid- to long-term natural gas planning of the national and provincial 13th Five-Year Plan. Some underground gas reserve and LNG reserve construction projects meeting certain conditions should be incorporated into the overall framework for national natural gas reserves.

10.4.2 Emphasise Gas Reserve Facility Legal and Regulatory Construction

Optimise as quickly as possible the nation's corresponding policies and legislation, ensuring that natural gas reserves have good external policy environments from a systemic perspective so as to encourage enterprises to construct reserve facilities, and carry out research on innovative mechanisms for the reserve operation model. Based on foreign experience, formulate China's natural gas reserve management statutes and reserve laws, clarifying reserve organisation and management institutions and their accountabilities and obligations. For example, in the United States' Natural Gas Act (1983), it was prescribed that the FPC regulates interstate natural gas pipelines, while each state government is responsible for overseeing state natural gas pipelines. China's natural gas authorities should oversee the formulation of plans and policies and carry out regulation of gas reserve facility third-party access, gas reserve prices, reserve usage, emergency mobilisation and corporate operational actions. Enterprise operations and user consumption should be lawfully operated within reasonable frameworks of law and regulation, avoiding abuse of market strengths and promoting effective competition within the market.

Use national laws to clarify the requirements for each level of government, upstream enterprise and fuel enterprise and their responsibilities and obligations in the natural gas reserve system. Implement stepped reserve management systems. The nation is responsible for the strategic reserves of natural gas. Natural gas upstream enterprises assume responsibilities and obligations for seasonable peak shaving and emergency response reserves. Reserve facilities are invested, built and operated by those enterprises, and each province and city's gas enterprises assume daily and hourly peak responsibilities and obligations, with the necessary peak shaving gas reserves being invested in, constructed and operated by those enterprises.

10.4.3 Accelerate Natural Gas Peak Shaving Emergency Reserve Facility Construction

The first task is to accelerate construction of large-scale gas reserves and LNG storage tanks that can satisfy seasonal peak shaving demands. In northern China, the north-east and the north-west, seasonal peaks and troughs are significant, and there are ample gas reserve locations for resources. First, underground gas reserves should be established, supplemented by LNG, small and medium liquefaction facilities and LNG receiving stations for seasonal peak shaving systems. In central China and the south-west, among other regions with good geographical conditions and nearby oil production areas, use exhausted oil and gas deposits to build underground gas reserves while at the same time using upstream gas fields, with a supplement from small and medium-sized liquefaction facilities. In eastern China, southern China and other locations with relatively poor environments for underground reserves, construction of LNG reserve storage should be a focus, supplemented by underground gas reserves and small and medium-sized storage for peak shaving systems.

The second focus is to construct small peak shaving facilities that can satisfy daily peak demand for gas. In central load gas-using cities, give full play to city fuel companies, accelerating construction of small LNG reserves, CNG tanks and accessory gas reserve facilities to resolve city daily peak shaving demands.

The third focus is to construct natural gas emergency response reserve systems. Clarify natural gas upstream production enterprise and city fuel company emergency reserve responsibilities. Rely upon and expand large reserves of gas, LNG storage and small tank reserves among other facility construction scope, accelerating the establishment of natural gas commercial reserves and satisfying issues of regional supply halt gas demand.

The fourth issue to address is to study the strategy for establishing suitably scaled natural gas reserves. From a global perspective, natural gas has not experienced a supply crisis on a global scale such as the oil crisis, and natural gas security issues are not yet pronounced. However, compared to oil, natural gas has a high degree of unification in the upstream, midstream and downstream. As part of this, if any stage has problems, it will affect the entire chain of natural

gas supply security, and there are major potential risks. The 2009 dispute between Russia and Ukraine is one such example of the danger. Therefore, China should develop research into natural gas strategic reserves as quickly as possible upon the basis of properly handling commercial reserves, so as to determine the layout in the mid- to long term for natural gas security issues, making plans early and establishing natural gas strategic reserve scale and models suited to China.

10.4.4   Formulate Proactive Tax and Price Policies

Establish reasonable gas reserve pricing mechanisms. The pricing stage in gas reserves must ensure that gas reserve investment and operation costs are recovered while guaranteeing that gas reserve enterprises obtain a reasonable return. On the other hand, standard service and fair competition must be promoted among gas reserve enterprises. The United States has developed peak period/off-peak period or seasonal gas reserve pricing based on service cost pricing approaches that has to a certain degree reduced the risk of imbalance in gas reserve services. In order to promote gas reserve service competition, the United States has also developed market demand pricing approaches. These pricing approaches have improved services to make them more suited to gas reserves, ensuring that gas reserve service providers obtain reasonable economic rewards. China should formulate pricing mechanisms suited to the current state of development of the natural gas industry at the gas reserve stage. Based on the operating costs for gas reserves/storage facilities, and based on government-determined internal rates of return calculations, the government can determine price levels and for each interval of time carry out an assessment and adjustment on gas reserve prices.

Commence differential pricing based on peak and off-peak periods, expressing natural gas value during peak periods. It is recommended that China implement differential pricing for various users during peak and off-peak periods, thereby guiding reasonable consumption, achieving a minimisation of peak and off-peak limits and improving system operation efficiency. The peak and off-peak pricing difference in the United States and France is generally 1.2–1.5 fold, determining January–February and November–December as the peak periods for gas usage. By implementing peak gas usage prices, price levels rise from base amounts.

Formulate discount policies and measures to encourage enterprises to construct reserve facilities. Because the construction expenditure for emergency reserve facilities can be enormous, along with major costs for the procurement of reserves and daily operating expenses, and because the main function is to satisfy seasonal peak shaving supply demand, it is recommended that the government put in place specific preferential policies for construction expenses related to facilities for peak shaving reserves. This could, for example, be similar to the approach taken with commercial product oil reserves, with upstream suppliers enabled to use 30% of LNG reserve costs to construct LNG peak shaving emergency response equipment as a rebate in payable income taxes. This is aimed at encouraging upstream suppliers to construct suitable amounts of LNG reserves so as to deal with temporary supply cuts or extra volume during peak demand periods.

10.4.5 Accelerate Gas Reserve Management System Reforms

System reforms include business unbundling, independent settlement and promotion of state-owned oil company gas reserve model reforms. Currently, the United States and European nation gas reserve operations have developed into entirely free-market independent business models, but the model must be within a competitive market environment, including for natural gas supply, shipment and storage stages, which must have many market participants who respect market rules and form a market environment with fair competition.

Considering the current circumstances of China's natural gas industry, future gas reserve operations could first employ a business model without full market liberalisation by establishing a gas reserve operating company within CNPC that is financially independent from the pipeline transmission business, thereby becoming an independent stage of the industry chain. As an independent profit-making entity, the gas reserves company can help facilitate the professionalisation of the gas reserve business and liberalise the market, while also helping to facilitate individual pricing at the gas reserve stage. As China's gas reserve business rapidly grows, it will be partitioned from the pipeline transmission business to become a stage of the industry chain that operates and develops independently.

Gas reserve peak shaving facility construction and operation diversification should be encouraged. Establish a diverse portfolio of entities as part of a gas reserve system, including state-owned oil companies, urban fuel companies and independently operated gas reserve companies. This will be beneficial in ensuring that natural gas supply security and gas reserve capabilities can rapidly grow. Looking at the American experience, this achieved natural gas reserve peak shaving commercialisation to a very high degree, with management and operation entities including interstate and intrastate pipeline companies, local distribution companies and independent gas reserve service providers. In order to encourage private enterprise and other social capital to enter gas reserve facility construction and operation, it is necessary to innovate in a commercial model, forming reasonable price mechanisms that achieve gas reserve facility third-party access and other measures for joint realisation.

10.4.6 Establish Prompt and Flexible Warning Systems for Emergency Response

Prompt and flexible warning systems for emergency response means active precautions and effective measures to relieve natural gas security risks. The aim is gradually to establish and optimise precautionary emergency response legal systems, organisational institutions and strategic mechanisms, information collection analysis and distribution systems, and various types and levels of emergency preparation and international assistance and co-operation agreement for natural gas emergency response systems.

1. **Establish a natural gas forecast warning system**

Use research to find favourable opportunities to build an indicator system for energy consumption, monitoring systems and evaluation systems to further standardise natural gas statistical systems. Establish statistical data as quickly as possible to monitor and assess as well as to distribute information, wholly integrating energy information channels to continually optimise energy statistics and information collection systems. Enhance forecasting and warning method research and develop forecast warning models suited to China's energy circumstances. Organise a professional team with strong capabilities to establish a timely and flexible forecast warning platform.

2. **Establish an emergency response system covering natural gas production, transmission, sales and all other stages**

Based on gas industry and regional characteristics, determine various levels of emergency response preparation. Establish emergency warning systems in the upstream and midstream focused on the central government and central government enterprises, and downstream plans focused on local governments and local fuel companies. Make full use of the co-ordinating role of the National Energy Council to establish handling for major energy incidents, to provide feedback and for information release review regimes as well as strategy mechanisms, ensuring that emergency response is prompt, strict and authoritative.

3. **Promote communication and co-operation**

China's natural gas forecast warning system construction should likewise be a broad international co-operation. First, strengthen government interdepartmental communication and contact. Strengthen the dialogue and co-operation between the National Energy Administration, National Bureau of Statistics, NDRC Operations Bureau and other relevant departments in the realms of mobilisation, data sharing, forecasting and warning, and other matters so as to establish a unified and well-prepared natural gas forecast warning and emergency response system.

At the same time, in terms of reserve project construction deployment, the National Energy Administration and other local governments should take a scientific approach to planning, with full proofs provided. For approved and authorised project construction sites, various levels of government should offer support to enterprises. In terms of peak shaving emergency response interregional deployment, the National Energy Administration and provincial energy bureaux as well as relevant enterprises should co-ordinate efforts.

Also, strengthen international dialogue and communication relating to energy. In terms of data and information, continue to strengthen data-sharing mechanisms with institutions such as the IEA. Promote dialogue with natural gas source nations including Russia, central Asia, Australia and Qatar, negotiating resolutions to major and outstanding problems. In terms of management experience, make full use of advanced international experience, with management systems and mechanisms to become more scientific in their handling. In terms of research results, fully absorb the research results from various nations and with relation to international organisations, so as to design forecast warning systems suited to China's specific circumstances.

The research in this chapter concentrates on the current state of China's natural gas market mechanisms and regulatory systems, on the nature of established natural gas market mechanisms in other countries and how they were created, and on the direction, underlying thinking, aims, main pathways, approaches and measures needed to ensure the success of reforms. For the purposes of this research, the natural gas industry is divided into upstream, midstream and downstream sectors.

The problems in China's current natural gas industry mechanism and regulatory system include: over-centralisation of mineral rights, a major lack of investment in exploration; bundling of infrastructure operation, with low usage efficiency; inappropriate natural gas pricing mechanisms which are detrimental to the development of downstream markets and the construction of associated natural gas storage facilities; regional monopolies controlled by downstream pipeline operators and cross-subsidies between different types of users; and incomplete supervisory systems.

It can be deduced from the experiences of mature overseas natural gas markets that diversification of the participants involved in the market is a precondition to the creation of a competitive natural gas market, while the opening of the upstream sector is absolutely essential. Implementation of reforms to pipeline access policy must progress gradually, with third-party access policies as the first step in infrastructure reform. Market-based pricing is a very important stage of the process of market liberalisation of natural gas, as natural gas pricing mechanisms are closely linked to the extent to which the market is able to develop. The creation of regional trading markets is essential for the market to be able to develop to any degree, while the creation of an independent, legally established regulatory system is the final guarantee for the effectiveness of natural gas market systems.

The basic concepts and overall aims of China's natural gas market liberalisation reforms include:

- the creation of a diversified, competitive, open, orderly modern natural gas market system;
- the establishment of pricing mechanisms that reflect factors such as scarcity of resources, the market fundamentals of supply and demand, external environmental pricing mechanisms and green taxation; and
- the creation of a legally established, service-oriented natural gas administrative system.

Based on the current level of development of China's natural gas market, proposals concerning natural gas market liberalisation reforms involve three stages:

- By 2020, a preliminary mineral rights primary market should be created in the upstream sector, in addition to the first steps being taken towards the creation of a secondary market. This will establish a primary market for preliminary midstream pipeline capacity trading and storage trading, including LNG facilities. Downstream, the natural gas terminal prices should be deregulated, to that they will then be based on market competition.

• By 2025, an upstream mineral rights primary and secondary market should be more or less complete. In the midstream, a pipeline capacity trading and storage trading primary market should be more or less complete, and this should be accompanied by the preliminary creation of a pipeline capacity trading and storage trading secondary market. A natural gas sales market system based on a spot-trading market consisting of manufacturers, independent traders, major users, local distribution companies and end users should be established downstream.

• By 2030, an upstream mineral rights primary market and secondary market should have been established, as well as midstream pipeline capacity trading and a storage trading primary market and secondary market. Downstream, a natural gas sales market system based on a combination of spot-trading and futures should have been established.

The main measures necessary to establish a natural gas regulatory system and to reform regulatory structures in China are:

• **establishment of a market system**: set up the Sichuan-Chongqing natural gas market liberalisation reform trial region;

• **completion of natural gas pricing mechanisms**: staged introduction of reforms to natural gas pricing, acceleration of the creation of a natural gas futures market;

• **clearing out pipeline transport bottlenecks**: extensive reform to the operation and regulation of natural gas infrastructure such as pipeline networks, clearly establishing functional roles, to ensure that pipeline transport services and sales services become separate from upstream and downstream activities and that third-party access occurs;

• **carrying out regulatory reform**: staged creation of a comprehensive, centralised, vertically unified, integrated, independent, specialised regulatory system;

• **stimulation of activity among market participants**: make reforms to state-owned oil companies more extensive;

• **setting up and broadening of the scope of the natural gas industry-related service market**: development of the market for natural gas will help broaden and deepen the ancillary gas-related services industry.

Finally, a comprehensive legislative system will need to be established as part of the reforms to China's natural gas institutions. The administration needs greater reform, and there should be financial and tax incentives to encourage the development and usage of non-conventional natural gas. A variety of measures can be employed to back up these goals, such as encouraging technical innovation, coordinated planning and greater international cooperation with commercial entities.

China's Current Natural Gas Market Mechanisms and Regulatory System

11

11.1 Current State of Natural Gas Market Mechanisms

11.1.1 Natural Gas Upstream Market

In 1998, as part of the separation of government functions from enterprise management, the former China Petroleum and Natural Gas Company and the China Petrochemical Corporation were restructured according to principles of fair allocation, interaction, preservation of advantages and orderly competition, resulting in the creation of the China National Petroleum Corporation and the China Petrochemical Corporation (Sinopec). In February 2000, Star Petroleum, which had originally come under the auspices of the Ministry of Geology and Mining, was completely absorbed into the China Petrochemical Corporation. China National Petroleum Corporation, China Petrochemical Corporation and China National Offshore Oil Corporation all became independent operations, responsible for their own profitability, each an amalgamated organisation that competed both upstream and downstream. According to the regulations, these three main oil companies were allowed to apply for oil and natural gas exploration plots extending throughout the whole country. Even though China National Petroleum Corporation and China Petrochemical Corporation were allowed to carry out marine exploration and China National Offshore Oil Corporation could also carry out land-based exploration, other companies were still not permitted to enter to the fields of conventional gas exploration and development. When it came to non-conventional natural gas, the Coal Industry Ministry issued the Coalbed Methane Exploration and Extraction Administration Temporary Regulations in 1994, then in 2012 the Ministry of Land and Resources issued a notice encouraging shale natural gas exploration and development monitoring and regulation, which permits other market participants access to the coalbed methane and shale natural gas exploration and development field. However, the proportion of total natural gas output that this accounts for is minimal.

Currently, the majority (97% or more) of oil and gas exploration rights and mineral rights are concentrated in the hands of the three large oil corporations. Based on Ministry of Land and Resources data, there were 1023 oil and gas exploration rights registered in China in 2012, covering 4,009,340 km^2, of which 942 belonged to the three main oil corporations, accounting for 3,906,848 km^2, or 97.44% of the total. There were also 671 oil and gas extraction rights registered, covering a total registered area of 118,949 km^2, of which 655 belonged to the three main oil corporations, covering 117,685 km^2, or 98.93% of the total (see Table 11.1 and Fig. 11.1).

* This chapter was overseen by Xiaoming Wang from the Development Research Center of the State Council and Mallika Ishwaran from Shell International. It was jointly completed by Yusong Deng, Jiaofeng Guo, Shouhai Chen from China University of Petroleum and Qingle Wu from Shell China. Other members of the topic group participated in discussions and revisions.

Shell International and The Development Research Center (Eds.), *China's Gas Development Strategies*, Advances in Oil and Gas Exploration & Production, DOI 10.1007/978-3-319-59734-8_11

Table 11.1 Oil and gas exploration rights and extraction rights registrations in 2012

	Exploration rights		Extraction rights	
	Number	Area (km^2)	Number	Area (km^2)
China National Petroleum Corporation	415	1,608,291	396	91,578
China Petrochemical Corporation	284	908,211	194	20,420
China National Offshore Oil Corporation	243	1,390,346	65	5687
Yanchang Oilfield	24	75,958	5	444
China United Coalbed Methane Corporation	24	18,019	2	193
Others	33	8515	9	627
Total	1023	4,009,340	671	118,949

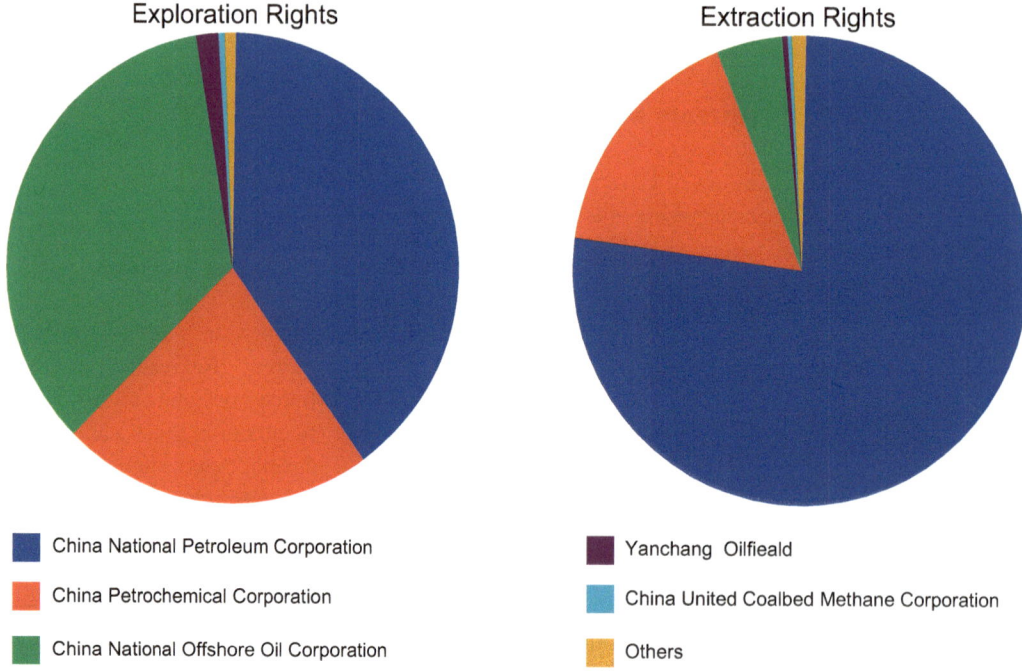

Fig. 11.1 Oil and gas exploration rights and mineral rights registered areas in 2012

Because the three main oil companies own the vast majority of natural gas exploration and mineral rights, in addition to owning all natural gas importing pipelines, LNG receiving stations and other infrastructure, the three main oil companies completely dominate natural gas supplies.

In 2014, natural gas consumption was 183 billion m^3, domestic natural gas output was 125.6 billion m^3, and coal-based gas output was 1 billion m^3. Imported natural gas accounted for 59 billion m^3, around one third of natural gas consumption. China National Petroleum Corporation was responsible for 95.5 billion m^3 of domestic gas output, 31 billion m^3 of imported pipeline gas and around 5 million tonnes of LNG (approximately 6.5 billion m^3), making up 72% of China's total natural gas consumption. China Petrochemical Corporation's domestic natural gas output was 20.2 billion m^3, with 100,000 tonnes of imported LNG (130 million m^3), accounting for 11% of China's total natural gas consumption. China National Offshore Oil Corporation's domestic natural gas output was 12.4 billion m^3, with 14.11 million tonnes of imported LNG (18.8 billion m^3), accounting for 17% of China's total natural gas consumption.

Overall, China National Petroleum Corporation had a three-quarter share of China's natural gas market, with China Petrochemical Corporation and China National Offshore Oil Corporation having somewhat smaller shares. In terms of regional markets, each of the three major oil companies have very high shares of the market in the regions that each supplies, in some cases reaching 100%. For instance, even though Beijing has seen diversification of gas sources,—with the natural gas being used including "Western gas going East", Central Asian gas, Inner Mongolian coal-based gas, imported LNG and Shaanxi-Beijing pipeline gas, this gas is provided by only one entity—China National Petroleum Corporation. However, as the Beijing Gas Group pointed out, China National

Petroleum Corporation was not actively seeking a monopoly in the Beijing market; the situation was brought about by Beijing's desire for China National Petroleum Corporation to act as the sole gas supplier, making it responsible for gas supply security. On the other hand, in south eastern coastal regions such as Guangdong and Fujian, it is China National Offshore Oil Corporation that has a relatively high market share.

11.1.2   Natural Gas Midstream Market

The natural gas midstream market encompasses natural gas transport and storage services, with the extent of development of the market being indicated by the scope of infrastructure present, including natural gas pipelines, LNG receiving stations and subterranean storage (see Fig. 11.2).

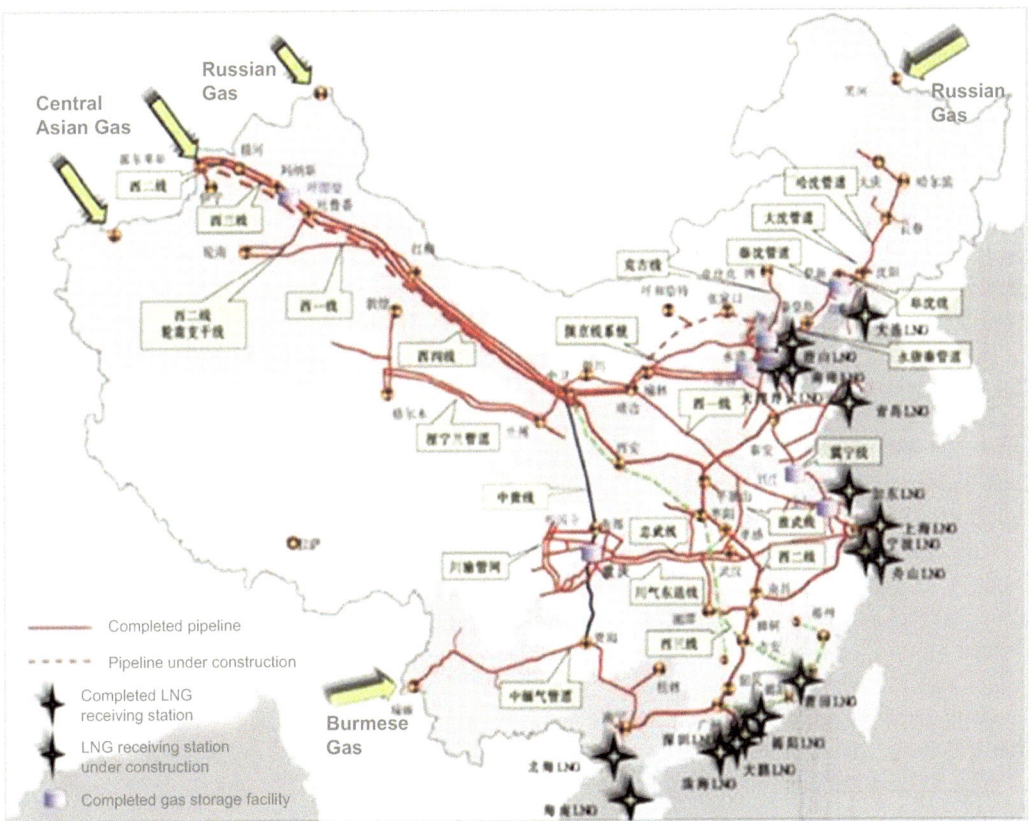

Fig. 11.2 China's natural gas infrastructure. *Source* China National Petroleum Corporation Economics and Technology Institute, 2014 Domestic and Overseas Oil and Gas Industry Development Report

1. Long-distance natural gas pipelines

The data and statistics for the current state of construction of China's natural gas pipelines are incomplete. By the end of 2014, the main natural gas backbone pipelines and branch pipelines covered around 63,000 km, of which around 46,000 km was owned by China National Petroleum Corporation, 73% of the total in China. Aside from the natural gas backbone pipelines owned by the three main oil companies, since 2003, as the natural gas market began to develop, local pipeline companies have begun to crop up, mainly owning provincial-level branch pipelines, and transporting natural gas from the backbone pipelines to municipal gate stations. In Zhejiang and Guangdong provinces there is a "complete province single network" policy, and here monopolistic provincial pipeline companies were created. In other provinces, where there were no provincial level-pipeline network companies, there are usually a number of pipeline companies; Shanxi province, for instance, has three main pipeline companies, while Henan province has almost 20. In general terms, the distances covered by local pipeline company pipelines are not great, for instance the length of pipelines completed by the Zhejiang provincial pipeline company is less than 1000 km. Details of the main natural gas pipelines in China built after 1996 can be found in Table 11.2.

2. LNG receiving stations

By the end of 2014, 11 LNG receiving stations had already been completed and brought online in China, with a combined receiving capacity of 40.8 million tonnes/year (see Table 11.3). Apart from this, there are 12 more LNG gas receiving stations under construction or planned for construction, with completion due in 2017, which will further increase China's receiving capacity of 42.4 million tonnes/year, after which the combined receiving capacity will be 83.2 million tonnes/year (see Table 11.4). The completed LNG receiving stations already online are all controlled by the three main oil companies. Of the 11 completed LNG gas receiving stations, China National Offshore Oil Corporation owns 7, China National Petroleum Corporation owns 3 and China Petrochemicals Corporation owns 1. However, as the state encourages access by private enterprise to the gas and oil infrastructure industry, diversification of LNG receiving station investors is gradually taking place, and both private capital and non-oil and gas companies are competing to build LNG shore-based discharging-receiving stations. The ENN Zhoushan LNG receiving station project is already under construction, while Xinjiang Guanghui Industry Investment (Group) Company Limited, China Huadian Corporation, Hanas Group and Beijing Jingneng Power Company Limited are all actively involved in the LNG receiving station development sector.

3. Subterranean storage facilities

A gas shortage occurred on a national scale in the winter of 2009, exposing the severe lack of natural gas storage facilities. As a result, the natural gas administration proposed a series of measures to accelerate the construction of natural gas storage facilities. During the 12th Five-Year Plan, China National Petroleum Corporation planned to build 13 oil and gas storage banks, with a total storage capacity of 453×10^8 m^3, the designed working gas capacity of which was 184×10^8 m^3, while it was expected that a total working gas capacity of 170×10^8 m^3 would have been created by 2015, equivalent to 10% of China National Petroleum Corporation's sales, which should be enough to satisfy downstream peak regulation requirements. Based on the "Sichuan gas going East" and Yuji pipeline projects, China Petrochemical Corporation is conducting a comprehensive selection of gas storage sites in relation to central and eastern region oil and gas fields and the western Sichuan gas region, and, based on this, construction of key gas storage facilities is planned during the 12th Five-Year Plan at Zhongyuan Wen 96, Zhongyuan Wen 23, Jintan in Jiangsu and Huangchang in Jianghan etc. Apart from China National Petroleum Corporation and China Petrochemical Corporation using depleted oil and

Table 11.2 Details of the main natural gas pipelines built in China, 1996–2014

Pipeline	Owner	Start point	End point	Length (km)	Gas transport capacity (100 million m³/year)	Came online in
Completed long-distance pipelines						
Ya- Gang Line	China National Offshore Oil Corporation	South China Sea Ya 13-1 gas field	Hong Kong, Hainan	778	34	1996-06
Shaan-Jing Line	China National Petroleum Corporation	Jingbian	Beijing	911	33	1997-09
Se-Ning-Lan Line	China National Petroleum Corporation	Sebei No. 1	Lanzhou	921	35.5	2009-11
Zhong-Wu Line	China National Petroleum Corporation	Zhong County, Chonqing	Wuhan	1364	70	2004-12
Western Gas Going East	China National Petroleum Corporation	Lun-nan, Xinjiang	Shanghai	3836	170	2004-12
Shaan-Jing No. 2 Line	China National Petroleum Corporation	Yulin, Shaanxi	Beijing	983	170	2005-07
Chang-Hu Line	Inner Mongolia Natural Gas Company Ltd.	Changqing, Jingbian	Hohhot	286	7	2009-01
Yong-Tang-Qin Line	China National Petroleum Corporation	Yongqing, Hebei	Qinhuangdao	320	90	2009-06
Chang-Chang-Ji	China National Petroleum Corporation	Changling, Jilin	Jilin Petrochemical Company	221	23	2009-12
Sichuan Gas Going East Line	China National Petroleum Corporation	Puguang Gas Field, Sichuan	Shanghai	1702	120	2010-03
Shaan-Jing No. 3 Line	China National Petroleum Corporation	Yulin, Shaanxi	Beijing	894	150	2010-12
West No. 2 Line	China National Petroleum Corporation	Huo-er-guo-si, Xinjiang	Guangzhou	9242	300	2011-06
Qin-Shen Line	China National Petroleum Corporation	Qinhuangdao	Shenyang	404	86	2011-06
Jiang-Ru Line	China National Petroleum Corporation	Jiangdu	Rudong	222	100	2011-06
Da-Shen Line	China National Petroleum Corporation	Dalian	Shenyang	423	84	2011-09

(continued)

Table 11.2 (continued)

Pipeline	Owner	Start point	End point	Length (km)	Gas transport capacity (100 million m³/year)	Came online in
Chang-Hu Dual Line	Inner Mongolia Natural Gas Company Ltd.	Changqing, Jingbian	Hohhot	518	80	2012-10
Yi-Huo Line	China National Petroleum Corporation	Yining	Huo-er-guo-si	64	300	2013-06
China-Burma Line	China National Petroleum Corporation	Ruili, Yunnan	Guigang, Guangxi	1727	100	2013-10
Fu-Shen Line	Datang International Power Generation Company Ltd.	Fuxin	Shenyang	125	40	2013-10
Ke-Gu Line	Datang International Power Generation Company Ltd.	Keshiketeng Banner, Inner Mongolia	Gubeikou, Miyun, Beijing	359	40	2013-11
Changning Region Shale Gas Trial Extraction Backbone	China National Petroleum Corporation	Ning 201-H1 well-group gas supply station	Shuanghe group distribution end station	95.6	15	2014-05
West No.3 Line Western Section	China National Petroleum Corporation	Huo-er-guo-si	Zhongwei	2445	300	2014-08
Completed connecting pipelines (1)						
Jing-Yu Line	China National Petroleum Corporation	Jingbian	Yulin	113	155	2005-11
Ji-Ning Line	China National Petroleum Corporation	Anping, Hebei	Yizheng, Jiangsu	1474	56.3	2006-06
Huai-Wu Line	China National Petroleum Corporation	Huaiyang	Wuhan	444	22	2006-12
Lan-Yin Line	China National Petroleum Corporation	Lanzhou	Yinchuan	460	19	2007-06
Zhong-Gui Line	China National Petroleum Corporation	Zhongwei	Guiyang	1613	150	2013-10

Source China National Petroleum Corporation Economics and Technology Institute, 2014 Domestic and Overseas Oil and Gas Industry Development Report

Table 11.3 LNG receiving stations online in 2014 in China ($\times 10^4$ tonnes/year)

Name of project	Location	Owner	Phase I	Total for Phases I + II	Phase I online	Phase II online
Dapeng LNG, Guangdong	Dapeng Bay, Shenzhen	China National Offshore Oil Corporation		680	2006-06	2011
Putian LNG, Fujian	Putian, Meizhou Bay	China National Offshore Oil Corporation		630	2008-04	2013
Yangshan LNG, Shanghai	Yangshan Deep Water Port	China National Offshore Oil Corporation		600	2009-10	
Rudong LNG, Jiangsu	Yangkou Harbour, Rudong	China National Petroleum Corporation		650	2011-06	2016
Dalian LNG, Liaoming	Dagushan Peninsular, Dalian	China National Petroleum Corporation		600	2011-07	2017
Ningbo LNG, Zhejiang	Zhongzhai, Baifengzhen, Ningbo	China National Offshore Oil Corporation		600	2012-12	2016
Jinwan LNG, Zhuhai	Gaolan Port, Zhuhai, Guangdong	China National Offshore Oil Corporation		700	2013-10	
Caofeidian LNG, Hebei	Caofeidian Port Zone, Tanghai, Tangshan	China National Petroleum Corporation		650	2013-12	
Fushi LNG, Tianjin	Nanjiang Port Zone, Tianjin Port	China National Offshore Oil Corporation	220	600	2013-12	2016
Qingdao LNG, Shandong	Dongjiakou, Jiaonan, Qingdao	China Petrochemical Corporation	300	300	2014	
Yangpu LNG, Hainan	Heiyan Port Zone, Yangpu Economic Development Zone	China National Offshore Oil Corporation	300	600	2014	

Source China National Petroleum Corporation Economics and Technology Institute, 2014 Domestic and Overseas Oil and Gas Industry Development Report

Table 11.4 LNG receiving station projects under construction or planned for construction in China ($\times 10^4$ tonnes/year)

Name of project	Location	Owner	Phase I	Phase I online	Status
Diefu LNG, Guangdong	Diefupian Zone, Dapeng New District, Shenzhen	China National Offshore Oil Corporation	400	2015	Under construction
Beihai LNG, Guangxi	Tieshan Port Zone, Beihai	China Petrochemical Corporation	300	2015	Under construction
Yuedong LNG, Guangdong	Huilai, Jieyang, Yuedong	China National Offshore Oil Corporation	200	2015	Under construction
Nangang LNG, Tianjin	Nangang Industrial Zone, Binhai New District	China Petrochemical Corporation	300	2016	Under construction
Zhoushan LNG, Zhejiang	Zhoushan Economic Development Zone	ENN	300	2017	Under construction
Binhai LNG, Jiangsu	Binhai Port Zone, Yancheng	China National Offshore Oil Corporation	260	2017	Permit received
Wenzhou LNG, Zhejiang	Xiaomen Island, Wenzhou	China Petrochemical Corporation	300	2017	Permit received
Zhangzhou LNG, Fujian	Xinggu Bay, Longhai	China National Offshore Oil Corporation	300	2017	Permit received
Lianyungang LNG, Jiangsu	Xuwei Port Zone, Lianyungang	China Petrochemical Corporation	300	2017	Permit received
Yuexi LNG, Guangdong	Bojia New Port Zone, Maoming	China National Offshore Oil Corporation	300	2017	Permit received
Yantai LNG, Shandong	Western Port Zone, Zhifu Port	China National Offshore Oil Corporation	300	Unknown	Early stage
Shenzhen LNG	Difupian Zone, Dapeng Bay	China Petrochemical Corporation	300	Unknown	Early stage

Source China National Petroleum Corporation Economics and Technology Institute, 2014 Domestic and Overseas Oil and Gas Industry Development Report

gas reserves to create subterranean storage facilities, China Petrochemical Corporation is working together with China National Salt Industry Corporation to construct gas storage facilities in subterranean salt caverns in the Jintan district of Jiangsu, while the first gas storage facility constructed with investment from a municipal gas company—Ganghua Jintan Gas Storage Facility—is already under construction. By 2013, the designed working capacity of gas storage facilities that had been completed or that were under construction had already reached 300 billion m³, but due to the need to extend subterranean storage facilities gradually, the actual completed working gas capacity is only 28.6×10^8 m³, which is only 1.7% of the current annual national gas consumption (see Table 11.5).

11.1.3 Natural Gas Downstream Market

The natural gas downstream market has a complex structure, and market participants are somewhat dispersed. Downstream users can be divided by gas supply method into users that are supplied directly and urban gas enterprises. Directly supplied users are those who purchase natural gas directly from upstream suppliers for the purposes of manufacturing or their own consumption, and that do not resell gas; urban gas enterprises buy gas from upstream suppliers, then resell gas to end users via their urban gas distribution networks. Direct users are generally industrial users that use large quantities of gas, but not all large industrial gas consumers are directly supplied customers. Due to the cross-subsidising that exists between residential users and commercial users in the pricing of gas supplied by urban gas enterprises, large-volume industrial users where the cost of supplying gas is relatively low are a major source of profit for urban gas enterprises. As a result, urban gas enterprises are violently opposed to upstream suppliers selling gas directly to industrial customers that fall within the geographical area to which their trading licences apply. Based on

research by the China Oil Economy Institute of Technology, the quantity of gas sold by urban gas enterprises in 2013 was around 90 billion m³, which accounts for 56% of China's natural gas consumption for that year.

In December 2002, the Ministry of Construction issued Suggestions Concerning Accelerating the Rate of Marketisation of Urban Utilities, which was followed by the publication of the localised legislation and policy-related Provincial and Urban Gas Administrative Regulations in various locations. This adopted the principle that "the party investing should make decisions on policy, the party benefiting should take on risks"; this was designed to actively attract state, private and overseas investment in urban gas infrastructure and management. Currently, China's urban gas industry has begun to convert from its previously ring-fenced situation, towards one with a new "four-pronged approach" consisting of centralised corporations, local state-owned enterprises, overseas investment and private enterprises, with over 800 gas companies throughout China in addition to the formation of five major transregional gas groups, including Towngas, China Resources Gas, PetroChina Kunlun Gas Company Limited, China Gas Holdings and ENN. In addition, there are relatively large local gas groups such as the Beijing Gas Group, Shanghai Gas (Group) Company Limited and Shenzhen Gas (Table 11.6).

11.2 Current State of China's Natural Gas Regulatory Systems

11.2.1 Upstream Market Administrative System

1. **Upstream natural gas administrative policy and legislation**

When compared to the midstream and downstream sectors, market regulation and legislation in China's upstream market is relatively complete; a preliminary core Mineral and Resources Rights Law has already been drafted,

Table 11.5 Construction status of main subterranean gas storage facilities in China

Gas storage facility	Owner	Location	Type	Designed capacity ($\times 10^8$ m^3)	Designed working gas capacity ($\times 10^8$ m^3)	Working gas capacity already created ($\times 10^8$ m^3)	Highest injection rate (10^4 m^3/day)	Date brought online
Lamadian	China National Petroleum Corporation	Daqing	Depleted reserve		1.00	Already abandoned	–	1975
Dazhangtuo	China National Petroleum Corporation	Dagang	Depleted reserve	17.8	6.00	18.4	320	Being brought online successively starting in 1999
Ban 876	China National Petroleum Corporation	Dagang	Depleted reserve	4.7	1.9		100	
Central–Northern Ban	China National Petroleum Corporation	Dagang	Depleted reserve	24.5	11.0		300	
Central–Southern Ban	China National Petroleum Corporation	Dagang	Depleted reserve	9.7	4.70		225	
Ban 808	China National Petroleum Corporation	Dagang	Depleted reserve	8.2	4.17		360	
Southern Ban	China National Petroleum Corporation	Dagang	Depleted reserve		5		240	
Ban 828	China National Petroleum Corporation	Dagang	Depleted reserve	4.7	2.57		360	
Jing 51	China National Petroleum Corporation	Huabei	Depleted reserve	1.3	0.60	4.2	–	2010
Jing 58	China National Petroleum Corporation	Huabei	Depleted reserve	8.1	3.90		–	
Yong 22	China National Petroleum Corporation	Huabei	Depleted reserve	6.0	3.00		–	
Liu Zhuang	China National Petroleum Corporation	Jiangsu	Depleted reserve	4.5	2.45	1.9	–	2011
Wen 96	China Petrochemical Corporation	Zhongyuan	Depleted reserve		2.95		–	2012-09
Shuang 6	China National Petroleum Corporation	Liaohe	Depleted reserve		16		–	2013-01
Hutubi	China National Petroleum Corporation	Xinjiang	Depleted reserve	117	45	2.6	1123	2013-07

(continued)

Table 11.5 (continued)

Gas storage facility	Owner	Location	Type	Designed capacity ($\times 10^8$ m^3)	Designed working gas capacity ($\times 10^8$ m^3)	Working gas capacity already created ($\times 10^8$ m^3)	Highest injection rate (10^4 m^3/day)	Date brought online
Xiangguosi	China National Petroleum Corporation	Chongqing	Depleted reserve		23		1380	2013-06
Suqiao	China National Petroleum Corporation	Huabei	Depleted reserve	67.4	23	0.2	1300	2013-06
Jintan	China National Petroleum Corporation	Jiangsu	Salt caverns	26	17.20	1.3	–	2007
Yunying	China National Petroleum Corporation	Hubei	Salt caverns		6		–	2015
Total				299.9	179.44	28.6		

Source China National Petroleum Corporation (CNPC)

and the natural gas upstream policy and legislation is based mainly on State Council regulations and rules implemented by the industrial administration.

In terms of oil and gas mineral rights management, the main legislation applicable is the Mineral and Resources Law (1996), the Rules for Implementation of Mineral and Resources Law (1994), the Mineral and Resources Exploration Plots Registration Administrative Measures, the Mineral and Resources Extraction Registration Administrative Measures and the Exploration Rights and Mineral Rights Transfer Administrative Measures. Regarding overseas co-operation, the main legislation applicable is the Regulations for Overseas Co-operation in the Terrestrial Extraction of Oil Resources and the Regulations for Overseas Co-operation in the Marine Extraction of Oil Resources. Regarding taxation and price regulation, apart from the general legal provisions in Enterprise Income Tax Law and the VAT Temporary Regulations, the main source of legislation is the Temporary Regulations For Resources Taxation, the Administrative Regulation for the Collection of Mineral Resources Compensation and the Exploration and Mineral Rights Usage Fees and Remuneration Administrative Rules, in addition to notices issued by the National Development and Reform Commission relating to adjustment of gas prices.

The sets of mineral administrative measures that apply to coalbed methane and shale natural gas differ from those that apply to conventional natural gas, and mainly consist of policy-type documents issued by the administration, including:

- Certain Suggestions in Relation to Accelerating Coalbed Methane (Coal-mine Gas) Extraction and Usage (2006) and Suggestions in Relation to Further Accelerating Coalbed Methane (Coal-mine Gas) Extraction and Usage, issued by the General Office of the State Council;
- Temporary Administrative Regulations for the Exploration and Extraction of Coalbed Methane, issued by the Coal Industry Ministry;

Table 11.6 Main provincial capital and urban gas enterprises in 2012

No.	Provincial capital/city	Gas enterprise	No.	Provincial capital/city	Gas enterprise
1	Beijing	Beijing Gas Group	17	Shijiazhuang	ENN
2	Shanghai	Shanghai Gas (Group) Company Ltd.	18	Jinan	Towngas, China Resources Gas
3	Tianjin	Jinran Huarun	19	Nanjing	Towngas, China Resources Gas, China Gas Holdings
4	Chongqing	Chongqing Gas	20	Zhengzhou	China Resources Gas
5	Guangzhou	Guangzhou Gas	21	Wuhan	Towngas, China Resources Gas, China Gas Holdings
6	Shenyang	Shenyang Gas	22	Changsha	ENN
7	Taiyuan	Taiyuan Coalgas	23	Fuzhou	China Resources Gas
8	Hangzhou	Hangzhou Gas	24	Nanning	China Gas Holdings
9	Hefei	Hefei Gas	25	Hohhot	China Gas Holdings
10	Nanchang	Nanchang Gas	26	Xi-An	Towngas
11	Haikou	Minsheng Gas	27	Harbin	Kunlun Gas
12	Guizhou	Guizhou Gas	28	Kunming	Kunlun Gas
13	Yinchuan	Hanas	29	Lanzhou	Kunlun Gas
14	Urumqi	Xinjiang Gas	30	Xining	China City Natural Gas Investment Co. Ltd.
15	Changchun	Changchun Gas	31	Lhasa	Kunlun Energy Co. Ltd.
16	Chengdu	Chengdu Gas			

- Notification Concerning Enhancing Administration of Overall Exploration and Extraction of Coal and Coalbed Methane and Notification Concerning Enhancing Supervisory and Administrative Work Relating to Shale Gas Resources Exploration and Extraction, issued by the Ministry of Land and Resources;
- Shale Gas Development Planning (2011–2015), Notification Concerning Various Issues in Relation to Standardisation of Approaches to the Development of the Coal Gasification Industry, issued by the National Development and Reform Commission;
- Notification in Relation to Subsidy Provision Policies Relating to Development and Usage of Shale Gas, issued by the Ministry of Finance; and
- Coalbed Methane Industry Policy, issued by the National Energy Administration.

2. **Natural gas upstream administrative bodies**

A multi-departmental shared administrative regime is employed. Apart from the Ministry of Environmental Protection and the State Administration of Work Safety, which are responsible for technical supervision of environmental protection and production safety, the main upstream administrative departments are the National Development and Reform Commission, the National Energy Administration, the Ministry of Land and Resources and the Ministry of Commerce, while the Ministry of Finance works together with the industrial administration to provide financial and taxation policy that promotes the development of the industry.

The National Development and Reform Commission (NDRC) is responsible for carrying out research into economic and social development policy on behalf of the State Council and is involved in maintaining a balance while carrying out macroscopic adjustments through reforms to the overall economic system. When it comes to drafting policy in relation to the natural gas industry and industry regulation, it mainly plays a role through:

- drawing up and organising strategy in relation to the national economy and social development, mid- to long-term planning and annual planning;
- prediction, forewarning and guidance information, making suggestions in terms of macroscopic policy, co-ordinating solutions for important issues affecting the progress of the economy, regulating economic activity, emergency interventions and transport-related co-ordination;
- involvement in establishing fiscal policy, currency policy and land policy, drawing up and organising the implementation of pricing strategy and the supervision and monitoring of pricing strategy implementation;
- organisation of drafting of comprehensive schemes relating to economic reforms, co-ordinating specific reforms to the economic systems, providing leadership in economic reform trials and in reform trial regional work; and
- authorisation, approval and auditing of major construction projects, projects with major foreign investment, major overseas resource development investment projects and projects requiring investment of large quantities of foreign currency.

The main duties of the National Energy Administration in terms of drawing up natural gas industrial policy and industrial supervision include:

- drafting of relevant supervisory and administrative legislation and its delivery for ratification and passing into law; drawing up and organising implementation of natural gas development strategy, planning and policy; advancing institutional reforms; drafting

relevant reform schemes; co-ordination in areas of development and reform where serious issues exist;
- organising and establishing natural gas industrial policy and related standards; approvals; checking and auditing of investment in fixed-asset investment projects;
- forecasts and warnings; dissemination of data relating to natural gas; involvement in operational adjustments and emergency measures in the natural gas industry;
- drafting of national natural gas planning and policy and management of its implementation; monitoring of changes in supply and demand in both domestic and international markets; proposals concerning ordering natural gas storage equipment and suggestions concerning rotation and usage in addition to organising the implementation of such activities; approval or auditing of natural gas storage equipment and installations and monitoring and regulation of commercial natural gas storage facilities;
- supervision of fair access to natural gas pipeline network installations;
- organisation and encouragement of international co-operation relating to natural gas; co-ordinating overseas resources development and exploitation; approval or auditing of major overseas investment projects;
- involvement in establishing policy relating to resources, finance and taxation; environmental protection and responses to climate change; proposals concerning pricing adjustments and suggestions concerning total import and export quantities.

The Ministry of Land and Resources is a major upstream administration for the natural gas industry, and its main responsibilities in relation to natural gas industrial policy and industrial regulation are:

- organisation of drafting of development planning and strategy;
- drafting of legislation, establishing regulations, policy, technical procedures and standards;

- leading local government law enforcement and investigation and handling serious contraventions;
- organisation, drafting and implementation of schemes relating to the application of mineral rights; handling rights applications; issuing exploration and development permits; auditing and authorisation of rights transfers; organisation and mediation in major rights disputes;
- standardisation and regulation of the rights market; investigation of rights holders; monitoring and supervision of extraction operations;
- regulation of the geological exploration industry and reserves of mineral resources; control over geological exploration accreditation; geological data; geological exploration results (it is legally entitled to gain from resources financially);
- authorisation of plots for foreign co-operation; supervision of foreign involvement in exploration and extraction activities.

The main role of the Ministry of Commerce in the regulation of the upstream natural gas industry is the drawing up of natural gas import policy, in the course of which they co-ordinate with other administrative departments regarding natural gas imports.

3. Main legislative regime applying to upstream natural gas regulation

(I) Rights management

Grade I national supervision applies to the exploration and extraction of conventional natural gas and shale natural gas, with unified planning applicable to the exploration and extraction of coalbed methane and administration at various levels, while there are differences in terms of admittance to the industry.

Applications for exploration or extraction of conventional natural gas are handled according to the Mineral and Resources Law and the mineral and resources exploration and extraction

registration administrative measures, and require submission of State Council documents approving the creation of an oil company or their approval for oil or natural gas exploration or extraction, in addition to submission of evidence of the identity of the legal representative of the exploration company. Exploration permits and extraction permits can only be issued after checking, after approval by an organisation stipulated by the State Council (the National Development and Reform Commission or the National Energy Administration) and after registration with the Ministry of Land and Resources. The threshold for admission to the conventional natural gas upstream field is therefore set very high, and requires "the approval or permission of the State Council".

While Grade I national supervision also applies to shale natural gas exploration and extraction, a notice issued by the Ministry of Land and Resources about the encouragement of shale natural gas exploration and development monitoring and regulation outlines a procedure for the sale of exploration rights based mainly on tendering and other such competitive approaches, with the aim of encouraging a variety of investment entities to enter the shale natural gas exploration and extraction field. In 1994, the Coal Industry Ministry issued the Coalbed Methane Exploration and Extraction Administration Temporary Regulations, and this had no specific accreditation requirements for companies involved in the development of coalbed methane, while also promoting the use of foreign capital and the adoption of advanced foreign technology for exploration and development of coalbed methane.

(II) Pricing regulation

Due to the monopolistic characteristics of the upstream natural gas market, the upstream natural gas price has been subject to long-term government control. In June 2013, the National Development and Reform Commission issued a notice relating to the adjustment of natural gas pricing. This notice mainly dealt with converting natural gas price regulation from an ex-factory-based approach to a gate station approach, with the introduction of gate station pricing regulation. The original "cost plus" approach to pricing was converted to a "market netback" approach.

Gate prices are government-guided prices, with a regulated maximum price, allowing suppliers and consumers to negotiate prices within a range beneath the maximum price. Government-guided pricing only applies to state-produced terrestrial natural gas and imported pipeline gas, while the prices of shale natural gas, coalbed methane, coal-based gas and LNG have been deregulated, now being determined purely by negotiation between suppliers and consumers. The natural gas gate price is pegged against the price of alternative energy sources. The alternative energy sources adopted for this purpose are fuel oil and LPG, with weighting of 60 and 40% respectively. Distinctions are made between stock gas and incremental gas, with an approach of "old approaches apply to old gas, new approaches apply to new gas", and the incremental gas gate price is no longer based on category of use. According to a notice issued by the National Development and Reform Commission, since the April 1, 2015, the pricing of stock gas and incremental gas has been combined, with one approach to pricing being applied, and with deregulation trials adopting a directly supplied gas-user gas-use gate price.

(III) **Fiscal and taxation regimes**

There are four categories of fiscal policy that apply to the upstream natural gas sector:

- taxation and administrative fees in relation to the extraction of natural gas resources—these are mainly a resource tax, a mineral resource compensation fee, a mineral rights usage fee, a mineral rights cost and plot use fee etc.;
- taxation and administrative charges in relation to imported natural gas—a VAT tax rate of 13% is applied to import, while some VAT refunds are available at import in relation to "state-approved natural gas projects";

- the general tax applicable to all companies, such as company income tax and VAT etc.;
- the fiscal subsidies provided by the state in relation to the development and usage of non-conventional natural gas resources; currently the subsidy provided by central government is 0.2 CNY/m^3 for coalbed methane and 0.4 CNY/m^3 for shale natural gas.

(IV) **Overseas co-operation**

Overseas co-operation is based on the Regulations for Overseas Co-operation in the Terrestrial Extraction of Oil Resources of the People's Republic of China and the Regulations for Overseas Co-operation in the Marine Extraction of Oil Resources of the People's Republic of China. China National Petroleum Corporation and China Petrochemical Corporation are responsible for overseas terrestrial co-operation, while China National Offshore Oil Corporation is responsible for overseas maritime co-operation. Based on a notice jointly issued in December 2010 by four ministries including the Ministry of Commerce and the National Development and Reform Commission, China National Petroleum Corporation, China Petrochemical Corporation, China United Coalbed Methane Corporation Limited and Henan Coalbed Methane Development Company Limited are allowed to develop overseas co-operation in the coalbed methane sector.

11.2.2 Midstream Market Administrative System

1. Midstream natural gas administrative policy and legislation

Regulation of the midstream natural gas market is divided into economic supervision and technical supervision. Economic supervision included pricing supervision and third-party access etc.; technical supervision includes environmental supervision and safety supervision. In terms of supervisory policy and legal documents, apart from the Oil and Natural Gas Pipeline

Protection Law issued by the National People's Congress in 2010, the main measures and regulations in force are:

- the Natural Gas Infrastructure Construction and Operation Administrative Measures, issued by the National Development and Reform Commission in February 2014;
- the Oil and Gas Pipeline Network Installation Fair Access Supervisory Measures (Trial Implementation), issued by the National Energy Administration;
- the Natural Gas Development Planning Under the 12th Five-Year Plan, issued by the National Development and Reform Commission; and
- Catalogue for the Guidance of Industries for Foreign Investment (revised 2015), issued by the Ministry of Commerce.

The Oil and Natural Gas Pipeline Protection Law's aim is to protect oil and natural gas pipelines, ensuring the security of oil and natural gas pipeline transport, and this mainly provides a legal framework for technical supervision. The Natural Gas Infrastructure Construction and Operation Administrative Measures provide fairly wide-reaching legislation regarding supervision of the midstream natural gas industry, and include natural gas infrastructure planning, construction and operation. They also include legislation covering natural gas market regulation and emergency guarantees. The Oil and Gas Pipeline Network Installation Fair Access Supervisory Measures (Trial Implementation) provide the legal framework in relation to the requirement for fair access to natural gas infrastructure, and establish the supervisory structures, supervisory duties and powers and modes of supervision that apply to ensuring fair access to natural gas infrastructure. In the Catalogue for the Guidance of Industries for Foreign Investment revised by the Ministry of Commerce in 2015, gas transportation pipelines and gas storage facilities are still listed as industries where foreign investment is encouraged. The purpose of this is to indicate openness to investment by foreign businesses, but it does not relate to

construction project planning or approval, and has even less weight when it comes to the supervision of natural gas pipeline operation or storage facilities. The Natural Gas Development Planning Under the 12th Five-Year Plan proposed certain infrastructure targets and guarantees, and presented planning for a number of pipelines and gas storage facilities and other such major infrastructure construction projects, and this is the main source of guidance in terms of development of the natural gas midstream sector.

2. **Natural gas midstream administrative bodies**

Based on integrated upstream and midstream operational modes, oil companies both construct pipelines and other such infrastructure and sell and transport natural gas that they produce themselves, so there is no conflict of interests between the pipeline company and pipeline users, and economic supervisory requirements are low. As a result, there is no specialist regulatory body. From the point of view of natural gas industry regulation, China adopts an approach whereby governing and regulation are combined, but there are a number of different bodies involved. The midstream natural gas administration includes the National Development and Reform Commission, the National Energy Administration and natural gas administration departments of People's Governments at the province, autonomous region and directly governed city levels, and these also take responsibility for economic supervision of the midstream natural gas industry. Technical supervision is carried out by geological, quality monitoring, environmental protection and production safety departments of local People's Governments.

The National Development and Reform Commission and the National Energy Administration are responsible for planning of national natural gas infrastructure construction. The natural gas administration departments of province, autonomous region and directly governed city People's Governments, on the other hand, act in line with national planning in drawing up natural gas infrastructure development plans for their administrative districts, submitting reports of such activities to both the National Development and Reform Commission and the National Energy Administration. Planning departments are responsible for approval, auditing and recording of construction of natural gas infrastructure. Authorisation documents relating to natural gas infrastructure projects authorised or approved by province, autonomous region and directly governed city People's Governments are submitted to the National Development and Reform Commission. Where a natural monopoly exists, natural gas infrastructure services are subject to government pricing control, with the National Development and Reform Commission and pricing departments at province, city and autonomous region level being responsible for determining the natural gas infrastructure service prices. The National Energy Administration and bodies under its auspices are responsible for supervisory work where it comes to fair access to natural gas pipeline networks.

3 Midstream market supervision

Midstream market supervision mainly involves establishing planning, project approval (authorisation), pricing control and enforcing the law etc.

The Natural Gas Development Planning Under the 12th Five-Year Plan issued in 2012 was the first-ever natural gas-specific planning in China. The main focus was to develop the natural gas industry by proposing development targets, policy assurance measures and major construction projects, and this provided guidance to lead the way for business investors. Based on the regulations, natural gas infrastructure construction projects required authorisation, approval or filing by industrial administration departments; if a project had not been authorised, approved or filed, then that project could not commence construction. Pricing for services in relation to naturally monopolistic natural gas infrastructure such as pipelines is controlled by the government, with service charges being collected based on prices set by government pricing departments.

Because "administrative measures" issued by the National Development and Reform Commission and the National Energy Administration fall into the category of ministerial regulations, they do not carry weight in terms of establishing administrative penalties. The punitive measures available to supervisory departments are therefore limited; apart from warnings and correction orders, they mainly rely on pricing law, anti-monopoly law and contract law to enforce the legal responsibilities of the relevant enterprises.

4. Midstream market regulation

In countries where reform of the natural gas market is advanced, such as the UK and the US, natural gas industry regulation mainly involves midstream regulation. This results in "supervision of the middle with freedom at both ends", thus creating an "X + 1 + X" market structure. In "X + 1 + X", the first X is the upstream supplier, the second X is the downstream user, and the 1 in the middle is the midstream pipeline network. "Supervision of the middle" refers to implementing strict supervision of natural pipeline network monopolies, enforcing a regime of third-party access and exerting control over pricing. Due to the long-term application of monopolistic operation to the upstream sector in China's natural gas industry and the unified mode of operation encountered in the midstream and upstream sector, a third-party access regime has never been established, and the majority of regulation relating to pipeline transportation prices stems from natural gas pricing requirements.

The first time that "fair access" was mentioned was in the Natural Gas Infrastructure Construction and Operation Administrative Measures, issued by the National Development and Reform Commission in February 2014, and the Oil and Gas Pipeline Network Installation Fair Access Supervisory Measures (Trial Implementation), issued by the National Energy Administration. The Natural Gas Infrastructure Construction and Operation Administrative Measures stipulate that natural gas infrastructure operators:

- should publicise information, including conditions of service provision, the procedures for obtaining such services and excess service capacity, and provide fair and just pipeline transport, gas storage, gasification, liquefaction and compression services to all users;
- may not use infrastructure in a manner that discriminates against other natural gas enterprises;
- where there is sufficient capacity, may not refuse to provide services to users who satisfy the necessary conditions; and
- may not make unreasonable demands.

The Oil and Gas Pipeline Network Installation Fair Access Supervisory Measures (Trial Implementation) defines the regulations relating to the conditions and procedures for handling user access applications and specific regulatory methods in more detail.

11.2.3 Downstream Market Administrative System

The downstream natural gas administrative system includes supervisory bodies, policies and legislation, in addition to main supervisory regimes.

1. Supervisory bodies

There are users in the downstream natural gas industry sector who are directly supplied by the upstream sector. There are also residential and commercial users, small industrial users, distributed power generation users and transport users who are supplied with gas via urban distribution networks. Where directly supplied users with larger gas consumption are concerned, such as gas for power generation, industrial fuels, gas for use in chemicals etc., the focus of government economic supervision is mainly the price of gas supplied and the fields in which gas is used. Currently, the National Development and Reform Commission is responsible for issuing natural gas pricing and natural gas use policy.

From the point of view of users supplied via the urban distribution networks, the main focus of government economic supervision is industry planning, control over admission to the gas industry, project allocation, price regulation, emergency assurance and so on, and the main object of such supervision is the gas companies. According to the regulations laid out in the urban gas administrative regulations, the State Council is responsible for the creation of administrative departments with responsibility for natural gas supervisory work at the national level. According to the relevant laws and regulations, People's Government gas administration departments at town level and above are responsible for determining and adjusting pipeline gas sales pricing, safety monitoring, quality inspections and provision of firefighting services, while other departments are responsible for specialist technical supervision.

2. Policy and law

Supervision of the downstream natural gas industry includes technical supervision and economic supervision. Technical supervision relates to supervision of product quality, of standards applied in engineering and construction and of safety in production. Economic supervision relates to supervision of admittance to the industry, of natural gas sales prices, of gas use sectors and of gas supply responsibilities. Generally, within policy or legislative documents, there may be both technical supervisory content and economic supervisory regulations. The Urban Gas Supervisory Regulations issued in 2010 are one such example of generalised administrative regulations, including both urban gas development planning and emergency assurances, and covering areas of economic supervision such as urban gas business and services, in addition to technical supervisory regulations relating to gas use, protection of gas installations, gas safety and prevention of accidents and their handling.

Certain specialist policies and legislation do exist in relation to downstream natural gas industry supervision. These mainly consist of natural gas usage policy and pricing regulation documents issued by the National Development and Reform Commission, and national urban gas development planning issued by the Ministry of Housing and Urban Construction, as well as urban gas supervisory regulations issued in 2010 by the State Council. Apart from this, contract law, emergency response law, product quality law, pricing law, fire prevention law, production safety law, metering law, special equipment safety inspection law, safe administration of dangerous chemicals regulations, housing construction and municipal infrastructure project completion and inspection filing administrative regulations and other such general law and regulations are also dealt with by downstream natural gas industry supervision.

3. Special business licence regime

In order to implement the reform demands expressed at the Sixteenth National Congress concerning "increasing reform to monopolistic industries, actively introducing competition mechanisms", in December 2002 the former Ministry of Construction issued Suggestions Concerning Accelerating the Rate of Marketisation of Urban Utilities, which suggested encouraging the involvement of social and overseas capital in the form of sole ownership, joint ventures and co-operation in the construction of municipal public utility infrastructure, to construct commercially operated municipal public utility facilities for water supply, gas supply, heating supply, waste water processing and waste treatment; tendering would allow selection of investment entities and would allow trans-regional and cross-industry involvement in the operation of municipal utilities enterprises, with special permits to be awarded by the government for such business operations. In 2004, regulations in the Municipal Public Utilities Business Licence Administrative Measures stated clearly that permitted municipal public utility operators are investors or operators of municipal

public utilities that have been selected by the government on the basis of relevant laws, statutes and regulations according to a market competition regime, and clearly outlines a regime whereby they may trade in certain public utility products or provide such type of services within a certain range and for a certain period of time. The special permits regime gives priority to appointments relating to specific business schemes, via a tendering process that allows selection of the best company, to which a special operators permit is awarded.

According to the Municipal Public Utilities Business Licence Administrative Measures, entities taking part in tendering for special business licences should:

- be a legally registered business entity;
- have corresponding registered capital, installations and equipment;
- have a good credit rating, be in good financial condition and be capable of servicing any outstanding debt;
- have experience in such operations and have a good business record;
- have sufficient key technical, financial and operational staff;
- present a realistic and feasible business plan;
- comply with any other conditions according to local laws and regulations.

After authorisation by People's Governments at the directly governed city, city and town level, government departments could then sign a business licence agreement with the successful bidder (the company that was awarded the special operators permit), the content of which included: the specifics of the permitted business operation, its geographical extent, its scope and duration; product and service standards; methods and standards for determining prices and fees and procedures for their adjustment; ownership and disposal of facilities; facility maintenance, update and conversion; safety management; performance guarantees; termination or alteration of terms of the permitted business operations;

liability for breach and methods of resolving disputes etc. Special business licence durations depended upon such factors as the type of industry, the scale of the operation and type of business and did not exceed 30 years.

4. Gas business licences

In July 2004, after the Administrative Licensing Law was issued, the original urban gas company accreditation administrative regime was rescinded, and the majority of provinces and directly governed cities drew up or revised local regulations and statutes to create a gas business licence regime. Clause 15 of the Urban Gas Supervisory Regulations of 2010 states that the state has adopted a business licence regime for gas businesses, and that companies involved in conducting gas business activities should satisfy the following conditions:

- they should satisfy the requirements of gas development planning;
- they should possess fuel gas sources and gas facilities that satisfy the national standards;
- they should have a fixed business address, a complete safety management regime and a complete business plan;
- the main person responsible for the business, the production safety manager and operational, maintenance and emergency repair staff should all have undergone professional training and have been found satisfactory;
- other conditions applicable according to legal and statutory regulations.

If these conditions are met, then the gas departments (at town level) and People's Governments (for larger areas) may issue gas business licences. It is clear from these regulations that the standard for the award of a gas business licence is "satisfactory" (satisfies the conditions), and as such, those that satisfy such conditions should all be awarded business licences.

11.3 Challenges Faced by the Current System

11.3.1 Over-Centralisation of Mineral Rights and Lack of Exploration

There are already oil and gas exploration rights registered in relation to 4 million km^2 across China as a whole. Of these, 3.9 million km (over 97%) belong to the three main oil companies, China National Petroleum Corporation, China Petrochemical Corporation and China National Offshore Oil Corporation. Oil and gas extraction rights already registered cover an area of 118,000 and 117,000 km^2 (99%) of these belong to the three main oil companies. The monopolistic position of the three main oil companies in terms of upstream exploration is hardly the result of free competition, but is rather due to legal restrictions and administrative monopolies. Based on the Mineral and Resources Law, the regulations governing mineral and resources exploration and the administrative measures governing the registration of extraction plots, applications to carry out exploration and extraction of oil and natural gas must be approved or agreed to by the State Council, but in actual practice only a very small number of companies ever receive such authorisation or permission.

Even though the exploration and extraction of shale natural gas, coalbed methane and other non-conventional types of natural gas is now open to other types of company, the number of plots actually accessible is limited, and the three main oil companies continue to dominate non-conventional natural gas extraction. The reason for this is that, in the majority of circumstances, non-conventional natural gas (mainly shale natural gas) plots coincide to a large degree with conventional natural gas plots. Apart from the legal restrictions, there are issues whereby it has not been possible to implement total third-party access to infrastructure, resulting in technical barriers arising in terms of extraction, and presenting difficulties for other companies trying to enter the upstream sector of the natural gas industry.

Due to the legal restrictions and the administrative protection of these monopolies, the three main oil companies virtually dominate all available oil and gas exploration plots. However, there is still a major lack of exploration. According to the Mineral and Resources Exploration Plots Registration Administrative Measures, owners of exploration rights should invest not less than 2000 CNY/km^2 in exploration within one year of the date on which the exploration permit was issued, rising to 5000 CNY/km^2 in the second year and 10,000 CNY/km^2 in the third. The Mineral and Resources Exploration Plots Registration Administrative Measures were issued in 1998 and were applied to all types of minerals, based on today's prices, and the level of investment that would be required for exploration for oil or gas (an annual exploration investment of 10,000 CNY/km^2) is unquestionably far too low.

In November 2014, the Ministry of Land and Resources subjected China Petrochemical Corporation and Henan Coalbed Methane Development Company Limited to fines of CNY 7,978,800 and CNY 6,035,500 respectively, and ratified reduction of the extent of their exploration plots. The reason for this was that these two companies had not been able to carry out exploration to the extent of their undertakings. Even though China Petrochemical Corporation and Henan Coalbed Methane Development Company Limited had only completed 73 and 51% of their undertakings in relation to exploration, their actual investment had already reached a level equivalent to 11.6 times and 3.6 times the level of investment required by law, respectively. According to China Petrochemical Corporation's annual accounts, China Petrochemical Corporation's investment in exploration in 2013 was only CNY 25.3 billion, which would be the equivalent of exploration of 1,600,000 km^2 of registered exploration rights, and based on this their average exploration expenditure was only 15,800 CNY/km^2. In actual fact, that CNY 25.3 billion also included the cost of overseas exploration, therefore it should not all be allocated to areas of domestic exploration.

11.3.2 Bundling of Infrastructure, Low Utilisation and Blocking Access to Upstream Markets

From the point of view of developing the natural gas market, building pipeline infrastructure is enormously important. Construction of natural gas pipelines and other infrastructure requires massive investment, with investment in backbone pipelines often reaching between tens and hundreds of billions of CNY, and branch pipelines costing between billions and tens of billions. The structure of every country's natural gas market is different, so variations in gas industry development and the types of infrastructure investment entities will naturally differ. In the early development stages of the US natural gas market, the upstream was very dispersed, and it was therefore often the downstream urban gas companies that were responsible for extending pipelines to oil and gas fields, where they bought gas for transport to, and sale in, downstream markets. In China, the development of the natural gas industry has followed the exactly opposite path; there is a relatively high level of monopoly upstream, while downstream it is relatively dispersed, and generally it is the upstream companies that are responsible for construction of pipelines and transport of natural gas for sale in downstream markets. In both cases, the approach adopted at the early stage involved bundling of pipeline transport and natural gas sales, with pipeline transport enterprises using their own pipelines to transport and sell natural gas.

Natural gas pipeline natural monopolies and bundling of business affects fair competition between upstream enterprises. In the United States in 1938, the Natural Gas Act introduced control over wellhead prices that continued for 40 years. The initial aim was to prevent pipeline transport companies from treating dispersed upstream producers unfairly. As described elsewhere, inappropriate wellhead price control eventually restricted the development of the natural gas industry in the United States, resulting in the severe gas shortages that occurred in the 1970s. In 1985 and 1992, the Federal Energy Regulatory Commission (FERC) issued Order 436 and Order 636, which introduced third-party access (TPA) and division of business, providing the basic conditions for the formation of a highly effective competitive natural gas market, while also acting as a stimulus for the shale natural gas revolution that has occurred in the US.

In February 2014, the Natural Gas Infrastructure Construction and Operation Administrative Measures and the Oil and Gas Pipeline Network Installation Fair Access Supervisory Measures (Trial Implementation), issued by the National Development and Reform Commission and the National Energy Administration respectively, stated that natural gas infrastructure operators should provide fair and open access to their services. However, without strict supervision, the "fair access" required by the Oil and Gas Pipeline Network Installation Fair Access Supervisory Measures (Trial Implementation) is actually "negotiated access" rather than fair third-party access. On top of this, neither of these two measures actually require the division of pipeline services from natural gas sales business. Although there is currently no actual evidence that bundled natural gas pipeline business operations act as an obstruction to entry to the upstream fields of exploration and extraction, in the second half of 2014, against the backdrop of a major drop in international oil prices, bundled operation of LNG receiving stations and long-term agreements in relation to gas prices were definitely responsible for obstructing the import of low-priced LNG from the international markets.

11.3.3 Irrational Pricing Mechanisms, Which Dampen Incentive to Build Gas Storage

The pricing reforms in 2013 were a major step forward in terms of liberalisation of the mechanisms that go to shape natural gas prices. However, prices are still not determined by the market; the transition from "cost plus" to "market netback" has only altered the method applied to pricing, rather than changing the actual price-determining mechanism. There are a lot of artificial factors at play in the choice of alternative energy source type, and these may not actually reflect the market value of natural gas. Taking the United States as an example, between 2000 and 2008, the price of fuel oil continued to rise while the price of natural gas virtually stayed the same. In 2010, with the arrival of the shale natural gas revolution, prices of fuel oil and natural gas began to deviate, with fuel oil increasing in price by one third, while the natural gas price dropped by two thirds. Against the backdrop of the widespread adoption of energy saving, emissions reduction and pollution control, the demand for coal dropped, as did the price, while the demand for gas continued to grow, as did the price, their prices changing in opposite directions. Therefore, adopting the price of coal to determine the price of natural gas makes no sense. Even if one tries adopting fuel oil and LPG as references, the changes in their prices still don't fully reflect the changes in supply and demand in the natural gas market, due to the differences in their uses and the difficulties in fuel conversion, and this may result in investors and natural gas users being exposed to erroneous market signals.

According to the pricing reforms of 2013, the natural gas gate station price was adjusted once a year based on the alternative energy source prices against which it is pegged, gradually transitioning to six-monthly and then quarterly adjustment. Regardless of whether adjustments take place once a year or at six-monthly or quarterly intervals, this year's natural gas price is still based on last year's alternative energy source prices. By making an adjustment once each quarter, the winter natural gas price is based on the autumn alternative energy source price. Warmer temperatures in autumn result in there being less need for heating and a relatively low demand for energy sources. Adopting the autumn alternative energy source price when deciding the price of gas for the peak gas use winter season is counterintuitive. So with a government-determined price based on the "market netback" approach, adjusted once a year or once each quarter, there is limited scope for adjustment and no capacity to reflect changes in supply and demand in the natural gas market up to the actual moment when the decision is made. Such an approach is not an attractive proposition to natural gas companies, given the massive levels of investment required to construct natural gas storage facilities. Unsurprisingly, gas storage facility working gas capacity in the US is between 18 and 20% of annual consumption, while in China it is only 2–3%.

11.3.4 Downstream Pipeline Operator Regional Monopolies and Cross-Subsidising Between Different Users

Although the reform of public utilities that started in 2002 resulted in competition in terms of market admission, it also had the effect of reinforcing monopolistic business operations. Companies that were awarded special business licences were allowed to carry out bundling of natural gas sales and network distribution over a period of time and within a certain range, thus excluding the possibility of competition occurring with other companies. Not only did these other companies encounter problems supplying gas to end users via urban gas networks, such activities as construction of pipelines to supply end users directly were also seen as an infringement of the special business licences awarded to urban gas companies. In addition to various urban gas companies being awarded special permits for supply of gas on the urban level via the tendering process, some provinces also set up provincial pipeline network companies, which then enjoyed provincial monopolies in purchasing and sales of natural gas. As a result, other companies were prevented from investing in and constructing natural gas pipelines in those provinces.

- Due to the monopolistic nature of urban natural gas networks and the bundling of sales and distribution, government-set pricing is generally applied to natural gas sold by urban gas companies. There are issues where pricing of urban gas is concerned.
- The pricing mechanisms are inflexible—since there is a mismatch between upstream and downstream pricing, the downstream price cannot immediately reflect changes in upstream gas source prices.
- It is difficult to monitor gas supply costs, and the pricing methods are somewhat opaque, so urban gas pricing departments find it hard to gain a clear understanding of the actual cost of gas supplied to urban gas companies, and there is no unified pricing method.

There is a mismatch between the natural gas sales price and the cost of gas supplied, with cross-subsidising occurring between different users, as industrial and commercial users (who have a low-cost of gas supply) pay a high price for their natural gas and a low natural gas price is applied to residential users (with a high-cost gas supply). In view of the cross-subsidising that exists between different users and the government control over residential user gas prices, it makes sense to some degree that urban gas companies then prevent upstream enterprises from developing direct supplies to users within their operational jurisdiction.

11.3.5 Incomplete Supervisory Systems and Insufficient Supervisory Capability

The laws that apply to natural gas are not complete, and this is not only reflected in the relatively few levels of legislation, with the majority of regulatory documents being ministerial regulations and there being no specific natural gas law, but also the fact that in certain sectors there are still no administrative regulations at all. The Mineral and Resources Law only requires submission for "authorisation or approval by the State Council", but does not actually define clearly what kinds of company can actually receive the "authorisation or approval" of the State Council. Although there is a fairly high hierarchical structure in Regulations for Overseas Co-operation in the Terrestrial Extraction of Oil Resources and Regulations for Overseas Co-operation in the Marine Extraction of Oil Resources, this is only applicable to foreign co-operation in extraction. The Oil and Natural Gas Pipeline Protection Law is the only administrative-type document dealing with the oil and natural gas industry at the legal level, but it was enacted to protect oil and natural gas pipelines and thereby ensure oil and natural gas transport security, and thus it mainly covers the legal issues of technical supervision. The Natural Gas Infrastructure Construction and Operation Administrative Measures provide a fairly comprehensive legal framework for supervision of the midstream natural gas industry, but lack force because they are merely ministerial regulations issued by the National Development and Reform Commission and thus lack hierarchical structure and are not particularly binding. The Oil and Gas Pipeline Network Installation Fair Access Supervisory Measures (Trial Implementation) provides the administrative framework to implement fair access to natural gas infrastructure; however, there are drawbacks in terms of the practicality of its application, with fair access actually being "negotiated access". In terms of the natural gas industry downstream sector, there are no legal documents providing an administrative framework, the legal document at the highest hierarchical level being the Urban Gas Administrative Regulations.

There is no specific natural gas industry regulator, and, apart from the areas of the environment, production safety and product quality supervisory departments, industry regulation adopts separate supervision of the upstream, midstream and downstream sectors, with co-operation and supervision divided between different departments, relying on a combined governmental supervisory approach. Under this multifaceted approach to regulation, natural gas supervisory functions are distributed between a number of departments, and inevitably there is a certain amount of duplication of roles as well as areas that are not fully covered. For example, the Ministry of Land and Resources, the National Development and Reform Commission and the National Energy Administration all have the power to decide industrial development planning and structure, to draft legislation and to establish administrative policy, while to some extent there are also jurisdictional conflicts between central government departments and local government departments.

With such a combined governmental supervisory regulatory system, there is a serious lack of supervisory input. For instance, the scheme regulation drafting team in the oil and gas section of the National Energy Administration whose area of responsibility is the "three establishments" (establishing functionality, establishing structure and establishing draft regulations) comprises less than 20 persons to cover the whole of China's oil and gas industry. Moreover, they are responsible not only for determining industrial development planning, drafting legislative framework and determining departmental procedures and industrial development strategy, but also for such supervisory functions as authorisation, surveying, co-ordination and imposition of penalties on specific individual projects, as well as the administration of work in relation to national strategic oil and gas storage.

International Experience of Liberalisation and Evolution of Natural Gas Markets

12

During the liberalisation of their natural gas markets, the United States and the United Kingdom both underwent a reform process that involved a transition from government supervision and gradual deregulation. This provides valuable experience to draw on in various areas, including natural gas exploration and development, storage and transportation, pricing mechanisms, market trading and institutional guarantees.

12.1 Incentivising New Entrants and Establishing Competitive Natural Gas Markets

From comparisons of the state of development of the natural gas markets of six countries and regions, it can be deduced that a competitive natural gas market requires the involvement of entities of a certain size (Table 12.1). The United States and Britain are the most typical competitive markets, and reflect the involvement of, and high levels of competition between, a multitude of different entities in many areas of the market, including exploration and extraction, storage and transportation, wholesale and retail, and trading

centres. By comparison, in China, Japan and South Korea there is a lack of competition; upstream and midstream business is mainly monopolised by oligarchs and there has been no division of production, storage, transportation or sales operations.

A competitive, reactive and fluid natural gas market should have the following characteristics:

- a greater number of entities, allowing competition in the upstream and midstream market sectors to provide diversified services to consumers, for which funds can be raised from investors;
- competitive prices at both the wholesale and retail levels;
- non-discriminatory access to pipelines, gas storage, LNG receiving stations and other similar infrastructure.

12.2 Opening the Upstream Sector to Competition

Availability of resources is a precondition of natural gas usage, so, looking at the natural gas industry as a whole, upstream production departments have virtual control over the operation of the whole industrial chain and the distribution of returns. As a result, opening up upstream exploration and development to new entrants is of major importance to the development of the natural gas market.

* This chapter was overseen by Xiaoming Wang from the Development Research Center of the State Council and Mallika Ishwaran from Shell International. It was jointly completed by Yusong Deng, Jiaofeng Guo, Shouhai Chen from China University of Petroleum and Qingle Wu from Shell China. Other members of the topic group participated in discussions and revisions.

© The Editor(s) and The Author(s) 2017
Shell International and The Development Research Center (Eds.), *China's Gas Development Strategies*,
Advances in Oil and Gas Exploration & Production, DOI 10.1007/978-3-319-59734-8_12

287

Table 12.1 Main characteristics of natural gas markets of various countries and regions

Natural gas index	Type	USA	EU	UK	Japan	S. Korea	China
Supply (bcm/year)	Domestic production	689	269	38	3	0.5	115
	Net imports	37	231	39	123	53	49
Consumption (percentage of total by sector %)	Power	40	30	30	65	50	15
	Industry	20	20	10	5	20	45
Transport pipelines (km)		500 k	200 k	8 k	5 k	4 k	50 k
Wholesale competition		✓	Limited (oligarchic monopoly)	✓	Limited (oligarchic monopoly)	X	X
Open access	Upstream	✓	✓	✓	✓	X	X
	Transportation	✓	✓	✓	✓	X	X
	Distribution	Diversified	Diversified	✓	X	X	X
Division of ownership of transportation and sales		✓	Diversified	✓	X	X	X
Independent (federal) market rights		✓	✓	✓	X	X	X
Fluid market centre		✓	✓	✓	X	X	X

Note Data is for 2013 (except natural gas consumption levels for China, which are based on 2011 data from power generation and industrial departments)
Source Vivid Economics, based on IEA, EIA, Chinese Government and ENTSOG data

In developed European countries and in the United States, the natural gas exploration and development field is for the most part a competitive market, and this is closely related to the system of access to mineral rights. For instance, the privatisation of land ownership (and resources that lie beneath it) has been effective in encouraging exploration and development of natural gas, while also creating the conditions necessary for the shale natural gas revolution. The approach adopted in The Netherlands of "50-50 state-private ownership" has acted as a major stimulus in encouraging exploration for natural gas resources. The government generally implements a system of business licences, including mineral rights release mechanisms, for any companies that have been awarded mineral rights, thus avoiding hoarding of rights and delayed development. In the UK, for example, companies involved in bidding for mineral rights are expected to strictly abide by exploration and development requirements, otherwise they are likely to lose their licences. However, such a system requires strict supervision by the regulatory body.

However, natural gas resources differ from the average commodity in that not only do they have a value as a raw material, but access to such resources is also accompanied by massive profits and, in addition, they have a strategic value that derives from their relatively large influence on the national economy as a whole. It is due to such factors that in recent years, resource holders such as Russia and Middle Eastern and South American countries have been exerting ever-tighter control over their oil and gas resources. In light of this, the problem of how to increase activity in the upstream exploration and development field, while also ensuring that the majority of profits remain within the country, in addition to ensuring greater energy security for China, deserves greater consideration.

12.3   Orderly and Gradual Implementation of Pipeline Access Policies

Between the 1980s and the present day, policy regarding the regulation of natural gas pipelines has been reformed in both the United States and the European Union, resulting in the creation of an independent, open natural gas transportation system. The increased level of supervision over the operation of natural gas pipeline networks (i.e. the aspects that usually attract natural monopolies) has ensured pipeline and network interconnectivity and access to services for third parties. The process by which this was achieved in general consisted of three stages.

In the first stage, pipelines were encouraged to provide transportation services to third parties, while government regulation was strengthened. This stage may be seen as a period of preparatory work for the encouragement of third-party access. However, governments do not force oil companies to provide non-discriminatory services in terms of access to their pipelines; incentives are provided to companies that provide access, and at the same time the gas end user is allowed to sign

contracts directly with producers. This procedure allows feasibility to be examined and enables the discovery of any problems that may exist in implementing third-party access to pipelines. At the same time, during this stage, the government continues to enhance its supervisory capabilities, with the gradual creation of an independent, unified, impartial and effective natural gas industry regulatory framework. This is in fact preparatory work that will allow the reform of supervisory capabilities and the formation of the teams necessary to carry this out.

In the second stage, pipeline third-party access is enforced. This effectively eliminates the bundling of natural gas sales and pipeline transport. The government issues the relevant regulations that make it obligatory for natural gas infrastructure operators to provide fair and just access to pipeline transport, gas storage, gasification, liquefaction and compression services to all users.

In the third stage, independence of pipeline transportation services is encouraged. The natural gas pipeline businesses of upstream/midstream/downstream integrated companies are split from them, creating a number of independent pipeline companies, after which creation of the oil and

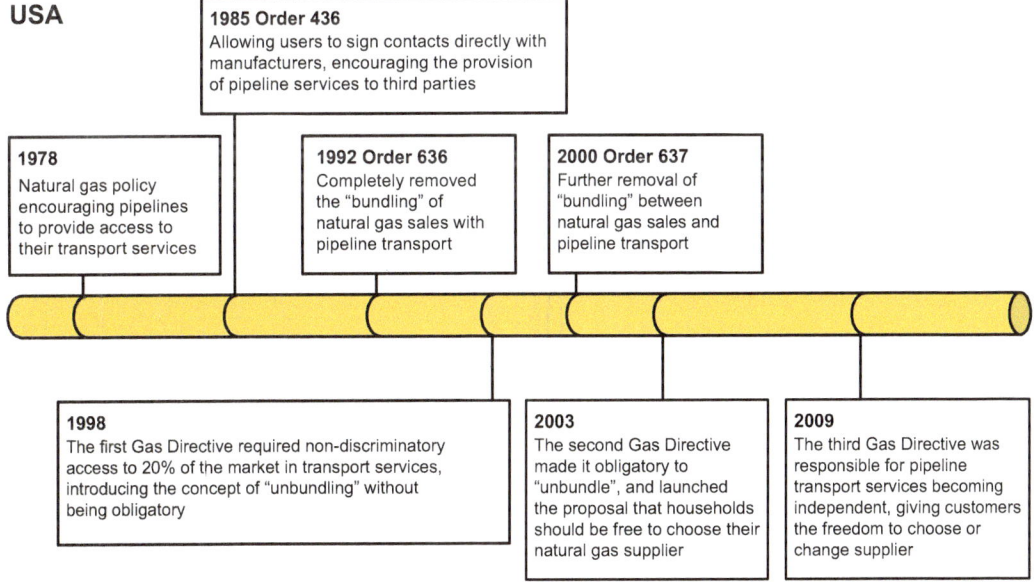

USA

1985 Order 436
Allowing users to sign contacts directly with manufacturers, encouraging the provision of pipeline services to third parties

1978
Natural gas policy encouraging pipelines to provide access to their transport services

1992 Order 636
Completely removed the "bundling" of natural gas sales with pipeline transport

2000 Order 637
Further removal of "bundling" between natural gas sales and pipeline transport

1998
The first Gas Directive required non-discriminatory access to 20% of the market in transport services, introducing the concept of "unbundling" without being obligatory

2003
The second Gas Directive made it obligatory to "unbundle", and launched the proposal that households should be free to choose their natural gas supplier

2009
The third Gas Directive was responsible for pipeline transport services becoming independent, giving customers the freedom to choose or change supplier

EUROPE

Fig. 12.1 Timeline of development of pipeline regulatory policy in the United States and Europe

natural gas pipeline supervisory system can be completed (Fig. 12.1).

Analysis of experience in the United States indicates that, from start to finish, reform of pipeline policy took more than 20 years, while in the EU the process has taken 10 years, so clearly this is a fairly slow process and cannot be achieved overnight. Moreover, pipeline reform in both the US and the EU involved a process of encouraging third-party access, followed by obligatory access, which was then followed by the splitting up of upstream/midstream businesses. This is the pattern of reform best suited to the natural gas industry, and deserves further study and emulation.

The development of US natural gas pipeline regulatory policy

In terms of structure, the structure of the early natural gas industry in the United States was similar to that encountered today in China: producers sold natural gas to pipeline transportation companies, and pipeline transportation companies then sold the natural gas to local distribution companies, with the natural gas being sold to the end user by the local distribution company. The natural gas sales price was controlled by federal government, and the price of natural gas sold to the end user by the local distribution company was controlled by local government institutions. The administration had virtual control over all parts of the natural gas industry, and the state was not only responsible for supervision of the naturally monopolistic pipeline transportation sector, but also for the competitive sectors of the natural gas industry (such as production and wholesale supplies). The strict regulation by the state resulted not only in natural gas companies being placed under a great deal of pressure, but also in abnormal development in terms of natural gas prices and consumption. The excessive regulation of natural gas producers led to the gas supply shortages encountered in the United States in the 1970s.

In order to remove the monopolies that had resulted from closed-type industrial

operations, and to encourage and co-ordinate development in the natural gas industry, the United States government began to adjust its policies towards natural gas industry regulation, and as a result natural gas pipeline regulatory policy underwent four main rounds of revision. In 1978, the Natural Gas Policy Act (NGPA) introduced a staggered process for relaxation of natural gas wellhead pricing controls, at the same time as encouraging pipelines to open up access to their transport services. In 1985, the Federal Energy Regulatory Commission (FERC) issued Order 436, which put pressure on pipeline transport companies of different states to separate sales and transport functions, while introducing competition mechanisms that gave natural gas end users and manufacturers more freedom of choice. In 1992, FERC issued order 636, which required state pipeline companies to separate natural gas sales from pipeline transport and establish separate companies to handle sales. By adopting pipeline usage rights, this order eliminated the widespread unfair competition largely associated with the supply of natural gas by state pipeline companies. It also introduced procedures whereby contracted transport could be sold on, allowing a "firm shipper" (any user of pipeline transport) to buy transport capacity from another firm shipper with excess transport capacity. In 2000, FERC then issued order 637, which went further, completing the unbundling of gas supplies and transport services.

The development of European natural gas pipeline regulatory policy

Europe is one of the three main global natural gas-consuming regions, and in the last 40 years, natural gas's share of overall energy consumption has continued to grow. At the same time, the business modes and regulatory systems applied to the natural gas industry in Europe have undergone continual adjustment. The natural gas industry regulatory systems and policies of each individual member country of the European Union are established with reference to "directives" provided by the European Union. If one looks at the development of the European natural gas industry over the last 10 years, the European Union issued three main directives, and a major component of these relates to encouraging the market liberalisation of natural gas pipelines.

In 1998, The EU issued the First Gas Directive, providing the impetus for the acceleration of the liberalisation of the internal European natural gas market. This directive proposed that in order to ensure the establishment and efficient operation of a European internal natural gas market, each country within the EU had a duty to ensure fair competition in terms of transport, storage and distribution, while requiring that all members should complete revision of their legislation within two years of the directive taking force. In terms of pipeline policy, the directive recommended the opening of a 20% non-discriminatory market access, with the introduction of the concept of "unbundling", without it being obligatory.

In 2003, the EU issued the Second Gas Directive, which required that all member states adhere to the regulations concerning the single market and adjust their own statutes, requiring that the natural gas market be opened to the non-residential users of all European countries before July 2004. Integrated companies had to complete the legal dismantling of their pipeline and marketing businesses, followed by the opening of natural gas markets to all users before July 2007, finally achieving the aim of allowing the consumer freedom of choice of gas supplier. Regarding pipeline policy, the enforcement of regulations relating to unbundling resulted in households being given the right to freely choose their natural gas supplier.

In 2009, the EU issued the Third Internal Energy Market Package (also known as the Third Package), which brought into force EU energy industry reforms. Based on these new reforms, large power and gas companies had to select one of the following three reform schemes:

- ownership unbundling, which requires companies to sell their distribution network, thus completing the separation of the manufacturer from the network;
- unbundling of business operations, which, while allowing retention of ownership of distribution networks, required the establishment of an independent company (known as an "independent system operator") who would be responsible for the operation of the gas distribution network;
- unbundling of managerial responsibility, allowing the ownership and management of gas distribution networks but requiring a subsidiary to manage the gas distribution network and entailed ensuring managerial and policy-making autonomy (known as an "independent transmission operator").

These reforms also required that each country establish an independent regulator, responsible for ensuring that large energy companies actually implemented this "division of production and supply" and for ensuring that the rules that apply to free competition in energy markets were put into play.

12.4   Third-Party Access: The First Step to Infrastructure Reform

In the natural gas industrial chain, upstream production and downstream sales are both sectors to which natural monopolies do not apply, and as such, efficiency can be improved by the introduction of market competition. Operation of midstream assets such as pipelines, on the other hand, exhibits the characteristics of a natural monopoly, with high fixed-cost investment and relatively low operational and management charges. There is thus a rationale for operation of pipeline services by a single company being the most effective approach to operation. This in turn is effective in reducing the overall operational cost to society. The special characteristics of midstream asset natural monopolies are one of the reasons that they are not exposed to the pressures of market competition, and to some extent they also dominate the markets. As a result of this, we recommend that greater regulation of midstream pipeline assets is required, to prevent companies that occupy a position that can be considered a "natural monopoly" from abusing their power to influence markets and extract ever-higher levels of monopolistic profits.

Generally, supervision of midstream assets consists of a number of approaches: regulation of fee standards applied to pipeline transportation and other such infrastructure, implementation of infrastructure third-party access and division of ownership. In order to prevent monopolistic enterprises extracting extremely high monopolistic fees for midstream services, the government needs to investigate actual operating costs and reasonable profit levels in order to establish reasonable charging standards. Apart from this, independent owners of midstream assets may also take advantage of their market dominance to increase pipeline transportation costs or restrict natural gas transportation, and this in itself is also responsible for reduced upstream and downstream competition, while also reducing pipeline usage. To prevent owners with natural monopolies over infrastructure taking advantage of their market position, and to ensure that all other parties have fair access to usage of pipelines and other infrastructure, the government is generally required to exert a supervisory influence, and this is exactly what is being referred to as the third-party access (TPA) regime.

International experience has demonstrated that third-party access policies balance out the economic interests of the owners of midstream pipelines and upstream and downstream production and consumption. This is not only beneficial in that it allows full use to be made of the potential of such infrastructure, but also in terms of attracting capital investment into the infrastructure construction sector

Regulatory policy and its effects in the UK provide excellent proof of this. Between the 1960s and the 1980s, when the North Sea oil fields were being developed on a significant scale, the production and transportation of British gas was led by only a few major companies. However, as the yield of these oil and gas fields has dropped, large companies have begun to find it hard to obtain satisfactory benefits, while receiving stations and pipelines have redundant capacity. At the same time, many small enterprises have entered the upstream field, developing small-scale oil and gas fields, using the pipelines of the larger companies to transport their oil and gas products. There is a lack of legal framework, and thus the potential exists for disputes to occur between the owners of infrastructure and the owners of oil and gas fields. Indeed, throughout the 1990s there was a worry that the cost of usage of infrastructure was becoming too high in relation to the costs and risks associated with the development of small oil and gas fields.

In order to resolve this situation, between the 1990s and the early 21st century, the British energy regulator (or the Department of Energy and Climate Change as it is now known) established a new access negotiation procedure, accompanied by a raft of legal and regulatory measures that reduced stakeholders' uncertainties at the same time as providing a regulatory framework with a legal basis and arbitration mechanisms that became a foundation via which negotiations could be completed in a manner that benefited both stakeholders. With the gradual appearance and introduction of this legal and regulatory framework, the natural gas markets were able to develop further, which benefited all the stakeholders concerned. This allowed smaller enterprises to enter the larger oil and gas exploration market, which in turn offered further benefits to the owners of infrastructure, reducing the cost of redundant infrastructure, while the government also benefited in terms of reduced unemployment and an increase in tax revenues.

A well-designed third-party pipeline access policy and regulatory framework should satisfy the following conditions:

- the policy should be effective, with the regulator being responsible for drafting and implementing policies relating to duties and standards;
- appropriate pricing, with the regulator providing a framework, allowing investors and users to reach agreement on pricing in terms of TPA and appropriate returns mechanisms;
- risk avoidance, requiring the regulatory framework to be stable and reliable, in a way that reduces the risks to the investor and the user to the minimum.

Third-party access policy frameworks will affect different types of natural gas markets in different ways. For instance, Japan is a typical natural gas importer, being incapable of achieving competition between upstream manufacturers, therefore TPA policy is limited. However, where China with its continuously growing markets is concerned, a regulatory framework offers significant benefits, providing a stimulus for investment in the midstream field.

12.5   Unbundling: An Important Element of Liberalisation of Natural Gas Markets

As mentioned previously, regulation that applies to the midstream sector generally involves approaches such as pricing standards applied being to infrastructure such as transport pipelines, the implementation of third-party access to such infrastructure and the unbundling or partitioning of ownership. In practice, even when the government establishes the principles of third-party access, monopolistic companies still

have many methods of ostracising third parties, for instance by a lack of transparency regarding available pipeline transport capacity or pricing, or by contractual obligations such as conducting obligatory technical research etc. Strict application of the principles of third-party access can suppress these anti-competitive market forces to some degree.

However, some experts are of the opinion that unbundling is a more effective approach to resolving such issues. Unbundling involves separating the factors that motivate operators at the midstream stage from those that motivate operators at the upstream and downstream stages, which reduces the likelihood of anti-competitive activities. Unbundling is not the actual objective, though; it is just an effective method of ensuring the feasibility of third-party access. Whether or not open access measures eventually result in an efficient actively competitive natural gas market is the final proof of whether mechanisms for unbundling have been successful or not. For instance, one of the benefits of unbundling is the additional upstream activity resulting from competition, encouraging greater manufacturing efficiency and expansion.

Where nations that have a domestic production capability are concerned, the earlier unbundling takes place and the more completely it occurs, the more effective it will be. Looking at the UK and The Netherlands and their domestic North Sea resources, and comparing them to nations such as France and Germany that have only limited domestic resources, unbundling occurred more rapidly. In the case of Japan, which has only minimal natural gas resources, only tentative unbundling occurred. In juxtaposition to this is the United States. The US has plentiful upstream resources, with a wide distribution. In fact, unbundling commenced in the US at a very early stage; in 1992 the United States had already required the introduction of legislation and other structural unbundling measures. One has to be cautious about making generalisations in the analysis of the factors that motivated this, but where nations that possess domestic resources were concerned, unbundling resulted in greater economic benefits, and this in turn has led to such nations being more willing to institute such reforms.

The unbundling process is of great importance, and depending on the extent to which unbundling takes place, it can be split into five categories: service unbundling, financial unbundling, legal unbundling, structural unbundling and unbundling of ownership. Unbundling at a higher level lays the foundations for unbundling at a lower level, lower-level unbundling being simply an extension of the unbundling occurring at higher levels. A good example of this is that financial unbundling must be implemented at the same time as service unbundling, and so on; the most thorough form of unbundling is ownership unbundling (see Table 12.2).

Analysis of midstream asset unbundling case histories from the United States and Europe indicates that, if regulators are given sufficient powers by the state, it is unnecessary to implement complete ownership unbundling. Among the nations analysed, only the UK and The Netherlands instigated complete ownership unbundling, while France and Germany carried out structural unbundling and never attempted complete ownership unbundling. Experiences in the US also demonstrated that, by applying a strict model to the operations of natural gas transport companies, it became unnecessary to apply ownership unbundling in order to ensure effective competition in the natural gas market. To summarise, the majority of the nations mentioned above instigated structural unbundling, without the need to carry out complete ownership unbundling. From this one may conclude that simply carrying out structural unbundling in conjunction with strict regulation provided the best results.

In contrast to pipelines, implementation of unbundling applied to infrastructure such as LNG receiving stations and peak gas storage installations has not been so strict. The reason for this is that, generally speaking, there is a large number of natural gas receiving stations and peak gas storage installations, therefore there is more opportunity for competition to arise; when compared to LNG receiving stations and peak gas storage installations, natural gas pipelines exhibit more pronounced natural monopoly characteristics. Where a pipeline network is connected to a fairly well developed market, LNG receiving stations take on the characteristics of supply resources, competition occurring with other receiving stations thus restricting market dominance. In a similar manner, in a diversified market with dense connections there will be many storage providers, while pipeline storage capacity, flexible natural gas production capacity and dynamic increase and decrease in demand can also be considered as providing storage services. In light of this, unbundling and third-party access policy should take this into account and treat such installations with more leniency.

12.6 Natural Gas Pricing Reform as Part of the Market Liberalisation and Development Process

Due to different natural gas markets having developed to differing degrees, there are three types of main pricing mechanism found in the main global importers of natural gas: the first type relies completely on market-driven prices, such as is encountered in the US and the UK. Here we have relatively mature natural gas markets, and a completely unregulated natural gas price, determined by market demand and supply; however, third-party access to pipelines is obligatory, resulting in a market where competition occurs directly between different gas companies. The NYMEX Henry Hub and the UK's NBP trading prices act as the reference prices for North American and European trading.

The second category is reliance on the netback pricing method, such as is adopted in European countries other than the UK. Market

Table 12.2 Five key unbundling models

Extent of unbundling	Model	Changes	Aims	Relationship to previous changes
I	Service unbundling	Midstream services (in particular pipeline services) must be unbundled from gas wholesale businesses, with service then being provided independently	From the point of view of the gas wholesale business, if transport services are not provided independently, this will have a negative impact on market competition, as the supplier is unlikely to sell gas to users other than the pipeline company	
II	Financial unbundling	The profits and losses of midstream operations will be borne by the company itself, having been isolated from the upstream and downstream concerns	This prevents reliance on midstream assets to cover losses or support upstream or downstream operations	If midstream operations do not become an independent service subject to individual taxation, then the midstream services are unlikely to possess a separate account for their own finances
III	Legal unbundling	The midstream business becomes an independent legal entity, while remaining within the vertically integrated structure, for instance existing as a wholly-owned subsidiary	The subsidiary must be in the form of a legal representative with its own legal standing. By introducing independent legal entities, it is possible to achieve more effective separation of managerial duties	An independent legal entity must have separate accounts for its finances and provide a separate service
IV	Structural unbundling	In order to isolate themselves from upstream and downstream motivations, many midstream services formulate their business policies in advance	This roots out any anti-competitive motivation; by removing such motivation from the midstream business of the entities concerned, it is possible to achieve more effective market liberalisation	A midstream company needs to be a legal entity; only then will it be able to respond appropriately to circumstances and fulfil its duties
V	Ownership unbundling	The midstream service must be split off from the vertically integrated structure and ownership must have been transferred to an independent entity	Comprehensive ownership unbundling completely separates the interests of midstream businesses from upstream and downstream businesses	If a company is under independent ownership, then it is intrinsically an independent legal entity, with independent accounts for finances and providing an independent service

Note In the EU Third Package, structural unbundling includes the Independent Transmission Operator (ITO) model
Source Vivid Economics

netback pricing is based on the market value of energy sources which are an alternative fuel to natural gas; this employs retrospective methods, involving deduction of transport, storage, distribution and other such costs to determine an upstream price. When compared to the US and UK markets, the European market is still in the stage of transition to full market competition. Natural gas consumption relies mainly on external imports, and most natural gas is traded in long-term contracts. The price is mainly pegged against oil products, but there is a certain amount of lag in the adjustment in contract prices.

The third type mainly relies on pegging against the price of imported oil, examples of which are Japan and South Korea. The development of the natural gas markets in both Japan and Korea have been somewhat delayed, with reliance mainly being on imports. At the same time, since there is a lack of a representative trading centre, there is little choice but to peg natural gas prices against crude oil. For instance, Japan's Chubu Electric Power agreed a pricing formula with the Qatar Gas Transport Company in 2001 which stated: $P = 0.1485 \times JCC + 0.08675 + S$. Here, S is an adjustment figure, which was used to moderate the effects of the violent fluctuation in the international oil price. Prior to 2005, the international oil price was relatively low, with the price of gas imported by Japan and South Korea being more or less the same as in Europe and the United States. In recent years, with further fluctuations in the international oil price, an "Asian Premium" phenomenon has arisen, with a major increase in the cost of gas use.

Reform of pricing mechanisms is a gradual process, with the final objective being a pricing mechanism which achieves completely open-market competition between different gas companies. If you take the historical situation in the US, the government controlled natural gas prices, applying a "cost plus" method to import prices, with the federal government controlling long-distance transport pipeline pricing and local government organisations being responsible for the price paid by the final user. Governmental pricing regulation resulted in the price remaining

relatively low over a long period of time, which led in turn to problems of supply being unable to satisfy demand. Subsequently, the government eased controls over the price of imported natural gas, and the creation of a trading centre allowed the gradual evolution of open-market pricing regulation. Historically, the reform of pricing mechanisms in the UK was similar to that of the US, involving a transition from prices being set by the government to prices being set by markets. Subsequently we find that the pricing regulation that was applied to these markets resulted in trading centres adopting pricing based on a reference market price model.

12.7 The Role of Natural Gas Trading Markets

Currently, global gas trading relies mainly on long-term contracts, and this plays a major role in balancing upstream and downstream interests and stabilising the relationship between supply and demand. Long-term contracts allow natural gas to be linked to the prices of alternative fuels, market trading prices and many other indices, thereby ensuring that upstream investors receive a stable return over a number of decades and protecting investment in exploration and production. This is particularly beneficial for stabilisation of upstream investment. Relative to long-term contracts, spot trading is more flexible and fluid and is taking on an ever-greater role in natural gas trading. Taking shale natural gas in the US as an example, due to uncertainties over yields, shale natural gas developers are unable to obtain long-term contracts on favourable terms and therefore the development of shale natural gas relies heavily on the spot trading market.

More important, though, is the fact that the creation of spot trading markets has provided a new pricing method for natural gas market liberalisation, which has also accelerated the creation and growth of trading centres. Take, for instance, US FERC order 636 in 1992, which made it obligatory for pipeline companies to provide access to their services, an event that resulted in a substantial loosening of regulation over gas markets, and a market competition-based pricing mechanism replacing the previous pricing model, a government-driven pricing mechanism. At the same time, FERC began to promote the concept of natural gas market trading centres, adopting the stance that natural gas market trading centres should be responsible for providing natural gas supplies, transport, storage, deployment, enhanced administration and other such services or combinations of services to the customers of pipeline transport operators. As a result, 36 internal natural gas market trading centres were created, covering all the US natural gas pipeline networks between 1993 and 1998. By 2003, 13 of these natural gas trading centres were closed down, either due to their trading base being insufficient, resulting in small trading volumes, or because they lacked competitiveness. Currently, there are a total of 24 natural gas trading centres, mostly in Texas and Louisiana, which provide hub services to the natural gas markets. The appearance of these natural gas trading centres accelerated the formulation and systematisation of open-market natural gas pricing mechanisms, streamlining the allocation of resources within the gas markets, increasing market efficiency and enriching the choice of investments available to those involved in these markets. This evolution also helped the US to establish itself as a leader in terms of energy pricing, as well as ensuring US national energy security.

Natural gas trading centres clearly offer both benefits and drawbacks. They have two major effects. The first is that there is a physical connection between the buyer and the seller, the second is that pricing is set based on market competition. As a result of this, allowing markets to provide pricing signals increases economic efficiency in terms of trading and investment decision-making, in addition to resulting in a relative drop in trading costs, giving greater returns to those investing in such markets. In addition, natural gas trading centres are able to balance problems in supply and demand due to their pricing mechanisms, thereby ensuring the security of natural gas supplies. A potential drawback is their impact on current market models. With pricing being determined by competition between different parties, rather than by the government or other leading influences within the market, it is possible that they can cause greater fluctuations in pricing in the short term.

From the global experience already gained by the main countries operating natural gas hub trading centres, we can conclude that, in light of the different functioning of natural gas trading centres, the success or otherwise of natural gas trading centres depends on three preconditions (see Table 12.3):

Table 12.3 Conditions for creation of a natural gas trading centre

Basic infrastructure conditions	A complete and open natural gas distribution network	Connections to both domestic and overseas gas-consuming and supplying enterprises in an open channel gas distribution network
Market conditions	A large number of different participants within the market	There must be sufficient numbers of entities within the market, including both buyers and sellers
	Relatively low level of market concentration	There should not be dominant suppliers or consumers
	Formalised trading activities	Natural gas trading and pipeline transport prices should be regulated by the government, with fair and reasonable prices; traded gas quantities should be able to enter pipelines freely; market prices should be made public and transparent
System conditions	Market liberalisation of prices	The wholesale trading price of gas along the supply chain should be determined freely by the market
	Stable regulatory framework	The government must create a fair commercial environment in order to allow competition policy and market models to take effect

Source Vivid Economics

- a natural gas transport network which is completely open to access, which is available for the use of market participants on a non-discriminatory basis;
- large numbers of independent buyers and sellers actively involved in arbitrage, none of whom possess the capability to dominate the market;
- government support for the opening of its natural gas wholesale markets accompanied by stable, transparent and reliable rules and regulations.

Trading centres also require enterprises to engage in trades based on arbitrage, which will improve market operating efficiencies, and arbitrage activities rely on the accuracy and high transparency of reports of natural gas prices.

By carrying out extensive natural gas sector reforms, China will possess the conditions that would allow it to become the premier natural gas trading centre in Asia. Looking at China's advantages, its natural gas output, the extent of development of its gas transportation pipelines, the scale of its LNG imports and the significance of its position in the Asian energy market are all clear. However, China's natural gas distribution network is not well developed, the market is still relatively concentrated, third-party access mechanisms are ineffective and pricing is still subject to government control, all of which are factors that are restricting the creation of a natural gas trading centre in China.

Experience gained in the United States and Europe since the 1990s in the creation and development of natural gas trading markets provides a valuable reference for China. The first thing to note is the extent to which the creation of a natural gas market will have a major effect on the development of trading centres. The development of natural gas hubs in Europe only started in 2008, and for a period of time they were not that successful, due to the slowness of market liberalisation in the European natural gas markets.

Second, government restrictions over trading activities can also have a major influence on trading centre development. Benefiting from the enormous supplies of the Groningen gas field, and recent increases in the number of LNG receiving stations and increased storage facilities, the Dutch Title Transfer Facility (TTF) is widely recognised as being the most successful European continental natural gas hub, and contract volumes indicated that by 2013/2014 TTF had already surpassed NBP to become the largest European fluidity hub (Natural Gas Daily 2014). Zeebrugge (Belgium), on the other hand, is viewed as the least successful example, due to its low level of trading activity, even to the extent that it is no longer viewed as a hub. Zeebrugge possesses suitable geographic conditions for it to become a natural gas hub, but commercial restrictions imposed by the local government have restricted the number of participants involved, reducing the trading platform's rate of growth.

Third, government impetus is capable of having a substantial effect in the short term. In the early 1990s, the United States had already achieved market liberalisation of its natural gas prices, but the link between natural gas prices of different regions (eastern region and western region) showed only a weak correlation. At the end of the 1990s, as the United States government promoted the adoption of the "law of one price" across the whole of the US natural gas market, after subtraction of transportation costs, a single transregional price was applied, at which stage price correlation had already developed to a very high degree (Cuddington and Wang 2006; Doane and Spulber 1994).

Fourth, formalised and orderly trading mechanisms also have a beneficial effect in encouraging the development of natural gas trading centres. Taking the development of the NBP in the UK as an example, the Uniform Network Code has played a crucial role. The Uniform Network Code provided the regulations and procedures for supervision of third-party

(TRA) use of the UK natural gas network. This introduced a daily balance system and short-term natural gas trading and one of the standardised contracts (NBP97) became the cornerstone of the UK direct trading contract.

Based on the experience gained in the development of natural gas trading centres in the United States and European Countries, and in light of the specific characteristics of China's natural gas industry, development of China's natural gas trading centres requires a 5- to 10-year plan to be drawn up. Trial trading centres should be established first. The government could establish natural gas pipeline networks to which third-party access applies in individual regions such as Shanghai, with the introduction of transparent natural gas pricing mechanisms. This would create the conditions for competition in natural gas demand and supply. Then, after conditions have developed over a certain period of time, a real-time gross settlement system should be introduced to trial trading centres. On the basis of this, the government can extend the areas covered by trials further, or increase the number of categories of buyers and sellers, at the same time as making further enhancements to market regulation. As the effects of these trials become more noticeable, they will attract the participation of greater numbers of producers and consumers in market trading. Once the price of natural gas in trading trials is less than the price of natural gas for traditional contracts, the number of traditional contracts traded by consumers will drop and further interest in purchasing natural gas from trial trading centres will develop. On the other hand, when the price of natural gas traded by trial trading centres is higher than that of traditional natural gas contracts, suppliers will be more willing to supply gas to these trial centres, thus also reducing the numbers of traditional contracts traded. As more and more diverse participants become involved in trading in these trial centres, the benefits of trading will increase, due to the network effect mentioned above. Hub-based pricing will be applied on an increasing scale and over a greater range, until partial market liberalisation is achieved or geographical limits become a factor.

12.8   Establishment of an Independent and Legally Protected Regulatory System

Major state regulatory bodies generally include an overall energy administration and an energy regulator, with powers being divided between these two bodies. The overall energy administration is mainly responsible for establishing energy strategy, planning and policy, playing a role in the balancing of total energy resources, with responsibility for energy security, modification of energy structures, promotion of energy saving and energy efficiency administration, collation and analysis of data, scientific innovation in relation to energy and international energy co-operation. It is also responsible for harmonising activities in conjunction with related departments. The energy regulator, on the other hand, is mainly responsible for ensuring order within energy markets, while promoting market competition within the industry and resolving disputes. It is also responsible for regulation of sectors in which natural monopolies exist, thus protecting the overall interests of the market participants. The overall energy administration is responsible for establishing policy; the energy regulator is responsible for implementing it (see Table 12.4).

These two bodies may be partially independent or completely separate from each other, and from the point of view of energy management they ideally act as a counterbalance to each other, which satisfies the requirements of modern administrative systems by separating the three powers of regulation-making, implementation and supervision. The regulator should be independent in terms of powers and positioning, to ensure that it is capable of reaching appropriate decisions which treat all market participants equally. Take the US Federal Energy Regulatory Commission (FERC) as an example. This is an

Table 12.4 The energy regulators of different countries

	Name	Function	Relationship to the overall energy administration	Funding source	Staff
United States	Federal Energy Regulatory Commission (FERC)	Regulation of interstate power, gas and oil transport, responsible for approval of LNG terminal and natural gas pipework construction, while also issuing hydroelectric power station permits	While in name it forms part of the Department of Energy, in actual fact FERC is an independent body	From fees charged for regulation of public utilities and service charges	Over 1200
Japan	Power market departments and gas market departments	Regulates distribution pricing, control over power station environmental, technical and safety standards, responsibility for encouraging market competition and resolving trade disputes	An internal branch organisation of the Agency for Natural Resources and Energy of the Ministry of Trade, completely subordinate to the Agency for Natural Resources and Energy	Government budgeting	In the region of 50 persons or more
United Kingdom	The Office of Gas and Electricity Markets (OFGEM)	Core principle is to encourage competition within the industry, protecting the interests of the consumer, including effective promotion of competition, regulation of monopolistic companies and pricing control	A government organisation, but completely independent from other energy-related departments	Provided by charges for regulation of public utilities	291 persons
Germany	Federal Cartel Office (FCO)	Regulation of the power and natural gas markets, resolving trade disputes, encouraging market competition in addition to carrying out surveys relating to pipeline charges	Subordinate to The Federal Ministry of Economy and Technology; this is an administrative body on the same level as other energy related departments	Government budget	The FCO office has around 300 persons, with 10 people allocated to the power section
Russia	The Federal Energy Commission of the Russian Federation (part of the Federal Commodity Pricing Bureau since 2004)	Regulation of delivery of electrical power, hydrothermal power, oil, natural gas and petrochemical products, regulation of wholesale and retail electricity pricing	A government organisation, independent from departments of industry and energy	Government budget	Probably around 200 persons
India	Central Electricity Regulatory Commission (CERC)	Mainly responsible for regulation of electricity prices of central power generating companies and interstate power transmission companies as well as issuing licences for power transmission and trading	A government organisation, but completely independent from the Ministry of Power	Funds provided by central government as well as charges made on licensed operators in addition to other income	68 persons
Korea	Korean Electricity Commission special task force, power regulator committee	Regulation of electrical power and other energy industries	Subordinate to the Ministry of Commerce, Industry and Energy (MOCIE), a semi-independent regulatory body	Government budget	Responsibility lies with an inspector, but exact numbers unknown

independent administrative body which is subordinate to the US government. It is responsible for the regulation of business activities involving gas, electrical power, hydroelectric power and oil, covering pipelines and distribution links; and policy-making and the resolution of disputes require a majority vote of the five committee members. There are five committee members, one of whom is the chair of the committee, and the names of whom are suggested by the President and ratified by the Senate (though not more than three fifths of the committee members may come from one political party). Policy decisions are issued by a court of law, and not by the United States Congress, and when cases are being heard, no private discussions are permitted.

The main regulatory role of FERC is to prevent companies from taking advantage of their monopolistic position, with the aims of the regulation including:

- prevention of discriminatory or preferential treatment;
- prevention of inefficient manufacturing and unfair pricing practices;
- ensuring excellence of service;
- prevention of infrastructure redundancy and wastage;
- where competition does not exist or is not able to exist, acting as an agent for the promotion of competition;
- promotion of safe, high-quality, environmentally friendly energy infrastructure via implementation of consistent policies;
- where possible, encouraging overall market competition in place of traditional regulation.

From the point of view of the natural gas industry, the main function of the Commission relates to pricing controls, interpretation of the relevant legal statues and service regulation, and the scope of regulation includes roughly 120 interstate gas pipelines. At the same time, the Commission has the power to authorise the construction and siting of gas delivery-related infrastructure as well as being responsible for the evaluation of the environmental impact that this may cause.

12.9   A Roadmap for Natural Gas Market Reform

Reform of market structures and the administrative regime applied to natural gas is a gradual process. Looking at the timeline of developments in the US natural gas market, this process involved a progression from complete government regulation and the government setting of prices to a situation where pricing controls were gradually withdrawn and bundling of gas sales was eliminated, followed by a fully developed natural gas market with pricing determined by the market itself (see Fig. 12.2).

The reform of the natural gas market is a long-term and difficult process, requiring a particular political environment, relevant infrastructure and associated government policy. Favourable political and market climates encouraged the reform of both the British and American natural gas markets. For instance, looking at the period from the 1970s to the 1980s, the oil crisis resulted in massive pressure being put on the government to introduce market reforms, while the extensive pipeline network infrastructure was also a driving force behind the natural gas market reforms in the US. In the UK, the Conservative party, which was in power at that time, under the leadership of Margaret Thatcher, was in the midst of carrying out privatisation of publicly owned industries, including vertically integrated power and gas companies, in order to relieve the financial problems faced by the government at that time; this also created political and market conditions favourable to market reform. In contrast to this, if reforms do not proceed effectively, there is also the possibility that they will fail. In the last 20 years, South Korea has encountered many difficulties in the reform of its natural gas market, in particular relating to the powerful unions of the publicly owned, vertically integrated monopolistic KOGAS, which objected to the surrender of KOGAS's global natural gas market purchasing capabilities. It is easy to see from this that the natural gas market reform process in all countries in question has been both slow and tortuous, and required tailoring to fit the current political and

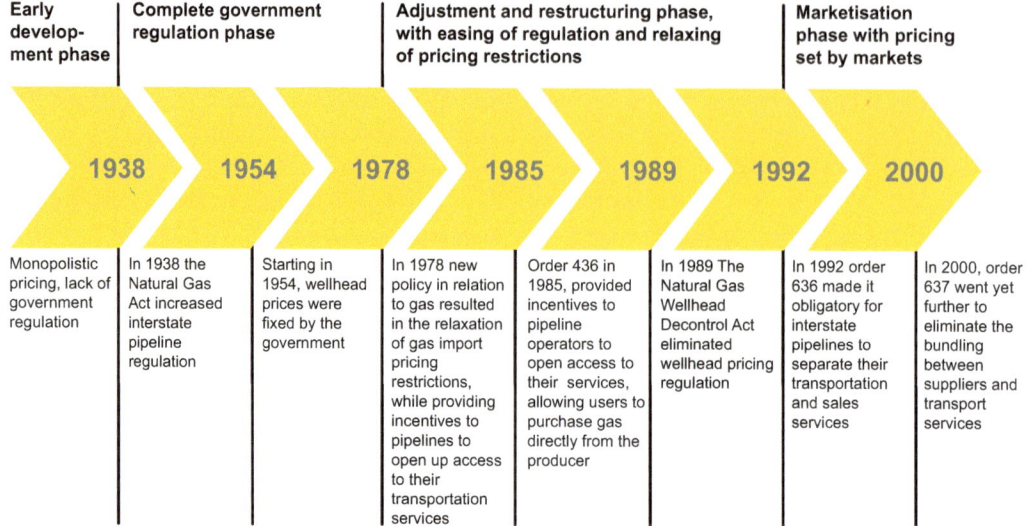

Fig. 12.2 Timeline in the reform of the US natural gas market

economic climate at the same time as implementation of the required policies and measures took place. As such, establishing a roadmap for reform and enforcing its implementation is crucial from the perspective of implementing natural gas market reforms.

Regulatory System Reform to Support Natural Gas Market Liberalisation

13

13.1 Direction of Reforms

Summarising international experience in relation to natural gas market reform, while giving special consideration to China's specific situation, the creation of a natural gas market system and reform of administrative systems should embody the principles described below.

The direction taken by reforms should be identical to the direction taken in the creation of a socialist market economic system. The creation of a market economy with specific Chinese socialist characteristics has been the overall direction adopted in the reform of Chinese economic systems. Statements issued during the 18th plenum of the People's Congress also made new proposals regarding the refinement of the socialist market economy system. The reform of the natural gas market must satisfy the requirements of a socialist market economy, ensuring that the relationship between the government and markets is dealt with effectively, in particular requiring the transformation of governmental functions. This applies particularly in the clarifying of the boundaries between the government and its markets; this will provide oil and gas companies with more opportunities to

develop. At the same time, in addition to providing the basic functions of a market system in optimising and allocating resources, it must improve the macroeconomic and market regulatory roles of the government, allowing implementation of effective regulation of naturally monopolistic markets while encompassing the important society-wide functions of safety, energy efficiency and emissions reduction regulation, in addition to evening out various flaws within the market.

Reforms must be implemented according to both long-term and short-term goals, taking place gradually. Natural gas is a basic element of our national economy; the reform of the natural gas industry must take into consideration both the positive and the negative impacts on socioeconomics, and this requires us to keep our eyes on the bigger picture, while establishing new system mechanisms. At the same time as we seek to implement these in the present, we also need to be able to solve current problems as they appear. This requires a unified design, which establishes the direction and objectives of such reforms, while allowing special emphasis to be placed where necessary, dealing first with the easier objectives and then the more difficult ones, first building up experience from trials, followed by gradual extension, followed by step-by-step implementation.

Reforms must ensure that they are all-encompassing and that the burden of them is shared evenly. Sufficient consideration must be given to the ability of different industrial sectors

---

* This chapter was overseen by Xiaoming Wang from the Development Research Center of the State Council and Mallika Ishwaran from Shell International. It was jointly completed by Yusong Deng, Jiaofeng Guo, Shouhai Chen from China University of Petroleum and Qingle Wu from Shell China. Other members of the topic group participated in discussions and revisions.

© The Editor(s) and The Author(s) 2017
Shell International and The Development Research Center (Eds.), *China's Gas Development Strategies*,
Advances in Oil and Gas Exploration & Production, DOI 10.1007/978-3-319-59734-8_13

to withstand adjustments in pricing, and to their different requirements in terms of the adjustment process. Special consideration must also be given to the harmonising of international and domestic market prices, and to the ability of both urban and rural residents to withstand such changes. The creation of new systems and procedures should ensure that reform of the market, the administrative system and pricing reform occur in step with reforms to taxation, while making active efforts to ensure the sustainable development of the natural gas industry.

The objectives of reforms should be closely linked to increasing industrial competitiveness. As globalisation takes on even greater economic significance, the rate at which China's oil and gas industries are becoming involved internationally, in addition to inviting international involvement in China, is increasing, and against this backdrop increasing the competitiveness of our energy industry is a matter of a major urgency. Reform of the natural gas industry must be linked to the creation of a more competitive industry. Whether or not China's oil and gas industries become more competitive internationally is a major benchmark against which we can measure the success of market reforms. Allowing competition to enter our markets will ensure that our oil and gas industries become more efficient and scientifically advanced, thus increasing the competitiveness of our industry.

13.2 Key Pillars of Reform

Reforms must follow the proposals aired at the 18th plenum of the People's Congress concerning completion of a market economy with specific Chinese socialist characteristics, which would allow markets to play a definitive role in the allocation of resources, accelerating the transformation of government functions, reducing bureaucratic red tape and interference in microeconomic activities and establishing a service-based natural gas administrative system. Reforms should be sufficiently market-oriented, while establishing pricing parity between natural gas and alternative energy sources, in addition to

creating pricing systems that reflect the extent of scarcity of resources, the market relationships of supply and demand and environmental externality in the creation of a green financial and taxation system. This will introduce the diversity of competition to the overall natural gas production chain, increasing external access yet taking care to ensure that the state retains control over the economy, enhancing regulation in terms of fair extension of access to services, while establishing a diversified, competitive, open and orderly modern market system. In overall terms, in order to guarantee that the strategic objectives regarding the development of natural gas in China are achieved, the natural gas market liberalisation reforms should accelerate the establishment of three main pillars, as outlined below.

13.2.1 Pillar I: Establish a Diversified, Competitive, Open, Orderly Modern Natural Gas Market System

The natural gas market liberalisation reforms mainly cover three areas. The first is the breaking up of monopolies, encouraging diversification and competition among participants. The introduction of private capital into the oil and gas exploration field should be supported, allowing its participation in conjunction with state oil and gas companies in oil and gas exploration and development. Reform of the current oil and gas mineral rights allocation regime, gradually allowing competitive tendering and bidding for quotas, based on implementation of a licensing regime, would then allow the structure of a competitive oil and gas resource exploration and development market to be established. Encouraging diversification of natural gas importers will attract private capital for the construction of oil and gas pipelines, which will then allow the separation of natural gas transport and sales. At the same time, it will make it possible to ensure that access to the midstream and downstream sectors is allowed on the basis of company safety, energy efficiency, environmental, quality and technical performance indices.

The second area is the acceleration of the creation of a natural gas trading market, involving training market intermediary organisations and ensuring an improvement in service. The construction of the Shanghai Oil and Natural Gas Trading Centre should be accelerated. In addition, at an appropriate time, research should be carried out regarding the creation of natural gas trading platforms for Beijing, Tianjin, Guangzhou, Chengdu and other trading centres. Encouragement should be given to energy efficiency conversion, financial, insurance, IT, consulting and other such participating professions, in order to improve the professional standards of all involved.

The third area is deepening the extent of reform of state-owned enterprises, encouraging the growth of modern energy companies. This can be accomplished by:

- promoting reforms such as mixed ownership, allowing private companies and overseas investors to hold shares in state-owned energy companies, as well as refining the process by which state-owned energy companies may list on the stock exchange, creating a diversified ownership structure while maintaining a state-owned controlling share over natural monopolies;
- optimising corporate governance structures in state-owned enterprises, introducing modern business regimes, creating boards of directors that are more than just symbolic and allowing shareholder meetings and other similar business administration and decision-making structures;
- tightening up company budgetary and expenditure management, ensuring cost-effective accounting, while giving full consideration to the social function of such enterprises, historical problems and future developmental requirements, increasing the proportion of profits being passed on, and ensuring that excess profits arising due to monopolies revert to the public coffers and go to benefit everyone.

13.2.2 Pillar II: Create a Pricing System that Reflects the Extent of Scarcity of Resources, the Market Relationships of Supply and Demand and Environmental Externality and a Green Financial and Taxation System

Natural gas pricing reform and financial and taxation reform must be dealt with in five ways:

- ensure that there is a mechanism by which competition in the natural gas industry results in market adjustments, and that natural monopolistic mechanisms are subject to legal regulation;
- internalise the costs of externalities such as environmental protection that required in the course of exploitation of energy resources and reduction of greenhouse gas emissions, which will go towards establishing energy pricing parity;
- bring about gradual integration with international natural gas markets;
- adopt natural gas as a clean and highly efficient energy type, with major encouragement being given to increased production capacity and importing;
- eradicate cross-subsidisation, which will ensure that "hidden subsidies" become "known subsidies".

In terms of pricing reform, it will be necessary to improve the "market netback" laws. This will allow regular dynamic adjustments correlating to energy resources, including pricing, category selection and weighting correlating to energy resource. Dynamic upstream and downstream linkage systems must be put in place, allowing for locally created price transmission systems, in addition to refining mechanisms via which low-income groups will receive financial subsidies. Creation of the Shanghai Oil and Natural Gas Trading Centre should be accelerated, while the scope of the trial should be extended. The introduction of differential pricing policies should be brought forward, encouraging the introduction of stepped prices and peak price adjustment etc. Once the market structure and other conditions allow, competitive systems should be introduced, allowing suppliers and users to determine prices freely. Once natural gas trading centres have been established, this will act as the basis for establishing and adjusting market prices and lead to the creation of a gas storage market and a pipeline market, allowing market forces to provide a solution to peak storage problems.

In terms of financial and taxation reform, it will be possible to create a green energy taxation system, which will act as a major financial stimulus in the development of natural gas. Then, at an appropriate point, carbon tax should be introduced, with "emissions charges" being converted to an "emissions tax" with the introduction of collection of an ad valorem natural gas resources tax, in conjunction with introduction of an energy efficiency/emissions reduction quota trading system. By increasing the financial support for infrastructure construction, scientific innovation, energy efficiency and safety assurances, it will be possible to accelerate reforms of the pipeline and storage infrastructure financing and investment systems. The duration of subsidies for shale natural gas and other non-conventional forms of natural gas should be extended and tax breaks offered to interruptible users, which in turn will increase research in terms of energy efficiency and emissions reduction within the industry as a whole. Strategic storage costs should be included in financial budgeting in addition to establishing a system of subsidies for low-income groups.

13.2.3   Pillar III: Establish a Service-Centred Natural Gas Administration with a Legal Basis

The transformation of government functions should be accelerated, reducing bureaucratic red tape and interference in microeconomic activities. Natural gas administrative bodies should mainly be concerned with macroeconomic administration relating to the field of natural gas. The formulation and realisation of oil and natural gas-related law and related policy strengthening will establish a developmental strategy that gives priority to energy efficiency, providing the capacity to respond to climate change and guide society as a whole in the direction of reduction of emissions and sustainable development. This should also entail enhanced macroscopic planning and industry guidance while reducing regulatory control over project approvals and pricing, thus reducing the level of interference in terms of market operations, production and other business activities. Enhanced demand-side management of gas-using industries is necessary, with greater attention being paid to long-term energy efficiency and greenhouse gas emissions management. At the same time, attention must be paid to ensuring the security of natural gas supplies; international co-operation should be encouraged, combining increased "Chinese external involvement" and "international involvement within China", improving reserve management capabilities and ensuring effective resource alerts and prediction.

Enhanced social oversight should be put in place through the creation of a service-centred natural gas administrative system. Natural gas regulatory bodies should mainly be concerned with market malfunction, with their main efforts being concentrated on their regulatory role. They must act appropriately in the regulation and encouragement of fair competition and the prevention of market monopolies. They need to extend their regulatory functions to include natural gas pipeline development and other areas where natural monopolies may exist, as well as the regulation of market prices and other aspects. In addition, they must be responsible for safety of operations, pollutant discharge and emissions as well as ensuring fairness of service and protecting the rights of consumers. They should be capable of ensuring compliance, strengthening production safety and improving levels of energy efficiency and environmental protection, ensuring that development is sustainable and making the social accountability of businesses a priority.

13.3 Reform Objectives

Full advantage should be taken of the strategic possibilities for reform that present themselves at this time of massive drops in international oil prices. Any actions should be taken according to the requirements of Chinese socialist market economic development, taking advantage of the definitive effect that markets play in the allocation of resources and making more effective use of the role played by the government, in order to support the establishment of "an effective government + an effective market". This approach will result in the creation of a mechanism for natural gas pricing mainly determined by market forces. The transformation of the government approach to regulation will result in the creation of a comprehensive oil and natural gas-related system of legislation, the founding principles of which are encouragement of independent control of businesses over their operations, fair competition, freedom of choice for consumers allowing them greater control over expenditure, free movement of commercial factors and equality in exchange. This will in turn ensure an effective, competitive market structure and market system, resulting in the overall advancement of reforms in the field of natural gas while both promoting and accelerating energy resource reforms in China.

Establish a comprehensive system of natural gas related legislation and standards: A "comprehensive legal system" means a national legal system which is complete, standardised, systematic, harmonised and unified. This should embody the legal principles of socialism with Chinese characteristics, providing a completely standardised legal system, with an effective system for the implementation of such laws. It should also be a system capable of rigorous monitoring of the law, which also consists of a system of legal guarantees, whereby national order is enforced according to the law, involving both governing by law and the government acting according to the law, in supporting the creation of a unified legal state, legal government and legal society. This in turn ensures that laws are scientifically established, strictly enforced, formulated fairly and obeyed by society, in a manner capable of modernising China's governance systems and capabilities. Regarding natural gas, legislative revision should occur rapidly, as there is a legal basis for the introduction of major reforms, while policy should be closely connected with such legislation. Moreover, there should be increased governance according to law, ensuring that industry administration occurs on a legal basis, ensuring fair legal process, enhancing the public's faith in the judicial process. Finally, increased work should occur in relation to the creation of legal teams and enforcing a restriction list, thereby ensuring that "no actions should be taken without a legal basis, while legal responsibilities must be fulfilled".

Create a new kind of integrated, open, competitive, orderly natural gas market system: The modern market systems must be integrated in nature, relying on the interaction and interconnection of a number of submarkets which are organically integrated. Integration, openness, competition and order are all essential characteristics of a market economy. Integration is the foundation of economic development, with products and factors able to flow between different industries, departments, regions, both domestically and internationally. Openness is a prerequisite in terms of economic vitality—in terms of the international division of labour, Chinese companies exhibit a major strength in that they are responsible for making products more competitive in the international production chain, increasing the efficiency of allocation of resources within specific sectors. Competition is the foundation of all economic efficiency, and fair competition with only the fittest surviving is crucial to ensure improved economic quality and efficiency. Order relates to maintaining economic order, requiring the establishment of strict, fair, open and transparent market regulations, which then maintain effective order within markets. Only by establishing an integrated, open, competitive and orderly market system is it possible to ensure that market factors are able to play a full role, which in turn will result in resources being allocated in the most efficient manner. This can be achieved via an overall increase in the extent of reforms extending as far as 2030, which will result in the creation of an integrated, open, competitive, orderly and modern oil and natural gas market system formed of a number of very large oil and gas companies in conjunction with other oil and natural gas production, transport and sales operators with different ownership structures and of differing size.

Creation of pricing systems and a financial and taxation system that reflect the extent of scarcity of resources, the market relationships of supply and demand and environmental externality: The system via which natural gas prices are set should be transformed from being government-dictated to being market-moderated, and this in turn will ensure that prices fall in step with those on the international natural gas markets. The pricing parity relationships between natural gas and alternative energy sources must be reorganised, and a financial and taxation system that takes into account ecology and environmental protection, reduction in greenhouse gas emissions and other such externalities must be established.

Encourage the creation of a modern natural gas market supervision system which is all-encompassing, where rights and duties are clearly delineated and which is both fair, transparent, efficient and capable of effective regulation: The manner in which the government implements supervision over the natural gas industry, which relied in the past mainly on an application and approval system, should be converted to one in which it has a strategic role in planning and guidance. This will encourage the creation and completion of a market regulatory system, improving the social accountability of government, the increased accountability encouraging fairer competition, preventing market monopolisation and protecting consumer rights at the same time as ensuring the safety of production operations; it will also provide encouragement to reduce the levels of pollutant emissions in addition to providing an impetus for sustainable social and economic development.

Roadmap for Natural Gas Market Liberalisation and Regulatory Reform

14

14.1 Roadmap

At different stages of the development of the natural gas industry, markets exhibit different characteristics depending on the different levels of competition and different problems encountered, and thus the focus of reforms should also differ. In general, no stage in the reform process can be left out, as each key issue must be dealt with effectively to ensure that policy and adoption of measures keep step. This approach should achieve a balance between the three aspects of timing, location and quantity. Attention should also be paid to the special characteristics of the natural gas industry sector and to the extent of the effects of the reforms and the uncertainties relating, for instance, to market conditions. Taking all these factors into account, the increase in the extent of reform occurring in the natural gas sector in China should take place at a gradual pace and to a moderate degree, so that the reforms are acceptable to society. Appropriate objectives and measures should be determined based on the stage of development reached by the natural gas market and according to its specific characteristics. For the purposes of this paper we envisage a gradual roll-out involving three stages.

* This chapter was overseen by Xiaoming Wang from the Development Research Center of the State Council and Mallika Ishwaran from Shell International. It was jointly completed by Yusong Deng, Jiaofeng Guo, Shouhai Chen from China University of Petroleum and Qingle Wu from Shell China. Other members of the topic group participated in discussions and revisions.

14.1.1 Upstream Sector Roadmap

Currently the vast majority of oil and gas mineral rights are concentrated in the hands of four major corporations: China National Petroleum Corporation, China National Offshore Oil Corporation, China Petrochemical Corporation and Shaanxi Yanchang Petroleum Group. Resolving this problem of excessive centralisation of mineral rights, resulting in control over land being awarded but no exploration taking place, is key to advancing reform in the upstream field. As a result, reforms need to start by looking at current legislative and administrative methods, concentrating on promoting diversification of investors, encouragement of effective competition and creation of a restriction list, in order to refine the system of market access, ensuring that a system for issuing tenders and bidding on quotas is implemented for natural gas mineral rights; this in turn will result in the orderly and effective encouragement of the creation of primary and secondary markets for exploration rights and extraction rights, at the same time as ensuring that basic data on natural gas resources is placed in the public domain.

1. By 2020, the preliminary creation of a mineral rights primary and secondary market should have taken place. This will involve the creation and refinement of a legislative system and framework with a "mineral resources law" at its centre. Restriction list, industry admission and competence baseline administrative systems will be established, allowing

© The Editor(s) and The Author(s) 2017
Shell International and The Development Research Center (Eds.), *China's Gas Development Strategies*,
Advances in Oil and Gas Exploration & Production, DOI 10.1007/978-3-319-59734-8_14

companies that are viewed as "competent" to import gas independently (pipeline gas and LNG) in addition to allowing freedom of choice of importer. This will also include the preliminary creation of a public-domain natural gas resources information management system, which will also include newly discovered Chinese gas hydrates mineral resources, while also involving introduction and refinement of a system for tendering and bidding in relation to conventional natural gas, shale natural gas, coalbed methane, gas hydrates and other such mineral rights and the preliminary creation of an exploration and extraction rights secondary market system and trading mechanism. The natural gas wellhead (ex-works) price will be deregulated and then decided by market competition, while the natural gas pricing method will be based solely on calorific value pricing. Amalgamation and integration of the main state oil companies will occur, according to experience gained overseas in terms of balancing development upstream, midstream and downstream, resulting in the creation of two or three vertically integrated large international oil companies, the upstream, midstream and downstream operations of which are equally balanced. This will allow the growth of an effective, competitive market structure, in addition to establishing and refining a specialised independent regulator.

2. By 2025, establish the basic elements of a primary and secondary mineral rights market. Establish a legislative system with an "oil and gas law" at its centre. Refine the restriction list and industry admission regime. Create a public-domain natural gas resources information management system. Create a regime for tendering and bidding on conventional natural gas, shale natural gas, coal bed gas, gas hydrates and other such mineral rights, including the preliminary creation of an exploration and extraction rights secondary market system and trading mechanism. Complete the honing and specialisation of the independent regulator.

3. By 2030, primary and secondary mineral rights markets should be fully established. A comprehensive legislative system based on an "oil and gas law" should have been established. A highly effective administrative implementation structure based on a restriction list and industry admission regime should have been created. A strict supervisory system relying on an independent specialist regulatory body should have been established. An integrated, orderly but competitive, legal, trustworthy, effectively regulated natural gas exploration and development market system should have been established.

14.1.2 Midstream Roadmap

Currently, Chinese oil and natural gas long-distance pipelines, branch pipelines and provincial pipelines are mainly concentrated in the hands of three major corporations (China National Petroleum Corporation, China National Offshore Oil Corporation and China Petrochemical Corporation). How to resolve the issues of the low numbers of participants involved in pipeline construction and operation, the bundling of transport and sales and a lack of third-party access are all key issues encountered in relation to overall reform of the midstream field. As a result of this, it is necessary to commence work regarding current legal and administrative systems, concentrating efforts in the areas of diversification of market participants (this in turn will encourage effective competition), creation of a restrictions list and improve market admission mechanisms. In this way, the complete separation of long-distance pipeline, branch pipeline and provincial pipeline transport services and sales operations will be ensured, so that the reform objective of third-party access to pipeline networks and open admittance to the industry can be achieved. Admittance to sectors where natural monopolies exist, such as the pipeline networks, charging and cost supervision should also be strengthened.

1. By 2020, allow preliminary pipeline capacity trading in a primary market, including the trading of storage capacity such as that within LNG installations. Ensure that reforms are implemented that make natural gas pipeline network finances independent, with the trial introduction of independence of production rights from the natural gas pipeline network. Enforce third-party access and open industry admittance. Completely separate long-distance pipeline, branch pipeline and provincial pipeline transport and sales services. Establish and refine an independent specialist regulator.

2. By 2025, the creation of a primary pipeline capacity and storage trading market should be basically complete, in conjunction with the preliminary creation of a secondary pipeline capacity and storage trading market. Ensure that the independence of the natural gas pipeline network from production rights has been accomplished and that the objectives of third-party access and open industry admittance have been achieved. In order to encourage the construction of pipelines and gas distribution networks and increase transport capacity, permit long-term natural gas supply and transport contracts (although gradual pricing regulation of transport and distribution networks would be necessary). Implement policies giving pipeline investment and construction projects that are of proven benefit to society and that have a major social influence immunity for limited periods (for instance by not enforcing third-party access to pipeline transport services for a specified period of time). Ensure that a specialised independent regulator is fully in place and refined.

3. By 2030, ensure the formation of both primary and secondary pipeline capacity and storage trading markets. Implement third-party access to distribution services, while allowing residential and commercial clients to seek out alternative suppliers or wholesalers other than their local distributor, either via the introduction of trials or as part of a compulsory process. Create a restriction list and an industry admission administrative

regime which will act as the foundation of an effectual administrative implementation system. Create a strict supervisory system based on a specialised independent regulator. Ensure that this results in an integrated, competitive, ordered, legal, trustworthy natural gas pipeline and storage market system which is effectively regulated.

14.1.3 Downstream Roadmap

China has already begun to establish downstream market structures, and in overall terms it is possible to see the existence of market competition to differing degrees. However, there are still a lot of problems, requiring solutions for issues such as irregularities in market regulation, incomplete pricing mechanisms, poorly developed market structures and localised monopolies, all of which are key areas of concern in ensuring thorough reforms in the downstream sector. As a result, work needs to begin on the existing legislative and administrative systems, concentrating efforts in the following areas: the formation of market systems; encouraging effective competition; the creation of a restrictions list and improving market admission mechanisms; and the deregulation of natural gas terminal pricing, with prices then being established by market competition. This should finally result in the formation at the national level of around 10 regional oil and natural gas spot trading markets, as well as establishing and refining a natural gas futures trading market, ending in the establishment of a fully featured, modern market and associated regulatory structures.

1. By 2020, ensure the deregulation of natural gas terminal pricing, which will subsequently be determined according to market competition. Establish natural gas spot trading markets and wholesale markets at the main natural gas trading centres and provincial distribution hubs. Establish a natural gas trading market within the Shanghai Futures Exchange. Create and refine an independent specialised regulator.

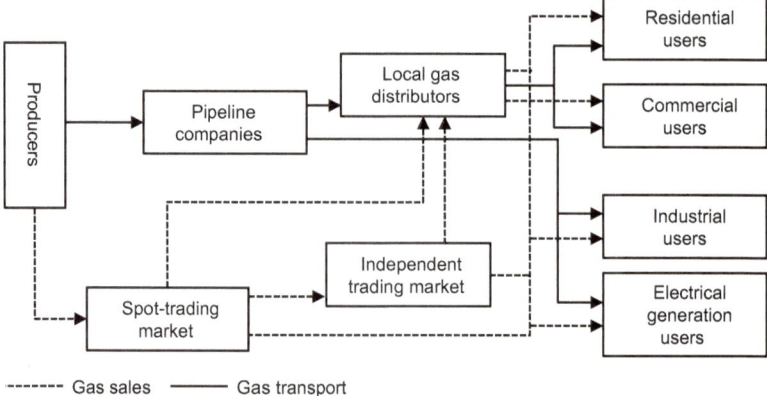

Fig. 14.1 Diagrammatic representation of the natural gas markets after complete separation of sales and pipeline transport. *Source* Xu (2015)

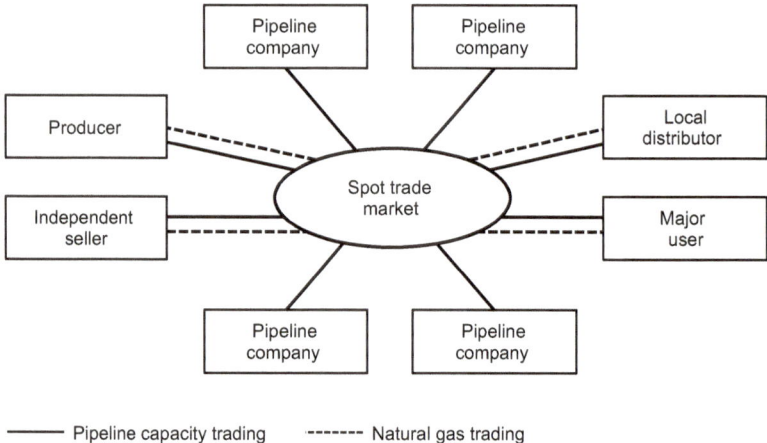

Fig. 14.2 Diagrammatic representation of a spot trading market system. *Source* Xu (2015)

2. By 2025, ensure the development of a natural gas sales market system based on a combination of producers, independent traders, large users, local distributors and end users that rely on a spot trading market. Create and refine an independent specialised regulator (Fig. 14.1).

3. By 2030, a natural gas sales market system based on a combination of producers, independent traders, large users, local distributors and end users that rely on a combined natural gas market system consisting of a natural gas spot trading market and a natural gas futures market should have been established. Create a specialised independent regulatory body as part of a strict regulatory system. Ensure that this results in an integrated, competitive, ordered, legal, trustworthy natural gas sales market system which is effectively regulated (Fig. 14.2).

14.2 Key Measures

The basic direction of China's natural gas policy it that it "supports energy efficiency, strengthens China's position, encourages diversification, is concerned about the environment, ensures thorough reforms, encourages international co-operation and improves the lives of everybody". This approach promotes transformation in terms of energy resources and modes of production and consumption, and forms the basis of a clean, efficient, safe and sustainable modern industrial system, placing complete reliance on the sustainable development of energy resources to support the sustainable development of the economy and society as a whole.

14.2.1 Creation of Market Systems

Efforts must be made to ensure that by 2020, an integrated, open, competitive and orderly natural gas combined production/transport/sales market system which encompasses "one network, diversified gas sources and 10 regions" has been created.

A "one network" pipework transport system is required in order to allow connection of gas resources from different sources and regions to the market, thus creating a national unified natural gas market system, which will encourage the flow of resources between the markets of different regions, and allowing effective competition, helping to ensure the security of natural gas supplies. To be specific, it will be necessary to create a pipeline backbone by perfecting the current "Western Gas Going East" system, the new Guangdong-Zhejiang and Shaanxi-Beijing systems, the "Sichuan Gas Going East" system and the China-Burma pipeline networks, while the Lanzhou-Yinchuan (Lan-Yin) line, Lanzhou-Yinchuan-Ningxia (Lan-Yin-Xia) line, Huaiyang-Wuhan (Huai-Wu) line, Zhongwei-Guizhou (Zhong-Gui) line and the Ji-Ning line act as connecting lines to create a "one network" pipeline transport system.

1. "Diversified gas source" gas supply framework

China's main natural gas market resource is state-produced gas from the Xinjiang, Sichuan-Chongqing, Qinghai and Shaanxi-Gansu-Ningxia regions as well as from maritime areas, in addition to gas imported from Central Asia via pipelines located in the west, gas imported via the northern China-Russia pipeline and imported LNG. Due to the need to concentrate on placing reliance on nearby supplies, more consideration needs to be given to a framework relying on "piping of western gas east, delivery of gas from the north to the south and bringing maritime gas ashore", ensuring that full use is made of a variety of market resources to satisfy the gas supply demands of various regions in order to ensure security of supply.

2. "Top 10" regional markets

Due to the factors affecting the distribution of natural gas resources, the layout of the pipeline network and existing socioeconomic conditions, the hope is to establish a seamless network in China connecting imported pipeline gas to the markets, connecting LNG receiving stations to the markets, connecting producing regions to the markets, connecting underground gas reserve storage to the markets and connecting producing regions with underground gas reserve storage. By around 2020, 10 large regional markets based on the Bohai Loop, Yangtze River Delta, Pearl River Delta, Sichuan-Chongqing, Yunnan-Guizhou-Guangxi, Central-Southern, Shandong-Henan-Anhui, Central-Western, North-Western and North-Eastern regions will be created, including:

- a locally supplied market to feed local production and consumption in Sichuan-Chongqing and the North-Western region, with Shanghai and Guangdong acting as production-transportation-consumer centres for the Yangtze River Delta and Pearl River Delta regional markets;
- Zhongwei in Ningxia, Wuhan in Hubei and Yongqing in Hebei acting as logistical hubs in the creation of Central-Western, Central-Southern and Bohai Loop regional markets, while gas from the China-Russia pipeline and pipeline gas produced domestically will come mainly from the gas source for a North-Eastern regional market;
- gas from the China-Burma pipeline and imported LNG acting as the gas source for the Yunnan-Guangxi-Guizhou regional market, with the underground gas storage groupings of the Yellow River Plain and northern China providing support to locally produced

pipeline gas and imported LNG and gas from other sources in the Shandong-Henan-Anhui regional market.

These "Top 10" regional markets are not completely independent but are interconnected and influence each other. This is achieved via a "single network" pipeline transport system connecting different pipelines and different entities in addition to effective market regulation, encouraging effective competition, and ensuring that market resources can spontaneously flow from high-value areas to low-value areas, thereby establishing a national pricing system based on different regional prices and a flexible natural gas market framework, in turn ensuring the optimised allocation of resources.

The creation and developmental progression of these "Top 10" regional markets should occur sequentially, by policy-led trial introduction. From the point of view of market creation and developmental progression, the relatively mature Sichuan-Chongqing, Yangtze River Delta and Pearl River Delta regional markets should be established within the time constraints of the 13th Five-Year Plan. When compared at the national level to other regional markets, the Sichuan-Chongqing region was the earliest regional natural gas market established in China, relying mainly on the Sichuan Basin, which is rich in oil and natural gas resources as well as having a relatively well established pipeline transportation network; this would make it much easier for a relatively mature market to become established. In light of this, it is recommended that a "Sichuan-Chongqing natural gas market reform trial zone" be established as soon as possible, with the results achieved in this trial allowing the creation of a mature regulatory system as far as market access and pricing mechanisms are concerned, which should be capable of overcoming inertia in the developing market and obstacles to effective competition, while allowing the market to play a definitive role. This in turn will allow the creation of a natural gas market with a 100 billion m^3 annual output capacity and a 70 billion m^3 annual sales volume.

Take full advantage of the economic advantages of the production-transport-sales centres of Shanghai and Guangdong and their ability to sustain prices and other such market advantages, which should allow for the rapid development and operation of a Shanghai oil and natural gas trading exchange. This will allow Shanghai and Guangzhou to act as pricing centres, making the creation of relatively mature regional markets in the Yangtze River Delta and Pearl River Delta districts possible within the time frame of the 13th Five-Year Plan. At the same time, by relying on the Ningxia Zhongwei, Hubei Wuhan and Hebei Yongqing natural gas distribution hubs and making full use of the advantages offered by distribution network nodes, it will be possible to gradually develop logistical systems so that they radiate out into surrounding areas, establishing these as oil and natural gas circulation and logistical centres, and allowing the formation by 2020 of relatively mature Central-Western, Central-Southern and Bohai Loop regional markets. Using this as the foundation, construction of the Yunnan-Guangxi-Guizhou, Shandong-Henan-Anhui, North-Western and North-Eastern regional markets will then be possible.

This activity should take place in conjunction with the construction of the "Top Six" subterranean gas storage groups. These rely on depleted oil and gas reserves, and depend on suitable storage conditions existing within the oil or gas deposits and create value from waste resources. The Central Plain, northern China, the North-East, Changqing, the North-West and the South-West oil- and gas-producing regions will form six major subterranean gas storage groups. These groups will gradually create and optimise a natural gas reserve peak adjustment system, which will be a major factor in the creation of China's "single network, with diversified sources and 10 regions".

By harmonising planning, construction and usage of natural gas markets, resources and infrastructure, it will be possible to create an integrated national upstream-midstream-downstream combined production-transport-sales market system, ensuring that development of the natural gas industrial chain occurs evenly at all levels.

14.2.2 Completion of a Natural Gas Pricing Mechanisms

1. Staged introduction of reforms to natural gas pricing

Currently, preliminary work has been carried out regarding the introduction of a dynamic mechanism for pegging natural gas prices against alternative energy sources. Considering the special circumstances of the natural gas industry, the widespread implications of reforms and the uncertainties concerning market conditions, it is recommended that, depending on the stage of development encountered in the natural gas market, reforms to natural gas pricing should advance steadily.

In the first stage, between 2015 and 2017, the emphasis should be on the creation of a pricing system. Application of the netback method should be introduced as soon as possible (including adjustment of the pricing coefficient K from 0.85 to 0.70–0.75). Determine the costs and pricing of long-distance pipeline, branch pipeline, provincial pipeline, urban pipeline and distribution network pipework transport in a scientific and acceptable manner. Establish residential gas pricing in the light of basic economic factors. Refine implementation methods to allow for seasonal price variations, peak and trough price variations, interruptible supply pricing and gas storage pricing. Revise measurement standards and valuation methods (conversion from a flow rate or mass valuation method to a calorific valuation method), and increase administrative and technical monitoring of illegal valuation activities. Bring an end to the charging irregularities occurring at local government levels (such as the current natural gas pricing adjustment fund, with local governments exacting this at rates that vary between 0.3 CNY and 1.0 CNY per cubic metre, which results in additional burdens being placed on the consumer).

In the second stage, between 2018 and 2020, remove all regulation of gas source pricing, including all provincial gate station pricing. Enhance regulation over residential use gas prices. Enhance the regulation of charging for use of such infrastructure as gas transport pipelines and gas storage, ensuring that there is effective regulation in relation to sector admission.

In the third stage, between 2021 and 2023, charging for all gas transport pipelines except gas distribution networks and for gas storage should be completely determined by the market. Complete introduction of supervisory and administrative systems with regard to pipeline networks and other infrastructure, ensuring that the objectives of allowing open, fair, orderly competition with regard to natural gas prices are achieved. In addition, policy should be formulated concerning the introduction of a financial and taxation system relating to resource tax, environmental tax and carbon tax. This will ensure that a suitable scientific financial taxation system and pricing mechanism is put in place, which will reflect the growing scarcity of resources, the relationships of market supply and demand and the costs of external environmental impact.

2. Accelerated creation of regional trading centres

In the short term, the Shanghai and Guangdong natural gas trading centres could be established. In the mid to long term, new regional trading centres could be established in Beijing, Sichuan, Hubei, Ningxia and Xinjiang. Efforts must be made to ensure that the Shanghai natural gas trading centre becomes both an Asian and international natural gas trading centre, resulting in the Shanghai trading centre system eventually becoming the centre for trading of both internationally priced and regionally priced gas, thus effectively creating a link between domestic and international supply and demand. The creation of natural gas trading centres should commence with spot trading, then the range and quantity of trading should be expanded, finally developing into futures trading, increasing the depth of the market.

14.2.3 Removal of Pipeline Transport Bottlenecks

The supervision of the operation of natural gas infrastructure such as pipeline networks should be extended, clearly establishing functional roles, to ensure that pipeline transport services and sales services are completely separate and that third-party access is facilitated.

The emphasis should be on reform to the natural gas pipeline network construction and operational systems, establishing functional roles within pipeline networks, and gradually resulting in a natural gas pipeline network to which fair access is provided, driven by supply and demand and which is both reliable and flexible. The formulation of policy in relation to market admittance in terms of pipeline network operation and services should be accelerated; this should involve the introduction of an admittance system of operator accreditation, ensuring the suitability of operators to undertake certain functions and duties. The basic requirements for operators will be the existence of an independent legal body corporate, and implementation of independent financial accounting systems. In light of the actual circumstances of China's natural gas industry, the separation of provision of infrastructure services could commence first with the relatively easily reformed long-distance pipeline network and LNG receiving stations. As for how to approach separation, this could begin with financial and legal separation. Then, as the extent to which reforms are adopted increases, it would be possible to institute trials in the eastern regions, where there is a greater diversity of gas sources, where competitive market structures are more or less already in place and where there is a relatively high density in terms of gas transport pipeline networks. This would be followed by a nationwide expansion. In addition to all this, an exploratory separation of the financial systems and legal status of gas storage services from gas distribution networks should be carried out. Such a change would encourage varied investment in the construction of gas storage capacity, allowing independent storage operators to participate in the natural gas market and profit from market peak and trough pricing differences.

Implementation of pipeline network interconnectedness allowing connection of third-party services should occur rapidly, and more effort must be put into the regulation of connection contracts, service pricing and quality of service —this will encourage the creation of the conditions necessary for market competition.

In the gas transportation and LNG receiving station sector, the scope of long-distance pipeline and LNG receiving station third-party access should be gradually expanded. Also, a permit administration system should be introduced, allowing any accredited operator of natural gas services to sign transport or storage contracts with pipeline networks and LNG receiving stations. Under such circumstances, when excess capacity in the distribution network systems arises, operators of pipeline networks and LNG receiving stations must offer this to any natural gas supplier or user who has need of such services, thus providing a non-discriminatory service based on fair pricing. Within the main pipeline backbone, all gas source input nodes and market terminals should be interconnected, removing any obstruction to the circulation of natural gas products. In terms of application, depending on the extent to which infrastructure has been developed and the degree of separation that has occurred between services and sales, it would then be possible to gradually introduce negotiable and subsequently obligatory third-party access. At this point, it would be essential to promote gradual deregulation (in terms of annual natural gas consumption), allowing large users to choose their natural gas suppliers directly ("large users" being urban gas companies, 200,000 kW power stations, combined cooling, heating and power generation stations, large-scale industrial companies (including for use as fuel and as a raw material) and LNG/CNG fuel suppliers, among others).

Regarding storage and urban gas distribution, third-party access mechanisms should be gradually introduced to users at different scales of consumption. Start with non-residential customers with fairly high annual consumption; this should result in a timetable and establish annual consumption scales. Then introduce freedom of choice of gas supplier to non-residential users according to different annual consumption levels while adhering to this timetable and depending on the extent of overall implementation of the previous measures. Alternatively, it would be possible to completely avoid urban pipeline distribution networks, or such urban networks would only be allowed to provide gas distribution services according to government regulated prices.

14.2.4 Further Regulatory Reform

Gradually set up a comprehensive, centralised, vertically unified, integrated, independent, specialised regulatory system, with fully established regulatory functions, and put more effort into establishing regulatory powers.

At different stages in their energy development, the United States and European nations each introduced an independent regulatory body. The US established both an energy administration and an energy regulator. The energy administration mainly has responsibility for fundamental policies relating to the development of energy resources and safety and for related policy research. The energy regulator mainly has responsibility for formulation and implementation of specific regulatory policy. In terms of its market regulation, this includes a staged upstream-midstream-downstream regulatory process. Although there is no regulatory committee for the upstream field, the federal government's Department of the Interior is responsible for regulation relating to production and use of energy sources, while individual states carry out the regulation of certain types of permit in upstream manufacturing. There are, however, independent regulators at both the midstream and downstream levels, with FERC being responsible for the regulation of state-level pipeline transport and sales regulation, while state public utility regulatory committees carry out regulation of downstream energy structures. These legally established regulatory bodies have adjudicative powers and are independent of the government, making them a more effective guarantee of the implementation of government policy regarding energy resources. Their open, fair, transparent,

legal principles make them more accessible to public scrutiny.

Establish independent regulatory bodies and clarify their regulatory responsibilities. Fully extend their regulatory functions, providing greater regulatory functionality. The energy-related regulatory system should be established as soon as possible. Enhance energy-related regulation. Create and complete the structure of regulatory organisations and related legislation, involving regulatory innovation, thus allowing improved regulation efficiency and ensuring the fairness and stability of the markets. This will help to create a healthy environment within which to develop the energy industry.

2015–2017: Complete and perfect harmonisation of the regulatory system. After establishing the relevant regulatory legislation, statutes and standards in conjunction with state energy development planning and strategy and energy industry policies, the different departments concerned should enhance communication in order to harmonise their work and co-operation. During the course of the implementation of the regulatory system, all departments should ensure enhanced sharing of information and co-operation. At the same time, it will be necessary to establish the functions and duties of each regulatory department in relation to central government and local organisations.

2018–2020: The State Council will deputise an energy regulation leadership group, which will be responsible for harmonising work between different departments and gradually establishing a relatively centralised regulation department. Each department will retain its original regulatory legislation, standards and functions, while the main burden of regulatory implementation will gradually be transferred to one organisation, which will then provide a relatively centralised regulatory service. At the local level, government will only establish a regulatory organisation consisting of a single level of administration, which will then dispatch specialist regulatory teams to various locations. In addition, the central regulatory department will be responsible for monitoring and supervision of local regulatory organisations.

2021–2023: Complete and refine the functions of a unified regulatory organisation. Apart from its role of regulation of the environment and national resources, this will gradually be endowed with full regulatory responsibility for the economic and social aspects of the entire energy industry chain. At this point a system of interaction between different departments will be clarified and their individual duties established and an independent, unified regulator will gradually be created. Finally, establish a system with an independent, unified, specialised regulator and vertical regulatory administrative organisations at different levels.

Enhance reforms in terms of social accountability and monitoring, with enhanced monitoring of present and future development and promotion of a system of market supervision based on standardised and procedural regulatory measures. Impose a system of unified market supervision, eliminating and discarding all regulations and practices that obstruct the creation of a national unified market and fair competition. Make better use of the government, in order to ensure that legal concepts and methods are applied to the adoption of market regulatory functions, with enhanced monitoring of present and future development and promotion of a system of market supervision based on standardised and procedural regulatory measures.

1. Enhance the role of government regulation

Completion and optimisation of standards and norms is vital. Accelerate the pace of work on the formulation and actualisation of regulatory legislation and technical standards and norms relating to all aspects of natural gas development, transport, storage and distribution. Complete and refine an environmental impact evaluation system, including evaluation of strategic environmental impact, evaluation of environmental planning impact and dynamic evaluation of the impact of construction projects related to development, transport, storage and distribution of oil and natural gas.

Enhance the overall regulation of the complete oil and natural gas development process, involving strict regulation in the pre-development stage, during development and in the post-development stage. Pre-development regulation should include planning and preparatory work in relation to oil and natural gas development, ensuring that all safety and environmental hazards are rooted out. At the stage when companies or contractors draft oil and natural gas development projects, they should also draft an oil and natural gas environmental impact report, which should make proposals relating to implementing measures to avoid or reduce all types of pollution. At the same time, companies undertaking development projects should be required to establish baselines in terms of the main environmental indicators (such as subterranean water quality, surface water quality, air quality and so on) prior to commencing work, and continuous monitoring should be carried out during the course of development. Well locations must be based on detailed geographical surveys, avoiding where possible densely populated areas, areas where environmental protection orders apply and so on, all the while reducing damage to the local land and using land more effectively. Where possible, pre-existing infrastructure should be relied on, thus reducing the need for the construction of new roads and infrastructure.

Regulation during development mainly relates to the drilling of oil and natural gas wells, their completion, gas extraction and normal operation. Contractors should be required to satisfy safety and environmental regulation standards and norms at all times and in all aspects of their work, adopting efficient, green, recyclable technology, all the while ensuring environmental protection and safety. At the same time, they should ensure that contractors are capable of responding effectively to prevent environmental pollution, that they have emergency protocols and that they provide anti-pollution equipment.

Post-development regulation mainly relates to the evaluation of the long-term risks resulting from the development of oil and natural gas. In the course of development of oil and natural gas it is possible that environmental factors which were not easily perceived pre-development and during development may arise. In the post-production phases it is necessary to carry out analysis of subterranean water, surface water, soil and air and other such aspects of the environment. These samples should be compared with pre-development baseline levels, allowing an appraisal of the environmental impact of oil and gas development. Post-development regulation must be strictly enforced, and companies that do not satisfy standards should be punished accordingly.

Special consideration should be given to safety and environmental protection as regards midstream pipelines:

- During pipeline construction, it is necessary to audit the legality and qualifications of prospective designers, suppliers and contractors. During the pipeline design, supply and construction process, there should be close monitoring for adherence to state safe production legislation and technical standards. Pipes and safety infrastructure should be scientifically selected and constructed, in order to root out any safety flaws that may exist in pipework. Companies responsible for distribution and delivery of oil and gas should have a safety executive in place, and ensure that suitable personnel and equipment is provided, ensuring the step-by-step creation and optimisation of a safe pipeline network and allocation of production safety responsibilities.

- Oil and gas pipeline safety work should be included within the scope of day-to-day government administration, establishing a comprehensive, scientific and sensible regulatory system, further enforcing specific responsibility for safety management in relation to pipeline installations. This should implement a system whereby responsibility for this lies with local leadership, thus establishing effective and enforceable procedures in relation to public safety. A system harmonising the work of various departments, the government and pipeline companies needs to be established, such as a system of joint progress consultations. The next level of government above will be responsible for

harmonising response in terms of cross-boundary safety issues, and this will include establishing a system of responsibilities for all related government departments and pipeline companies, fully implementing a safety information sharing system and rapid safety event response system.

- Those responsible for illegal appropriation and excavation and other actions that represent a serious threat to the safety of oil and gas pipelines will be held accountable, and any organisations or persons found responsible will be dealt with.

2. **Enhanced enforcement of the safety and environmental responsibilities of enterprises**

Establish a comprehensive safe production supervisory and administrative body and regime. Enterprises must set up safe production committees and safety management departments, which should be staffed with appropriate managerial staff. In addition, enterprises must establish a comprehensive production safety management system, including a system of production safety duties, a system of safety instruction, a safety meeting system and a safety equipment management system, including a safety incentive and penalty system and emergency response system.

Clarify production safety duties, providing a comprehensive production safety duty regime. Enterprises must establish a comprehensive production safety duty system, based mainly on job responsibilities, with the person with overall responsibility at its centre, which establishes who has overall responsibility for production safety and consists of a system of production safety duties applicable to staff involved in all departments and activities, thus forming part of the overall production safety monitoring and duty system.

Strengthen safety management in relation to specialist technology. Strengthen safety management in relation to specialist technology on the basis of production safety legislation, standards and norms.

Institute a safety troubleshooting regime, thereby enhancing preparation for emergencies. Enterprises must carry out intermittent potential accident troubleshooting on the main aspects of the production process, the main points of activity and main items of equipment used. Where any potential accident areas are found, these should be subjected to specialist evaluation, categorisation and pre-emptive restrictions and special precautions. It is necessary to establish a strict potential accident reporting regime, standardising potential accident control work. Where accidents occur, it is necessary to have an effective emergency response organisation and specifically formulated responses, which should be practised in the course of day-to-day operations.

Introduce a health, safety and environment (HSE) management system and a pipeline integrity management system. HSE management is particularly concerned with the safety of operative activities, safeguards and operational safety, while pipeline integrity management systems are specifically concerned with equipment and technical management and preventative maintenance in order to ensure the safe and reliable operation of equipment. By effectively integrating the functions of these two roles, allowing them to complement each other, it is possible to create a new pipeline management system. This not only enhances HSE management systems, but also increases the level of safety management, supervision and the professional skills of auditing personnel and is an extremely effective approach to safety.

3. **Increased openness of information and public scrutiny**

Openness of information and public scrutiny should be standard throughout the entire process of oil and natural gas production, transportation and usage. Government departments should provide effective guidance and carry out effective supervision of work on the openness of company data and public scrutiny. This work should concentrate on providing open, honest descriptions of any environmental, safety or health issues

related with production, transport and usage processes, and the response of the company concerned to such risks. Furthermore, the government should rapidly publish safety and environmental testing data, and should make communication with local residents a part of all stages of the development process, ensuring that residents are fully aware of any challenges, risks or benefits that may arise as a result.

NGOs and other civil society organisations should be active in carrying out public scrutiny, to establish whether or not the companies concerned respect environmental standards, while also holding regulatory bodies accountable. The public should be aware of environmental issues, and have a full appreciation of the safety issues and environmental risks associated with oil and natural gas production, transportation and usage, thus allowing them to play a role in the public scrutiny of oil and natural gas production, transport and usage. The media should play a role in the dissemination of environmental knowledge and the exposure of illegal activities. Where media reports are critical, they should require companies to carry out investigations into the veracity of the claims and resolve the issues in question. In addition, companies should then provide reports within two weeks to the relevant departments and news organisations regarding the results of rectification work they have undertaken or the progress of their investigations.

14.2.5 Further Reform of the State-Owned Oil and Gas Companies

State-owned oil and gas companies played a dominant role in The Netherlands before reform, making the situation very similar to that encountered in China. Despite having created a unified energy market according to EU requirements, due to path dependence, The Netherlands adopted a different development approach. This approach involved the retention of a number of major state-controlled upstream, midstream and downstream enterprises, while EBN, a completely state-owned company with special characteristics, stepped in and took control over operation of the oil and natural gas markets, a step which proved to be an effective measure.

The evidence proves that the presence of state-owned enterprises does not necessarily mean the presence of a monopoly or low efficiency. However, there does need to be specialised supervision and management in order to improve operational efficiency, which under fair market conditions will effectively promote the entry of social capital. The background of China's development from a planned economy to a market economy as well as features such as the strategic position of gas resources and their exploration and development, the high-risk nature of pipeline transport and delayed returns on investment have determined that in order to ensure the security and supply of oil and gas, involvement and investment by state-owned enterprises is, at present, essential. Especially from the point of view of natural monopolies and important public goods and service sectors, dominance by state-owned enterprises should continue for a certain period of time. However, at the same time it is important to encourage effective competition through diversification of investment and improvements in tendering and bidding structures.

There is a need to optimise the allocation of resources afresh, while remodelling the core competitiveness of the enterprises concerned, in order to create oil companies that satisfy basic rules of structural proportions in terms of the upstream and downstream sectors that apply throughout the global oil and gas industry. This will result in the creation of international oil and gas companies that are more competitive internationally, which will in turn improve the standing of China's oil and natural gas industry, as well as providing greater national energy security. Between two and three international oil and gas super-companies which have balanced development in terms of the upstream, midstream and downstream sectors should be established. At the same time, in order to ensure that suitable conditions exist for effective market competition to occur, after merging, such unified international super-companies should then hive off part of their midstream and downstream businesses.

Simultaneously with this restructuring, if state-owned oil and gas companies are to converted into satisfactory market participants, then there will need to be a deepening of reforms. This process would mainly require the completion and refinement of state-owned asset supervision and administration systems and operational systems, so that government's various functions can be separated: social and economic, asset ownership management, state-owned asset management and state-owned asset operation. This would achieve separation between government and business enterprises and between government and assets, resulting in the creation of efficient operational mechanisms and governance structures, which would encourage the management of state-owned enterprises to conduct managerial activities and operations based on market forces, thus encouraging greater activity by state-owned enterprises.

14.2.6 Establish and Improve the Services Market

The experience of the United States in its successful development of shale natural gas shows that multiple investors and combined development systems for specialised services can mobilise the positive aspects of venture capital, technology research and development, upstream extraction, infrastructure, market development and end applications, as well as put in place a system for the implementation and improvement of the regulatory system to ensure the rapid and orderly development of the shale natural gas industry. In an open and competitive environment, there are advantages that may be exploited, such as increased specialisation and access to technology within the technological services industry, providing horizontal drilling, well completion and cementing and multi-stage hydraulic fracturing services to rights owners etc., as well as professional and technical services for engineering, logging and experimental testing. After a company completes a service in relation to a specific aspect, it can then be replaced by another company that specialises in the next aspect. A high degree of division of labour allows for splitting of shale natural gas extraction operations into individual aspects, requiring less investment, with shorter operating cycles and a faster recovery of funds, attracting a large amount of venture capital and private capital into the shale natural gas field.

Market mechanisms must therefore be allowed to play a complete role. This will provide a strong impetus for the development of oil and natural gas-related professional services and technology companies, in line with the principles of "production requirements, advanced technology and a good reputation". Market mechanisms and qualifying constraints can be used to regulate service enterprises and their activities, organising specialist construction teams, while establishing new systems founded on "exploration,

production, site operations, costing, safety and the environment". All the while, procedural regulation of production processes and standardised operations will ensure effective and orderly exploration and extraction.

At the same time, the training of professional personnel needs strengthening. Sustainable development depends on the personnel of the oil and gas industry, so competition between enterprises is ultimately competition using human resources. It is therefore necessary to establish professional resources for China's oil and gas market—and, indeed, for the international market. Not only is there a need for experts with a great deal of management and technical expertise, they also need to be encouraged to expand their proficiency to legal matters and international trade, resulting in a relatively stable human resource system which encompasses all the professions required.

Policy Measures and Safeguards to Support Natural Gas Market Liberalisation and Regulatory Reform

15

To promote the rapid and healthy development of the natural gas industry, completion and refinement of policy safeguards must take place as soon as possible.

15.1 Create and Complete a Natural Gas Legislative Framework

The fourth plenary session of the Communist Party Central Committee proposed the creation and completion of a legislative system, constituting a historical move towards the creation of the conditions necessary for legislative reform, laying the foundation for future legislative work. Thus legislative work should take a leading role in promoting economic and social development, which will involve a transition from "policy-guided action" to "legally guided action", ensuring that reform and legislation keep step, and that reforms occur according to the law. Only by ensuring that development of the natural gas industry is realised according to the law, and by increasingly establishing scientific and democratic legislation for the natural gas industry, can good legislation and governance be achieved, and a basic system for the

legal regulation of the natural gas industry put in place.

Although the initial framework for a legislative and regulatory system within the natural gas sector in China is already in place, the existing legal system lacks co-ordination, consistency and systematisation. For example, China still lacks a specialist and comprehensive "oil and natural gas law" that completely covers the entire upstream-midstream-downstream production chain. In terms of exploration and development of natural gas, the existing mineral resources legislation is based on solid minerals, and deals with problems that are common to the rights issues encountered with mineral resources, making it unsuitable for resolving the specialised problems arising in relation to natural gas, a gaseous mineral. Moreover, most of the existing legislation was created for a planned economy and restricts the rules of acquisition and transfer of mineral rights and is unable to fulfil the particular needs of natural gas and market liberalisation reform. In terms of natural gas pipeline networks, the end consumer sector and environmental protection, the current regulatory legislation is incomplete, while the responsibilities and obligations of local governments and related businesses are not sufficiently clear in terms of transportation, gas storage and gas distribution. The technical specifications and standards of the modern gas industry are still not sufficiently normalised and complete.

The law is the most important tool for governing a country, and good laws are a prerequisite of good governance. To form a complete system of legal standards, first, key areas of

* This chapter was overseen by Xiaoming Wang from the Development Research Center of the State Council and Mallika Ishwaran from Shell International. It was jointly completed by Yusong Deng, Jiaofeng Guo, Shouhai Chen from China University of Petroleum and Qingle Wu from Shell China. Other members of the topic group participated in discussions and revisions.

legislation must be strengthened, the systematic nature of laws and regulations must be enhanced and weaknesses in the legal system causing the laws to contradict each other in terms of logic and value must be eliminated. Therefore, the legal framework of the natural gas industry should be improved as soon as possible. There should be a comprehensive and effective natural gas legal framework that consists of legislation specific to the natural gas industry, with an "oil and natural gas law" at the core.

Improving the application of such legislation will make it better suited to the characteristics of natural gas exploration, production, transportation, storage, distribution and usage. Completion of improvements to relevant implementation rules and supporting regulations should come first. Gas-specific legislation should be established as soon as possible, consisting of the formulation or revision of the "oil and gas mining rights regulations", "natural gas midstream and downstream administrative regulations", "gas extraction environmental protection regulations", "offshore oil and gas pipeline protection regulations", "natural gas reserves regulations" and other administrative regulations. Improvements to legislation in relation to resource rights, exploration and development contracts, infrastructure and operational management, storage, sales and usage, safety alerts and emergency response, production safety and compensation for environmental impact, cross-border investment and import and export trade are all vital as well.

Administrative regulations for natural gas midstream and downstream should be researched and drafted. The regulations should include, at the very least: supervisory participants and duties at different levels of government; the organisation, position, powers, duties, regulatory principles and operating mechanisms of a natural gas regulatory body; natural gas transportation administration modes and operational mechanisms; the rights, obligations and responsibilities of businesses in relation to natural gas; the principles and mechanisms which will determine the prices and distribution charging rates for natural gas; pipeline construction and authorisation; maintenance, security and open access to pipelines; and resolution of disputes.

Environmental protection regulations for gas extraction should be researched and formulated. Special provisions must be made for the following aspects of natural gas exploitation: pollution prevention planning; environmental impact assessments; pollution permits; drilling fluids and exhaust gases; the recycling or processing of waste water and toxic gas; noise control; management of radioactive sources; environmental monitoring; excessive pollution; and emergency response to pollution incidents.

Areas of conflict between the oil and gas pipeline protection law and other laws should be dealt with through by legal interpretation. There are various urgent needs: co-ordinating pipeline development planning and other specific planning; dealing with the convergence between pipeline construction planning and urban and rural planning; legal determination of underground pipeline passage rights; restrictions on land use in relation to pipeline safety; and conflicts between pipeline safety and road and rail safety requirements.

15.2 Deepen Reform of Oil and Gas Regulations

The existing energy administration systems must be rationalised, and this will involve a gradual transition to a high-level, centralised energy administration. Industry administration should be enhanced by placing general industry-focused planning, market access approval and legislative functions under the control of one overall department. More emphasis should be placed on strategic planning in relation to energy development, implementing comprehensive planning, policies and standards in the administration of the industry. Increasing the rate at which related governance is simplified and deregulation occurs, further removing and deputising various administrative duties and application approval functions, and making a clearer distinction between the role of government and the role of business, will reduce the extent of government interference in microeconomic affairs. The separation of government and business functions should be promoted further by separating naturally monopolistic businesses from competitive businesses, and opening up specific areas to competition, encouraging orderly access for all types of investor to all areas of the energy industry, to allow fair access and encourage effective competition.

Admission management should be implemented using a restriction list. A "restriction list" is a government-compiled list of fields, businesses etc. to which prohibition or restrictions apply. Access to fields that are not included on the list is completely open, as a principle of "permissible unless prohibited" applies. Give full play to the market's decisive role in resource allocation, and ensure that where control is released this occurs effectively and reduces admission thresholds. If not prohibited by law, activities should be open to market participants. If not authorised by law, government activity in activities is not permitted. This will provide greater space for the market to play a determining role. A unified market admission system should be implemented, founded on a restriction list regime, encouraging and guiding the access of various types of market entity to participate in fields other than those listed in the "restriction list" in accordance with the law. This will promote diversification of investors in the energy sector.

However, in addition to relaxing market admission, there should also be certain large, significant changes involving deregulation of both large and small enterprises. For example, in the oil and natural gas development and infrastructure fields, where problems related to administrative monopolies are particularly pronounced, allowing access to one or two major competitors will result in more effective competition. Consideration could be given to tendering and bidding for conventional oil and gas mineral rights for the Sichuan Basin pilot scheme, to the release to competition of construction and operation of the provincial pipelines and to the complete deregulation of the oil and gas markets for Sichuan, Chongqing and other cities. Deregulation of oil and gas importing would open up both domestic and foreign markets and allow private enterprises to grow in influence.

15.3 Deepen Reform of the Fiscal and Tax Systems

1. Further improve fiscal policy

Action should be taken to take advantage of market effects so as to expand funding for geological exploration, with a focus on supporting and promoting unconventional and deep water natural gas resource development and international co-operation. Government support mechanisms for fundamental, strategic and cutting-edge scientific research and research into generic technology and mechanisms for funding major equipment should be improved. The peak regulation, frequency modulation back-up and compensation policies should be refined.

The manner in which taxes are divided between central government and local government needs adjusting. When compared to thermal power generation and other such traditional energy sources, the financial discounts applied to natural gas and other clean energy sources offer few benefits in terms of local fiscal income, and as a result there is little stimulus for activity in this area in local government. On the other hand, when compared to central taxes, the sources of local taxes are fairly dispersed, while collection and administration presents greater difficulties, with unstable income affecting funding for the development of natural gas by local governments. We suggest an appropriate adjustment of the proportions of VAT, company income tax and other taxes allocated from the natural gas sector, increasing local government fiscal income, which will act as an encouragement to local government to develop natural gas.

2. Extension of financial subsidies for shale natural gas

Efforts to provide policy support for coalbed methane as soon as possible should be intensified. Financial subsidies to coalbed methane development companies should be increased to 0.60 CNY/m^3. The corporation tax exemption policy should be extended to 2020, and there should be strict implementation of the VAT refund policy or implementation of an "immediate refund" policy.

Between 2012 and 2015, China's central government gave a standard subsidy of 0.4 CNY/m^3 to shale natural gas businesses who satisfied the relevant conditions. The period to which the 13th Five-Year Plan applies will be critical for the launch, development and growth of China's shale natural gas industry. Encouragement of exploration and development of shale natural gas will continue, despite the sharp drop in international oil prices and the likelihood that these will remain at between 60 and 80 $/barrel for the next 3–5 years and the associated reduction in global investment in shale natural gas. In light of this, the continuance and implementation of further shale natural gas exploration subsidies is of major importance if shale natural gas is to continue to be developed on a large scale in China. The standard subsidy for 2016–2018 is currently 0.3 CNY/m^3; the standard subsidy for 2019–2020 is 0.2 CNY/m^3.

3. Accelerate resource tax reform

The objective of resource tax reform should be the promotion of the rational development and utilisation of natural resources, the promotion of ecological and environmental protection and the acceleration of transformation of development modes. It should be introduced in three stages and cover three areas. In 2015–2016, preliminary completion of a resource tax system; in 2017–2018 the resource tax system should be more or less completed; and in 2019–2020 the objective of the resource tax reforms should have been fundamentally achieved, leading to an emphasis on establishing the application of resource tax, determining appropriate taxation levels and extending the scope of tax collection, improving calculation and collection methods, and refining a system of resource commodity pricing, this being achieved via integrated measures.

Research should be carried out on adjusting petroleum product consumer taxation systems and tax rates, promoting petroleum product consumer tax reform using multifaceted taxation schemes. On the basis of improving the existing tax system, multifaceted collection schemes are recommended, promoting petroleum product consumer taxation reform. First, strict accreditation of wholesale businesses should be applied, through channels such as media monitoring and public oversight, to ensure that access is provided openly, fairly and justly to approved wholesalers. Second, tax departments should focus on a new tax collection scheme, adjusting the regulatory focus to ensure that the transition between the old tax system and the new one is smooth and successfully implemented. Third, the proportion of oil consumption tax to be shared between local and central government should be determined in a scientific manner, encouraging the active involvement of local government in collection and administration. Fourth, thorough monitoring and improvement systems should be put in place in order to deal with new problems and situations which arise with the new tax system.

4. Establish and improve environmental tax and carbon trading policy systems

Environmental taxes are taxes levied for environmental purposes and include special taxes imposed or levies imposed as a form of preventing pollution or damage to the environment. The scope of such taxes ensures that they are levied against and specifically target behaviour harmful to the environment. This form of taxation (also known as "independent environmental taxation") is thus intimately connected with environmental protection, and generally includes pollutant emission taxes, contaminant taxes and carbon taxes. Pollution emission taxes are environmental taxes levied against pollutant emissions (such as waste gas, contaminated water and solid waste). Contaminant taxes are environmental taxes levied against potentially contaminating products (such as energy fuels, motor vehicles, ozone-depleting substances, fertilisers, pesticides, detergents containing phosphorous, mercury-cadmium batteries). Carbon taxes are environmental taxes levied against carbon dioxide-generating fossil fuels such as coal, oil and natural gas.

During the period of the 12th Five-Year Plan, encouragement was given in particular to the breakthrough in carbon trading which encouraged pilot schemes to realise a move from "voluntary trading" to "mandatory trading". In the long term, the aim is to establish a comprehensive primary and secondary market which covers the entire economy and which integrates international markets, with market mechanisms playing a decisive role in the optimised allocation of all kinds of energy efficiency and emissions reduction resources, in order to achieve maximum emissions reduction for minimum cost. Regarding carbon tax, carbon tax legislation should be accelerated, making it possible to think about ways in which carbon trading can be used in conjunction with carbon taxes, thus allowing implementation of a variety of policies. At the same time, enterprises which actively use emissions reduction and carbon dioxide recovery technology and which reach certain standards should be given tax breaks.

Legislative work in relation to the transformation from an environmental charge to an environmental tax system and the introduction of pollutant emissions taxes should be accelerated. The objects of existing charges levied on pollutant emissions should all be included in the scope of the pollutant emission taxation, and comprehensive reform of the pollution charging system should be implemented. Further refinement of

energy efficiency and emissions reduction taxation policy is required, as is the creation and refinement of an environmental compensation mechanism, while the possibility of implementing a green tax system should be explored.

15.4   Establish and Complete a Natural Gas Data Management System

Refinement of a unified information management system should occur as soon as possible. Submission of oil and gas data should be linked directly with mineral rights administration, allowing comprehensive integration and data sharing in the mineral rights administrative processes. Accelerate the establishment of a national database. The main areas covered by this should be the results of domestic oil and gas geographical surveys and data relating to exploration and development, covering both domestic and foreign oil and gas resource data. Establish a national public information network for oil and gas resources, implementing concentrated and dynamic, publicly accessible data, mineral rights administration and raw data in an integrated oil and gas data management, publication and sharing service platform, in order to effectively promote an open-door policy on mineral rights, with truly equitable access, thus encouraging effective competition.

15.4.1   Data Submission

Various types of data related to the exploration and development of conventional natural gas, shale natural gas and coalbed methane, including original geological data, outcome data and relevant geophysical information from geological work, should be submitted to the departments responsible for management of state-owned land resources, which will then be filed with the natural gas mineral rights administration. Original geological data and results data to be submitted should include specialised information of aspects such as field geology, geophysics, geochemistry, geological remote sensing, experimental testing, engineering geology, drilling engineering and information technology. The complete dataset for the previous year should be submitted by the second quarter for the previous year.

15.4.2 Standardised Data Management

To strictly regulate the resource-based integrated data management of conventional natural gas, shale natural gas and coalbed methane, submission of natural gas resource raw data will be linked with gas mineral rights administration. This will establish mechanisms for natural gas resource data acquisition, processing, management and storage, promote the digitisation of natural gas resource raw data and establish and improve new mechanisms for natural gas resource raw data management and services, creating an integrated management and shared services platform for natural gas resource raw data.

15.4.3 Database Development and Information Disclosure

Resource information covering domestic conventional natural gas, shale natural gas and coalbed methane should be set up, adopting China's natural gas resource survey and exploration and development data as the basis of the main content of a national gas resource database. Implement comprehensive integration and data sharing in relation to the natural gas mineral rights administrative processes. Create a national public information network for oil and gas resources, implementing unified natural gas resources information management of concentrated and dynamic data, public data, mineral rights administration and raw data.

15.5 Increase Reform and Technical Innovation in the Natural Gas Sector

State investment in gas development and utilisation technology should be increased. The focus should be on key technological breakthroughs, increasing investment in the areas of key technology common to unconventional gas exploration, deep water gas development and LNG storage and transportation. Strategic planning and expansion of research and development should be instituted in the following eight key technology areas: shale natural gas exploration, deep water gas development, coalbed methane extraction, gas hydrates extraction, coal-based methane, gas-powered vehicle (and ship) engine manufacturing, combined cycle gas turbine generators and carbon capture and storage, with a focus on achieving key technological breakthroughs and early deployment of fundamental research in the relevant fields.

The administrative system in relation to existing technological investment needs to be reformed. The national energy departments should organise and co-ordinate scientific, territorial resource, industry and information, finance, environmental protection and standardisation departments, defining the tasks and responsibilities of each different department with respect to support for natural gas technology innovation. With leadership from the State Council, uniform supportive policies should be developed according to the shared features of technology. Recommendations for improving the support system for natural gas technology innovation include:

- Establish a national major technology laboratory for unconventional gas, providing manpower, resources and finance for research breakthroughs relating to key technology, should be established.
- Support business innovation and enhancing industrial technological capabilities.
- Support businesses to transform themselves and upgrade to clean, gas-based energy.

- Improve production efficiency through technological innovation and investment.
- Assist backbone enterprises in adopting core technology, thus enhancing their technical capabilities and establishing their core competitiveness.
- Strengthen the links between producers and academic research.
- Support research into shared technology and develop and implement independent projects promoting autonomous, localised projects relying on clean gas based energy equipment.
- Encourage localised manufacturing of clean, gas-based energy technology.
- Intensify efforts to train personnel, supporting the establishment of corporate research and development centres and post-doctorate research centres.
- Strengthen shale natural gas geological research, and accelerate the development and application of "factorisation" and "complete sets" in the adoption and development of technology.
- Explore the formative process involved in the development of advanced practical shale natural gas exploration and development technologies and business models.
- Foster in-house innovation and equipment manufacturing capacities.
- Encourage joint scientific advances and international co-operation, ensuring the introduction, digestion, absorption and innovation of advanced technology.
- Adopt shale natural gas exploration, development and production technology suited to China's surface and subsurface geology.
- Accelerate the formation of environmentally friendly and cost-effective key technology and equipment systems specific to China and implement these in order achieve widespread adoption.

In terms of equipment, the first step should be to increase the power rating of technology, reducing the number of devices necessary and reducing the footprint of well-sites, making them more suitable for mountainous settings. Next, the aim should be to bring about equipment modularisation, miniaturisation and portability so as to facilitate work in complex surface configurations. To achieve cost reductions, design process optimisation and research into low-cost hydraulic fracturing options should be actively pursued. As regards environmental issues, environmental protection and the safety of additives should be ensured, as should co-ordinated resource development and environmental protection. Over the next 3–5 years, at the same time as digestion, absorption and innovation technologies are introduced, key shale natural gas engineering technology installations systems with Chinese characteristics should be adopted on a wider scale, and these should be environmentally friendly, economical and effective. After 2020, efforts should be made to ensure that localised and mature proprietary shale natural gas technology and equipment is available.

Large-scale coal gasification and methanation equipment, energy efficiency cascade technology improvements, microalgal biodiesel technology as well as key sewage treatment technologies should be given priority and included in key national basic research projects, while there should be increased investment in the development of technology and applications. Relying on existing model coal methane projects, further innovation and research should take place concerning the suitability of coal types and high-pressure gasification, increasing production capacity, improving gasification, reducing water consumption, promoting environmentally friendlier new-generation coal gasification techniques and key equipment and other such breakthroughs in proprietary innovations. In addition, research and development of high-temperature methanation synthesis and reactor technology and equipment should take place, gradually resulting in the creation of core production technology and automated integrated systems suited to the characteristics of Chinese coal.

Research efforts into gas hydrates exploration and development technology should be ramped up, enhancing the research and development capabilities in relation to core technology, thereby encouraging the development of core technology intellectual property.

- Investigation and evaluation of gas hydrates will provide a basis on which the orderly development of gas hydrates can take place. Further improvements in resource surveying, exploration and evaluation techniques should be applied to gas hydrates, prioritising prospecting criteria and methods, allowing the distribution of China's maritime and tundra gas hydrates resources to be plotted— this will establish which gas hydrate land areas are commercially viable.
- Strengthening scientific and technological research into extraction technology will promote a process by which gas hydrates could be developed in China. This should mainly include:
  - research into methods that result in rapid and efficient decomposition of gas hydrates, such as depressurisation, heating, chemical inhibition and carbon dioxide substitution;
  - research into key technology such as drilling, completion, cementing, horizontal wells and fracturing and improved gas collection, storage and transportation technology, which will establish a relatively complete gas hydrates extraction technological system;
  - research and development into extraction equipment and other key equipment, encouraging the introduction of locally manufactured gas hydrates extraction equipment, enhancing the technical level of China's gas hydrates extraction equipment, strengthening the integration of extraction techniques and methods, optimising extraction schemes and equipment portfolios, and formulating practical gas hydrates development technological systems.

- Strengthening research into extraction safety and environmental assessment, in order to establish the likely impact of gas hydrates extraction on global climate change, seabed geological disasters and deep-sea ecosystems as soon as possible, will allow the introduction of appropriate dynamic monitoring, disaster warning and control systems.

15.6 Expand International Energy Co-operation

Co-ordinate the use of domestic and international resources and markets. Particular attention should be paid to ensuring that investment and trade go hand in hand, while making use of both land-based and maritime channels, in addition to accelerating the development and implementation of mid-term to long-term planning in relation to overseas energy resources, focusing on expanding import channels, building the Silk Road Economic Zone, the 21st Century Marine Silk Road, the BCIM Economic Corridor and the Pakistan Economic Corridor, and actively supporting energy technology, equipment and engineering teams to "look outside".

Strengthen the construction of the five key energy co-operation regions of Russia and Central Asia, the Middle East, Africa, the Americas and Asia-Pacific, and deepen bilateral and multilateral international energy co-operation to establish regional energy markets. Actively participate in global energy governance. Strengthen co-ordination and encouragement of enterprises to "look outside" and to acquire or obtain shares in foreign companies dealing with advanced shale natural gas and other unconventional gas development technologies in a systematic and targeted manner, allowing China to accumulate technological experience relating to unconventional gas development.

- Further optimise the authorisation process to improve authorisation efficiency. Competition for international oil and gas resources in the mergers and acquisitions market has been relatively intense. Good projects often receive a lot of attention, while requirements relating to deadlines have become stricter. In light of this, it is recommended that the relevant departments tighten up their overseas investment authorisation procedures, reducing risks where possible and increasing authorisation efficiency in order to shorten the time required for authorisation to be granted.

- Take effective measures to solve the problem of the shortage of investment and financing for overseas corporate mergers and acquisitions. First, policy banks can give the necessary support to enterprises for acquisition financing. Second, China Investment Corporation may be able to act as an investor for certain projects. Third, some China-Africa development funding and other such equity investments can play a role in overseas investment.

- From the long-term perspective, it is necessary to build a policy system which will encourage oil companies to "look outside". First, co-ordinate the development of national oil security strategy and energy diplomacy policy, creating an international environment which is conducive to overseas investment in oil and gas, such as signing of investment protection agreements, securing relatively reasonable fiscal terms on resources in regions like Latin America and Central Asia and refinement of the government's mechanism for rapid response to overseas emergencies. This will minimise the political, security and policy risks faced by oil companies due to overseas investment. Second, establish overseas venture exploration funds for projects approved by the state for which oil companies can apply; after high-risk exploration pays off, a share of the proceeds is returned, or partly written off in the case of failure.

At the same time, the oil and gas futures market being created in China must take into consideration the status and trends of the distribution of global resources, production development, trade flow and increasing consumption. A perfect, modern, international oil and natural gas market system with foundation like these will help China attain a relatively favourable position in the process of using global oil and natural gas resources to develop its economy. Therefore, on the one hand China should respond to the developments and trends in the market while improving the domestic oil circulation system and price formation mechanisms, in order to establish an oil and natural gas industrial system based on diversified natural gas resources, diversified participants and open-market prices. On the other hand, though satisfactory market admission, foreign currency and customs policies must be put in place, in order to attract a large number of international investors to participate in building a global international oil and natural gas futures market.

The establishment of the China (Shanghai) free trade test area (the FTA) provides the conditions necessary to develop a Chinese oil and natural gas futures market. Organisations such as the People's Bank of China have issued policies such as Suggestions on the Financial Support of the Shanghai FTA, under the principles of "first release, then regulate" and "controllable risk, steady progress". Capital account convertibility, CNY cross-border use and interest rate market liberalisation have already been achieved in the FTA, indicating that China has entered a new stage of financial innovation and that the establishment of an offshore financial market has already begun. Moreover, the Shanghai international energy trading centre will be specifically responsible for the construction of an international oil and natural gas futures platform, contract design and supporting mechanisms (such as foreign investor participation in transactions and bonded delivery), thus establishing a market trading system based on three main principles: value-changing mechanisms, fair trading rules and good credit systems. Solid progress on this work indicates that China expects to launch international oil and natural gas futures trading in the near future, an event which will attract the participation of large numbers of international investors, which will allow us to take better advantage of both domestic and international resources and markets. This will in turn help to improve the international competitiveness of Chinese oil and gas companies on the open market, promoting the establishment of a Chinese oil and natural gas market system that is able to meet the ever-growing economic and social development needs for energy while ensuring national energy security.

PART IV

International Experience from the Development of Gas Markets Globally

Looking at the global energy system, natural gas consumption is, on average, growing as a share of the energy mix. This growth is driven by gas market liberalisation, a switch from other fuel sources and improved access to unconventional natural gas reserves. At the same time, the world is increasingly being divided into regions that consume natural gas and regions that supply it. The imbalance in natural gas supply and demand has in part driven the sharp increase in trade in liquefied natural gas (LNG) over the past decade. In this context, as an important part player in the global energy markets, China's policy framework has a major influence on global natural gas markets.

Global natural gas pricing is in the process of evolution, with oil-indexed long-term contracts increasingly being replaced by market-based pricing. For example, the US Henry Hub price reflects current supply and demand conditions, seasonal variations, the impact of major disruptions, as well as long-term trends. In markets with oil-indexed gas pricing, prices are less sensitive to and reflective of market conditions. Historical links between natural gas and oil prices are no longer relevant in many regions, and as the fundamentals of natural gas and oil markets have diverged, so have natural gas prices in regions with competitive pricing compared to regions continuing with oil indexation. The objective of market-based pricing, such as through natural gas hubs, is to create a liquid market where natural gas is competitively priced.

A competitive natural gas market is key for achieving a more affordable, secure and environmentally sustainable energy system. In turn, this requires five core elements of a liberalised market: creating institutions, ensuring open access to infrastructure, deregulating prices for the competitive segments of the value chain, setting standards and ensuring market transparency and protecting end users.

The natural gas value chain can be broadly divided into three segments: upstream markets, midstream infrastructure and downstream markets. The upstream segment primarily refers to domestic exploration and production of natural gas, where licensing regimes and fiscal policies are important for incentivising and supporting production. The midstream segment refers to infrastructure for the transport of natural gas—domestic and imported—through natural gas transmission networks and local distribution companies. The downstream segment refers to wholesale and retail markets, including natural gas trading hubs. A liberalised natural gas market requires deregulation of the upstream and downstream segments, and third-party access to midstream infrastructure.

As natural gas demand rises globally, security of supply is becoming an increasingly important issue. Key elements of natural gas security include: diversity of natural gas supply sources and availability of alternatives, sufficient quantity of good quality supply and transmission infrastructure such as storage facilities, pipelines and

LNG terminals, and institutions and policies to support the development of natural gas sources and infrastructure.

The energy system has implications for environmental pollution—for example, reducing local air quality, which in turn produces negative health outcomes and lowers the productivity of the economy. Substituting natural gas for other fossil fuels in industrial applications and residential heating can reduce local air pollution and provide significant benefits for the economy.

International Development Trends   16

16.1 Global Natural Gas Markets

Throughout the world, natural gas is gaining prominence in national energy systems. In many economies its share of the energy mix is growing. Major factors behind this global growth include market liberalisation, a switch from other fuel sources and improved access to unconventional natural gas reserves. Environmental and energy security considerations have added impetus to the trend.

In this chapter, we examine developments in natural gas supply and demand globally. We look at developments and trends in international natural gas markets in terms of supply, demand and prices. We also consider the experiences of many developed economies of liberalising the natural gas value chain. Given the increasing importance of natural gas as an energy source, we examine the effect of natural gas on national energy security. Finally, we consider the environmental benefits of a shift to natural gas, in terms of boosting local air quality, improving health outcomes and increasing the overall productivity of the economy.

As global energy demand has grown, so has the importance of natural gas in the energy system, both in absolute terms and as a share of the global energy mix. In 1980, natural gas accounted for 19% of global energy consumption, and by 2010 that figure had risen to 23%. During these two decades, natural gas consumption rose on average by 2.6% per year, compared to 2% for overall energy consumption, and is expected to sustain a 2% annual increase until 2030. While much of the current demand for natural gas comes from developed markets, demand in coming years is expected to be driven by increased consumption in emerging markets, particularly for power generation. Indeed, the bulk of this new demand is expected to come from Asia, and particularly China.

Demand for natural gas is driven primarily by switching from other fuel sources, such as oil and coal. In addition, increased economic intensity—which is pushing demand for all fuels—is contributing to growing consumption of natural gas. Reduced energy intensity (the amount of energy needed for a given output) is mitigating these growth pressures slightly. Perhaps surprisingly, our research finds that price differentials between natural gas and other fuels have a relatively small effect on fuel switching and demand growth.

The development of unconventional natural gas has further stimulated natural gas demand growth. In the wake of recent success in the United States in exploiting shale natural gas, a larger portion of the world's natural gas supply is

* This chapter was overseen by Jigang Wei from the Development Research Center of the State Council and Taoliang Lee from Shell Eastern Trading Corporation. It was jointly completed by Yaodong Shi, Zifeng Song, Ren Miao from the Energy Research Institute of the National Development and Reform Commission and Cindy Wang from Shell China. Other members of the topic group participated in discussions and revisions.

Shell International and The Development Research Center (Eds.), *China's Gas Development Strategies*,
Advances in Oil and Gas Exploration & Production, DOI 10.1007/978-3-319-59734-8_16

expected to come from unconventional natural gas reserves. In 2030, shale natural gas and coalbed methane are expected to account for more than half of North America's natural gas production. Further, the US Department of Energy has estimated that China holds the world's largest reserves of technically recoverable shale natural gas, almost twice as much as found in the United States. However, cost, the geographic distribution of the natural gas fields, and the need for the appropriate capabilities could be obstacles to exploiting these reserves.

The global natural gas trade, and especially the flourishing LNG trade, has supported the continued growth in natural gas demand. Global trade in LNG doubled between 2003 and 2013. The increase in global LNG trade is expected to continue, leading to an increasingly linked and integrated global natural gas market. Partly as a result of the increased production of unconventional reserves in the United States and elsewhere, countries exporting LNG are likely to become more diversified in coming years, creating a supply network with links to a broader range of demand centres. LNG shipping is also more flexible, adding further momentum to global LNG trade. Because they do not rely on immovable pipeline networks, uncommitted LNG shipments can respond to price and be quickly directed to the best markets, based on price. Energy security advantages and arbitrage opportunities also help LNG shipments to compete with natural gas delivered over pipelines.

Pricing systems for natural gas are gradually shifting away from oil-indexed contracts, especially outside Asia. While natural gas contracts have traditionally been linked to the price of oil, the underlying economic fundamentals for these two fuels have diverged. Oil-based fuels are primarily used for transportation, with power and heat generation becoming a much smaller contributor to demand, while the use of natural gas in power and heat generation is increasing. In addition, the creation of natural gas hubs—notably in Europe and the United States—has enabled natural gas contracts to be linked to transparent and readily available natural gas price benchmarks, such as Henry Hub in the US and the British National Balancing Point (NBP).

As part of this study, we examined the impact of different Chinese natural gas demand scenarios on global energy markets. The analysis suggests that an increase in China's natural gas demand is not likely to have a significant impact on global natural gas prices, as global supply is able to respond to increased Chinese demand. However, Chinese coal demand displaced by the increase in natural gas demand is likely to have a significant impact on global coal prices. China currently accounts for a large share of global coal demand, and the decline in demand, combined with global coal supply being slow to respond to the new market conditions, is likely to depress global coal prices. In the absence of any policies to restrict coal use, regional economies such as India and Indonesia are likely to be the main beneficiaries of lower coal prices.

16.2 International Experiences of Liberalising the Natural Gas Value Chain

As part of our study, we examined the liberalisation experience of natural gas systems in five markets that have witnessed significant increases in consumption in recent years. In all of these markets—the United States, the European Union, the United Kingdom, Japan and South Korea—market liberalisation and increased competition played a significant role in increasing domestic demand for natural gas. The market development in each of these cases was unique, but there are nonetheless positive and negative lessons that China can learn from.

When it comes to increasing the share of natural gas in the energy mix, three key energy-related concerns for policy-makers are:

- providing affordable natural gas for business, residential and industrial end users;
- enhancing national energy security by diversifying supply networks and improving the country's negotiating position in international natural gas markets; and

- delivering ancillary benefits, such as improving air quality or supplying energy to poor households and regions.

Liberalising the natural gas value chain can help to balance and support these objectives. For example, increased competition in the extraction and sale of natural gas can drive down costs, and therefore reduce the price paid by consumers and end users. It can also help to diversify sources of supply, increasing energy security.

The countries we studied followed similar paths to liberalisation, although they differed in the details and the level of success. The process of liberalisation helps to deliver competitive market outcomes—to provide end users with the greatest choice at the lowest price. This can be achieved by opening markets to competition or through regulatory measures to limit or cap prices where natural monopolies exist.

The natural gas value chain can be divided into three segments: the upstream exploration and production segment; midstream infrastructure, including transmission and distribution pipelines and LNG terminals; and the downstream wholesale and retail markets for natural gas that bring the fuel to consumers and end users. Competitive markets can exist at many points in this value chain.

Competition in the upstream segment can drive greater efficiency and innovation in the exploration and production of natural gas, as a way to drive down costs and develop greater volumes of domestic supply. For example, before the liberalisation of natural gas markets in the European Union, state-owned, vertically integrated natural gas companies had little incentive to improve the efficiency of their operations, leading to increases in the price of natural gas. Similarly, greater competition in the downstream segment can increase choice and reduce the price paid by the end users of natural gas, for example by increasing competition in the shipping and sale of natural gas. These segments can be opened to competition, within a regulatory oversight framework that ensures and supports competitive markets and without the need for further regulatory interventions.

However, opening the midstream segment to market competition is typically not desirable, and is likely to fail. Midstream infrastructure, such as pipelines, is highly capital-intensive. For example, in 2013, onshore oil and natural gas pipeline construction in the United States cost $4.1 million per mile, while offshore pipelines cost $7.6 million per mile. High fixed costs and relatively low operation and maintenance costs point to large economies of scale: the costs of delivery decline rapidly with volume. This fits the definition of a natural monopoly, where the lowest long-run average costs are realised when production and ownership are concentrated in a single firm. Countries have dealt with the natural monopoly characteristics of midstream infrastructure in a number of ways: by mandating third-party access to pipeline infrastructure; by regulating the tariffs that pipeline infrastructure owners can charge for access; and/or by separating transport services from upstream and downstream businesses (a process known as unbundling).

To be more specific, countries that have successfully liberalised the natural gas value chain began by creating the necessary institutions, particularly a strong and independent natural gas regulator. They then moved to ensure open access to the natural gas network by, for instance, requiring large integrated companies to unbundle their holdings, regulating network charges and overseeing third-party access to transmission and distribution networks. These countries also deregulated prices where appropriate, especially in the upstream and downstream segments of the value chain. They set market standards and mandated transparency, for example by standardising trading agreements and supporting the development of natural gas hubs. And finally, they strived to protect end users by overseeing competition and managing adverse market impact.

16.3 Liberalisation of Different Segments of the Natural Gas Value Chain

In addition to examining the evolution of natural gas systems in selected markets, our study also looks at international experiences of liberalising specific parts of the natural gas value chain.

In the upstream segment, liberalisation has focused on encouraging new entrants and increasing competition through appropriate fiscal and licensing regimes. International experience suggests that these measures are necessary to meet national objectives, as well as to ensure the quality and quantity of natural gas production.

In the midstream segment, the natural monopoly characteristics of natural gas transmission and distribution infrastructure have meant that liberalisation efforts have focused on providing third-party access and unbundling. In setting the framework and terms for third-party access, governments need to balance the need for fair and open access with incentives for further investment in midstream infrastructure. This balance varies by country and circumstance. For example, regulators in Japan have favoured midstream investment over access. Thus, while provisions for third-party access exist in law, enforcement has tended to be lax.

In addition to third-party access, unbundling is an important consideration with regard to midstream assets. Often, as a country begins the process of liberalisation, the national market is dominated by one or a few companies that own assets throughout the value chain. These companies have a strong incentive to restrict access to midstream infrastructure in order to benefit their own upstream or downstream businesses. Because of this natural tendency, countries seeking to liberalise their natural gas markets often require these vertically integrated companies to separate—or unbundle—midstream assets into autonomous businesses. Unbundling has taken various forms. While countries like the United Kingdom and The Netherlands have achieved full ownership unbundling, others like the United States, Germany and France have chosen to stop one step short, at structural unbundling. The experience of these latter countries indicates that full ownership unbundling is not required, and that structural unbundling under a strong regulator can capture many of the benefits of full ownership unbundling.

Finally, with respect to the downstream segment, our study examines the development of competitive natural gas wholesale markets and the establishment of natural gas hubs. Wholesale markets connect upstream sellers and retail buyers and establish a basic price point for natural gas in a country or region. Regulatory efforts here have often strived to encourage the development of natural gas hubs, national or regional centres that ease transactions between buyers and sellers. Hubs can push natural gas pricing away from oil-indexed contracts, creating a more competitive pricing mechanism that is based on the value of natural gas and is more responsive to changes in market conditions.

Successful hubs enable competitive pricing of natural gas, more efficient market co-ordination and increased market transparency, with consumers and end users benefiting from the more efficient pricing of gas. By reflecting the true cost and value of natural gas in the economy, hub pricing also ensures that trade and investment decisions are made on the basis of economic benefits and efficiency. Hubs also create powerful and self-reinforcing network effects, which help to lower transaction costs and increase competition in wholesale gas markets. Finally, a hub also supports energy security objectives by supporting the diversification of sources of supply. Diversified and liquid hubs are better able to respond to changes in demand compared to the inflexibility of long-term contracts, for example by procuring LNG cargoes on the spot market during peak demand periods.

16.4   Natural Gas Energy Security and Social Influence

As global energy demand has grown, so has policy-maker interest in the security of energy supply. Until recently, concerns about energy security centred mostly on safeguarding oil supplies, but as the global energy mix has diversified, the scope of energy security has broadened as well. Our study finds that while there are a wide range of approaches that a country can use to achieve energy security, they can be distilled down to three critical aspects of the supply chain: sources, infrastructure and institutions. Combining imports, domestic supplies and access to alternative fuels results in a diversified supply with readily available fuel substitutes. Ensuring the quantity and quality of physical assets such as storage facilities, pipelines and LNG terminals means that sources of supply can be accessed reliably. Finally, proper market and political organisations, as well as underlying incentives, are necessary along the supply chain in order to develop these sources and infrastructure.

Assessing China against this framework, our study finds that the country is in a good position with respect to natural gas energy security. In terms of sources, rising LNG trade, future pipeline imports and significant availability of alternative fuels provide a diversified source of supply compared to other large natural gas-consuming countries and regions such as Japan, South Korea and the European Union. Rapid infrastructure development has meant that China has capacity (especially storage) to match the levels of other major gas-consuming countries.

Since China became a net importer of natural gas in 2007, import sources have rapidly diversified. Figure 16.1 shows, compared to the EU, Japan, and Korea, China's rapid increase in the diversification of its natural gas imports, thereby increasing its energy security. In 2007, China's only import source was Turkmenistan. The Herfindahl Index was 1. Following the development of LNG trade, China's import sources have diversified rapidly, and its Herfindahl Index has

correspondingly dropped. Today, China's natural gas import diversification is comparable to that of the EU. In 2010, it was even comparable to nations with the most diversified sources of natural gas—Japan and Korea.

Compared to nations with high natural gas energy ratios that rely more on natural gas imports, China's imports are more diversified. China, the EU, Japan, Korea, and Turkey are the top five importers of natural gas. These five nations have very diversified sources for their imports, and their Herfindahl Indexes are all very low. However, as shown in Fig. 16.2, compared to other countries, China's natural gas energy ratio is lower, and its reliance on natural gas imports is lower. In fact, other nations do not have the import diversification that China enjoys, and their reliance on imports and on natural gas is significant. However, these nations rarely face major natural gas security issues, which indicates that China could safely raise the proportion of natural gas in its energy mix as well as its proportion of natural gas imports.

LNG enables not only China but also the entire world to achieve greater supply diversification. From 2003 to 2013, global LNG trade doubled, with a huge influx of new buyers and sellers. Figure 16.3 shows Japan's LNG trade network. Japan's LNG sources are in the inner circle, while the outer circle shows other countries that Japan's import sources supply. The arrows show LNG trade direction. The coloured

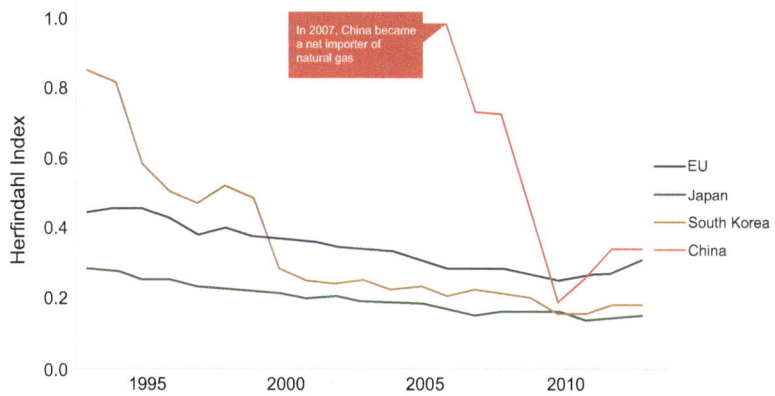

Fig. 16.1 China's natural gas import diversification. *Source* Vivid Economics, based on EIA data

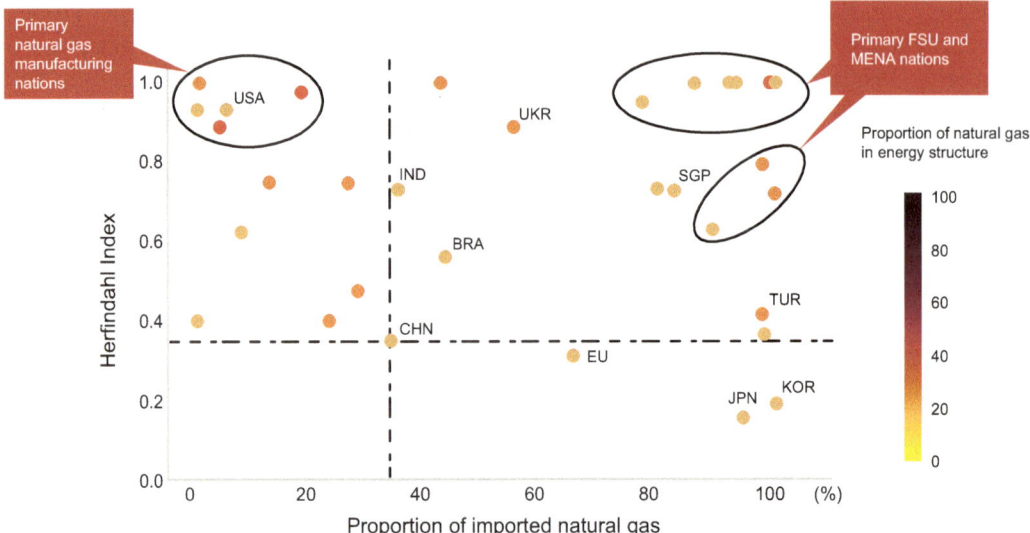

Fig. 16.2 China's import security as compared to major nations. *Source* Vivid Economics, based on IEA and EIA data; natural gas proportions are from 2011, with remaining data from 2013

arrows indicate how reliant each exporting nation is on Japan. Red arrows show relatively high reliance on Japan and yellow arrows show low reliance. In 2003, there were relatively few nations supplying LNG to Japan, and Japan was one of the few consuming nations. By 2013, Japan had added more nations from which it imported LNG, diversifying its import sources. Moreover, these exporting nations had grown their client base since 2003, though many still relied heavily on Japan for their LNG sales.

Figure 16.4 shows China's natural gas trading network in 2013. In addition to LNG trade during this period, there was also one Turkmenistan natural gas pipeline. Compared to other nations, China's imports of LNG were more diversified. However, compared to Japan in 2013, there is a marked difference. For exporting nations, Japan is a more important export destination.

In the future, new pipeline natural gas will improve the security of natural gas for China. Figure 16.4 shows that currently China's pipeline trading has negotiating power. Turkmenistan has only Kazakhstan and Iran as its other export options. China, on the other hand, also plans to engage in pipeline trade with Uzbekistan, Kazakhstan and Myanmar (the options for these

countries are limited as regards pipeline destinations) as well as Russia (East-West Siberian oil and gas field natural gas export destinations are limited). Therefore, even though Japan's LNG supply is more diversified, China (leaving LNG to one side) still has the option to use pipeline natural gas and can also engage in domestic production. This means that China does not need to have a strong relationship with natural gas-exporting nations as Japan does, but can instead use natural gas hubs to make exporting nations compete with one another.

Domestic natural gas production can greatly improve China's natural gas security. Domestic production is a reliable option to replace imports, and can improve negotiating power. China's domestic production has great potential, and features some of the world's major unconventional natural gas reserves. However, current production levels are low, and future production remains uncertain. This uncertainty is a key factor in China's natural gas security, since international experience shows that natural gas security of producing nations is entirely different.

There is a close relationship between natural gas market liberalisation and whether a nation has domestic production. Nations without

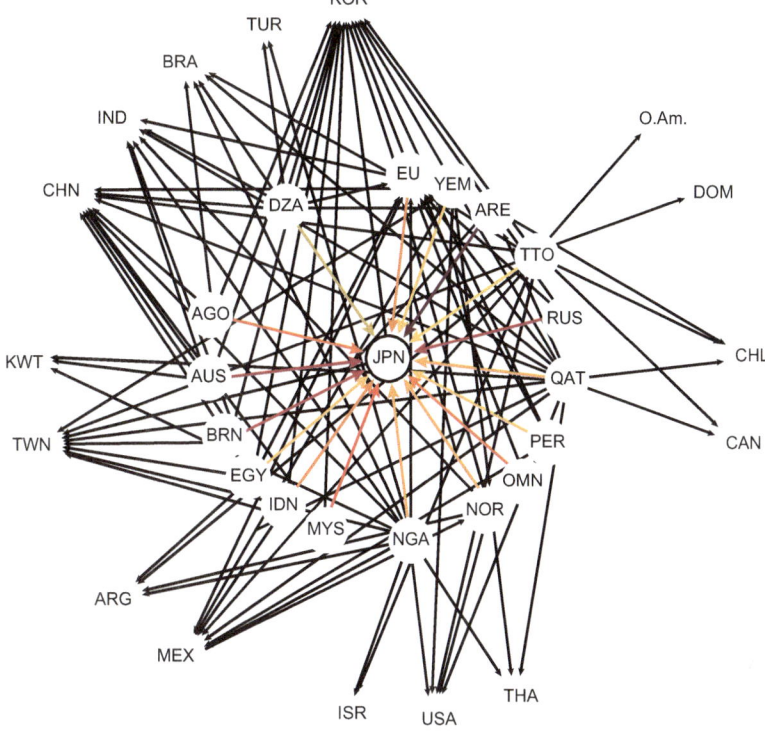

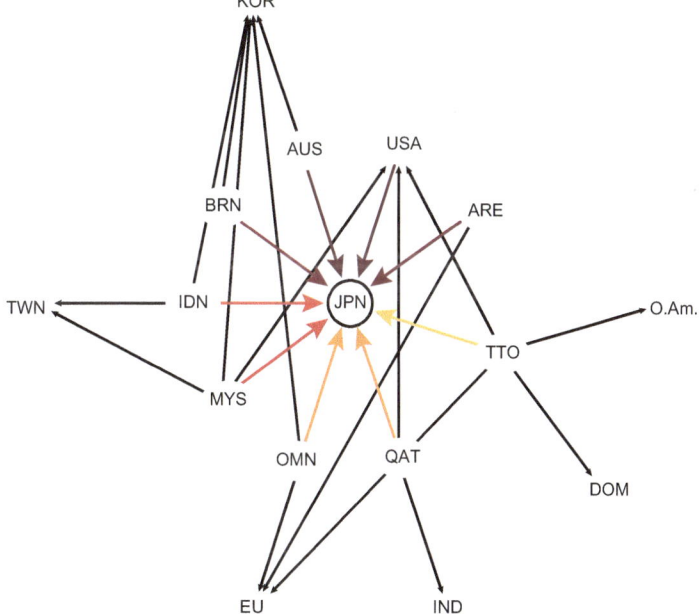

Fig. 16.3 Japan's LNG supply sources are diversified. *Source* Vivid Economics, based on IEA data

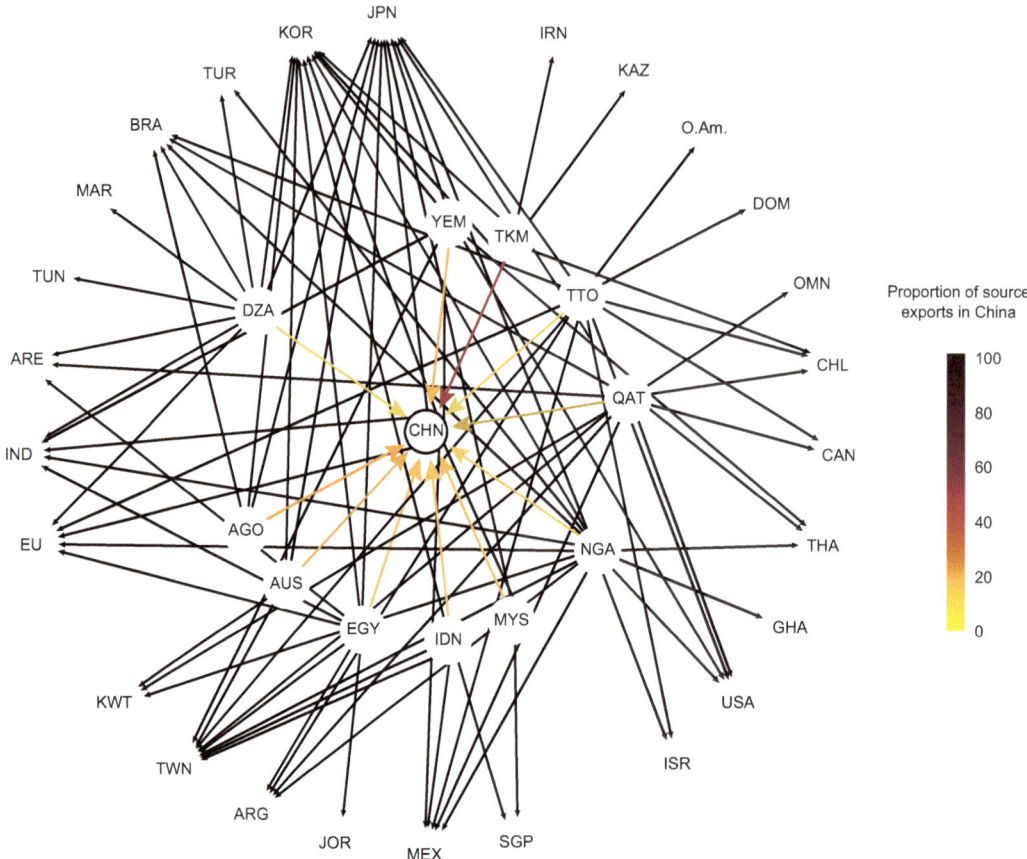

Fig. 16.4 Sources of China's natural gas. *Source* Vivid Economics, based on IEA data

domestic production such as Japan and Germany and nations with large volumes of domestic production such as the United Kingdom and United Sates prove this point. For example, Japan's market is more co-ordinated and there are large companies with bargaining power with exporting nations. However, the United States and United Kingdom have more open markets and their competitive pressures stimulate more domestic production. Another factor is that domestic production improves bargaining power with exporting nations, since when import supply is cut off, the nation has a reliable substitute source. Figure 16.5 illustrates the consistent relationship between domestic production and market liberalisation.

Currently, China's natural gas market is a co-ordinated one. However, there is still major potential to develop domestic production.

International experience has shown that further deepening of market liberalisation will increase domestic production, which then improves natural gas supply security. This would compensate for any reduction in bargaining power that might arise from moving away from the current co-ordinated market system. Therefore, out of consideration for energy security, China could strive to increase domestic production, and market liberalisation is one means to stimulate domestic production.

China has another abundant substitute for natural gas that can further improve source security. Natural gas can be used for power generation and supply of heating—and these are energy services that China's abundant coal or renewable energies can provide. To a lesser extent, natural gas is also used for transport, where oil products can potentially be substituted.

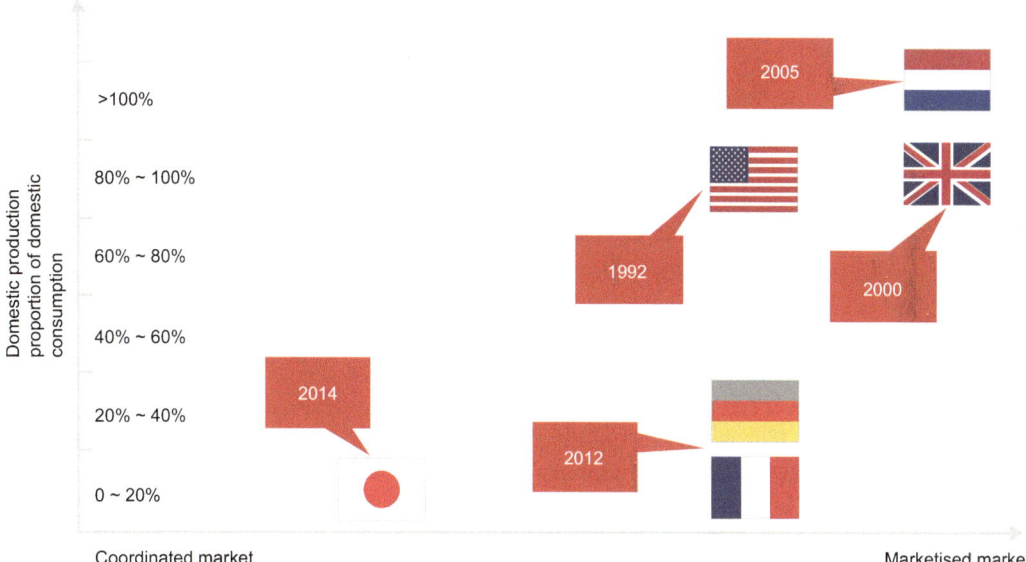

Fig. 16.5 Time points when major nations promoted market liberalisation reforms

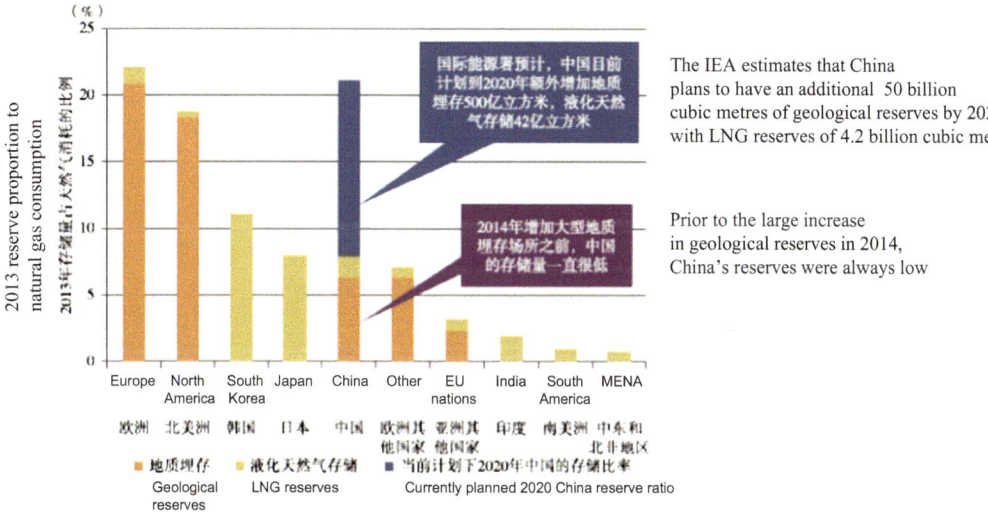

Fig. 16.6 Gas reserve volume as a proportion of consumption, showing planned increase for China. *Note* 2013 China data includes the 10.7 billion m³ of China National Petroleum Corporation Hutubi. *Source* Vivid Economics, based on IEA data

if the natural gas supply were disrupted, switching over to other fuels could limit the negative repercussions. However, using other fuels requires a variety of technologies, not all of which are available to consumers, such as being able to switch smoothly from natural gas to fuel oil or other sources for heating. Moreover, using coal and petroleum as substitutes will have a greater impact on the environment and climate. For these two reasons, the priority should be to

improve energy security through domestic production and diversification of imports.

China has the world's sixth-largest natural gas
import capacity, and through the West-East pipeline it also has major pipeline import capabilities.
This infrastructure was developed after 2005, and
China is continuing to develop its infrastructure,
with plans for major expansions in the future.
Current plans indicate that by 2020, China will
have increased its pipeline import capacity by
114 billion m^3 over 2012, with 40 billion m^3 of
LNG import capacity, which is forecast to satisfy
planned natural gas import demand.

In terms of reserves, China's reserve volume
is on par with other major natural gas-consuming
nations. For example, in 2014 storage facility
construction brought consumption reserve

volume on a par with Japan, as shown in
Fig. 16.6. In 2020 it is planned that reserve
capacity will increase, even if demand grows as
expected, to the highest level worldwide, the
level currently held by Europe.

Based on international experience, a lack
of infrastructure, maintenance and prompt
post-incident repairs are the main reasons for
natural gas supply cut-offs. Moreover, in
many nations, these problems are often
overlooked, viewed as operational issues
rather than policy issues. As China's natural
gas industry develops, continued investment
in infrastructure and the establishment of
sufficient and reliable accompanying technologies and incentive measures will be
especially important.

The Global Natural Gas Market

17

Global natural gas consumption, both in absolute terms and as a share of the energy usage mix, is continuing to grow. This growth has been driven primarily by changes in patterns of economic activity, energy intensity and energy substitution. Of these, the primary driver has been the switch to natural gas from other sources, primarily oil. Natural gas prices have consistently remained high in recent years, but despite this and the fact that gas has not been given any major preferential advantage compared to other energies, there has been no marked dampening of demand.

As global consumption volumes grow, the divisions between regions that consume natural gas and regions that supply natural gas will become more marked. The Asian OECD member nations[1] and Europe have become the major natural gas importers, with the former Soviet Union states, the Middle East and North Africa dominating natural gas exports. The growth of the gap between natural gas production and consumption is continuing to widen, and greater volumes of trade are necessary to balance supply and demand.

As natural gas supply hubs rise in prominence, natural gas prices are gradually breaking free from their link to oil prices. In the United States and Europe, because the natural gas market has matured, supply hubs have sufficient liquidity for competitive trading, and natural gas is priced for its own value rather than being linked to oil prices. However, an Asian natural gas trading hub is unlikely to develop soon, as this region's markets are still controlled by a small number of larger buyers and sellers, and market liquidity is low. Given the imbalances in regional supply and demand, shipping costs and trade restrictions, natural gas prices are likely to continue to vary by region rather than to converge on global unified price levels.

17.1 An Overview of the Global Energy Market

Natural gas is a growing player in the global energy mix (Fig. 17.1). In 1980, natural gas provided 57 EJ or 19% of the global total primary energy consumption volume of 300 EJ. In 2010, natural gas provided 124 EJ, representing 23% of the 525 EJ global total. Natural gas grew its proportion of primary total energy at a time when global energy consumption was itself rising rapidly. So natural gas demand grew at a faster rate than total energy demand—since 1980, average annual growth rate for natural gas has been 2.6%, compared to 2.0% for total energy.

Natural gas is increasing its share in global and regional energy mixes by displacing oil products

---

* This chapter was overseen by Jigang Wei from the Development Research Center of the State Council and Taoliang Lee from Shell Eastern Trading Corporation. It was jointly completed by Yaodong Shi, Zifeng Song, Ren Miao from the Energy Research Institute of the National Development and Reform Commission and Cindy Wang from Shell China. Other members of the topic group participated in discussions and revisions.

[1]Japan and South Korea.

Shell International and The Development Research Center (Eds.), *China's Gas Development Strategies*,
Advances in Oil and Gas Exploration & Production, DOI 10.1007/978-3-319-59734-8_17

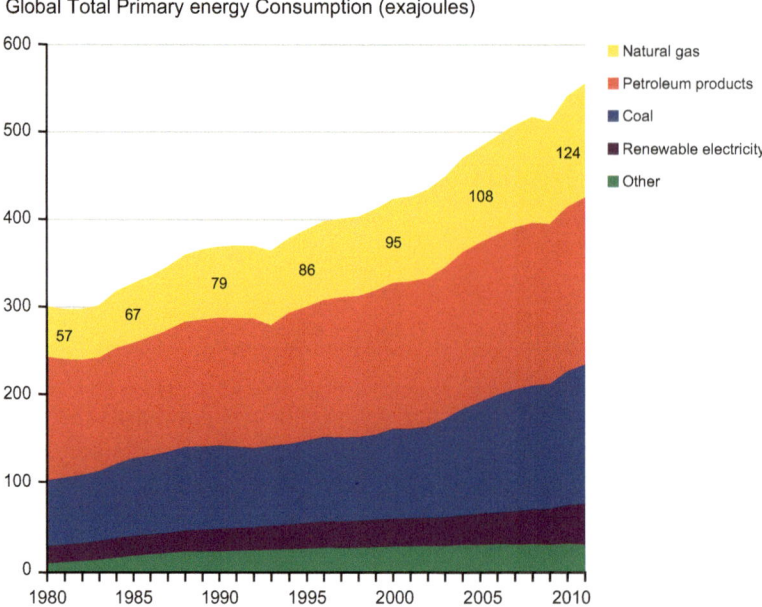

Fig. 17.1 Natural gas's share of global energy consumption. *Note* Data is global total primary energy consumption. *Source* Vivid Economics, based on EIA data

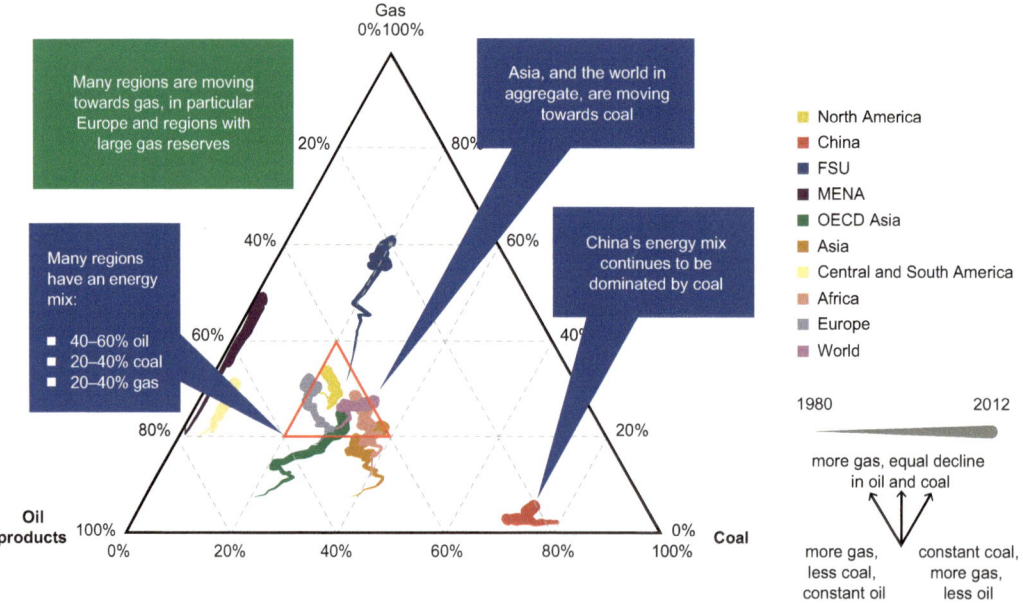

Fig. 17.2 Changes in share in the energy mix between oil, gas and coal. *Note* Data is share of fossil fuel energy; each country's coloured trail along the matrix gets thicker over time; movement toward the apex illustrates increased share of natural gas at the expense of oil (*lower left*) or coal (*lower right*). *Source* Vivid Economics, based on EIA data

and, sometimes, coal (Fig. 17.2). Each region uses differing amounts of oil products, natural gas and coal to supply its primary fossil fuel energy needs each year, and the balance of these energy sources changes over time. In the Middle East and North Africa and countries of the former Soviet Union, for example, natural gas has tended to supplant oil, while in North America and Europe it has taken some of coal's share. In general, the global trend has been toward natural gas consumption in lieu of oil use, with a parallel movement toward coal as well in some regions.

Looking to the future, global energy demand will continue to grow, and demand for natural gas is expected to increase faster than demand for other fossil fuels, continuing the trend of recent years. The IEA has noted that rapid population growth, increasing prosperity and improved access to reliable electricity are driving this trend. Average annual growth rate for natural gas is generally forecast to be 2% from 2012 to 2040.

The growth in natural gas demand in the period to 2040 is expected to be widely dispersed geographically, with Asia and the Americas playing an important role (Fig. 17.3). Asia is expected to account for the largest proportion of global natural gas demand growth during this period (as well as production). In particular, natural gas demand in China is expected to grow by 5.2% a year, accounting for 56% of Asia's natural gas consumption growth. Natural gas demand is expected to grow steadily in other

emerging economies as well, with India rising 4.6% a year and Brazil 4.0%. However, OECD member country growth should be slower, averaging about 1% a year.

Power generation will play a significant role in increasing natural gas consumption, accounting for 36% of total growth from 2012 to 2040 (Fig. 17.4). Driven by the petrochemical industry, the industrial sector is also expected to contribute strongly, with an annual growth rate reaching 1.9%. Although natural gas use in transportation is expected to grow strongly at 3.3% a year, this sector will still continue to account for only a small share of total demand, about 9% by 2040.

During the same period, unconventional natural gas (such as shale natural gas) will come to account for an increased proportion of the overall supply mix. Global natural gas resources are widely dispersed geographically, and at the current rate of production, there are 60 years of proven reserves (Table 17.1). Even though US shale natural gas development has seen strong growth, there remain significant uncertainties about the development of shale natural gas in other regions. Currently, there are essentially no other regions outside of the United States that have successfully developed shale natural gas. In Poland, Sweden and Ukraine, resource discoveries have been disappointing, while developments in Algeria, France, South Africa and other countries have met with public opposition.

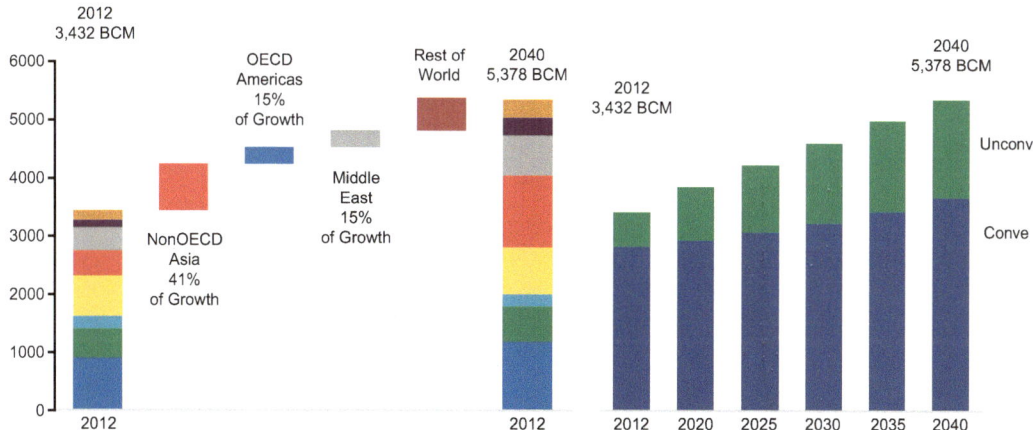

Fig. 17.3 Rising diversity in natural gas supply and demand up to 2040. *Note* bcm is billion m^3. *Source* IEA

Fig. 17.4 Sources of demand for natural gas 2012 versus 2040. *Source* IEA

17.2 Factors Driving Demand

Driven by a number of factors, global natural gas demand has continually risen in recent years, and is expected to maintain its upward momentum. The three major drivers for this growth are economic activity, energy intensity and switching to natural gas, of which the last has played the biggest role in most countries. On the supply side, the availability of domestic natural gas resources has played an important role. Consistently high natural gas prices have not prevented the fuel source from gaining substantial ground in recent years, albeit from a low base.

17.2.1 The Main Factors Driving Demand

Based on research analysing the natural gas markets in seven different countries (see Chap. 2), it is apparent that economic activity, energy intensity, switching to natural gas and the availability of domestic natural gas resources are the four major factors influencing natural gas demand.

As economic activity increases, so does energy demand, and thus natural gas consumption increases in line with its share of the overall energy mix. From looking at the experiences of various nations, it appears that residential users and power utilities have seen the highest increases

Table 17.1 Remaining technically recoverable natural gas resources (end 2013)

	Conventional	Unconventional				Total	
		Coal gas	Shale natural gas	Coke gas	Sub-total	Reserves	Proven reserves
Eastern Europe/Eurasia	143	11	15	20	46	189	73
Middle East	124	9	4	–	13	137	81
Asia-Pacific	43	21	53	21	95	138	19
Americas OECD member countries	46	11	48	7	65	111	13
Africa	52	10	39	0	49	101	17
Latin America	31	15	40	–	55	86	8
Asian OECD member countries	25	4	13	2	19	45	5
Global	465	81	211	50	342	806	216

Note Data is in trillions of cubic metres
Information source IEA data

in usage, probably because rising household incomes allow people to spend more on heating and power companies in turn increase their use of natural gas to meet the greater general demand for electricity. In addition, natural gas demand from manufacturing industry has also increased, primarily due to a rise in production volume and the corresponding energy demand.

Under certain economic conditions, energy demand will fall in response to a fall in overall energy intensity of economic activity, and natural gas demand will decrease in line with its share of the overall energy mix. A drop in energy intensity is closely related to structural adjustments to industry. The experience of several major natural gas-consuming nations shows that when the service sector rises as a proportion of a nation's output or when a move from heavy industry to light industry occurs, energy intensity will exhibit a downward trend.

When a given industry chooses to replace oil and coal with natural gas, its natural gas consumption rises significantly. Global trend analysis shows that using natural gas to replace oil or coal is the major reason for the rise in natural gas demand. In most markets studied, the change in natural gas demand between 1982 and 2012 linked to switching fuel sources to natural gas was equal to or greater than that attributed to increased economic activity. Research into major natural gas-consuming industries also shows that a large proportion of consumption growth is accounted for by switching from other fuels to natural gas.

As nations develop, the service industry begins to play a larger role in the economy, urbanisation increases, and controls on air pollution become more stringent, leading to a marked tendency towards replacing oil and coal with natural gas. Australia, Europe and the United States have shifted a large portion of their energy mix from coal to natural gas, and China is currently considering similar measures. Many other countries have switched from oil to natural gas. These switches are significantly linked to the proportion represented by the service industry in each country's economy, with growth in the service industry directly affecting demand, especially driven by the need for office heating.

Also, legislation aimed at greater urbanisation and improvement of air quality leads to natural gas (thanks to its cleanliness) often being first choice as a substitute energy source.

Finally, the availability of domestic natural gas resources is another factor that prompts energy transition. Domestic reserves are generally the cheapest source of natural gas for a country, and analysis has shown that, regardless of how big the reserves are, they always have an influence on the proportion of natural gas in the country's energy mix. This is partly because nations without natural gas reserves generally lack the infrastructure and systems to support natural gas imports. In addition, countries with domestic supplies and existing infrastructure and institutions are also more open to trade in natural gas, which may further increase the share of natural gas in the energy mix.

China currently is in a period of rapid development, becoming more urban, becoming more concerned with air quality and developing domestic natural gas reserves. These are all characteristics of countries with a high proportion of natural gas in the marketplace. However, if China wishes to promote natural gas, it will need to act to put measures in place that stimulate the transition from coal to natural gas. The experience of other nations suggests that this is more difficult to achieve than a transition from oil to natural gas, and currently only Europe has accomplished it.

17.2.2 Natural Gas Price Elasticity and China

International experience has shown that demand for natural gas is not sensitive to price changes, and that rises in natural gas prices do not necessarily cause a reduction in levels of natural gas consumption. From 1987 to 2000, global natural gas prices remained stable at low levels and natural gas demand in OECD countries grew rapidly. Beginning in 2000, natural gas prices began to rise, but demand did not react to the price changes, remaining stable in the majority of countries, and growing in others.

China's market performance is consistent with international experience. Since the 1990s, Chinese natural gas consumption has grown rapidly. The rise in natural prices that began in 2000 has not had a marked effect, and growth rates have remained relatively subdued.

Price competition between fuels appears to have only a modest effect on natural gas demand. From 1987 to 2000, the natural gas price remained a relatively constant fraction of the electricity price and the coal price. During that period, natural gas demand rose substantially. After 2000, even though natural gas prices rose substantially, there was no large decline global natural gas demand. During the period 2000–2012, OECD member country residential and industrial natural gas demand only dropped by 8%, despite a 60% price increase in real terms.

The main reason that price competition between fuels has only a minor effect on natural gas demand is that natural gas commands loyalty among its users. Once natural gas demand has been established, it can adapt to a changing economic environment, and this is especially true in terms of residential usage. Once consumers start to use natural gas, they tend not to switch away, because of its quality. Natural gas is a relatively clean source of controllable heat that is easily delivered and easy to use. In comparison with other fuels, such as coal, the characteristics of natural gas may mean that once the infrastructure is in place, residential and industrial users are less troubled by the price.

In summary, the non-price characteristics of natural gas appear to be important drivers of demand, so high natural gas prices in China may not hold natural gas demand back. However, China's energy prices are controlled, which means that China's situation could be different from other nations. In the majority of cases, major natural gas-consuming nations have already implemented energy price market liberalisation, with prices following changes in supply and demand. In China, however, some energy prices are controlled, a situation which lacks responsiveness to changes in supply and demand. The synergy between demand and price in other nations could result in different scenarios in China.

17.3   Supply and Demand Imbalances

Globally, natural gas-producing regions are gradually diverging from demand centres. This imbalance has triggered a boom in global natural gas trade, with a proliferation of international pipeline and LNG projects.

17.3.1   Summary of Global Resources

Unconventional natural gas resources are likely to account for a larger share of total gas supply as development continues. The IEA estimates that global remaining conventional technical recoverable natural gas resources have reached 465 trillion m^3 and unconventional resources, 342 trillion m^3. Between 1994 and 2013, global proved reserves increased by 56%, and available resources are expected to continue to grow. Even though unconventional gas production volumes are expected to increase, the majority of natural gas production is likely to continue to be from conventional resources.

According to current estimates, global natural gas remaining proven recoverable natural gas reserves are approximately 186 trillion m^3. Over the past 30 years, proven recoverable natural gas reserves have grown by 3–4 trillion m^3 a year, and the reserve-to-production ratio has stayed steady at close to 60 years.

For the most part, these reserves are in three countries: Iran (18%), Russia (17%) and Qatar (13%). Turkmenistan also holds a significant share of the world's proven recoverable natural gas reserve, with the other 11 countries among the 15 with the largest reserves holding considerably less (Fig. 17.5).

1.   Centres of demand

In 2013, global natural gas consumption, including LNG, rose by 1.1% from 2012. During the same period, natural gas international trade volume increased by 2.3%. Global international trade in oil for the same period rose by only 1.7%. The top 10 natural gas-consuming nations

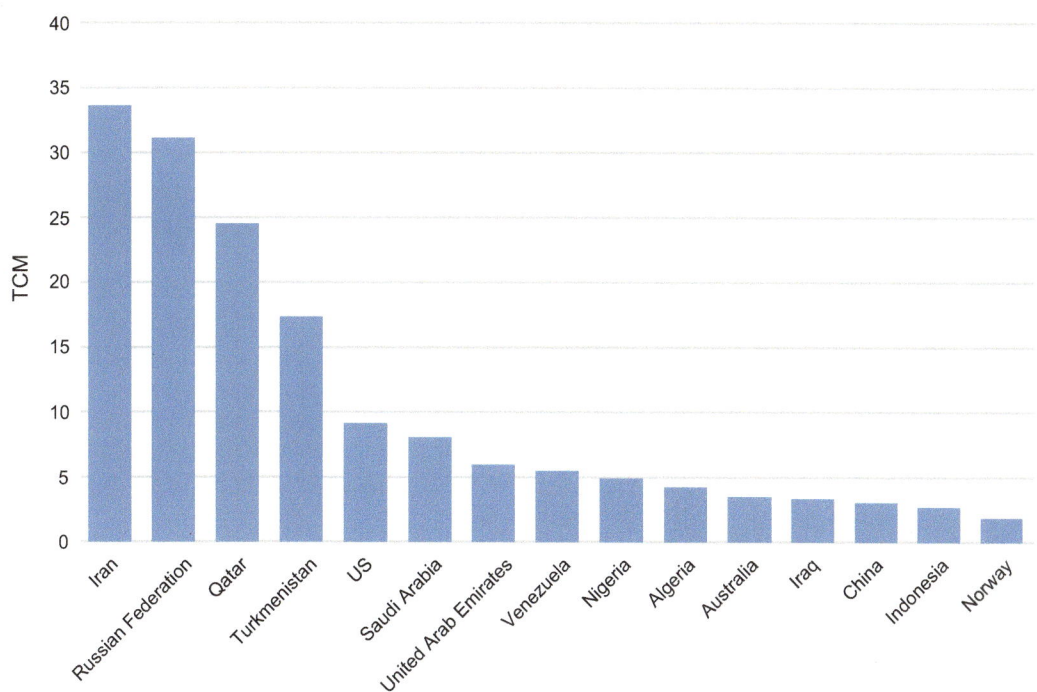

Fig. 17.5 The 15 countries with the largest proven recoverable natural gas reserves. *Source* BP Statistics 2014

consumed a total of 2 trillion m³, accounting for 61% of total global consumption. These largest consumers were: the United States (22%), Russia (12%), Iran (5%), China (5%), Japan (4%), Canada (3%), Saudi Arabia (3%), Germany (3%), Mexico (3%) and the United Kingdom (2%).

The United States is the largest natural gas consumer, and in 2013 consumed a total of 737 billion m³. The largest natural gas-consuming industry was power generation, accounting for 33% (this proportion was 25% 10 years ago). The second largest user was industry, accounting for 31% of total volume in 2013 (its share fell below that of power generation in 2008). Shares taken by other uses in 2013 included residential (21%) and commercial (14%). In the next 10 years, with the completion of various LNG liquefaction projects, primarily around the Gulf of Mexico, US natural gas production and the country's LNG exports to Europe and Asia are expected to further increase.

In the United Kingdom, natural gas consumption in the last decade has been more volatile after a sustained period of growth, led largely by increased residential and power generation use of gas. From 1990 to 2003, the share of natural gas used for power generation in the United Kingdom rose from less than 1 to 38%, but this had dropped to 27% by 2013.

Prior to the 2011 Fukushima reactor incident, Japan was already the world's largest importer of LNG. The Japanese government's response to the incident, especially its closure of all nuclear power plants, pushed natural gas consumption in the country even higher. When nuclear power, which had previously accounted for 31% of power generation, was entirely halted, Japan's LNG imports rose from 69 million tonnes to 82 million tonnes. Even though Japan's nuclear power plants are expected to resume operations, continued domestic opposition and new standards make the timing for this uncertain.

Consumption in Russia, the world's second-largest natural gas market, has been volatile in recent years. Use dropped sharply after the 2008 global financial crisis, rebounded strongly, and then softened again going into 2012 and 2013, echoing the country's lower economic growth. In Europe, Germany's 2012 and 2013 natural gas consumption continued to grow despite a reduction on average in EU member state consumption. In the Middle East, Iran, another large consumer, has seen natural gas consumption growth moderate in recent years. Finally, China has seen a brisk rapid growth in natural gas consumption since 2000, placing it now among the world's 10 largest natural gas-consuming countries.

While India is not yet among the world's top consumers of natural gas, policy initiatives—especially in the power generation sector—have led to remarkable increases in use. Domestic consumption nearly doubled from 1995 to 2005, reaching about 37 billion m^3 or 1.3% of total global consumption. However, price controls, a drop in domestic production and issues surrounding LNG infrastructure access have dampened the natural gas market in recent years. As these problems are resolved, India's natural gas consumption will likely resume growth.

2. Centres of production

World natural gas production was 3.4 trillion m^3 in 2013, an increase of 1.1% from 2012. Between 1970 and 2013, global natural gas production grew nearly 3.5 times. The largest natural gas-producing countries and their proportions of total global production volume were as follows: United States (20%), Russia (18%), Iran (5%), Qatar (5%), Canada (5%), China (5%), Norway (3%), Saudi Arabia (3%), Algeria (3%), India (2%) and Malaysia (2%). However, from the perspective of export totals for natural gas pipeline and LNG, Russia was the largest exporter, accounting for 22% of the global total, primarily in pipeline natural gas. Ranking second was Qatar (12%), primarily LNG, followed by Norway (10%), primarily in pipeline natural gas.

The United States is already the world's largest natural gas producer, and production is set to increase significantly as several LNG projects

come online. The success of US shale natural gas production will soon turn the country from a net importer to a net exporter. LNG exports are expected to start in 2016 as Cheneire's Sabine Pass facility begins operation. By 2020, the United States is expected to have 40 million tonnes in annual liquefaction capacity. There are yet more LNG exports projects in planning, but not all of them are expected to be completed.

The world's second largest natural gas producer, Russia, has been operating a large-scale pipeline gas and LNG export business for a long time. In 2013, Russia produced 605 billion m^3 of natural gas, and pipeline exports to Europe were about 211 billion m^3, with the main destinations being Germany (40 billion m^3), Turkey (26 billion m^3) and Italy (25 billion m^3). Russia also exported 14 million tonnes of LNG, primarily to Japan. It is expected that as the natural gas pipeline between China and Russia enters operation, and with new LNG facility operation commencement, Russia will see further growth prior to 2020 in natural gas production volumes and export volumes.

A significant amount of pipeline natural gas is also exported from Norway, The Netherlands and Algeria to European markets. Dutch gas production is underpinned by production from the Groningen Field, northwest Europe's largest natural gas field and one of the largest in the world. The Netherlands exported 53 billion m^3 by pipeline in 2013, but in 2014 the government lowered the production limit following an earthquake in the region linked to natural gas extraction.

Qatar was the world's largest LNG exporter in 2013, exporting over 78 million tonnes of LNG. The world's largest non-associated gas field straddles the Qatar–Iran border, with some estimates placing the recoverable reserves at 900 trillion cubic feet in the North Field (the Qatari portion) and 500 trillion cubic feet in the South Pars (the Iranian portion). Since production from the North Field began in the early 1990s, Qatari natural gas production has increased from 6.3 billion m^3 in 1991 to 158 billion m^3 in 2013, with most destined for export. Qatar initially developed its LNG export capacity to supply significant volumes to each major market—Asia, Europe and North America—but the North American shale natural gas boom and strong LNG demand in Asia resulted in almost 70% of its LNG going to Asia in 2013. Iranian gas production also increased significantly over the same period, but all of its natural gas is consumed domestically.

Australia is expected to become an important natural gas producer and exporter soon. Although production volume in Australia was 1% lower in 2012 than in 2012, a series of LNG liquefaction projects will become operational over the next few years that will bring Australia's production capacity up to 88 million tonnes, overtaking Qatar to become the world's largest LNG exporter. However, some of these facilities will be the first in the world to use coalbed methane to produce LNG, so some uncertainty remains about these figures.

Even though 2013 production volume was not significant, Papua New Guinea began natural gas production in 2014 as the PNG LNG project came online. This project's annual production capacity is 6.9 million tonnes of LNG, and production capacity expansion plans are being assessed.

Canada is already a major producer of natural gas, and even although its exports are currently limited to pipelines to the United States, the country plans to export more widely. Canada's west coast has a significant number of LNG projects in planning, hoping to take advantage of the region's abundant unconventional natural gas and its access to large LNG importing countries like Japan, South Korea and China.

Based on their abundant natural gas resources, Tanzania and Mozambique are two more countries that could potentially become major LNG exporters. For them, natural gas is a brand new industry, and it is likely to take longer to begin operations. For these—or indeed any—new natural gas projects to attract the necessary financing, they will need long-term commitments from high-quality buyers at prices sufficient to provide adequate returns to both the host country and the developers.

17.3.2 Regional Imbalances

Globally, supply and demand is unevenly distributed between regions. The main producing regions are countries of the former Soviet Union, North America and the Middle East and North Africa (Fig. 17.6). Countries of the former Soviet Union and the Middle East and North Africa are the world's largest exporters of natural gas. The Asian OECD countries, China and Europe are the main importers. In other regions, production and consumption are basically balanced, with the exception of the rest of Africa, which is a relatively small, but growing, exporter.

Because global natural gas consumption has grown, and regional supply and demand are unbalanced, import and export trade volumes are also growing. From 1990 to 2013, global natural gas consumption volumes grew by approximately 70%, but there were relatively large disparities in each region's supply and demand growth rates (Fig. 17.7). Nations exporting natural gas expanded exports, and importing nations likewise increased imports. From 1990 to 2013, approximately 20% of the newly added consumer demand was realised through inter-regional import and export trade.

In recent years, production increases have been supported by large discoveries of conventional, and more recently unconventional, natural gas. Since the late 1980s, there have also been major discoveries of conventional natural gas, especially in the Middle East and North Africa, causing these regions rapidly to become major producers of natural gas. At the same time, production volumes even rose in regions where there had been no major increases in reserves, indicating a preference for domestic sources over imports to meet demand.

While reserves of unconventional natural gas reserves, including tight natural gas, shale natural gas and coalbed methane, are largely unconfirmed, they are potentially huge. Their exploitation in the United States since 2002 has fundamentally changed the country's natural gas market, although unconventional production remains minimal in all other regions. However, the potential for unconventional natural gas to

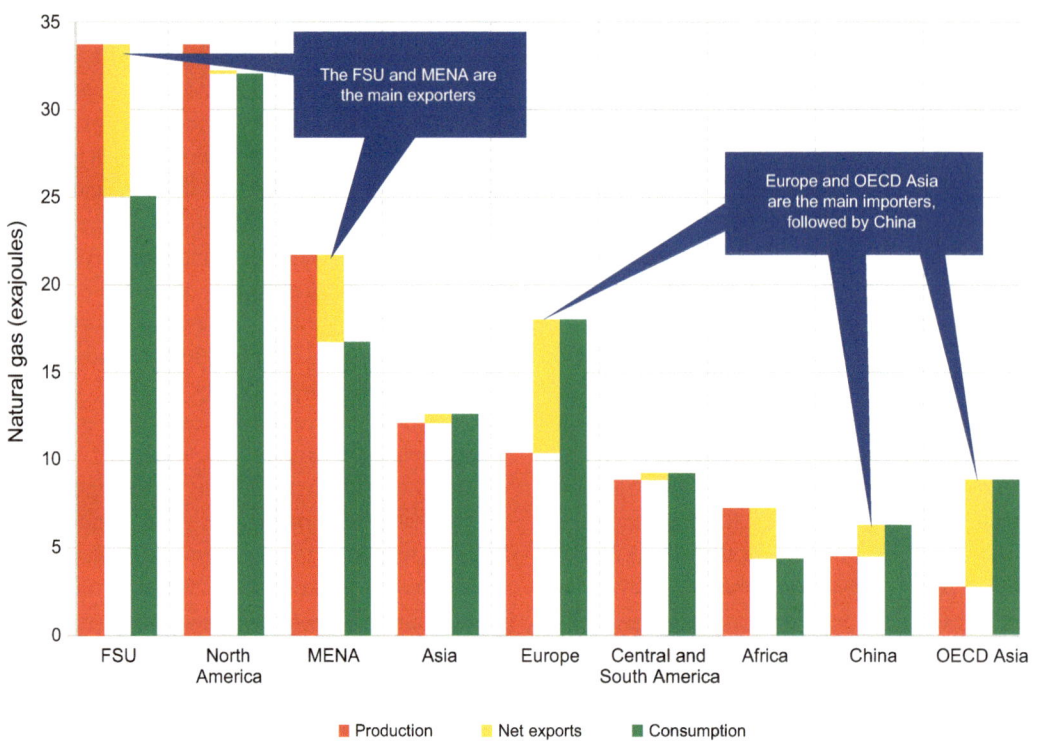

Fig. 17.6 The major natural gas producing and consuming regions. *Note* Data for 2013. *Source* Vivid Economics, based on IEA data

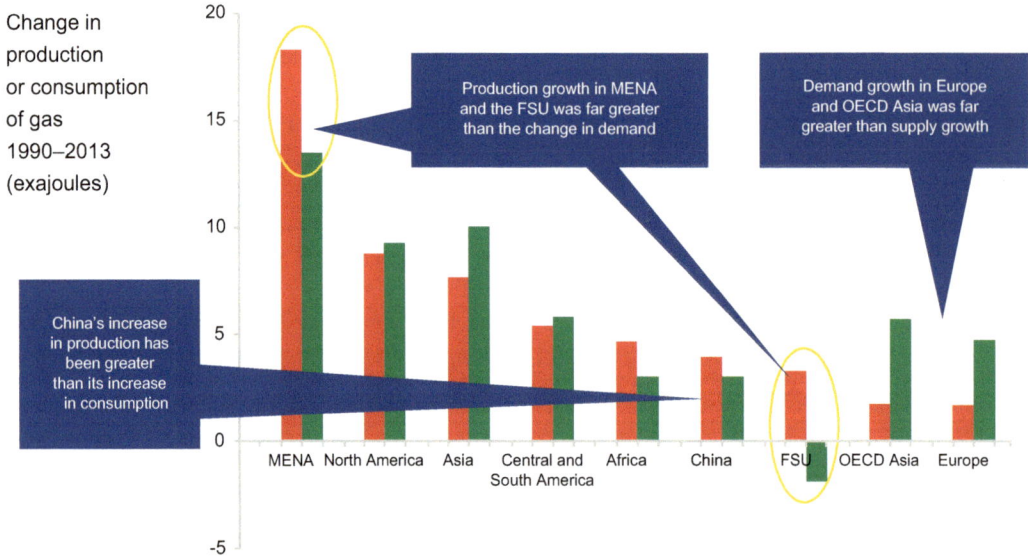

Fig. 17.7 Natural gas supply and demand growth rates of various regions, 1990–2013. *Source* Vivid Economics, based on IEA data

disrupt other regions is large. Unconventional natural gas reserves are mostly distributed in countries without major amounts of conventional natural gas (Fig. 17.8), and if these reserves could be commercially exploited, then the inter-regional imbalance between natural gas producers and consumers would probably be reduced. Such a change would reduce demand for natural gas from countries of the former Soviet Union, the Middle East and North Africa.

As a result of the developments discussed above, the world is increasingly becoming divided into regions that supply natural gas and regions that consume it (Fig. 17.9). The natural result of this situation has been an increase in inter-regional trade.

17.3.3 Inter-regional Natural Gas Trade

With the regional imbalances in natural gas supply and demand, inter-regional trade,

especially LNG trade, has been increasing. From 1993 to 2013, natural gas trade over pipeline almost doubled, while LNG trade—albeit from a lower starting point—quadrupled (Fig. 17.10). Natural gas trade over pipeline, however, still dominates the market, accounting for about two thirds of total trade.

1. **Pipeline gas trading**

Since 1993, the volume of natural gas traded over inter-regional pipelines has almost doubled, though the network has not significantly increased its connectivity. In 1993, inter-regional pipeline natural gas trade volume was 470 trillion cubic feet, and 86% of this was exported by the former Soviet Union, with 94% imported to Europe. At the time, the natural gas pipeline network was limited, with most natural gas flowing from the former Soviet Union, the Middle East and North Africa (Fig. 17.11). By 2013, 8.5 trillion cubic feet was being traded annually over pipeline, with the share exported

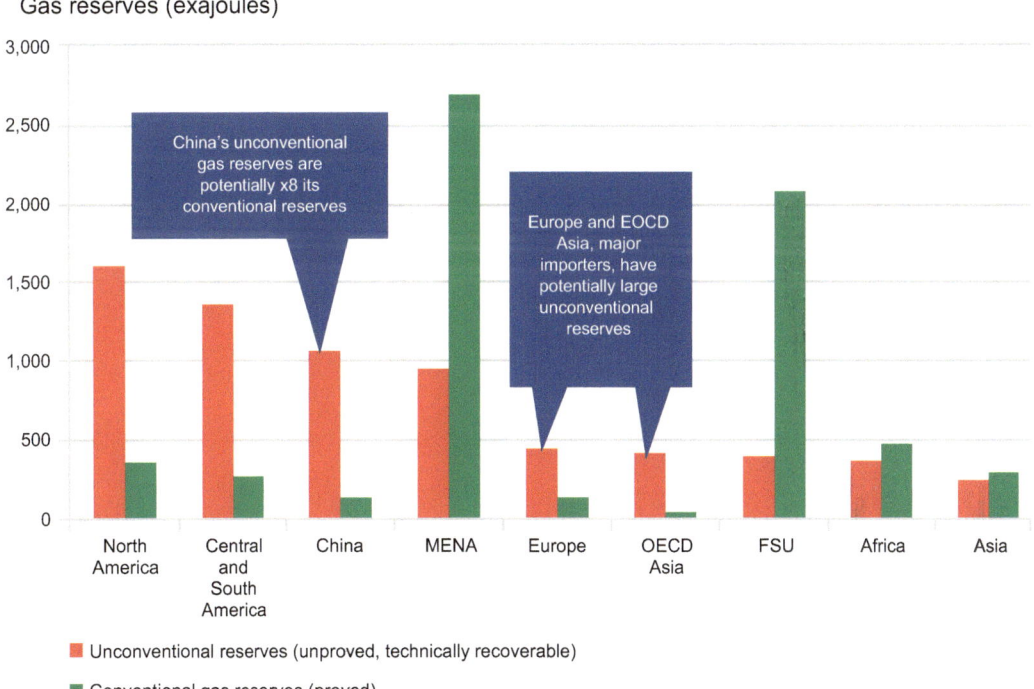

Gas reserves (exajoules)

Fig. 17.8 Unconventional natural gas reserves by region. *Note* Data is for 2013; unconventional reserves include tight natural gas, shale natural gas and coalbed methane. *Source* Vivid Economics, based on EIA and IEA data

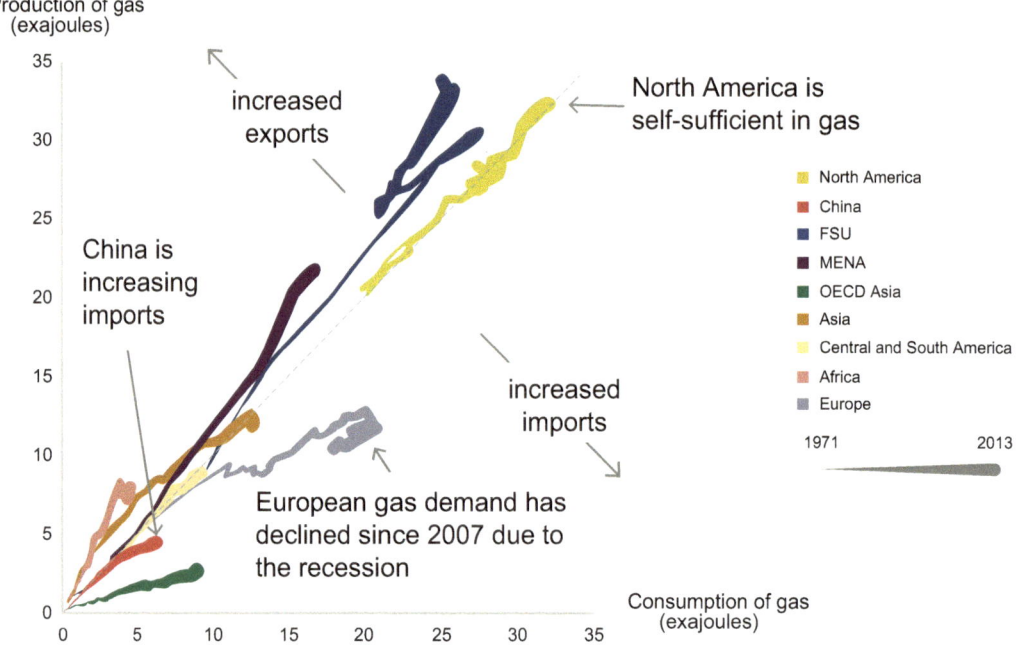

Production of gas (exajoules)

increased exports

North America is self-sufficient in gas

China is increasing imports

increased imports

European gas demand has declined since 2007 due to the recession

Consumption of gas (exajoules)

North America
China
FSU
MENA
OECD Asia
Asia
Central and South America
Africa
Europe

1971   2013

Fig. 17.9 The increasing polarisation of regions into suppliers and consumers of natural gas. *Note* Data is on a gross calorific basis; each country's coloured trail along the matrix gets thicker over time; the dotted midline represents equilibrium between production and consumption, with greater production above the midline and greater consumption below it. *Source* Vivid Economics, based on IEA data

Fig. 17.10 Types of global natural gas trade. *Note* Data is a global aggregate of international trade. *Source* Vivid Economics, based on IEA data

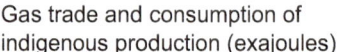

Gas trade and consumption of indigenous production (exajoules)

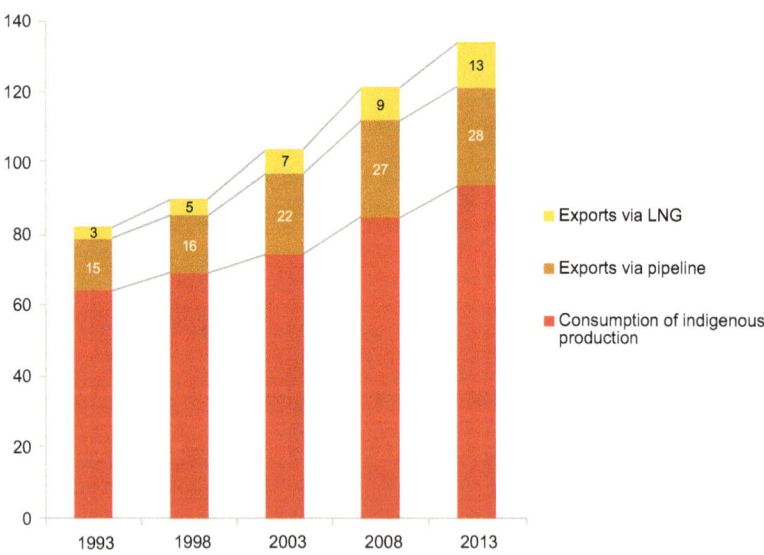

Exports via LNG

Exports via pipeline

Consumption of indigenous production

by countries of the former Soviet Union falling to 75%, and European imports to 65%. In the same year, Chinese imports accounted for 11% of pipeline natural gas, exclusively from countries of the former Soviet Union. Even though there had been some small changes between 1993 and

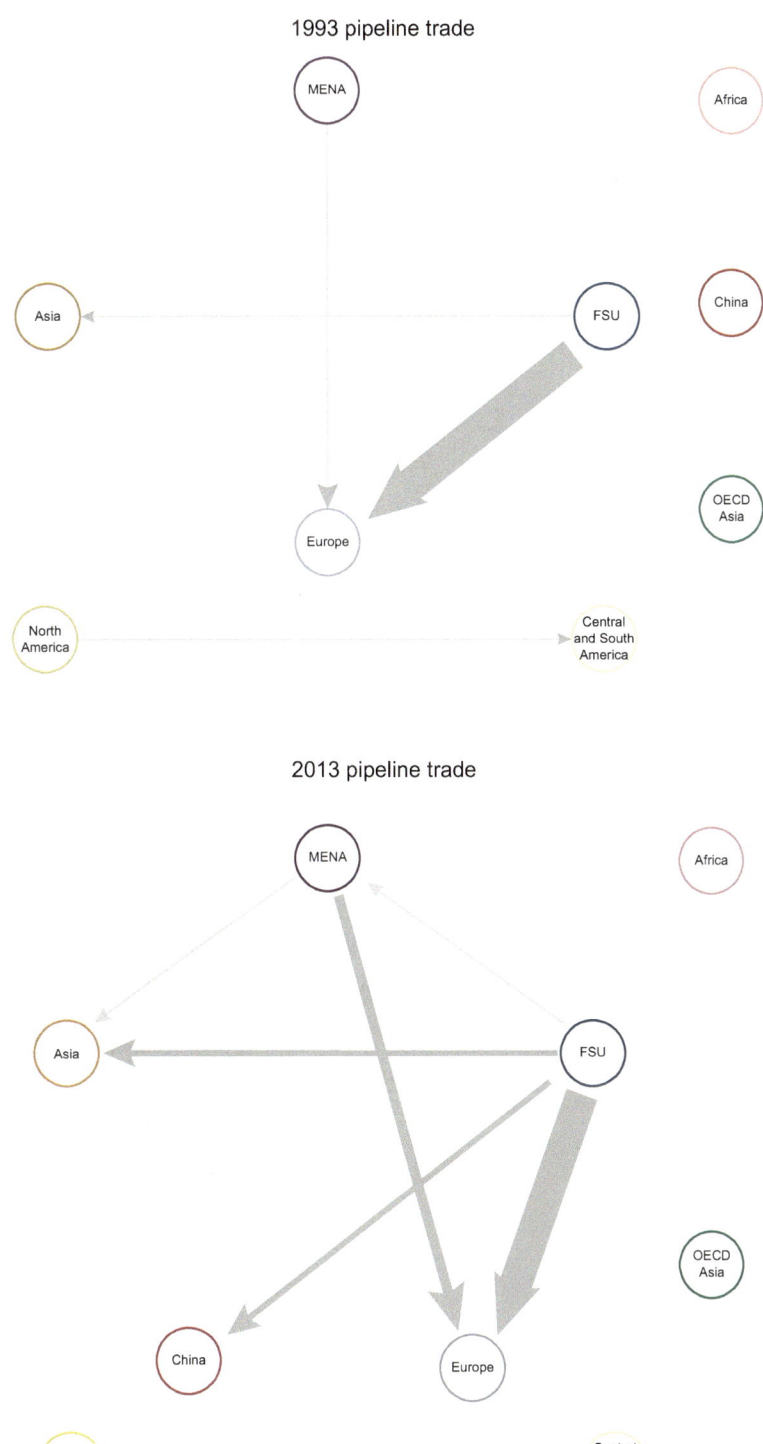

Fig. 17.11 Global inter-regional pipeline natural gas trade. *Note* The thickness of the arrow represents the percentage of global pipeline trade flowing between two regions. *Source* Vivid Economics, based on IEA data

2013, the inter-regional natural gas pipeline network remained relatively limited, especially in light of the global pattern of reserves, production and demand.

2. LNG trading

The volume of natural gas traded through LNG has more than tripled since 1993, and the network has significantly increased its connectivity. In 1993, inter-regional LNG trade volume reached 2.7 trillion cubic feet, of which 64% was exported from Asia, and 70% was imported by Asian OECD member nations and 25% by Europe. Though networks facilities were limited in 1993, primarily concentrated on trade between Asia and Japan and between Africa and Europe, by 2013 they had expanded significantly (Fig. 17.12). In 2013, inter-regional LNG trade volume had reached 9.1 trillion cubic feet. Of this, 57% was exported by the Middle East and North Africa, while the major importers were Asian OECD member nations (56%), Europe (15%) and China (9%). LNG trade in Asia,

excluding China and OECD countries, has become more complex, handling imports and exports, and accounting for just 10% of net inter-regional LNG trade volume, compared to 64% in 1993.

The degree of connectivity in LNG markets has begun to form a global natural gas market. By 2013, the network of LNG trade had become much denser, with many regions connected to each other. The flexibility of LNG delivery means that LNG can be used to exploit arbitrage opportunities across a range of regions, connecting markets together. However, LNG trade is currently dominated by a small number of participants in the Middle East and North Africa, and trade volume on some routes is still low. In addition, there are legal restrictions on exports, such as in the United States, and limitations due to infrastructure capacity. Furthermore, LNG developments remain capital-intensive and risky, and as a result interregional price differences need to be wide to motivate trade. Faced with such problems, even though regional markets are far more connected now than in the recent past,

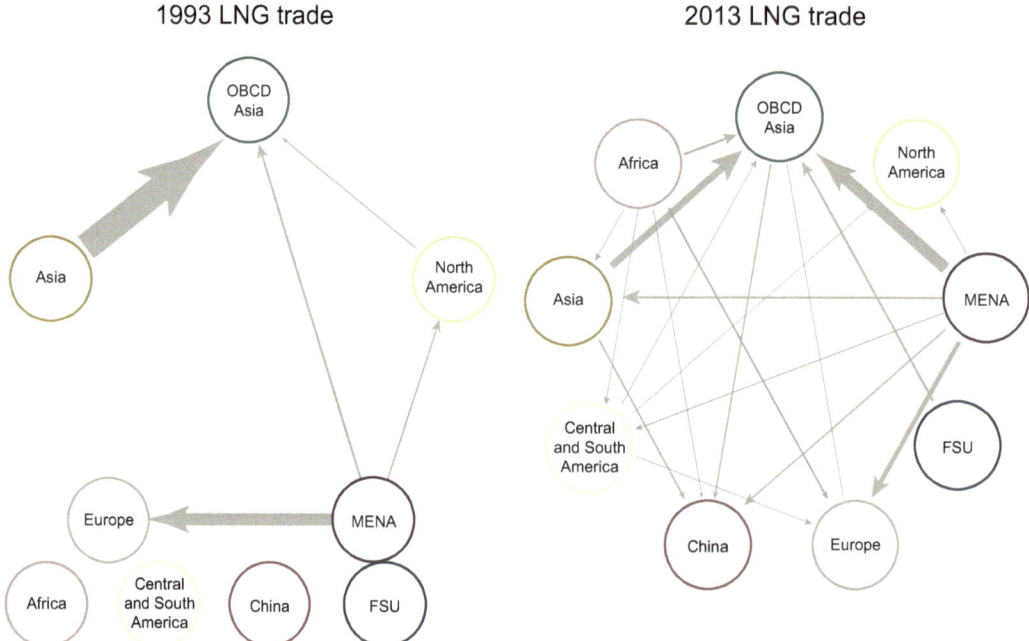

Fig. 17.12 Global LNG trade *Note* The thickness of the arrow represents the percentage of global pipeline trade flowing between two regions *Source* Vivid Economics, based on IEA data

the degree of connectivity is still insufficient to support the formation of a global market.

3. **The problems facing LNG trading**

LNG capacity has increased rapidly, and the expansion is expected to continue. In 2013, LNG liquefaction capacity, which is required for export, was 2.25 times greater than in 2003, while regasification capacity, an import requirement, was twice that in 2003. By 2023, liquefaction capacity is expected to grow almost threefold, while regasification capacity is expected to nearly double. This will bring global liquefaction capacity to almost 40 trillion cubic feet/year by 2023, and regasification capacity to almost 60 trillion cubic feet/year. (Regasification capacity is greater than liquefaction capacity because regasification capacity is distributed across more locations and countries build import capacity to meet peak demand, whereas the flow of exports tends to be smoother, requiring a lower capacity for the same volume.)

However, LNG costs are volatile, and may not fall significantly in the future. In general, the cost of liquefaction plants has increased by 50% over the last decade, and some projects, primarily in Australia, have cost at least twice normal levels. Costs have risen because of higher commodity prices, such as steel, and, in Australia in particular, higher labour costs. This pattern of high costs could persist. Even if commodity prices fall, LNG is likely to remain an expensive process, because the scope for technological breakthroughs is limited, and the large amount of expensive capacity recently added will lock in higher costs. Furthermore, LNG is only competitive with overland pipelines over long distances, such as from the Middle East to Japan (Fig. 17.13).

These issues suggest that LNG is economically limited. This fact could constrain investment, or may require long-term contracts to manage risk. Moreover, it is likely to mean that LNG remains an expensive fuel, serving as a marginal source of supply in markets where supply fails to keep up with demand.

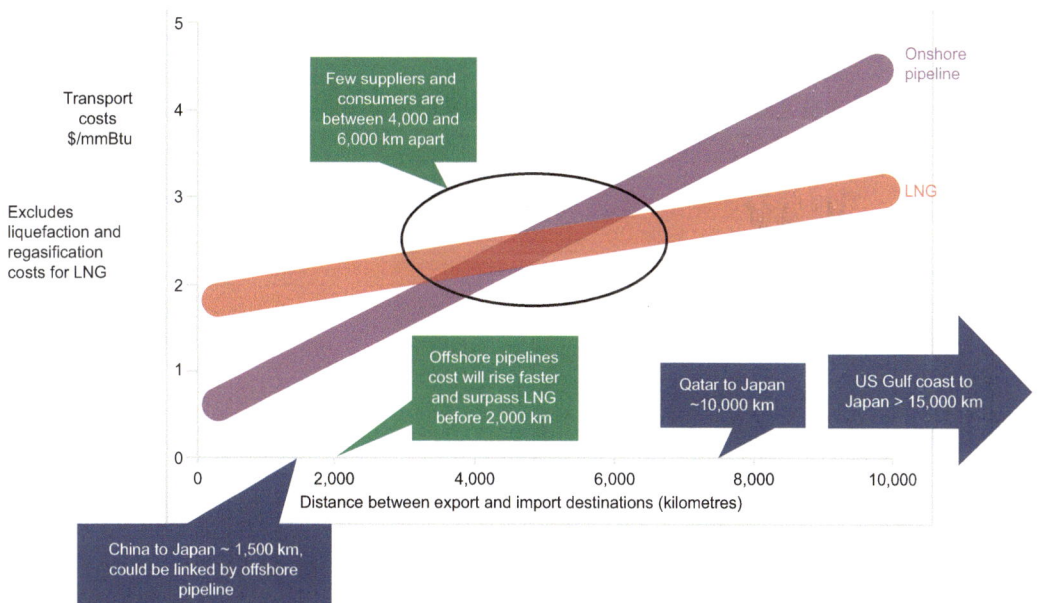

Fig. 17.13 Comparison of competitiveness between pipeline natural gas and LNG. *Source* Vivid Economics based on SBI Energy Institute (2014)

17.3.4 Unconventional Natural Gas Resources

Confirmed unconventional natural gas reserves account for approximately three-quarters of global technically minable reserves. In a 2012 report, the IEA estimated that global confirmed technically minable natural gas reserves amounted to 420 trillion m^3, with 331 trillion m^3 recoverable with unconventional technology. The unconventional resources were 208 trillion m^3 in shale natural gas, 76 trillion m^3 in tight natural gas and 47 trillion m^3 of coalbed methane. In a separate report, the EIA assessed the world's recoverable reserves of shale natural gas at 7299 trillion cubic feet, of which China's share was the largest, followed by Argentina, Algeria, the United States and Canada (Table 17.2).

Unconventional natural gas has transformed the natural gas market in North America. In the United States alone, shale natural gas production is expected grow from 9.7 trillion cubic feet in 2012 to 19.8 trillion cubic feet in 2040, bringing its share of total US natural gas supply from 40 to 53%. This increase in natural gas supply is expected to give US manufacturers an added advantage over foreign competitors and has

already resulted in a significant number of projects that seek to liquefy natural gas for export.

North America's success in exploiting unconventional natural gas has inspired other countries. According to IEA estimates, global unconventional natural gas total production volume will reach 928 billion m^3 in 2020, including 454 billion m^3 of shale natural gas, 148 billion m^3 of coalbed methane and 294 billion m^3 of tight natural gas.

Despite the optimism, there nonetheless remain significant uncertainties about how quickly unconventional resources can be brought online outside the US, especially in countries where little or no production has been taken place. Although China is estimated to have unconventional resources totalling about 32 trillion m^3, the government recently reduced its near-term outlook for reaching these reserves. Problems cited included that the resources were spread across more than 500 basins and the geography was difficult, as well as costs, inadequate infrastructure, water disposal concerns, lack of channels for introduction of international companies and a lack of innovation as a result of the small number of participating companies. In another example, Argentina, with about 23 trillion m^3 in unconventional resources, has

Table 17.2 Top 10 countries with technically minable reserves of shale natural gas

Ranking	Country	Shale natural gas reserves
1	China	1115
2	Argentina	802
3	Algeria	707
4	United States	665
5	Canada	573
6	Mexico	545
7	Australia	437
8	South Africa	390
9	Russia	285
10	Brazil	245
	Global total	7299

Units trillion cubic feet
Source US Department of Energy

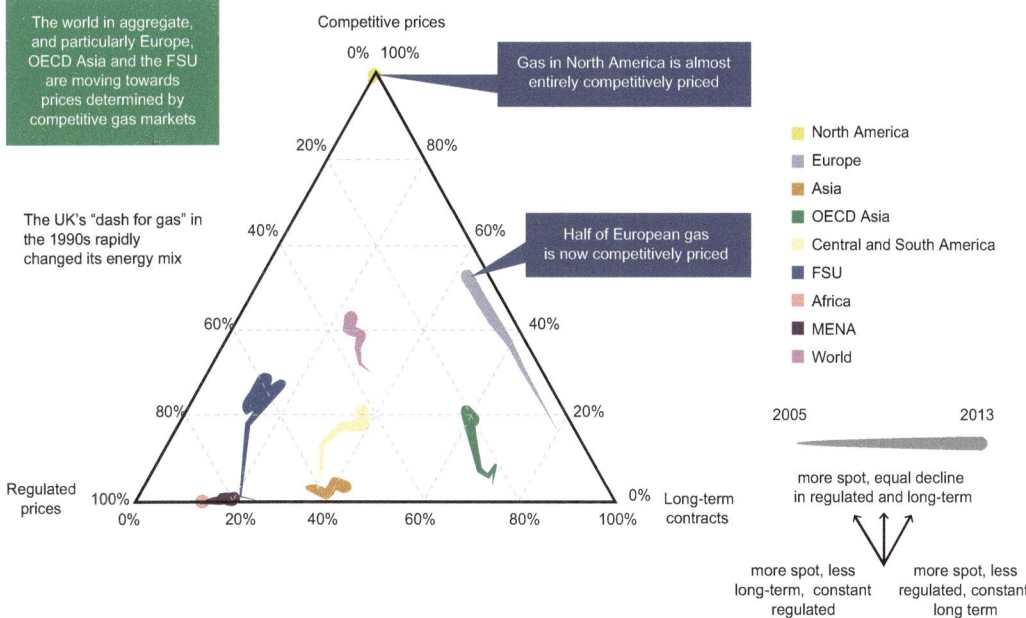

Fig. 17.14 Changes in natural gas contract types. *Note* Each country's coloured trail along the matrix gets thicker over time; movement toward the apex illustrates increased share of natural gas at the expense of oil (*lower left*) or coal (*lower right*). *Source* Vivid Economics, based on IGU data

solid pre-conditions in place, but is constrained by funding and non-technical risks.

17.4   Pricing

Natural gas prices and price-setting mechanisms have evolved over time. While long-term contacts linked to oil prices were once dominant, now price setting is taking different forms across markets, and prices vary by region. A 2014 report from the International Gas Union (IGU) found that 43% of global wholesale volume was based on competitive natural gas pricing, also known as gas-on-gas pricing, and was not indexed to oil prices,[2] while 19% was indexed to oil. The term of contracts is also becoming more diverse, compared to oil-indexed contracts, which tended to be long-term agreements. Prices are being set more often based on

competition among natural gas suppliers, on hubs or on spot markets (Fig. 17.14).

In China, where energy price controls are relatively stringent, the mechanisms of natural gas pricing differ greatly from the way they function internationally, with different pricing methods depending on gas source and on use. However, China's natural gas pricing is undergoing reform, and is gradually moving toward national natural gas pricing rules.

Natural gas hubs are an important factor in natural gas pricing mechanisms, given that their core function is to provide a physical connection within the natural gas system and to facilitate competitive pricing. Natural gas hubs break the link between the price of natural gas and oil prices. The competitive pricing that the formation of hubs allows becomes a kind of substitute plan for controlled prices linked to oil prices. In addition, natural gas hubs also form an important component of natural gas downstream markets. See the special discussion in Chap.19 for the principles, effects and practical cases of natural gas hubs.

---

[2]Contracts connected to oil price can also be competitive, but gas-on-gas pricing contracts are normally more competitive, and are more reactive to changes in the market fundamentals.

Due to the great size and rapid development trends of China's natural gas market, as well as previous experience of large-scale international trade in other commodities, policy adjustments for China's natural gas market are likely to have a major impact on global natural gas markets. This study established a model to analyse China's natural gas demand growth in various scenarios and its effects on global natural gas markets, including price, energy mix, and other aspects.

17.4.1   Current Pricing Regulations

The development of a global natural gas market is limited by geography, with most international trade being over natural gas pipelines or by LNG shipping. Geographical limitations and high shipping costs—the construction of international long-distance pipelines, as well as the costs of shipping and storing LNG—restrict trade between different regions, causing the natural gas market to develop distinct regional characteristics, particularly regarding how prices are established (Table 17.3).

Price levels across the regions have also varied significantly, reflecting the changes to the supply and demand for each market (Fig. 17.15). In North America, the influential Henry Hub price generally reflects the supply and demand dynamics in the United States, for example by reflecting seasonal variations, major incidents (such as hurricanes Katrina or Rita) and

Table 17.3   Regional market pricing characteristics

Region	Market description	Method of price formation
North America	Natural gas market with competition-based natural gas pricing. Interconnected infrastructure linking storage, supply and demand hubs	Multiple natural gas indices, with Henry Hub the dominant openly-traded LNG index. Natural gas index reflects North American natural gas supply and demand
Europe	Multiple natural gas markets with varying degrees of competition-based pricing. Markets operate and regulations are developed under a framework established under the European Union, but strong national interests remain. Infrastructure is primarily interconnected, with some bottlenecks	Long-term contracts connected to oil price or oil product prices are being increasingly challenged by competitive pricing, for example from NBP in the UK and Title Transfer Facility (TTF) in the Netherlands
Japan, South Korea, and Taiwan	Markets primarily based on national monopolies and supply in the region primarily under long-term contracts, with some active spot LNG buying to manage supply and demand or some portfolio optimising. Customs data for LNG imports publicly available and can be used to determine the average import price	Strong oil indexation for long-term contracts to the Japan Crude Cocktail (JCC), which is generally defined within individual contracts and lags current oil prices because they are typically based on recent average prices of crude imports into Japan. These are prevalent in Asia and may include price review clauses. Spot cargoes are primarily based on a fixed price, typically negotiated bilaterally or based on tenders. Surveys by various reporting agencies seek to capture this through the Japan Korea Marker (JKM). JKM is not a price for spot cargoes, but is currently the best estimate in the industry
China	Market dominated by state-owned enterprises. Supply based on a mix of domestic production, pipeline imports from central Asia, Myanmar and, soon, Russia and LNG imports. Infrastructure development continuing. Market reforms ongoing	Natural gas market pricing reform ongoing. Natural gas supplied under a mix of cost-plus for domestically produced natural gas and oil-indexed pricing, primarily for imports. LNG price formation similar to Japan, South Korea and Taiwan. Natural gas sold at regulated prices set by a National Development and Reform Commission formula linked to LPG and fuel oil

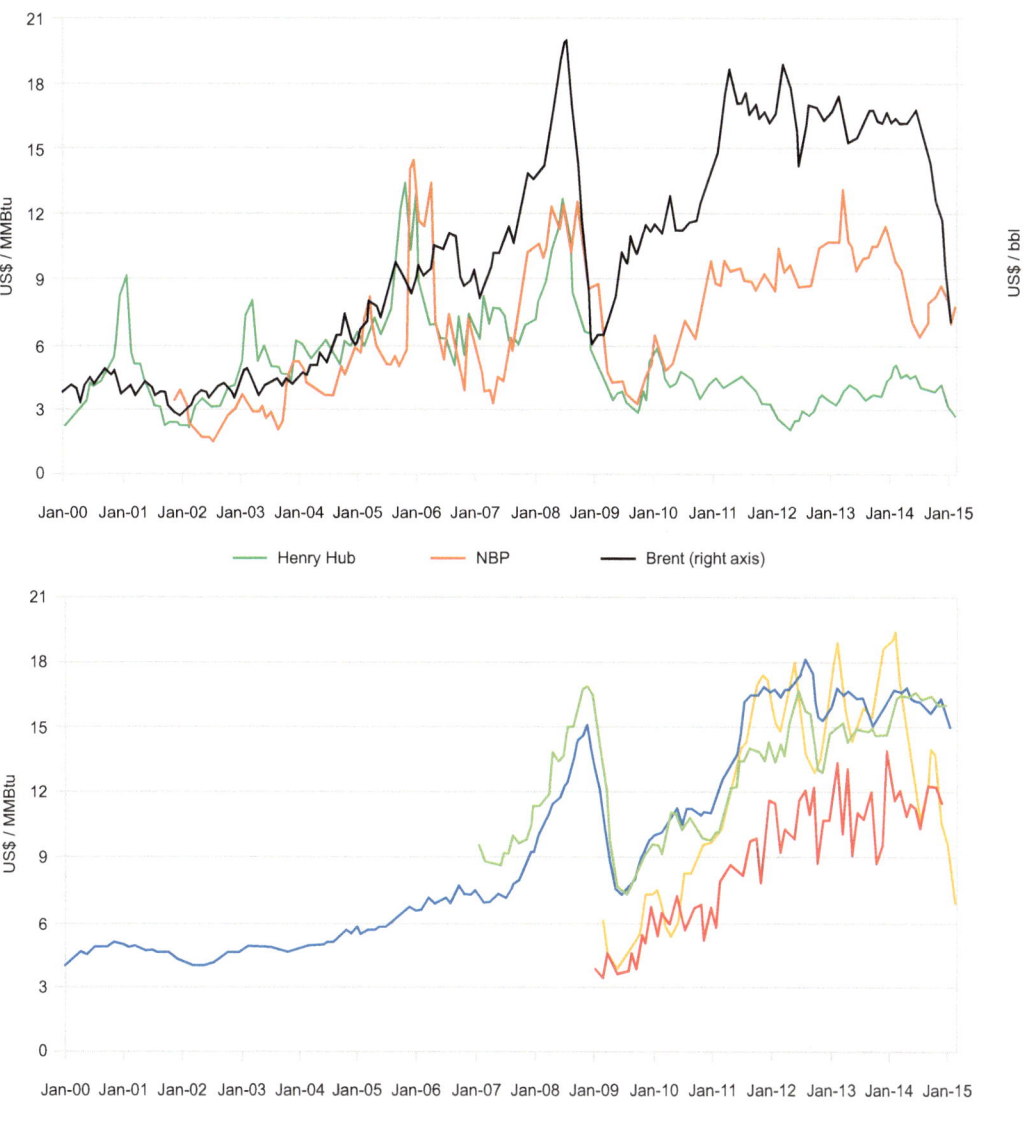

Fig. 17.15 Key price indices and Japan's average import price. *Source* Platts, Energy Intelligence, Intercontinental Exchange, and Heren

longer-term trends (such as the shale natural gas revolution). Indeed, there were periods when US natural gas prices were higher than the average LNG import price in Japan, for instance when the market expected the United States to need substantial LNG imports.

In the past, the Henry Hub price was widely seen as a benchmark for the US market, and many natural gas liquefaction projects around the world were begun targeting exports to the United States based on these prices, relying on the Henry Hub price for their export plan pricing, along with the belief that the United States would be a long-term LNG importer. A significant number of regasification terminals in the United States were also proposed (Fig. 17.16).

The United States was expected to be a long-term LNG importer, and developers were

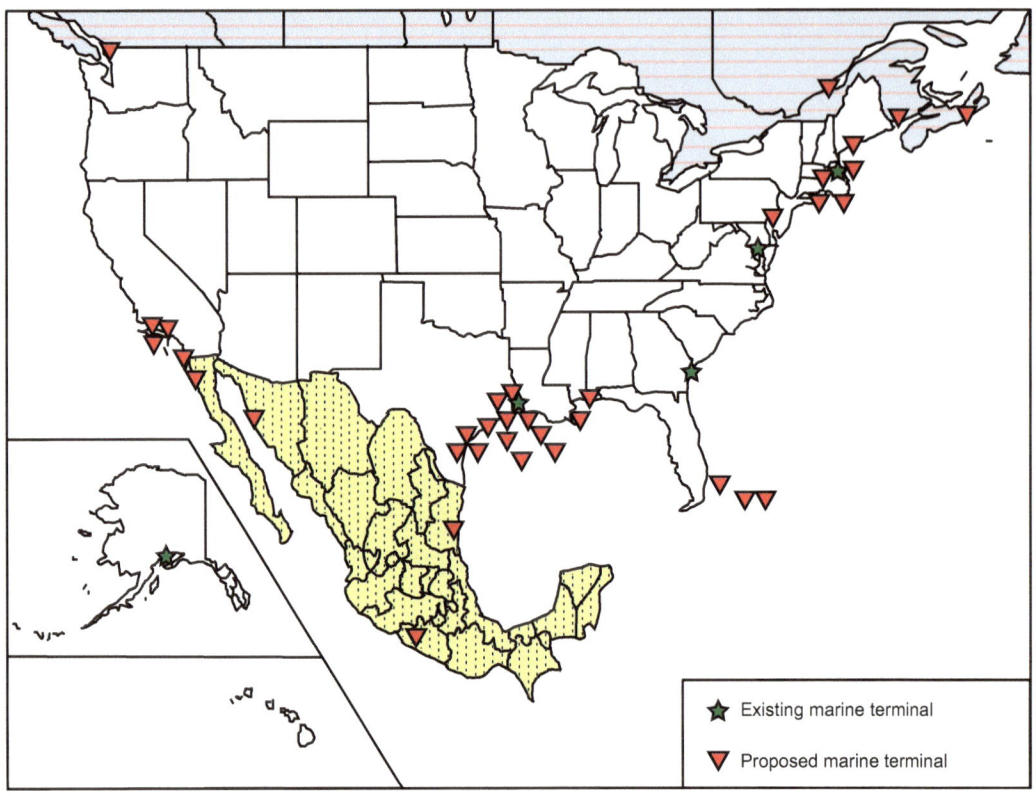

Fig. 17.16 Current and proposed LNG regasification terminals in North America, 2004. *Information source* EIA

willing to price their export plans based on the Henry Hub largely because it was widely accepted as reflecting US supply and demand fundamentals. A significant number of regasification terminals were also proposed in the United States.

In markets in which pricing is predominantly linked to oil indices, prices also respond to changes in supply and demand, but not as efficiently as in competition-based pricing. The mechanism is also complicated by the supply and demand of oil.

One example of this follows new LNG sales and purchase agreements in Asia-Pacific, where prices gradually diverged from oil until around 2005 and slowly retracted (Fig. 17.17). In the early 2000s, project developers in the region had to enter markets relatively new to LNG imports and had the alternative of exporting to the United States. Costs for building LNG projects were lower than current levels, and their oil price outlooks were anchored around historical levels, which had generally been below $60 a barrel for 15 years. After 2005, higher project costs and greater demand for natural gas led to higher prices, in some cases reaching parity with crude oil. The 2011 nuclear power plant accident in Fukushima, Japan put additional upward pressure on natural gas prices in Asia. More recently, however, natural gas prices have begun to soften, partly a result of lower oil prices, weaker global economies, additional capacity (both completed and proposed) and new projects, such as US LNG exports to Asia.

Looking ahead, natural gas prices, especially LNG exports to Asia-Pacific markets, could become particularly volatile with the onset of LNG exports for the United States, which are primarily indexed to Henry Hub prices. Companies are hoping to capitalise on the US natural gas boom by liquefying the fuel and trading it on international markets. Many projects have been

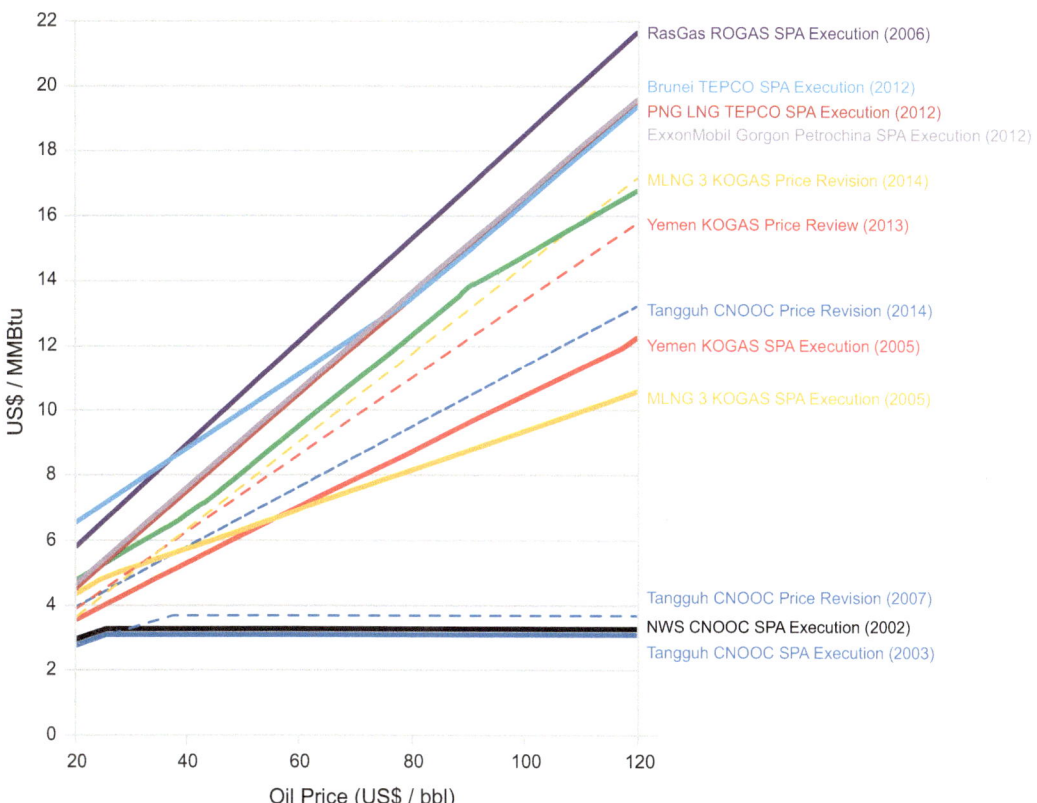

Fig. 17.17 Progressive divergence by different degrees from oil prices of Asian oil-linked natural gas prices. *Source* Wood Mackenzie and public information

announced, but not all will be completed (Fig. 17.18).

In general, pricing of American LNG exports has two components: one linked to Henry Hub prices representing the feed gas and fuel costs, and an annual fixed component representing investment in the liquefaction facility (Fig. 17.19). Based on currently available purchase and sale agreement terms, the "Henry Hub" component is typically 115% of the Henry Hub price, while the second component is based on a fixed dollar amount, with about 15% of this linked to US inflation rates.

Buyers gain supply diversity by procuring US LNG exports, but they also acquire some risk. Coupled with contracted destination diversity, which allows shipments to be rerouted to the most favourable markets, US LNG exports can contribute to a beneficial diversified energy

supply network. However, US LNG supplies can be less competitive than oil-indexed supplies, depending, among other factors, on movements of global oil prices and Henry Hub natural gas prices (Fig. 17.20).

Compared to Asia, where natural gas contract prices are linked to oil, US natural gas exports have a different risk profile. Many US LNG export sales and purchase agreements include a tolling agreement section, which means that the buyer is responsible for procuring the natural gas to be liquefied, and lower plant utilisation rates would lead to higher unit prices because the buyer continues to pay the fixed component, except in extreme cases. The buyer is responsible for purchasing the natural gas to be liquefied, with the unit price rising with lower liquefaction facility usage rates, because the buyer continuously pays a fixed fee portion, except in

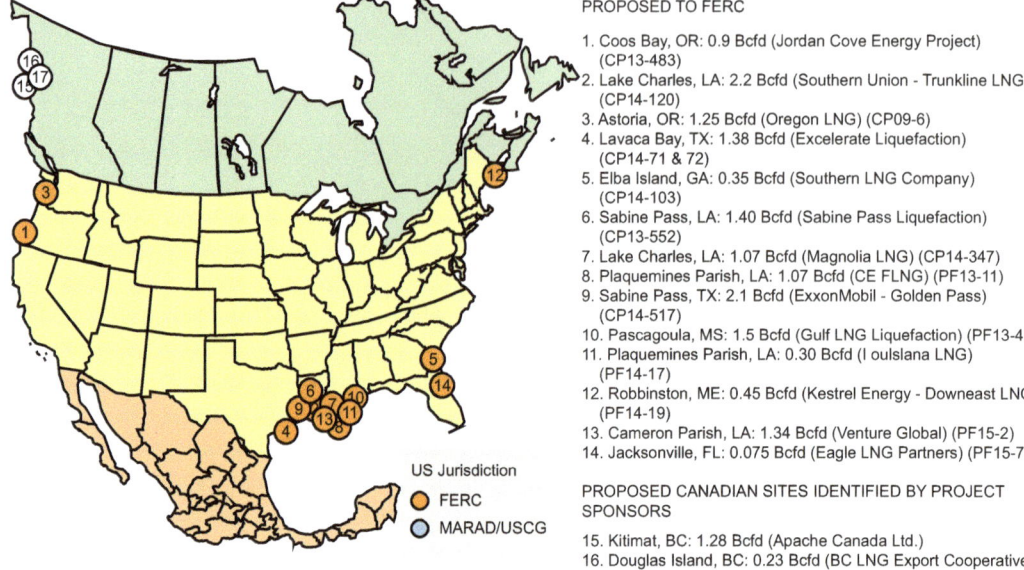

Fig. 17.18 Proposed LNG export terminals in North America, February 2015. *Source* US Federal Energy Regulatory Commission

exceptional cases. In recent periods, oil prices have dropped, and long-term oil price trends are uncertain, with recent Henry Hub prices seeing multiple record lows, highlighting the relative risk of US LNG supply and supplies whose prices are linked to oil.

Even though natural gas pricing mechanisms are trending further toward competitive pricing, the formation of a global unified price is still impossible. However, the growth in inter-regional natural gas trade volume, as well as rises in trading hub or spot market natural gas sales volumes, are prompting the formation of regional markets based on the original national foundations, for example in Europe.

Even if trade and competitive pricing were to promote the formation of a globalised natural gas market, regional prices are likely to continue to exist, and inter-regional price variance could even exceed shipping costs. The reason for this is that natural gas regional demand and supply is often imbalanced: large consumers are not large producers, resulting in the signing of long-term contracts to reduce energy security issues and the potential concentration of market forces.

Likewise, formation of a global unified price is not possible, since the flexibility of global trade capabilities is likely to be insufficient, and will be unable to satisfy trade demands that can change at any time. Finally, the potential for economically recoverable unconventional natural gas remains unknown, but if it were to be unlocked at a low cost, as seen in the United States, the volume generated would be more likely to influence prices in the immediate region of the supply rather than in other regions.

17.4.2 The Relationship Between the Price of Natural Gas and the Price of Oil

Natural gas, whether delivered over pipelines or as LNG, has historically been priced based on competing fuels in the receiving markets and supplied under long-term agreements. Examples include exports from the Groningen natural gas field in The Netherlands, where pricing adjustments are tied to the market price of three major types of fuel, or LNG imports to Japan, which are

LNG Sale and Purchase Agreements (SPAs)
Sabine Pass Liquefaction

~20 mtpa "take-or-pay" style commercial agreements
~$2.9B annual fixed fee revenue for 20 years

	BG Gulf Coast LNG	Gas Natural Fenosa	Korea Gas Corporation	GAIL (India) Limited	Total Gas [6] & Power N.A	Centric plc [6]
Annual Contract Quantity (MMBtu)	286,500,000 [1]	182,500,000	182,500,000	182,500,000	104,750,000 [1]	91,250,000
Annual Fixed Fees [2]	~$723 MM [3]	~$454 MM	~$548 MM	~$548 MM	~$314 MM	~$274 MM
Fixed Fees $ / MMBtu [2]	$2.25 - $3.00	$2.49	$3.00	$3.00	$3.00	$3.00
LNG Cost	115% of HH	115% of HH	115% of HH	115% of HH	115% of HH	115% of HH
Terms of Contract [4]	20 years	20 years	20 years	20 years	20 years	20 years
Guarantor	BG Energy Holdings Ltd.	Gas Natural SDG S.A	N/A	N/A	Total S.A.	N/A
Corporate / Guarantor Credit Rating [5]	A- / A2 / A-	BBB / Baa2 / BBB+	A+ / A1 / AA-	NR / Baa2 / BBB-	AA- / Aa1 / AA	A- / A3 / A-
Fee During Force Majeure	Up to 24 months	Up to 24 months	N/A	N/A	N/A	N/A
Contract Start	Train 1 + additional volumes with Trains 2,3,4	Train 2	Train 3	Train 4	Train 5	Train 5

(1) BG has agreed to purchase 182,500,000 MMBtu, 36,500,000 MMBtu, 34,000,000 MMBtu and 33,500,000 MMBtu of LNG volumes annually upon the commencement of operations of Trains 1, 2, 3 and 4, respectively. Total has agreed to purchase 91,250,000 MMBtu of LNG volumes annually plus 13,400,000 MMBtu of seasonal LNG volumes upon the commencement of Train 5 operations.
(2) A portion of the fee is subject to inflation, approximately 15% for BG Group, 13.6% for Gas Natural Fenosa, 15% for KOGAS and GAIL (India) Ltd and 11.5% for Total and Centrica.
(3) Following commercial in service date of Train 4. BG will provide annual fixed fees of approximately $520 million during Trains 1–2 operations and an additional $203 million once Trains 3–4 are operational.
(4) SPAs have a 20-year term with the right to extend up to an additional 10 years. Gas Natural Fenosa has an extension right up to an additional 12 years in certain circumstances.
(5) Ratings are provided by S&P/Moody's/Fitch and subject to change, suspension or withdrawal at anytime and are not a recommendation to buy, hold or sell any security.
(6) Conditions precedent must be satisfied by June 30, 2015 or either party can terminate. CPs include financing, regulatory approvals and positive final investment decision.

CHENIERE

Fig. 17.19 Sabine Pass LNG sale and purchase agreement structure. *Source* Cheniere Energy Inc

priced based on Japan Crude Cocktail prices, an amalgamation of oil prices that is generally defined within individual contracts.

For many years, natural gas contracts have been indexed to the oil price; when the oil price changes, the price paid for natural gas changes as well, using a preset formula. Until about 2008, the link has led to a close correlation between natural and oil two prices (Fig. 17.21). The link loosened in periods when oil price showed high volatility, such as the oil crises, and natural gas

contracts tended to be renegotiated. Anticipating a need to rebalance the long-term agreements, some natural gas contracts included clauses allowing price formulas to be reviewed based on an established set of factors.

The connection between oil and natural gas prices started to loosen in 2008, typified by the divergence of Henry Hub prices from oil prices. The separation was partly a result of the US recession and a boom in unconventional natural gas supplies in the United States, which meant

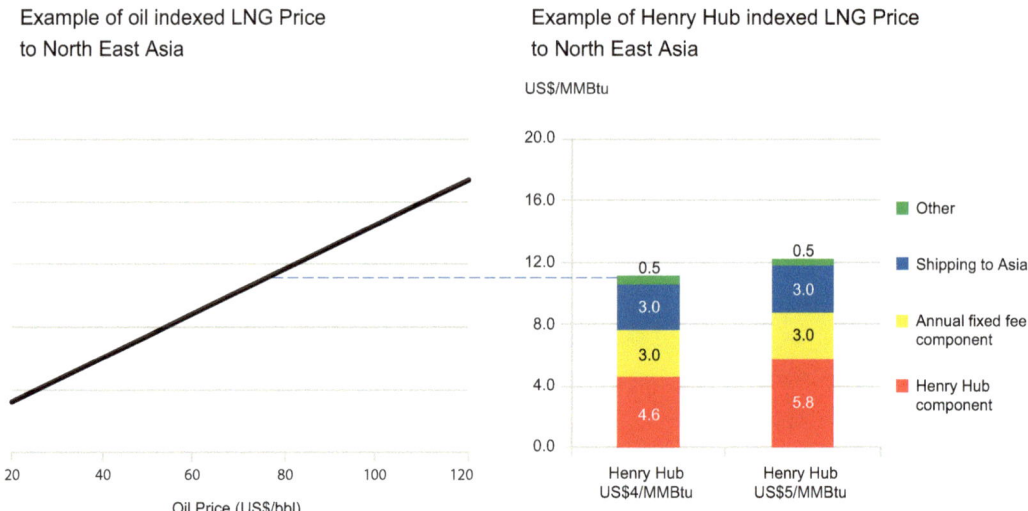

Fig. 17.20 Henry Hub natural gas prices could be higher than natural gas prices linked to oil prices

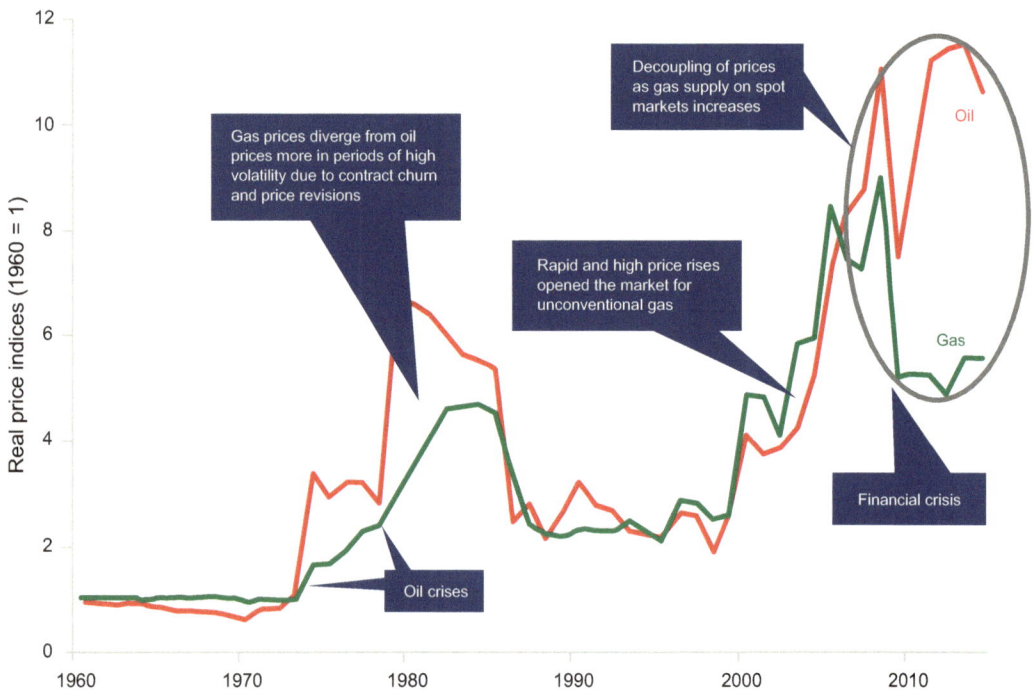

Fig. 17.21 The relationship between natural gas price and oil prices. *Note* Data are indexes of global average prices. *Source* Vivid Economics, based on World Bank data

that the country no longer needed significant LNG imports. In Europe, an economic downturn also triggered decreased natural gas demand, leading to ample supplies, which were increasingly purchased through the natural gas hubs without any link to oil prices.

Recently, the possibility of Russian natural gas supplies that pass through Ukraine being disrupted added market volatility, but prices recovered even as the supply to Europe was reduced. Interestingly, the supply reduction was linked largely to LNG imports to Europe, rather than Russian pipeline deliveries. Strong demand for LNG in Asia-Pacific created arbitrage opportunities, and LNG delivered to European buyers was reloaded by the buyers for sale in Asia-Pacific. Faced with low gas index prices, low demand, or both, buyers in Europe chose to profit by selling to Asia-Pacific buyers at higher prices. The arbitrage opportunity disappeared as oil prices and European natural gas hub prices fell.

Historically, four rationales have supported the need for linking natural gas contracts to the price of oil:

- **Benchmark price**: Natural gas exploration, production and transport are capital-intensive and the industry often requires long-term contracts. To share the risk between buyer and seller, the contract prices generally vary according to a benchmark, and the oil price has provided a transparent and robust benchmark. Also, many of the costs of natural gas development—for example, drilling rigs and skilled labour—are comparable in both oil and natural gas production, and these costs are affected by oil prices. However, in some markets the fundamentals for pricing oil and natural gas have diverged. For instance, in the United States and parts of Europe, sufficiently deep and liquid natural gas benchmarks have become available.
- **Co-production**: Natural gas has been a by-product of oil extraction and sold on similar terms. Exclusive natural gas extraction, however, is becoming more common, even though co-production remains important. In the United States, for example, natural gas is a by-product of shale oil production. Low natural gas prices in the US market are partially a result of high oil prices, which have offered significant incentives to shale projects. The low natural gas prices may not be sustainable or easily replicated outside the United States.

- **Similar routes to market**: Natural gas and oil have similar transport infrastructure, and the same companies have tended to deliver both because of their expertise, market dominance or both. However, natural gas infrastructure has increasingly become more independent.
- **Competition for customers**: Natural gas and oil products were used for power generation and heating, and natural gas was often priced just lower than oil to compete. However, oil products are increasingly used only used for transportation, with other fuels used for power generation.

As markets have evolved, however, only one of these reasons—the need for a transparent benchmark—remains relevant, but even here in some markets natural gas hubs are now more able to provide a natural gas benchmark.

The fundamental economic rationale for the link between natural gas and oil prices is whether the economic drivers behind the value of oil are similar to those behind the value of natural gas. When oil and natural gas were substitutes—for example when they were both used in power generation—this was often true, and their value would change at the same rate. Now, however, oil is primarily used for transportation, and natural gas is not. The value the market puts on these two fuels is often driven by different forces, so the price of oil no longer reflects the value of natural gas as accurately. Because the uses for natural gas and oil and their production costs have largely diverged, the fundamental economic rationale for a price link is unlikely to be re-established, especially in markets in which natural gas benchmarks are emerging.

17.4.3 Chinese Pricing Mechanisms

China's natural gas is derived from domestic natural gas, imported LNG and imported pipeline natural gas from countries of the former Soviet Union. The different treatment of these resources from different source results in a three-strand pricing mechanism:

- **Domestic natural gas**: Price is determined on a cost-plus basis.
- **Imported LNG**: Contract prices and linked to Japan Crude Cocktail prices, with an S-curve-based ceiling mechanism that limits exposure to oil price movements.
- **Imported pipeline natural gas**: In a pricing mechanism that the IGU has called a "bilateral monopoly", price is based on intergovernmental negotiations and entails significant uncertainty.

The differences inherent in China's pricing structure have already caused a series of market convergence problems, especially in periods when oil prices are high.

Experience from North America and Europe suggests that prices for imported natural gas or LNG can be based on a natural gas index that reflects the supply and demand of the recipient region or country. A relevant natural gas index is only possible, however, if the participants, infrastructure and regulations are in place to form a natural gas hub that is liquid, transparent and widely used. The natural gas index must also be acceptable to financial institutions, which supply credit to developers and buyers. Once in place, a Chinese natural gas index could provide the necessary price signals to attract additional imports as required. Although natural gas market reforms in China continue to gather momentum, international experience suggests that a natural gas hub is not likely to develop soon and intermediate measures may be needed in the interim.

As China continues to reform its natural gas market, pricing mechanisms will need to be allowed to evolve. As an interim measure, the current practice of offering end users natural gas at prices based on the prices of competing fuels can continue. However, this approach exposes importers, especially LNG importers, to risk. LNG imports to China would continue to be driven by global supply and demand factors, and China would have to compete with other buyers on the market. During this time, oil-indexed contracts would continue, at the same time as LNG deals linked to Henry Hub prices gain momentum. If the market is allowed to open

further to competition, the mismatch between imported natural gas prices and end-user prices should narrow.

17.4.4   The Influence of Chinese Demand on the World Market

This section analyses the influence of China's natural gas consumption on global energy markets. We will explore three scenarios for Chinese demand: the first where there are no new policies to reduce consumption or stimulate natural gas consumption, and then two scenarios that feature increasing substitution of natural gas for coal, both in the power sector and in the wider economy.

The key findings from the modelling are:

- **Natural gas prices**: Flexible global natural gas supply would dampen any increase in domestic natural gas prices if higher natural gas demand in China put upward pressure on domestic prices.
- **Coal consumption**: Lower Chinese coal consumption would be largely offset by increased coal use in other countries, unless those countries adopt similar policies.
- **Domestic coal**: Policies that encourage only the power sector to reduce coal consumption could result in lower coal prices and a switch to coal from natural gas in industries not covered by the low-carbon policies. Policies aimed at reducing domestic coal production should have broad applicability in order to have the greatest impact.

1. **The three policy scenarios**

Three scenarios for China were developed to illustrate the effects of a reduction in domestic coal consumption—for power generation and for industrial use more widely—and to analyse how readily natural gas could be substituted for coal and what impact these changes would have on global energy markets. These three scenarios (Baseline, Power Generation Sector and All

Sectors) simulate the effects of varying degrees of restriction on coal consumption in China. Each scenario differs in the stringency of the restriction, as well as the sectors that are targeted.

- **Baseline scenario**: China does not introduce any new policy measures aimed at reducing coal consumption or stimulating natural gas consumption. Coal-fired generation capacity in the power sector continues to grow at historical rates for the duration of the 13th Five-Year Plan, from 2016 to 2020. It continues to grow steadily, albeit at a slightly slower rate, after that. This scenario is broadly consistent with recent IEA and EIA analysis. However, given the momentum for change in China's energy policy, it should be viewed as a useful baseline against which to compare the impact of the other two scenarios, rather than a realistic future pathway. For example, the scenario does not take into account current commitments to LNG projects.
- **Power Generation Sector scenario**: China restricts coal consumption in the power sector, reducing investments in coal-fired generation capacity and increasing the effective price of coal for power generation. This scenario is very similar to current policy trends in China, and thus is considered to be the most realistic.
- **All Sectors scenario**: China restricts coal consumption in the power sector, similar to the Power Generation Sector scenario, but extending the restrictions to other sectors of the economy, reducing investment in all coal-intensive capital stock and raising the effective price of coal economy-wide.

2. The impact on Chinese energy supply and demand

Restrictions on coal use would raise the effective price of coal and reduce incentives for coal-intensive investments. Coal consumption would peak in 2020 under the Power Generation Sector scenario and in 2018 under the All Sectors scenario. The later peak in the Power Generation Sector scenario reflects the leakage of coal use from the power sector to other sectors of the economy, most notably manufacturing, that would result from restricting coal use only in the power sector.

The decline in coal consumption would be accompanied by an increase in the consumption of natural gas and electricity. Under both the Power Generation Sector and All Sectors scenarios, the share of natural gas in China's energy mix would rise by 2030—to 17% in the Power Generation Sector scenario and to 21% in the All Sectors scenario (Fig. 17.22). The share of natural gas in the power sector would rise to around 15% in both scenarios. Against a backdrop of rising energy demand overall, this would translate into a 70% increase in natural gas consumption in the Power Generation Sector scenario over the Baseline scenario and a 100% increase in the All Sectors scenario.

China's domestic natural gas production would remain below consumption levels in both the Power Generation Sector and All Sectors scenarios, despite an assumed tripling in domestic production by 2030. As a result, the share of natural gas imports in total consumption is estimated to rise to around 50% by 2025 in both scenarios. Given current and planned future pipeline capacity, the LNG share of natural gas imports is estimated to rise to more than 50%. The consumption/production gap would be larger in the All Sectors scenario, where LNG imports are estimated to account for 73% of total natural gas imports.

Restricting coal use only in the power sector (as in the Power Generation Sector scenario) would lead to a leakage of coal use to other sectors, leading to natural gas-to-coal substitution in these sectors. In turn, this would benefit other coal-intensive sectors of the economy, such as manufacturing, as lower coal demand in the power sector would place downward pressure on the price of coal paid by other sectors. On the other hand, an increase in natural gas demand for power generation would place upward pressure on the domestic price of natural gas, putting other gas-intensive sectors, such as residential heating, at a disadvantage because their main substitutes for natural gas are electricity and oil.

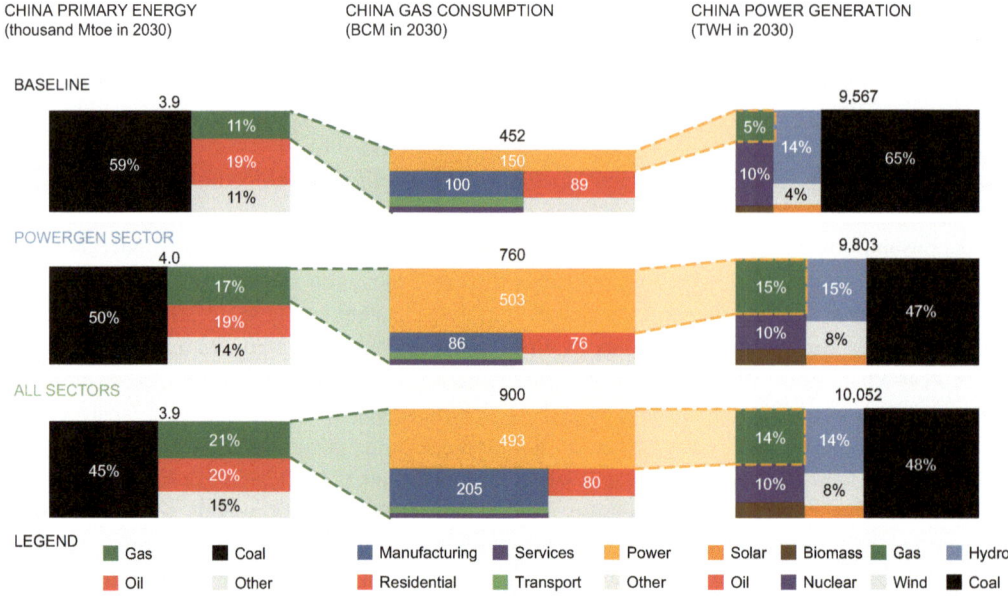

CHINA PRIMARY ENERGY
(thousand Mtoe in 2030)

CHINA GAS CONSUMPTION
(BCM in 2030)

CHINA POWER GENERATION
(TWH in 2030)

LEGEND

| | Gas | | Coal | | Manufacturing | | Services | | Power | | Solar | | Biomass | | Gas | | Hydro |
| | Oil | | Other | | Residential | | Transport | | Other | | Oil | | Nuclear | | Wind | | Coal |

Fig. 17.22 China's 2030 energy mix, natural gas consumption and power generation energy mix, by scenario. *Source* Aurora Energy Research, Global Energy Model

3. **The impact on the global energy market**

The IEA estimates that emerging economies like China will account for most of the growth in global energy demand over the next two decades, and changes to the energy policies and the energy mix in these economies will have implications for global energy markets. Overall, the modelling indicates that while global coal consumption remains relatively unchanged, natural gas consumption would increase in the Power Generation Sector and All Sectors scenarios (Fig. 17.23).

(I) **Impact on coal consumption**

In the Power Generation Sector and All Sectors scenarios, decreased coal use in China depresses the coal price enough to stimulate coal consumption elsewhere, leaving total coal consumption largely undiminished at the global level. Both scenarios result in a decrease in Chinese coal consumption, with a greater reduction in the All Sectors scenario than in the Power Generation Sector scenario. However,

total global coal consumption by 2030 remains relatively undiminished and roughly the same in both scenarios. This is because global coal supply is relatively price-inelastic, that is, supply is relatively unresponsive to falling demand and hence falling prices. While the reduction in Chinese coal demand would put downward pressure on the coal price, this would not lead to a significant reduction in price. Cheaper coal prices, on the other hand, would lead to increased coal consumption in other countries without similar restrictions on coal use.

Overall, in the Power Generation Sector scenario 96% of the decrease in China's coal consumption is offset by an increase elsewhere, while in the All Sectors scenario 89% of the decrease is offset. One reason for the difference between the two scenarios is the nature of trade flows in the coal market in each scenario. In the Power Generation Sector scenario, some of the coal that would have been imported by China compared to the Baseline scenario would no longer be needed and instead would be consumed closer to its source. In the All Sectors scenario, however, both domestic coal production and coal imports would

Global gas demand (thousand Mtoe)   Global gas demand (thousand Mtoe)

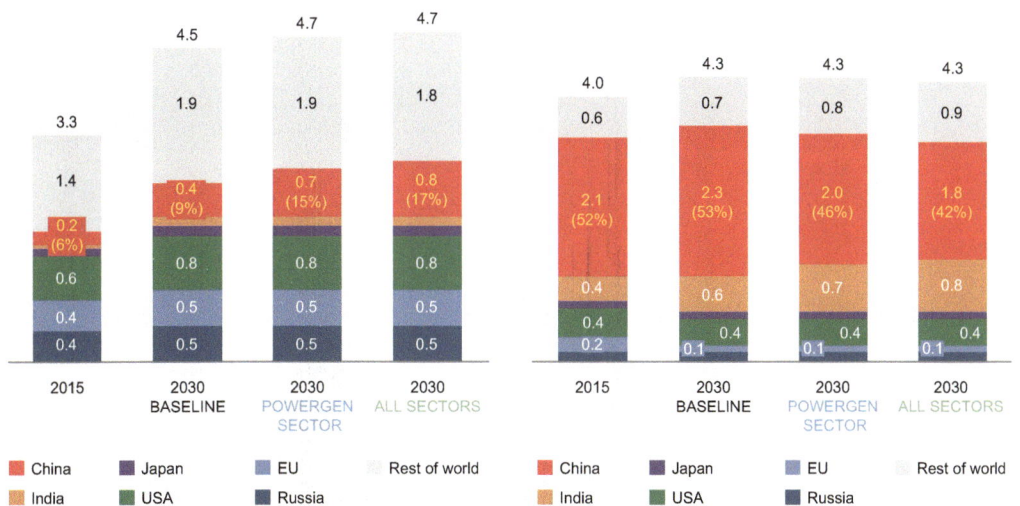

Fig. 17.23 Global natural gas and coal demand, by region and scenario. *Source* Aurora Energy Research, Global Energy Model

be displaced, and China would become a net exporter of coal by 2025. However, countries that consume China's surplus coal production would incur transportation costs, leading to a smaller share of China's coal consumption leaking to other countries.

The coal leakage effect would be driven purely by the economics of lower coal prices—neither the Power Generation Sector nor the All Sectors scenario assumes comparable coal reduction policies in other parts of the world. This effect would be limited if comparable coal restrictions were also in place in other Asian countries. A key implication of this result is that, given the interconnected nature of global energy markets, a meaningful reduction in global coal consumption can only be achieved through a concerted international policy effort.

(II) **Impact on natural gas consumption**

In contrast to coal, natural gas supply is relatively price-elastic, that is, global natural gas supply increases in response to rising demand and rising prices. The supply response mitigates some of

the increase in price, resulting in only a modest increase in natural gas hub prices.

Restrictions on coal use in both the Power Generation Sector and All Sectors scenarios would drive substantial substitution away from coal and towards natural gas in China. The higher demand would be met by production from many different geographical regions, including a significant increase in domestic production (Fig. 17.24). The rising share of LNG to meet domestic consumption would push domestic natural gas prices higher, nearing Japanese LNG prices, which would stimulate domestic production.

The modelling indicates that, while increasing Chinese natural gas demand would put some upward pressure on natural gas prices, natural gas-intensive countries like the United States would not decrease consumption substantially in response. Thus global natural gas demand overall would be higher under both the Power Generation Sector and All Sectors scenarios, and the consumption decrease outside China would only be enough to offset about 42% of the increase seen within China.

Global gas production (thousand bcm)

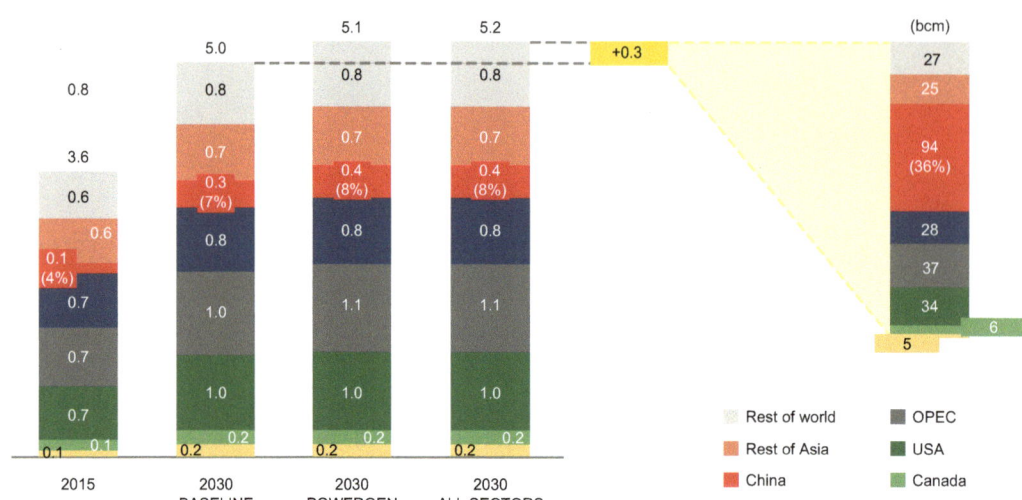

Fig. 17.24 Trends in global natural gas production by region and scenario. *Source* Aurora Energy Research, Global Energy Model

(III) **Impact on primary energy consumption**

Both the Power Generation Sector and All Sectors scenarios indicate a net increase in global energy consumption (Fig. 17.25). The world's overall energy bundle would effectively become cheaper as the coal price would drop by more than the increase in the natural gas price. The decline in coal consumption in China would be offset by increased coal consumption elsewhere and the increase in natural gas consumption in China would not be offset. Moreover, in China, higher total primary energy consumption would be driven by substituting coal with electricity.

(IV) **Changes in energy prices**

Substituting natural gas for coal in China would have a large effect on global coal prices and a relatively small effect on global natural gas prices. Global coal prices in 2030 would be lower in both the Power Generation Sector and the All Sectors scenarios compared to the Baseline scenario. In the Power Generation Sector scenario, the difference would be 15% or $12/tonne and in the All Sectors scenario, 26% or $20/tonne (Fig. 17.26). The main drivers for the significant drop in global coal price would be China's exceptional position as the largest consumer in the international coal market—China currently consumes more than half of the world's coal—and the high degree of global integration in the coal market.

Natural gas prices everywhere would increase as a result of rising Chinese natural gas demand, but this increase would be very modest at the major international natural gas hubs: a 3% price increase on average in 2030 across three major natural gas hubs: the NBP in the United Kingdom, the Henry Hub in the United States and the LNG import price in Japan (Fig. 17.27). In all three scenarios, there would be considerable convergence between these hub prices, as global LNG trade increases substantially and average LNG transport costs fall. For instance, the difference between Japanese LNG prices and the Henry Hub price would fall from just more than $12/MMBtu in 2015 to $7–8/MMBtu in 2030, as long-run East Asian LNG prices become broadly equivalent to the Henry Hub price, plus transport costs.

Global primary energy demand (thousand Mtoe)

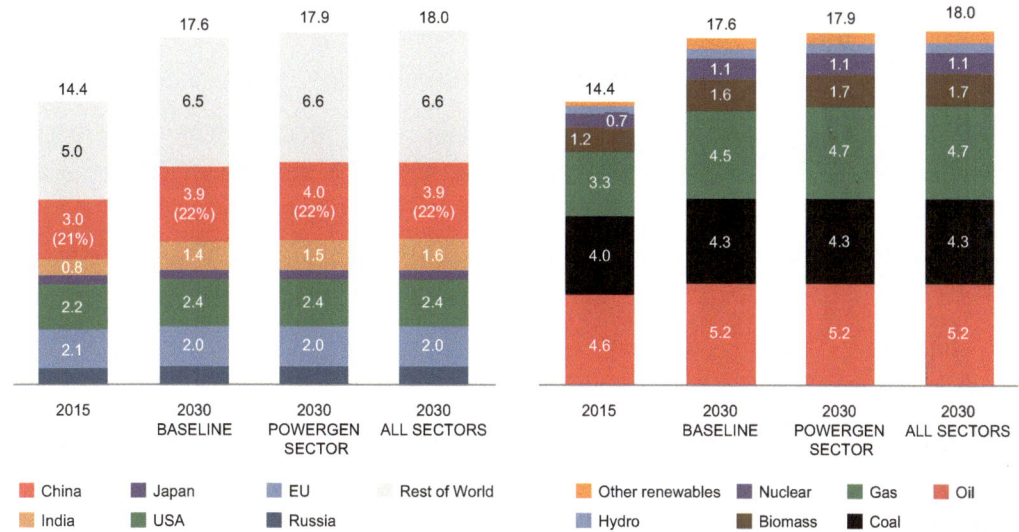

Fig. 17.25 Global primary energy demand 2015–2030 by region and scenario. *Source* Aurora Energy Research, Global Energy Model

Coal price (2014 USD/tonne)

Gas & coal consumption in 2030 (△Mtoe, relative to baseline)

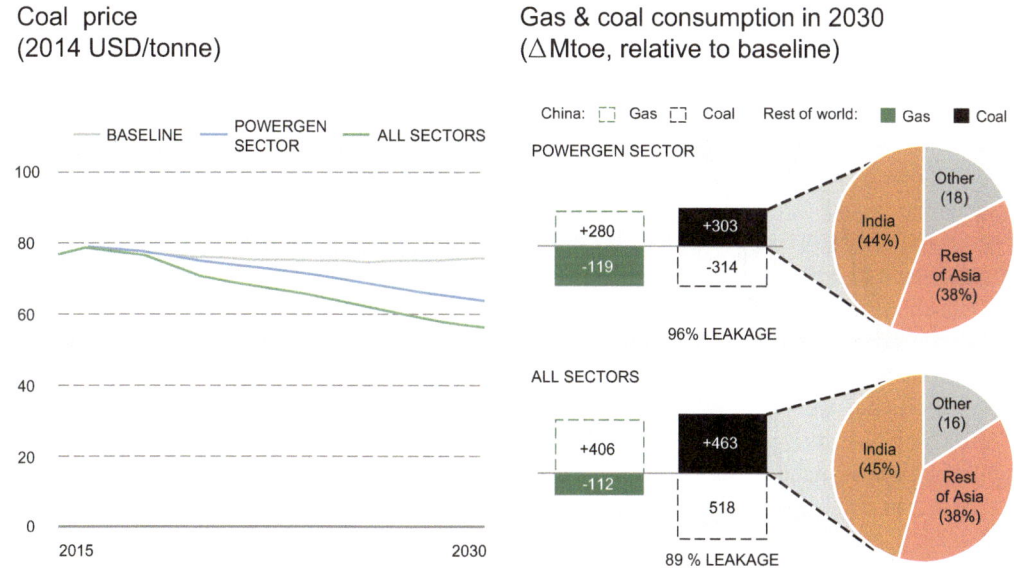

Fig. 17.26 Decline in China's coal demand and leakage of capacity. *Note* Price references ARA 6000 kcal per kg coal. From 2015 to 2018, displayed prices reference current futures prices. *Source* Aurora Energy Research, Global Energy Model

Global gas prices (2014 USD/MMBtu)

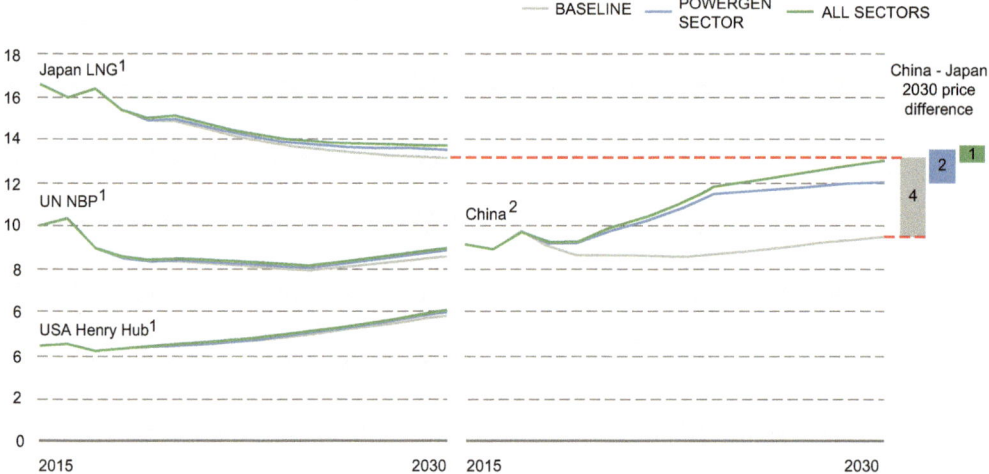

Fig. 17.27 Global natural gas hub price change trend forecasts. *Notes* (1) For 2014–2018, prices shown take into account current futures prices (as of December 12, 2014, converted to real 2014 US$). (2) For China, this is a weighted average producer gas price index (domestic, pipeline, LNG), which equates supply and demand in the AER-GLO model. It includes regasification costs for LNG. Weightings are based on the standard Armington assumption governing trade in CGE model. As a consequence, this figure will not necessarily match wholesale prices at any particular region in the Chinese market. *Source* Aurora Energy Research, Global Energy Model

The increase in global natural gas prices would be much smaller than the decrease in global coal prices for several reasons:

- The increase in Chinese natural gas consumption would be slightly less than the decrease in Chinese coal consumption because there would be some substitution from coal to non-natural gas fuels in China.
- China is a larger player in global coal markets than in natural gas markets, and its decisions linked to coal would have greater implications for global markets.
- Global trade in natural gas is much smaller than that in coal, partly because the rigidity of pipeline supply and the high transport costs for LNG restrict global trade flows.
- Global natural gas supply is more price-elastic than global coal supply, and natural gas production would increase to meet higher demand, while coal production would not fall significantly, creating excess supply and lower prices.

However, domestic natural gas price in China would rise more substantially in both the Power Generation Sector and All Sectors scenarios. By 2030, domestic prices would reach $12/MMBtu in the Power Generation Sector scenario and $13/MMBtu in the All Sectors scenario, compared to between $8/MMBtu and $10/MMBtu in the Baseline scenario. The reason for the disproportionate effect on prices in China is that as China starts substituting natural gas for coal, the increase in natural gas consumption would be made possible by increased natural gas imports, LNG in particular. As a result, natural gas prices in China would converge towards the East Asian LNG price.

4. **A description of the model**

The simulations of three possible energy scenarios for China were based on the Aurora Energy Research global energy model, which is a hybrid of the computable general equilibrium

Modelling approach

1. GGEM solves for prices given fossil fuel and electricity production

2. Fossil fuel and electricity production are optimised at given prices

3. Iterates until an internally consistent solution is found

4. Moves on to next year

Aurora's global energy model (AER-GLO)

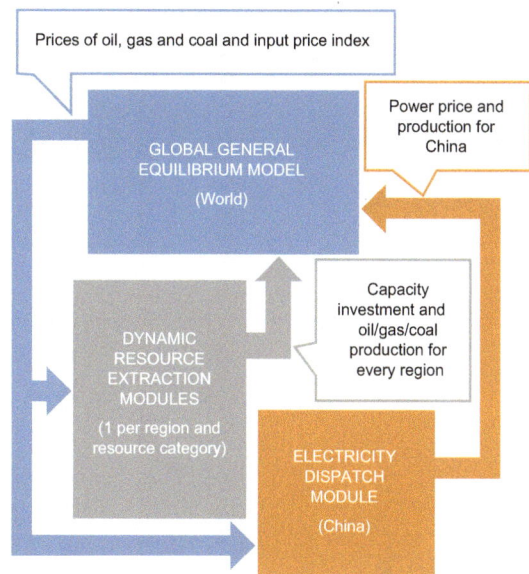

Fig. 17.28 Hybrid computable general equilibrium (CGE) model used by Aurora for global modelling. *Source* Aurora Energy Research, Global Energy Model

model that the company had created to analyse global energy markets.

The model drew from three key building blocks: a global general equilibrium module to represent energy demand and economic activity in all countries, a number of dynamic resource extraction modules to model the supply of oil, natural gas and coal in a dynamic financial setting; and an electricity dispatch module, which emulates the Chinese power sector (Fig. 17.28). The model lets all three modules solve iteratively in each year until an internally consistent solution across all three is found in every year. The hybrid structure combined the benefits arising from the robust structure of global general equilibrium models with the added detail of partial-equilibrium models of fuel extraction and electricity dispatch. Both types of modelling approaches are essential for understanding global energy markets.

The model was used to forecast the impact of energy policy changes on the global economy, as well as individual sectors. In addition, with

129 regions globally and 57 sectors within each region, the model could produce outputs that included country- and industry-specific forecasts of demand and supply of all goods and services, changes in imports and exports, gross regional product, consumption, investments, returns to capital and emissions of greenhouse natural gases and particulate matters.

For this study, the model was calibrated against recent economic and energy data, and envisioned three illustrative energy policy scenarios, in which key Chinese policy parameters relating to coal consumption were changed. These scenarios simulate plausible changes in China's energy policy environment and the effects these are likely to have on global coal and natural gas markets.

A global general equilibrium structure was necessary for modelling the global markets for fuels because of the intimate linkage between energy and future developments in modern economies, the substitutability between different fuels in final energy use, and the uneven distribution of resources across the world. As a result, the impact of China's energy policy on global energy markets provided by the model were driven to a large extent by the interaction of different markets under equilibrium and the resulting substitution between fuels in all geographical regions in the model.

An Overview of International Regulatory Experience

18

As part of our study, we looked at the evolution of natural gas markets in five selected countries that experienced significant increases in gas demand over the last few decades. In general, these five markets liberalised their natural gas systems in phases, allowing increased competition in the upstream and downstream segments, ensuring fair access to midstream infrastructure such as pipelines and LNG terminals, and striving to ensure that market changes supported overarching energy goals and benefited end users. To varying degrees, the experience of the United States, the European Union, the United Kingdom, Japan, and South Korea shows that natural gas market liberalisation plays a key role in developing and maintaining well-functioning natural gas markets.

Three key factors that determine the development and evolution of natural gas markets are: fundamentals, market regulation and ancillary policies:

- **Market fundamentals**: The fundamentals of the natural gas market include demand for natural gas in the economy, the availability of sources of supply, and the extent to which sources of supply compete.

- **Energy market regulation**: Natural gas market regulation is a crucial factor in the development and evolution of natural gas markets. A clear and consistent regulatory framework requires a strong and independent market authority and includes key provisions for the regulation of natural monopolies, safety and environmental rules, particularly for exploration and production, security of supply and market transparency, where competitive markets can exist.

- **Ancillary policies**: These include policies and regulations that affect natural gas markets indirectly, for example, air quality and climate change legislation that influences the cost-competitiveness of substitute fuels for power generation, policies to encourage and support greater energy access for low-income households, and subsidies for energy-related research and development.

Market fundamentals, energy market regulation and ancillary policies interact to determine market composition. In addition, the entire energy system and even other sectors in the national economy can influence natural gas market regulation and ancillary policies. Likewise, the market fundamentals can be influenced by technological development or by geopolitical events that reduce the relative costs of indigenous and foreign sources of supply.

Turning for a moment to natural gas regulations, these generally have three goals:

---

* This chapter was overseen by Jigang Wei from the Development Research Center of the State Council and Taoliang Lee from Shell Eastern Trading Corporation. It was jointly completed by Yaodong Shi, Zifeng Song, Ren Miao from the Energy Research Institute of the National Development and Reform Commission and Cindy Wang from Shell China. Other members of the topic group participated in discussions and revisions.

- **Affordable supply**: Providing economical natural gas to end users.
- **Secure supply**: Making gas delivery more secure, especially in the segments of the natural gas value chain where natural monopolies exist.
- **Secondary benefits**: Formulating policies that provide knock-on benefits for society as a whole, for example, restricting air pollution and carbon emissions, and providing energy to poor families.

The market liberalisation of natural gas can achieve these three goals. First, market liberalisation can provoke competition, which helps to reduce consumption costs and thus provide affordable energy to end users. Second, market liberalisation opens the market to a more diverse range of suppliers, which can improve security of supply. Third, because market participants will react promptly to any incentive measures contained in secondary policies, the entire natural gas value chain will be able to respond more flexibly to fulfil these policy objectives.

However, market liberalisation also brings greater risks to investment. For example, under free market conditions, domestic natural gas enterprise buying power could be reduced in international markets. Market liberalisation will also make it harder for poor families to obtain energy. Case studies show that governments often use targeted market liberalisation that is restricted or controlled, so as to rein in these adverse influences.

Natural gas market development does not solely rely upon regulation, though. The fundamentals of a market, such as the ready availability of natural gas supplies and the level of natural gas demand, are primary forces in natural gas market outcomes. For example, regardless of the regulatory regime, a country like Japan, with negligible indigenous production, will always have different natural gas market fundamentals than the United States with its large indigenous production. At the same time, competition between natural gas and other fuels will also influence development of natural gas markets, as will a wide variety of other factors such as air

quality, climate change, energy sufficiency and electricity market structure, all of which will influence natural gas market development.

International experience suggests there is a core set of actions required to deliver competitive natural gas market outcomes, across a range of different market fundamentals and ancillary policy priorities. The analysis of gas market evolution in different countries seeks to draw general lessons for natural gas market liberalisation, noting contrasts in each region that force different regulatory approaches. The Chinese context will be different again, but the common core of actions across the case studies is likely to be relevant to China, although modified for the country's unique characteristics. In short, international experience offers important guidance to China as it designs the composition and operation of its natural gas market.

The political sequencing of regulatory reforms is as important to successful market liberalisation as correct implementation. Natural gas market regulation impinges on a number of politically important issues, in particular energy affordability, the funding model for national infrastructure and energy security. When political momentum is supportive of liberalisation, regulatory reform has delivered competitive markets within 10–15 years; for instance, in the United States starting in the late 1970s or the United Kingdom starting in the late 1980s. When political momentum is not supportive, the process of regulatory development could require decades.

In countries with significant domestic natural gas resources, such as Norway, The Netherlands, the United Kingdom, and the United States, market liberalisation has driven increased competition and greater development of these resources. At the same time, the experience of these countries shows that the course of market liberalisation requires favourable upstream licensing, fiscal or environmental policies. In the United States, private ownership of land and the resources underneath it enabled rapid development of shale natural gas exploration and production. Where resources are owned by the government, such as on the outer continental shelf of the North

Sea in Norway, The Netherlands and the United Kingdom, processes are in place to enable effective and competitive leasing of exploration and production rights.

Another lesson emerging from the case studies is that interactions between ancillary policies are often unpredictable, such as environmental and technology policies on the one hand and liberalised gas markets on the other. Environmental policies—relating to carbon emissions and local air pollutants—can increase the costs of emissions-intensive coal relative to natural gas. While ambitious policies on climate change and air quality favour natural gas over coal in electricity generation, the design of these policies could produce some unintended consequences. This has been the experience in the European Union, where the interaction between the two has undermined environmental objectives and led to the coal-renewables "energy paradox". Similarly, the United States has supported energy-related research and development programmes for decades, such as on unconventional natural gas and microseismic fracture monitoring, which eventually proved instrumental in driving the US shale natural gas boom of the last decade.

In summary, market fundamentals, energy market regulation and ancillary policies interact, and shape a country's natural gas market over time. International experience has shown that technical developments and geopolitical events can affect domestic and international supply costs, and can influence the market fundamentals. Likewise, the market fundamentals have an influence on natural gas market regulation and ancillary policies, and together they all form the natural gas market. Natural gas market liberalisation is thus ultimately dependent on the history of the natural gas market, geography and the political environment, and these major influences are pivotal in the formation of natural gas markets. For example, natural gas market regulation and ancillary policies are determined by political trends and events. This chapter will introduce international experiences from natural gas market regulation, market liberalisation and other perspectives, analysing case studies of foreign natural gas market liberalisation in depth in the hope

of providing a reference point for, and insight into, China's natural gas market liberalisation reforms.

18.1 Regulatory Reform

18.1.1 The Reasons for Regulation

Policymakers regulate the natural gas value chain in order to balance three major policy goals:

- **Economic supply**: Providing affordable heating to households and competitive natural gas prices for power generation and energy-intensive industries.
- **Security of supply**: The maintenance of a high-quality supply of natural gas to end users. Policymakers also often consider natural gas supply infrastructure to be part of critical national infrastructure required for the functioning of the country and the delivery of the essential services.
- **Safe and clean supply**: This objective presupposes that natural gas supply should pose minimum risk to public health and safety. It also captures the notion that increased natural gas supply can replace alternative fuels that have a higher direct environmental impact and higher greenhouse natural gas emissions.

These policy goals can conflict on multiple levels, and are often referred to as the energy policy "trilemma". For example, within economic supply, ensuring affordable natural gas prices for end users has often led countries to subsidise end-user prices, either through cross-subsidisation of one class of users by another or directly from the state budget. Further, delivering economic supply may not always be compatible the with goals of safe and clean supply.

Liberalising natural gas markets can help deliver across the energy trilemma objectives, but may need to be supplemented by a broader policy framework to manage some of the remaining trade-offs. For example, market liberalisation increases competition, which in turn helps to drive down costs for consumers over the

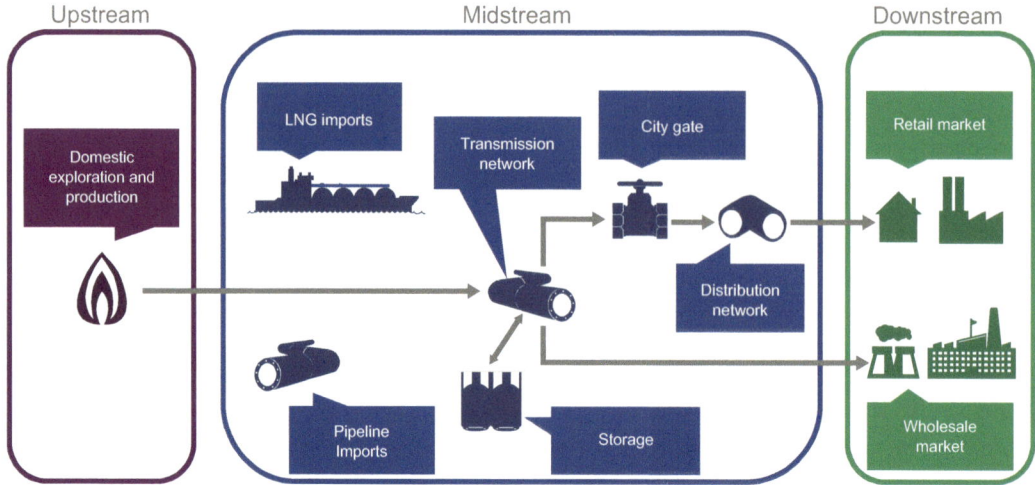

Fig. 18.1 The natural gas value chain. *Note* Natural monopolies, where markets fail, are mainly in the midstream segment

long term, delivering an economical supply. It also opens the market to a diverse range of suppliers, increasing security of supply. Finally, policies to encourage natural gas use and increase its share in the energy mix can reduce the risk to public health and safety, for example by reducing indoor and outdoor air pollution and greenhouse gas emissions caused by coal use.

However, liberalisation can also make investment riskier, and fragmented domestic natural gas firms can face reduced buying power on international markets. It can also raise costs in the near term, reducing the ability of low-income households to access energy, unless other policies are put in place to compensate. The five case studies below demonstrate that gas market liberalisation is often limited, or complemented by other policies, in order to manage these impacts.

18.1.2 The Process of Market Liberalisation Across the Natural Gas Value Chain

Liberalisation refers to the process by which competitive market outcomes are delivered: providing end users with the greatest choice at the lowest price. This can be achieved either by

opening markets to competition or through regulatory measures to limit or cap prices where natural monopolies exist. To achieve the goal of natural gas market liberalisation, the structure and economics of the natural gas value chain need to be considered (Fig. 18.1). With the exception of some segments where natural monopolies exist, market competition exists throughout the value chain.

The natural gas value chain can broadly be divided into three segments—upstream markets, midstream infrastructure and downstream markets. The upstream segment refers to domestic exploration and production of natural gas. The midstream segment refers to infrastructure for the transport of natural gas—domestic and imported—through transmission pipelines, LNG terminals and local distribution networks.[1] The

---

[1]For the purposes of this study, LNG terminals are classified as midstream infrastructure rather than upstream assets. Whether to treat LNG terminals as upstream assets or midstream infrastructure depends largely on the nature of competition in a market. In the United States, LNG terminals compete with domestic production, and are hence considered equivalent to domestic well heads and classified as upstream assets. In Europe, most of the competition is with imported natural gas, and hence LNG terminals are considered part of the transmission network and classified as midstream infrastructure. The appropriate treatment of LNG terminals in China is not yet clear

downstream segment refers to wholesale and retail markets supplying gas to end users.

Competitive markets can exist at many places across the natural gas value chain. Competition in the upstream segment can drive greater efficiency and innovation in exploration and production of natural gas, as a way to drive down costs and develop greater volumes and sources of domestic supply. For example, before the liberalisation of natural gas markets in the European Union, vertically integrated and state-owned natural gas companies had little incentive to improve the efficiency of their operations, leading to increases in natural gas prices. Similarly, greater competition in the downstream segment can increase choice and reduce the price paid by the end users of natural gas, for example by increasing competition in the shipping and sales of natural gas. These segments can be opened to competition, within a regulatory oversight framework to ensure and support competitive markets and without the need for further regulatory interventions.

Opening markets to competition is typically not desirable and can fail in the midstream segment. Midstream infrastructure, such as pipelines, are highly capital-intensive. For example, in 2013, onshore oil and natural gas pipeline construction in the United States cost $4.1 million per mile, while offshore pipelines cost $7.6 million per mile. High fixed costs and relatively low operation and maintenance costs point to large economies of scale: the costs of delivery decline rapidly with volume. This fits the definition of a natural monopoly, where the lowest long-run average costs are realised when production and ownership are concentrated in a single firm.

However, enterprises in positions of natural monopolies tend toward using their market position to harm consumers and will collect prices higher than those formed by market competition. In addition, natural gas transmission pipelines and other midstream infrastructure owners are generally vertically integrated, and they provide various services throughout the

value chain. Such enterprises will generally take actions to restrict competition, for example collecting excessively high fees from third-party natural gas suppliers and shippers using the infrastructure so as to maintain their leading position in upstream sectors, or else raise their profits in downstream sales business. This results in US and European natural gas market regulatory bodies requiring transmission pipeline owners to open their facilities to third parties, or to regulate the fees collected from customers by the pipeline infrastructure owners. They also insist on the separation of transmission services and upstream or downstream business—this process is called deregulation.

Liberalisation of natural gas markets requires identifying the competitive and natural monopoly segments of value chain, and applying appropriate regulatory approaches. For example, between 1954 and 1978, the United States in effect treated the entire value chain as a natural monopoly, with prices from the wellhead onward fully regulated to protect consumers. As a consequence, US natural gas investment failed to respond adequately to the 1970s oil crisis. The recognition in 1978 that only pipelines are natural monopolies, requiring regulation to ensure open access, marked the start of a period of liberalisation. The reforms unbundled the natural gas industry, separating transportation and sales businesses, and ensured open access to interstate pipeline infrastructure for third parties. By 1992, after 14 years, the fundamentals for a competitive and efficient market had been established, and they supported a 40% increase in natural gas consumption from around 500 billion m^3 in 1990 to 700 billion m^3 by 2010.

A review of international case studies shows that a competitive, responsive, liquid natural gas market is characterised by:

- **Critical mass**: Many players that compete for substantial upstream business and access to midstream infrastructure, serving many buyers with large volumes and financed by responsive investors.
- **Competitive pricing**: Competitive price formation at the wholesale and retail levels.

(Footnote 1 continued)
because a competitive natural gas market has yet to develop.

- **Open access**: Non-discriminatory open access to natural monopolistic midstream infrastructure, as well as regulated tariffs.

Overall, a competitive market can optimally allocate resources in many stages of the value chain, within an oversight framework to ensure well-functioning competitive markets. However, pipelines are natural monopolies and require more direct regulation. Getting the right combination of competitive markets, regulatory oversight and direct market regulation is essential for unlocking the benefits of liberalisation.

18.2 Market Liberalisation

18.2.1 Core Initiatives for Liberalisation

Market regulation is crucial to achieving and securing the liberalisation of natural gas markets. A regulatory framework, codified in law, defines the principles of regulation along the natural gas value chain. It determines which segments of the value chain should be opened to competition and establishes the institutional arrangements to ensure free and fair competition. It also determines which infrastructure is to be treated as a natural monopoly and sets the basis for access to the infrastructure, as well as the level of unbundling for vertically integrated players. The regulatory framework also includes rules around safety, security of supply and environmental standards, which apply throughout the value chain. Implementing the framework requires a strong, independent regulatory authority.

The natural gas market liberalisation experience of the United States, the European Union, the United Kingdom, Japan and South Korea has lessons that might be relevant to China. Each country has different characteristics. For example, the United States has by far the greatest indigenous supply, while the European Union is the biggest importer (Table 18.1). Understanding

Table 18.1 Market traits of five natural gas markets plus China

Natural gas indicator	Type	US	EU	United Kingdom	Japan	Korea	China
Supply (bcm per annum)	Indigenous	689	269	38	3	0.5	115
	Net imports	37	231	39	123	53	49
Consumption (% of total)	Power	40	30	30	65	50	15
	Industry	20	20	10	5	20	45
Transmission pipelines (km)		500k	200k	8k	5k	4k	50k
Wholesale competition		✓	Limited (oligopolies)	✓	Limited (oligopolies)	✗	✗
Open access	Upstream	✓	✓	✓	✗	✗	✗
	Transmission	✓	✓	✓	✗	✗	✗
	Distribution	Varies	Varies	✓	✗	✗	✗
Ownership unbundling of transport and sales		✓	Varies	✓	✗	✗	✗
Independent (federal) market authority		✓	✓	✓	✗	✗	✗
Liquid market centres/hubs		✓	✓	✓	✗	✗	✗

Note Data for 2013, except China's shares of power and industry in natural gas consumption for 2011. bcm is billion cubic metres

Source Vivid Economics, based on IEA, EIA, Chinese government and ENTSOG data

these differences can help identify experiences that best fit China's circumstances.

The case studies suggest five core actions for liberalisation that can deliver a competitive outcome in the natural gas market, as shown in Fig. 18.2. Creating institutions allows for a fundamental natural gas legal framework and a strong, independent regulatory agency. Enabling open access focuses on assuring fair and reasonable access to infrastructure owned by a natural monopoly. Deregulating prices enables market forces to determine prices and provides clearer signals about supply and demand. Setting standards and ensuring transparency offers regulatory mechanisms to ensure fair market practices. And finally, protecting end users safeguards the interests of commercial and residential natural gas users and seeks to manage any negative consequences of market liberalisation.

These measures take into account the broader context of natural gas market liberalisation: the need to balance the positives and negatives of liberalisation; interactions with fundamentals and ancillary policies; the importance of natural monopolies in the value chain; and the legacy of state intervention in the natural gas market.

1. Creating institutions

Institutions, including a natural gas law and a strong, independent regulator, form the fundamental building blocks of liberalised natural gas markets. Regulators are most effective when they have statutory independence from political, government and industry influence to ensure appropriate decision-making and equal treatment of market participants, as well as to avoid a myopic approach to long-term infrastructure investment decisions. The regulator's authority and tasks should be laid down in a regulatory framework codified in law.

The case studies below underscore the importance of regulatory institutions being independent of political influences and operating transparently. They illustrate that decisions should be made in a consultative and transparent

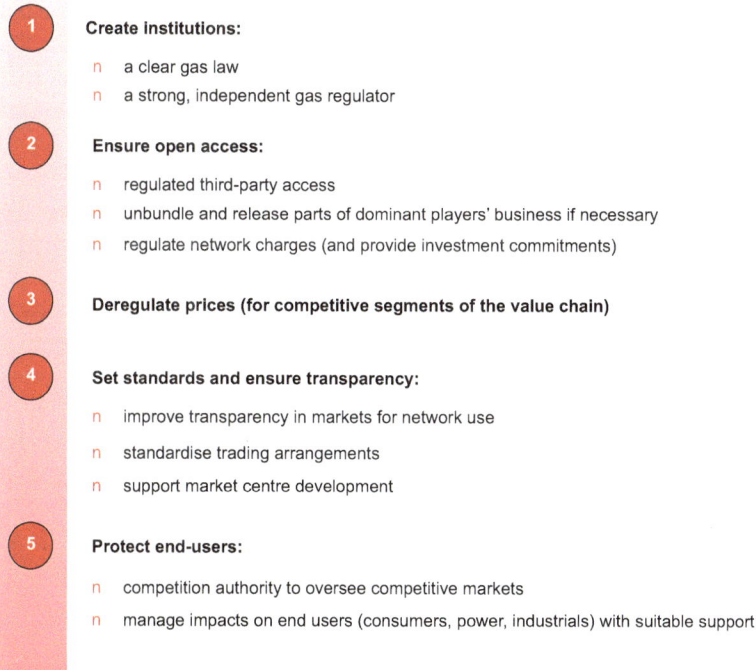

Fig. 18.2 Five steps to successful natural gas market liberalisation. *Source* Vivid Economics

manner by involving stakeholders, publishing evidence and final decisions, and allowing appeal in court. The regulator should have access to adequate financial and human resources to carry out its tasks satisfactorily.

The US Federal Energy Regulatory Commission (FERC) is a classic example of a natural gas market system foundation operating well. FERC was established in 1997 and is charged with overseeing industries including electricity, hydroelectricity, oil and natural gas. In all of these industries, participants often hold significant market power. FERC regulates the markets to ensure that companies do not misuse their monopoly positions, and its regulatory objectives range from preventing discriminatory service and unfair pricing to promoting environmentally sound infrastructure. FERC's main responsibilities in the natural gas industry are regulating the rates and services offered by interstate pipeline companies, certifying and permitting new pipeline construction, enforcing competition and preventing market manipulation where markets exist, and handling related environmental issues.

2. Ensuring open access

Owners of natural monopoly infrastructure, such as pipelines, have an incentive to misuse their market power, particularly when they are vertically integrated across the natural gas value chain. Regulatory action is generally required in five specific areas:

- **Third-party access**: This requires owners of networks with natural monopoly characteristics to grant access to their assets to interested third parties. (A detailed review of third-party access experience is presented in Sect. 19.2.)
- **Regulating network charges**: To prevent monopolistic pricing of transportation services, the tariffs that a network owner and operator can charge for transportation should be regulated. The allowable tariff should be set at a level that encourages investment, but does not price shippers out of the market.
- **Unbundling ownership**: A vertically integrated company that has natural gas

extraction, sales and transportation businesses has an incentive to engage in anti-competitive behaviour, such as charging excessive rates for transportation services, to prevent third parties from competing with its upstream extraction, downstream sales units or both. To enable non-discriminatory open access to networks, the interests of any extraction or sales business should be separated from the transportation services business. This is ideally achieved by unbundling or separating businesses operating in different segments of the natural gas value chain. (A detailed review of unbundling experience is presented in Sect. 19.3.)

- **Transparency on capacity availability and the terms and conditions of infrastructure use**: Open access and unbundling alone are not sufficient to deliver liberalised markets, and provisions to ensure transparency around midstream transport infrastructure use are vital in enabling competition, both upstream and downstream.
- **Release parts of dominant players' business**: Dominant players tend to have long-term contracts with suppliers and customers. The regulator should intervene to break up these contracts so that sufficient market volume is available for entrants.

Granting open access to networks is often initially attempted through negotiated third-party access or voluntary arrangements, leaving network owners free to negotiate terms and conditions for access. All five jurisdictions that were reviewed adopted this approach at first. However, negotiated agreements consistently fall prey to market power abuse by network owners that are vertically integrated with interests throughout the value chain, for example in production, import capacity or local distribution. Such firms have a clear incentive to use their power to prevent entry of competitors.

Third-party access should be regulated carefully, with provisions for allowable network use charges that provide incentives for investment, terms and conditions of use, and transparency around capacity availability, tariffs and terms of use. Even with regulated third-party access, incumbent market supply dominance may need to be proactively broken by the regulator. For example, the United Kingdom regulator forced British Gas, the dominant incumbent in the UK natural gas market, to release parts of its long-term supply contracts in the 1980s so that new entrants could start to compete and serve the market instead.

In terms of establishing fee rates, regulatory bodies should satisfy the requirement to permit open usage of gas transmission infrastructure while also ensuring that gas transmission infrastructure is able to receive ample investments. There are essentially two methods of establishing fee rates: one is the sum of costs method, setting total revenue as equivalent to investment and operating costs plus a reasonable return. Another approach is the incentive-managed pricing utilised by the European Union and United States in recent years, in which pipeline operator costs and permitted revenue are separated and, based on cost baselines or cost analysis models, a baseline revenue is established, along with an additional permitted revenue to incentivise supply quality and other set goals. Together with these two goals, the controlled rates of return always promote investment as an important factor, with set levels both encouraging investment and also not leading to excess investment or insufficient investment. Sometimes, in order to encourage investment, infrastructure need not be required to provide open usage, for example with LNG terminals in the United States and Britain.

Unbundling is a key element in ensuring non-discriminatory access to infrastructure. Owners of pipelines have an incentive to offer favourable terms to their own upstream extraction and downstream sales businesses over other third parties. Authorities observed this anti-competitive behaviour by incumbent firms in the United States in the 1930s and in the United Kingdom in the 1980s. The response in both countries was to require significant unbundling in addition to open access, which facilitated the entry of gas shippers and brokers into wholesale and retail markets. The European Union allows for different forms of unbundling. It allows

member states to keep networks under incumbent vertically integrated ownership, with strict arrangements to ensure independent decision-making about transmission system operations. This reflects the desire of EU member states to retain control over who owns and operates vital infrastructure and to maintain a certain degree of buyer power with major import sources. It also reflects the reality that some member state markets may be too small to achieve real competition, and the costs of liberalisation could outweigh the benefits.

Provisions to ensure transparency around midstream transport infrastructure use are vital in enabling competition. Open access and unbundling, even when required by law, are not in themselves sufficient to deliver liberalised markets. In the 1980s, the liberalised United Kingdom natural gas market was slow to attract entrants and generate competition because market participants were not able to acquire essential information on capacity availability and the terms and conditions of infrastructure use. Since then, the United States and United Kingdom have mandated that information about storage availability and capacity and rates for pipeline transport be published on bulletin boards. Compliance is monitored and enforced by the relevant market regulators and competition authorities.

In Japan, open access to LNG terminals has not yet resulted in third-party use, partly because market participants do not have sufficient information. Moreover, following the Fukushima nuclear plant disaster, liberalisation lost momentum in Japan as the country moved to secure increased LNG imports. Japan now encourages collective LNG purchasing by its natural gas importers to protect the economy against volatile LNG prices, although the debate around further natural gas market liberalisation, including third-party access to LNG terminals, resumed in 2014. Similarly, in South Korea, winter peak natural gas demand concerns and volatile LNG prices have led to the protection of its vertically integrated and publicly owned monopoly supplier, Korea Gas (KOGAS), to sustain its buying power on global markets.

Finally, it is important to recognise and address any existing long-term contracts held by the incumbent that create barriers to entry by limiting available capacity and restricting access to midstream infrastructure. These legacy contracts are generally long-standing, reducing capacity available for third parties and therefore preventing competitive forces from taking effect. Regulators can intervene to break up long-term contracts incumbents may have with suppliers and end users, to reduce incumbents' market share, and to open the market for new entrants. Breaking up existing long-term contracts was a key step in liberalising markets in the United States and United Kingdom: dominant market players were forced to release capacity from contracts to enable market entry and competition.

3. Deregulating prices

Prices should be deregulated where conditions exist for the establishment of competitive markets, for example in upstream exploration and production and in the downstream wholesale and retail markets. Deregulating prices is necessary to ensure that economic signals about supply and demand fundamentals are transmitted between market participants. Markets tend to fail when they exhibit natural monopoly characteristics, and in these stages of the value chain—namely transmission and distribution—prices need to be regulated. Elsewhere, however, participants are able to allocate resources efficiently when prices are left to respond freely to changes in supply and demand fundamentals.

Wholesale prices are key transmitters of economic signals because they form the link between upstream exploration and production and the rest of the value chain. Achieving a competitive natural gas market involves a phased liberalisation of wholesale prices. The United States evolved from cost-plus wellhead prices, using netback[2] and oil-linked contracts, to a system in which spot and futures contracts are the dominant types, traded by a diverse set of

participants. The UK market evolved from a price-setting single buyer—British Gas—linking contracts to substitute fuels and inflation to regulated caps and, finally, to more competitive hub-based trading. The end point of price deregulation in these markets has been a liquid and responsive wholesale market.

These market-based prices imply a move away from long-term contracts, which tie the price of natural gas to that of another commodity, or regulated prices, which are set by the government. This has been changing as the volume of natural gas traded, particularly LNG trade, has expanded. In the United States, natural gas is now almost entirely based on market prices, whereas in the European Union it is split evenly between market-based prices and long-term oil-indexed contracts.

Long-term contracts linked to prices of substitute fuels, spot markets prices or other indices offer a degree of certainty for investors to support upstream development. For example, when combined with netback pricing, a take-or-pay clause, which provides a penalty payment if natural gas deliveries are cancelled, can ensure a steady income stream for decades and enable investment in exploration and production. Such clauses are common in the contracts for the Groningen natural gas in The Netherlands. However, long-term contracts may lead to low spot market liquidity and price uncertainty. Moreover, as discussed previously, breaking up existing long-term contracts is crucial to ensure open access, enabling greater third-party access to midstream infrastructure and market entry and competition in the upstream and downstream segments.

Reaching a stage of highly liquid wholesale spot markets can also support upstream development. For example, in the European Union, once natural gas hubs gained sufficient liquidity, contracts were increasingly linked to natural gas spot markets rather than oil prices. In the United States, shale development relied on the existence of liquid spot markets. Developers of US shale plays depended on spot markets because they were unable to secure attractive long-term contracts because of the high uncertainty around field production.

---

[2]Cost back value pricing uses a price equation based on the actual price collected from end users.

4. **Setting standards and ensuring transparency**

To enable new entrants to enter the market, transparency around market rules is essential, as is a regulator with appropriate legal authority and independence from political parties, government branches and industry in order to enforce transparency requirements. This includes transparency around the terms and conditions for third-party access to networks and disclosure of available capacity on networks. It may also include measures to enhance competition and increase market liquidity in the upstream and downstream segments and, as a result, improve transmission of economic signals between market participants. For example, the regulator can increase upstream competition through the design and implementation of licensing and fiscal regimes using measures such as allowing trade in upstream licences or setting beneficial fiscal terms for exploration and production. The regulator can also standardise trading arrangements in downstream wholesale markets, for example through a network code or other guidelines that help build trust among market participants, or by standardising contracting arrangements to support the creation and development of a natural gas hub.

The regulator can also promote the use of market centres, such as in the case of Henry Hub in the United States and the National Balancing Point (NBP) in the United Kingdom. In the United States, FERC has supported the development of market centres by requiring market players to provide certain services and by standardising trade arrangements. In the United Kingdom, a virtual hub, the NBP, emerged as a result of the Network Code in 1996, which required a location where its balancing mechanism could be implemented by the regulator. The Code was complemented by a set of natural gas trading arrangements that established contract standards, allowing parties to trade natural gas with confidence in contracts.

5. **Protecting end users**

To maintain a competitive natural gas market, regulatory oversight needs to include a competition authority and provisions for managing impacts on end users. A competition authority would be responsible for ensuring well-functioning markets that provide end users with the greatest choice of suppliers at the lowest price. It could also protect consumers, power generators and industrial users that may be disproportionally or inequitably affected by any market reform or resulting market outcome.

Consumer and end-user protection requires the appropriate legal, financial and political support. In the United States, from the 1950s until the late 1970s, consumer protection and, by extension, welfare distribution influenced price setting for natural gas. However, policies to protect consumers and other end users should be disconnected from the gas market liberalisation process. In the long term, liberalised markets generate competition that drives down costs for consumers and delivers sustained low prices. During the transition period, as natural gas markets liberalise, social welfare and equity considerations may result in additional support being provided to the most exposed sections of society and sectors of the economy. The most economically efficient approach would be to allow natural gas markets to liberalise, including price deregulation, and to provide income or other support separately to consumers and end users, for example using social welfare payments or lump sum subsidies. Such an approach would not interfere with the efficient functioning of the natural gas market, while ensuring that the most vulnerable businesses and individuals are protected.

18.2.2　Political and Economic Factors of Market Liberalisation

The sequencing and timing of liberalisation is significant, and international experience suggests a sequence of reforms that includes key milestones and potential points of tension (Fig. 18.3).

Initially, vertically integrated incumbents tend to supply all or most natural gas, acting as the monopoly producer and supplier of natural gas. Supply and demand are unresponsive as there is no free price formation that enables transmission of economic signals. The first step of the liberalisation process is to create a natural gas law and establish a natural gas regulator to implement market reforms, including open access to the existing infrastructure and networks for third parties.

Alongside measures to encourage new entrants and facilitate greater competition in the upstream and downstream markets, the liberalisation process requires that a start is made on deregulating prices to that they reflect market supply and demand conditions, for example wholesale markets moving from simple cost-plus pricing to more responsive approaches, such as netback pricing. For midstream infrastructure, network charges need to balance increased third-party access with incentives to invest in

maintenance and new network capacity. In the initial stages of liberalisation, incumbents tend to battle to retain their market power, by putting up barriers to entry, contesting regulatory interventions and retaining their long-term contracts. They may also withhold crucial information from the market, such as the capacity available on their networks and information on terms and conditions of third-party access.

The next step in the liberalisation process is to increase market transparency, for example by requiring network owners to disclose network capacity and the terms and conditions for access. In addition, at this stage the regulator may intervene to break up long-term contracts held by incumbents with suppliers and customers. A mature market is one with a strong regulator that can ensure and secure competitive market outcomes and a diverse set of players across the value chain. These players include some integrated firms, but also new entrants, such as pure natural gas shippers, marketers and brokers. If liberalisation is extended to cover distribution networks, natural gas retailers may also emerge to compete for smaller end users. Building on this, the next step would be the establishment of standardised trading arrangements, a balancing mechanism and service requirements to support the development of market centres or hubs as platforms to

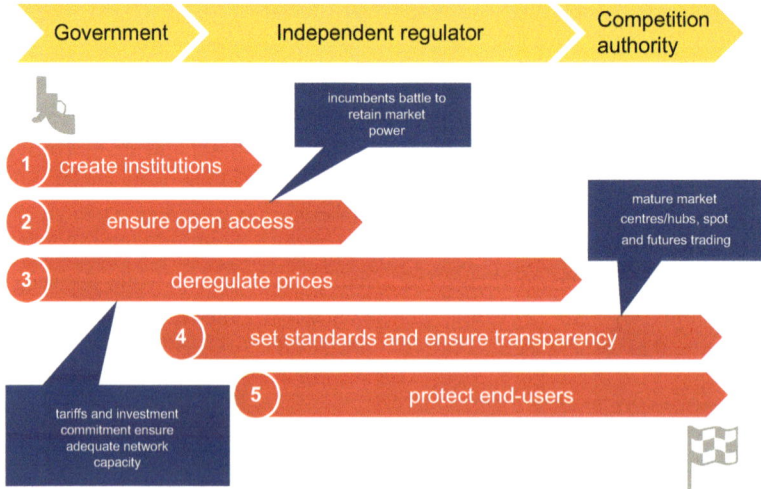

Fig. 18.3 The sequence of five regulatory reforms and potential market reactions. *Source* Vivid Economics

trade short-term contracts. This allows market liquidity to build and economic signals to be transmitted between the demand and supply sides of the market.

As the market matures, the emphasis of regulatory oversight shifts to end-user protection and the competition authority takes on the role of ensuring fair competition. The focus of policy also shifts to mitigating undesirable impacts of the reforms on end users.

However, a key observation from the case studies is that the reform of natural gas markets is a lengthy and difficult process and is likely to fail if the political context, fundamentals and ancillary policies are not aligned favourably. The United States in the 1970s and 1980s exemplifies a context conducive to natural gas market reform. At the time, there were many upstream producers and network infrastructure was well developed. In addition, there was strong political pressure to make the market more efficient, especially downstream wholesale and retail markets, following the failure of natural gas supply to respond during the 1970s oil crisis. Similarly, in the United Kingdom in the 1980s, the political and market context were conducive to reform. The conservative government of Margaret Thatcher strived to resolve poor government finances by privatising publicly owned industries, including vertically integrated electricity and natural gas companies. Further, the government allowed natural gas to be used in power generation, an opportunity for investors because of the emergence of cost-effective combined-cycle natural gas turbine technology.

On the other hand, liberalisation processes in the European Union, Japan and South Korea provide examples of protracted and incomplete efforts to open up the natural gas value chain. Even today, the European Union faces political opposition in its attempt to liberalise continental natural gas markets, a process that began in 1998. Opposition focuses on concerns over supply security as a result of greater import dependence, as well as over ownership of infrastructure that is in the national interest. Increased global tension centred on Russia, Europe's main natural gas supplier, and the Ukraine illustrates the risk.

Recently announced proposals for a European Energy Union seek to balance these concerns while continuing to push for greater liberalisation of natural gas markets.

Similarly, in Japan, energy security considerations following the 2011 Fukushima nuclear disaster have caused a setback to the liberalisation process, driving towards more co-operation, rather than competition, among the regional monopoly suppliers in securing long-term LNG contracts. However, in 2014, further liberalisation appears to have returned to the agenda. South Korea has also faced difficulties over the past two decades as it tried to liberalise its natural gas markets, largely because of the power of the KOGAS labour union and reluctance by KOGAS to give up its buying power on world LNG markets.

18.2.3 The Impact of Market Liberalisation on Domestic Mining

The experience of countries with significant domestic natural gas resources, such as Norway, The Netherlands, the United Kingdom and the United States, has been for market liberalisation to drive greater development of these resources. These countries have tended to liberalise earlier and more comprehensively than countries without significant domestic natural gas resources. Before liberalisation, upstream exploration and production tended to be dominated by state-owned vertically integrated monopoly producers. Opening the sector to competition helped drive greater efficiency and innovation in exploration and production of natural gas, reducing costs and facilitating the development of greater volumes and sources of domestic supply, as seen in the European Union. In addition, access to midstream infrastructure and competitive wholesale and retail markets have attracted new entrants and fostered greater competition in upstream exploration and production. With such access, new upstream entrants can be confident of being able to transport natural gas to wholesale and retail suppliers, and eventually to end users.

Liberalisation of the natural gas value chain has also been supported by favourable upstream licensing, fiscal or environmental policies in these countries. In the United States, private ownership of land and the resources underneath it enabled rapid development of shale natural gas exploration and production. Where resources are owned by the government, such as on the outer continental shelf of the North Sea in Norway, The Netherlands and the United Kingdom, processes are in place to enable effective and competitive leasing of exploration and production rights. For example, licensing arrangements in the United Kingdom are designed to encourage exploration and production and penalise hoarding; UK licences can be taken away if companies do not follow the work programme proposed in their bid for exploration and production rights. The Netherlands developed its large Groningen natural gas field through a 50:50 public-private partnership arrangement, which provides a steady income stream to encourage new entrants and greater investment in upstream exploration and production, while also ensuring that the investments were economic and competitive. Specific fiscal incentives and other support policies are also provided, such as in the United Kingdom, to incentivise exploration and production in more-challenging areas.

18.2.4 Market Liberalisation and Ancillary Policies

There are many interactions between liberalised gas markets and ancillary policies. Ancillary policies, relating to the environment or to technology development, affect the evolution of natural gas markets. Analysis of the case studies shows that while some of these interactions have been anticipated, others have been unpredictable.

Environmental policies—such as the US Clean Air Act Amendments in the 1990s, the more recent US Environmental Protection Agency carbon emission standards, the European Union's emission trading system to price in carbon emissions, and the EU air quality directives and regulations—can increase the costs of emissions-intensive coal relative to natural gas. While ambitious policies on climate change and air quality favour gas over coal in electricity generation, the design of these policies could produce some unintended consequences. This has been the experience in the European Union, where the interaction between climate policy and energy markets has undermined environmental objectives and led to the coal-renewables "energy paradox".

While the European Union has an overall target to reduce greenhouse gas emissions by 20% by 2020 over the 1990 level, it also has a separate target for a 20% share of renewables in the energy mix by 2020. The separate renewables target and the large subsidies provided to achieve it have dampened wholesale energy prices because of the low—and at times negative—marginal cost of renewables. This has, in turn, reduced incentives to invest in back-up fossil fuel capacity, which is required to compensate for the potential of intermittent supplies from renewable sources. Moreover, flaws in the design and implementation of the EU Emissions Trading Scheme—the European Union's primary mechanism for meeting its overall greenhouse gas emissions target—combined with recent weak economic conditions and the relatively low price of coal, have shifted the energy mix towards coal and renewables at the expense of natural gas.

These unintended consequences of energy and climate policies have undermined the profitability of power suppliers and destabilised their business models. The problem is illustrated by the challenges faced by companies like E.ON and RWE in Germany, whose marginal cost pricing-based business model has been put at risk as subsidised and distributed sources of renewable energy of significant size enter the energy mix. Going forward, Europe renewables are likely to remain a significant part of the energy mix to 2030 if Europe is to meet its climate policy ambitions, and electricity market regulation will need to evolve alongside climate policies. In turn, this will have knock-on impacts for natural gas demand and markets.

The natural gas market has also benefited from ancillary interventions, such as energy-related research and development support and subsidies. This has been particularly significant in the context of US shale natural gas development. Various research and development programmes related to unconventional natural gas were set up by the US Energy Research and Development Administration in the 1970s and continued by the Department of Energy. These programmes, along with other research and development programmes, such as those centred on microseismic fracture monitoring, have proved instrumental in driving the US shale natural gas boom of the last decade.

18.3 Case Studies of Natural Gas Market Liberalisation

Careful study of individual national natural gas markets makes clear the precise nature of the market liberalisation that has occurred as the natural gas value chain has developed. Each market's liberalisation has its own characteristics, but the ultimate goal is always to establish a more vibrant natural gas system that supports the nation's strategic goals. Understanding the development of these markets can provide valuable reference points as China formulates its energy system objectives.

18.3.1 Case Study 1: United States

The development of a dynamic and competitive natural gas market in the United States (Fig. 18.4) reveals three main phases of regulation. These phases are set against a backdrop of events, such as the shale natural gas boom and key regulatory changes, as well as varying levels of gas supply and gas prices between 1960 and 2013.

The pre-liberalisation phase was marked by gradually increasing levels of regulation over

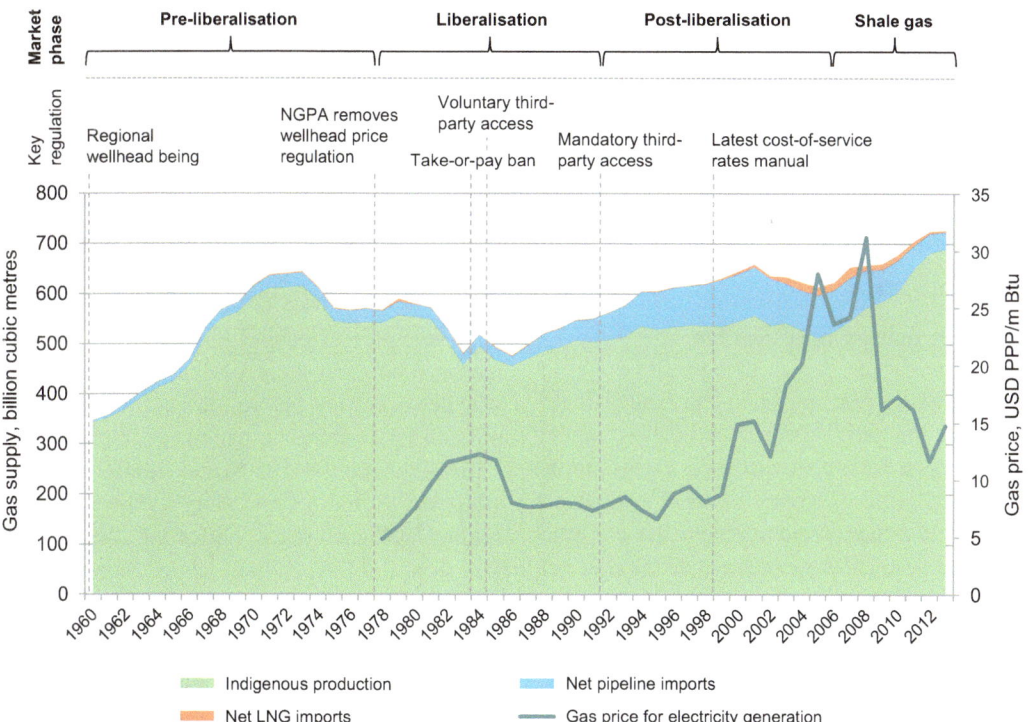

Fig. 18.4 Phases, development and key regulations in the establishment of a liberalised natural gas market in the US, 1978–1992. *Note* Nominal natural gas prices paid at power plant, national currency converted to average GDP purchasing power parity. *Source* Vivid Economics, based on IEA data

15 years, which began as early as 1938 and reached full wellhead price regulation in 1954. These price controls were justified on the basis of the perceived natural monopoly characteristics of the natural gas value chain, warranting heavy state intervention to protect consumers from abuse of market power. These interventions echoed earlier interventions in the oil value chain to counteract the anti-competitive practices of Standard Oil and other companies. The price controls suppressed investment and led to significant supply shortages, particularly during the 1970s oil crisis, which were only ameliorated by the subsequent recession.

In response, a new market authority, FERC, was established and several orders to reform the market were enacted from 1978, marking the start of a period of liberalisation. The reforms unbundled the natural gas industry, separating transportation and sales businesses, and ensured open access to interstate pipeline infrastructure for third parties. By 1992, after 14 years, the liberalisation programme had largely been completed. The post-liberalisation phase saw the enactment of various orders to refine the unbundling and open access arrangements, but the fundamentals for a competitive and efficient market had been established and they supported a 40% increase in natural gas consumption from around 500 billion m^3 in 1990 to 700 billion m^3 by 2010.

1. Before market liberalisation

The natural gas market in the United States started in the early 1900s. Before then, natural gas produced from coal was used locally. In the early 1900s, the discovery of large natural gas basins in the Southwest triggered the construction of large-scale natural gas transmission networks to reach the population and industrial centres of the United States. Vertically integrated companies started selling natural gas from basins in the Southwest to end users on the East Coast, the Mid-Atlantic Coast and the West Coast.

A 1935 Federal Trade Commission investigation found high levels of market concentration and abuse of market power by vertically integrated companies in the exploration, production and transportation of natural gas. In response, the Gas Act of 1938 established the Federal Power Commission (FPC) to regulate interstate pipelines and ensure "just and reasonable" wholesale prices. Regulation of intrastate pipelines was left to state regulators. Any purchase of natural gas for interstate transport and sale to a local distributor now required a certificate from the FPC, which set a maximum price for the natural gas. This price allowed pipeline companies to recover some of the costs incurred in pipeline construction and operation. The exploration and production of natural gas remained exempt from federal regulation: wellhead prices remained unregulated. The final sale price to end users was also unregulated. The FPC further gained authority to approve the construction and operation of facilities and the provision of services used in interstate natural gas transmission.

The industry defended deregulated wellhead prices provided for under the 1938 Gas Act. However, by the 1950s, consumer groups were championing a system of regulated prices for both producers and pipelines based on the perceived natural monopoly characteristics of natural gas supply and fear of market power abuse among vertically integrated companies. This debate culminated in the Phillips Decision, which established full wellhead price control by extending the powers of the FPC to "the rates of all wholesales of natural gas in interstate commerce, whether by a pipeline company or not and whether occurring before, during, or after transmission by an interstate pipeline company."

Throughout the 1950s and 1960s, the FPC struggled to find a satisfactory methodology for setting wellhead rates. Following the Phillips Decision, the FPC initially aimed to establish a cost-of-service rate for wellhead natural gas for each producer. It was administratively challenging and a large backlog developed, complicated by the wide variety of producers, contract types and cases. In 1960, the FPC adopted an averaged approach based on a regional pricing strategy, setting provisional price ceilings based on the average costs of exploration and development in

24 producing regions. There were many hurdles, including a wide variety of production cost profiles within the established regions, and as a result the process was protracted and it was not until eight years later, in 1968, that the first permanent price was established.

This approach to setting prices led to significant natural gas shortages in the 1970s. Rising energy prices, particularly during the oil crisis, led to increased demand for comparatively inexpensive natural gas. Unfortunately, as a result of wellhead price regulation, there was no incentive to invest in exploration and production and, as a result, the supply response was very limited. Significant shortages prevailed by the mid-1970s, and often supply was curtailed for some customers and new customers were refused connections. The Southern states were the exception to these shortages: consumers in these natural gas-producing states remained able to source the natural gas they wanted because these intrastate markets were not subject to interstate wellhead price controls.

2. Market liberalisation

The Natural Gas Policy Act of 1978 sought to improve the functioning of natural gas markets by phasing out price controls and establishing a new natural gas market regulator, FERC. The United States had found itself vulnerable to the oil crisis, partly because of the way it regulated upstream wellhead and interstate pipeline prices of natural gas. The new act sought to increase security of supply by improving incentives to invest in indigenous natural gas production. It also aimed to establish a single national natural gas market, with producers allowed to choose their own wellhead prices. Producer market power would be addressed through competition within this new single market.

The act also established a complex transitional system of escalating wellhead price ceilings across more than 30 categories of "old" and "new" natural gas. To encourage new supply, but prevent producers of already-committed natural gas from benefiting from increasing prices, the act established higher initial price ceilings and an

accelerated schedule of wellhead price decontrol for new contracted natural gas, which did not apply to old natural gas supplies. The act scheduled the full phase-out of all price controls by the end of the century. But after just 10 years, regulators reviewed the wisdom of market and regulated prices coexisting for such a long period, and the Gas Wellhead Decontrol Act of 1989 hastened the phase-out schedule, dictating that all price ceilings were to be removed by January 1, 1993.

A single market required a single price, so the differential inter- and intra-state pricing had to be ended. The act replaced the FPC with FERC, which was granted authority over interstate natural gas trade (see box "Overview of the US Federal Energy Regulatory Commission").

Overview of the United States Federal Energy Regulatory Commission

The US Federal Energy Regulatory Commission (FERC) was created under the 1977 Department of Energy Organization Act as a government agency independent from political party influence or affiliation, other branches of government and industry participants. It has a staff of about 1500 and an annual budget of $300 million. It is governed by five commissioners, nominated by the president and confirmed by the US Senate. Each commissioner serves a five-year term, and no more than three can come from the same political party. The commission operates by majority rule.

In a testament to FERC's independence, the commission's decisions are reviewed by a court, rather than Congress or any other branch of government, and private discussions during case proceedings are prohibited. In addition, there is a clear distinction between FERC authority, comprising interstate business and infrastructure, and authority held by state governments on intrastate business and infrastructure.

FERC oversees electricity, hydro power, oil and natural gas. Its activities centre on ensuring that companies with natural monopolies do not misuse their market positions, and its regulatory objectives include:

- preventing discriminatory or preferential service;
- preventing inefficient investment and unfair pricing;
- ensuring high-quality service;
- preventing wasteful duplication of facilities;
- acting as a surrogate for competition where competition does not or cannot exist;
- promoting a secure, high-quality, environmentally sound energy infrastructure through the use of consistent policies;
- where possible, promoting the introduction of well-functioning competitive markets in place of traditional regulation;
- protecting customers and market participants through oversight of changing energy markets, including mitigating market power and ensuring fair and just market outcomes for all participants.

The primary responsibilities of FERC in the natural gas industry are regulating the rates and services offered by interstate pipeline companies, certifying and permitting new pipeline construction and handling some closely-related environmental issues.

The 1978 Natural Gas Policy Act led to rising prices, and an increase in long-term take-or-pay contracts. This moved FERC to intervene in 1984 as market conditions drove natural gas prices down, but left buyers paying high prices under long-term contracts. Natural gas contracts in the 1970s were typically multi-year purchasing agreements with take-or-pay clauses obliging buyers, including interstate pipelines and local distribution companies, to pay for a contract

volume, whether or not they took the natural gas. The contracted sales allowed upstream companies to recover their investments in exploration and production and provided buyers with a hedge against further expected price increases as a result of shortages.

For a few years following passage of the act, many contracts tied prices to the escalating regulated price ceilings. This became problematic in the early 1980s, when natural gas demand and petroleum market prices declined under prevailing market conditions, diverging from the price ceilings. As a result, some buyers were locked into long-term contracts under which they were paying prices far above spot market prices. In 1984, FERC Order 380 alleviated the burden of take-or-pay contracts by eliminating minimum bill obligations for local distribution companies. The minimum bill represented the amount of natural gas that had to be paid for whether or not the natural gas itself was received. However, although local distribution companies were released from this requirement, pipeline companies remained locked into take-or-pay contracts with natural gas producers.

In 1985, FERC moved to unbundle pipeline and sales services with Order 436, introducing voluntary third-party access to interstate pipelines. This unbundling of transportation and sales services enabled alternative sales arrangements to the merchant package, or the sale of natural gas plus its transportation, which had traditionally been offered to local distribution companies. The order provided incentives to interstate pipeline companies to offer standalone transportation services, in the form of blanket certificates that allowed them to engage in new activities such as opening new facilities without prior authorisation from FERC. Third-party access to pipelines remained voluntary and was to be offered on the basis of negotiated rates. The order did not address open access to other services, such as storage facilities. The changes to voluntary third-party access did not resolve the high cost of take-or-pay contracts of the 1970s, which still haunted pipeline companies. These were addressed by FERC two years later, by Order 500, which allowed interstate pipeline companies to shift some of the take-or-pay liabilities to other businesses in the value chain, leading to the voluntary renegotiation of most remaining liabilities.

FERC Order 636 of 1992, known as the Restructuring Rule, marked the final stage of natural gas market liberalisation, fully implementing the third-party access regime. The order required interstate pipeline companies to unbundle their sales and transportation services, ensuring that natural gas of other suppliers could receive the same level of service as previously enjoyed by the pipeline company's own natural gas sales. It diminished the market power of pipeline companies and increased competition among natural gas sellers.

The major provisions of Order 636 were:

- A requirement that interstate pipeline companies provide open access to transportation and storage services that are equal in quality, regardless of whether the natural gas is purchased from the pipeline company or elsewhere. This included a provision for companies to create an internal firewall to prevent exchange of information between marketing and transportation divisions. It also included provisions for legal unbundling, requiring companies operating pipelines, storage and LNG facilities to restructure production and marketing branches as arm's-length affiliates and cease all-merchant services.
- Encouragement of market centres where several pipeline systems interconnect, including mileage-based rather than fixed-tariff or "postage stamp" rate setting for natural gas transportation.
- Establishment of a capacity release market in transportation and storage, with a requirement for pipeline companies to maintain electronic bulletin boards to provide information about availability of services on their systems.
- Redesign of regulated rates for pipeline transportation, shifting all fixed-cost recovery onto firm customers with a daily capacity reservation and not onto interruptible customers without a daily capacity reservation.

Variable costs were to be recovered through a usage fee based on the natural gas transported. This was intended to eliminate any distortions in purchasing behaviour in response to the previous design, which allocated certain fixed costs to the usage fee.
- Requirement for pipeline companies to offer "no notice" firm transportation services to local distribution companies, allowing customers to receive natural gas on demand up to their maximum contract level.

Order 636 represented a watershed moment in US natural gas market development. It introduced competition among natural gas suppliers and greater efficiency in the use of natural gas industry infrastructure. The industry restructuring which began with voluntary third-party access under Order 436 and led to mandatory unbundling under Order 636 fundamentally changed natural gas transportation rates and patterns. According to the US EIA, the higher level of flexibility and resulting increased competition among natural gas suppliers "has contributed to changes in regional production, transportation, and consumption patterns, and to greater efficiency in the use of the natural gas industry infrastructure."

As a result of the new regulatory regime, entities purchasing at the wellhead can negotiate for spot, short-term and long-term contracts at various rates. The market can operate with firm and interruptible delivery of spot natural gas on a contract-by-contract basis rather on the previously prevailing long-term contract obligations. The electronic bulletin boards facilitate offers and bids for natural gas and for pipeline space. Marketers and brokers trading on spot and futures markets collectively clear the market, with higher prices leading to increased wellhead and pipeline capacity supply.

3. The impact of market liberalisation

Liberalisation of the US natural gas market has created a system with unique characteristics.

In the upstream market, regulation of exploration and production is mostly within the jurisdiction of states, and therefore highly diverse. At the end of 2013, the United States had 9.3 trillion m^3 of proved natural gas reserves, and a wide range of state regulatory provisions covering licensing, ownership and safety and the environment of exploration and production across the country.

Key elements of upstream regulation in the United States are:

- Ownership of resources resides with ownership of land. In the United States, land is mostly owned by private entities. Onshore and offshore resources owned by the federal government are governed by the Department of the Interior and the EPA, which control and administer leases, revenues, land use planning and environmental standards.
- States generally have regulatory commissions that deal with all aspects of the energy value chain within the state. Development of privately owned and federal resources needs to comply with state regulations on safety and the environment to ensure that resource development is safe and environmentally responsible and therefore in the public interest.
- Exploration and production rights for federal resources are auctioned to private entities on an open and competitive basis once land use planning has been determined. Prospective lease holders bid for exploration and production rights. They pay an initial bonus, and then rent for the right to develop the resources. Leases are valid for 10 years or as long as there is at least one producing well.

In terms of regulation of midstream assets, FERC has continued to refine the regulatory framework since 1992 to promote market efficiency and increased levels of competition. Notable further improvements in setting rates and regulating unbundling include:

- In 1999, FERC published a cost-of-service manual, listing transparent guidelines for FERC's determination of allowable project cost recovery by pipeline companies on the basis of a reasonable rate of return, operating and maintenance costs, depreciation charges and tax recovery mechanisms. This rate of return regulation is a general model for the treatment of natural monopoly utilities, including pipelines.
- The 2000 FERC Order 637 refined and updated pipeline transport regulations relating to rates, including the suspension of price caps on capacity release sales of less than one year, to achieve a greater transparency and efficiency in the use of pipeline services.
- FERC's Hackberry Decision of 2002 established that LNG import terminals would be exempt from open access requirements to encourage more LNG site development. It allowed the owner of the Hackberry LNG terminal to agree to terms of use with its affiliates rather than under regulated cost-of-service rates. This effectively defined LNG terminals as an upstream supply source, part of the competitive sector, rather than part of the natural monopoly transportation chain, with individual project approval depending on FERC assessment of the general need for the project as well as its economic and environmental impacts.
- The 2003 FERC Order 2004 introduced new functional unbundling regimes applicable to both electricity and natural gas industries, requiring that marketing, transmission and affiliated energy employees of electricity and natural gas companies function independently. The order established the Office for Market Oversight and Investigations to monitor the implementation of functional unbundling, which has since been absorbed into FERC's Office of Enforcement.
- The 2008 FERC Order 717 further reinforced functional unbundling. It included specific provisions for the physical separation of facilities and staff, for example the separation of occupational functions, which meant that employees can be responsible for work on marketing or on transport, but not both.

Downstream policy is generally subject to local regulations, and remains inconsistent across states. For example, intrastate natural gas commerce is subject to state regulation, and retail unbundling is at the discretion of the state regulator. Only four jurisdictions—New Jersey, New York, Pennsylvania and Washington, DC—have unbundled retail markets. Unbundling is delivered by means of natural gas residential choice programmes in these four jurisdictions. The residential choice programmes allow consumers to purchase the components of natural gas supply separately, rather than relying on the bundled products offered by a single local distribution company.

The regulatory framework has ensured that pipeline companies transport natural gas, but no longer buy and sell it, so they cannot abuse market power based on the natural monopoly characteristics of pipeline infrastructure. This has enabled the construction of 305,000 miles of transmission pipelines and the establishment of a high number of competing market participants. The US interstate and intrastate transmission network comprised more than 210 separate

pipeline systems (Fig. 18.5). This system is supported by ancillary infrastructure, including 1400 compressor stations, 24 hubs or market centres, 400 underground storage facilities, 49 pipeline import and export locations, eight LNG import facilities, and 100 LNG peaking facilities. This has also enabled competitive markets in the upstream and downstream segments. It has led to 6300 natural gas producers producing from more than 480,000 wells, including 21 companies considered major producers. There are 160 different pipeline companies, 123 storage operators and 1200 local natural gas distribution companies.

The liberalisation process, particularly the 1992 FERC Order 636, created the conditions for the rapid growth of natural gas trade through market centres such as Henry Hub as pipeline infrastructure improved and competition increased. Henry Hub, a physical trading hub in south Louisiana, has emerged as the most prominent market centre, with spot and future contract prices based on competitive natural gas

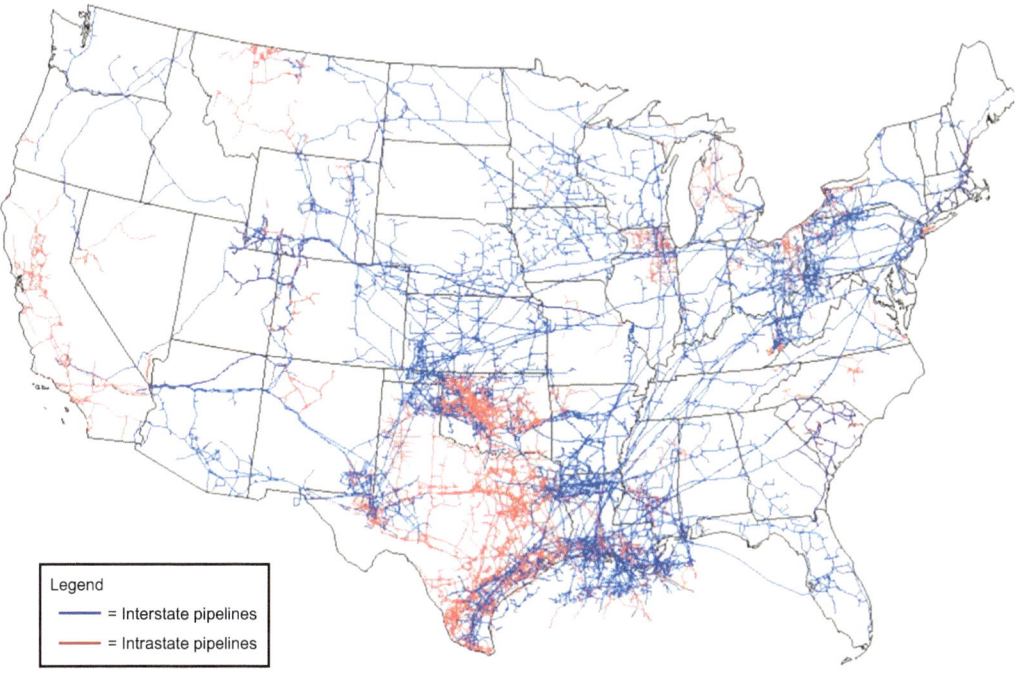

Fig. 18.5 US natural gas transmission pipelines. *Note* Data for 2009. About two thirds of the length is interstate transmission pipelines. *Source* US EIA

pricing rather than oil indexing. It was established by Sabine Pipe Line LLC, a subsidiary of Chevron, in 1989. In 1990, the NYMEX selected Henry Hub as the basis of its natural gas contracts because of its central location and liquidity. Henry Hub now sets the benchmark price for the entire North American trading area. It has emerged as the most liquid natural gas market in the world, trading more than 100,000 natural gas contracts a day, compared to 918 contracts on its first trading day. The scope of NYMEX futures contracts has extended from terms of up to three years in 1997 to up to 10 years at present.

Other hubs have emerged since, such as Opal in Wyoming, where price differentials compared to Henry Hub are determined by regional disparities in production and transport costs. The arbitrage opportunities between regional hub prices drive investment in pipeline capacity. Notably, the recent shale natural gas boom has led to trade volumes shifting to hubs that are closer to shale natural gas fields, such as the Dominion South hub, a key supply point in the

Marcellus shale in southwest Pennsylvania. This could result in a fundamental shift of the main price reference point in the United States away from Henry Hub.

The liberalised regulatory framework has enabled the market to transmit economic signals, and encouraged the rapid expansion of shale natural gas production once extraction technology became competitive. On the supply side, price signals led to a surge in shale natural gas production since the mid-2000s. Once drilling and hydraulic fracturing technologies were cost-effective, many firms adopted the technology and secured acreage. Economic growth and price hikes of substitute fuels led to rising natural gas prices in the 2000s and, in response, heavy investment in shale natural gas drilling by incumbents and new entrants alike. On the demand side, market liberalisation coincided with increasing levels of natural gas consumption throughout the economy (Fig. 18.6). The increase was greatest in electricity generation, where natural gas competes directly with

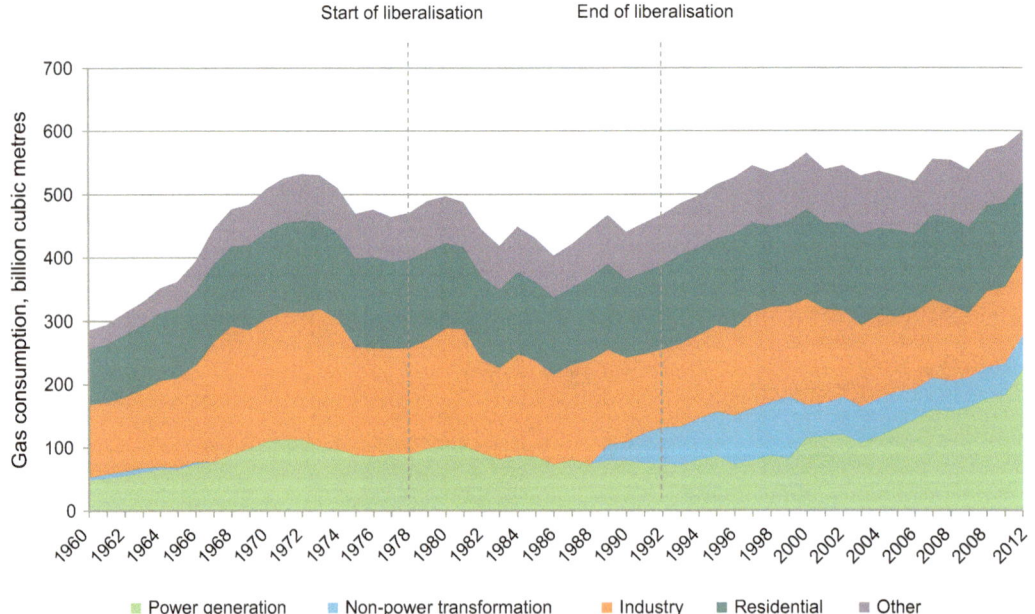

Fig. 18.6 US natural gas consumption before, during and after the period of market liberalisation. *Note* Excludes transport, non-energy use, energy industry own use and losses; non-power transformation includes CHP and natural gas to liquids plants; other includes the services, agriculture and fishing sectors. *Source* Vivid Economics, based on IEA data

alternative fuel inputs, especially coal, but also nuclear and renewables. Increased shale natural gas supply and falling natural gas prices led to a rapid increase in natural gas use for electricity in the last decade.

4. **Ancillary policies**

The natural gas market has not only benefited from the regulatory framework directly governing it, but also from ancillary intervention, including research and development subsidies and environmental policy.

On the supply side, various research and development programmes related to unconventional natural gas were set up by the US Energy Research and Development Administration in the 1970s and continued by the Department of Energy. These have been important for shale natural gas development. Other research and development programmes, such as those centred on microseismic fracture monitoring, have also contributed to shale natural gas development. Fiscal incentives including tax credits for the production of unconventional fuels under the Crude Oil Windfall Profits Tax Act of 1980 further enhanced the financial viability of shale natural gas extraction.

On the demand side, the competitiveness of natural gas compared to substitute fuels—mainly coal—is a key driver. Emission standards, such as those imposed under the Clean Air Act Amendments ("CAAA") of 1990, improved the competitiveness of natural gas relative to coal. With effect from 1995, 110 power plants were required to surrender CAAA emissions allowances. The programme was extended in 2000. More recently, in June 2014, the EPA proposed a new plan to cut carbon emissions from power plants, which is likely to further improve the competitiveness of gas.

18.3.2 Case Study 2: Europe

Liberalisation of the natural gas market in the European Union began as part of the creation of an internal energy market with the First Gas Directive in 1998. Regulatory efforts in the European Union have evolved over decades, passing through various milestones (see Fig. 18.7).

In 1988, the European Union articulated the goal of creating an internal energy market. It then took a decade for the European Commission (EC) to draft legislation that was acceptable to member states. The First Gas Directive passed by the European Parliament in 1998 included requirements for third-party access and unbundling, although these provisions had been much diluted in negotiations during the preceding decade. The Second Gas Directive, adopted in 2003 and complemented by further regulation in 2005, focused on implementation of third-party access and unbundling provisions introduced in the First Gas Directive. This was achieved through strengthening market oversight across all member states, allowing market authorities to establish cost-of-service tariffs for monopolistic infrastructure, and other mandates. The Third Gas Directive, commonly known as the Third Package, was passed in 2009 and focused on incentivising infrastructure development, providing for stricter unbundling, network planning and supply security. Despite the provisions of the Third Package, the European natural gas market —apart from the United Kingdom market—is not yet fully liberalised.

1. **Before market liberalisation**

Similar to the US experience, before liberalisation most EU member states had blanket regulatory regimes that treated the entire natural gas value chain as a natural monopoly. The natural gas industry structure in each country was similar across the bloc, with the upstream segment of the value chain often dominated by one or a few state-owned, vertically integrated natural gas companies engaged in exploration and production as well as transportation and sales. The downstream segment was dominated by local distribution companies, which controlled distribution and retail monopolies.

As state-owned enterprises, these vertically integrated national natural gas companies had

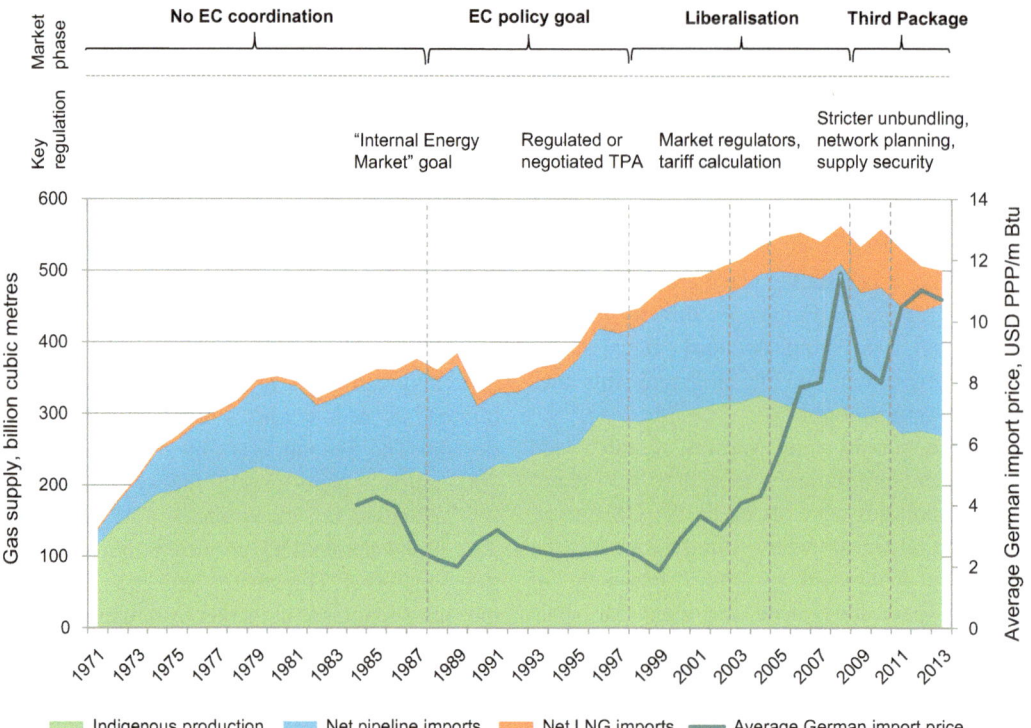

little incentive to improve the efficiency of their operations, which led to increases in natural gas prices. Natural gas markets were governed by national law and were generally exempt from EU competition law, which granted special status to public natural gas companies on the basis that application of competition rules might obstruct the supply of natural gas. The national laws governing public natural gas companies within member states were generally based on similar principles, including exclusive rights to build and operate networks, prohibition of entry, vertically integrated operations, compensation on the basis of historical costs and a high degree of central planning.

2. Establishing energy policy objectives

The EC established the creation of an internal energy market as a policy goal in 1988. The internal market is a foundational principle of the EU, codified in the 1986 Single European Act, and aims to facilitate the free movement of goods, persons, services and capital, with competitive markets in all sectors. Although not explicitly part of the Single European Act, a competitive energy sector was deemed a pivotal element in a well-functioning internal market: a competitive energy sector, and the associated energy cost reductions, contributes directly to improved   competitiveness   of   European industries. This position was elaborated in a 1988 EC green paper, *The Internal Energy Market*, which set out procedures for the creation of an internal energy market, including harmonisation of rules and technical norms, the opening up of public procurement markets and the removal of fiscal barriers.

The EC then set out to develop several directives[3] targeting a common carrier system, under which consumers could purchase natural gas from any supplier, regardless of the ownership of pipeline infrastructure. However, the directives faced strong opposition in the Council of Ministers, and the EC had to change its approach to a more staged, bottom-up approach. Complex negotiations ensued, and in 1993 the European Parliament took an active role in finding a compromise that was acceptable to both the EC and the Council. The development of the first Electricity Directive was given priority over a directive covering natural gas. The Electricity Directive was adopted in 1996, with provisions for electricity market liberalisation.

3. The beginning of market liberalisation

The First Gas Directive was adopted in 1998. It established a framework for liberalisation for member states, with provisions for third-party access and account unbundling. The directive sought to create a single internal market for natural gas in the European Union, but was a significantly diluted version of the initial EC proposals for a common carrier system. Member states were left free to choose their preferred pace and regulatory measures for natural gas market liberalisation.

The directive required owners of natural monopoly infrastructure, including transmission networks, storage and LNG facilities, to provide open access to third parties. However, member states could still choose between negotiated and regulated third-party access.[4] Vertically integrated companies were required to separate the accounts of natural monopoly infrastructure and sales businesses, but not necessarily to realise full functional unbundling. The directive further provided that power generators and end users consuming more than 25 million m^3 of natural gas a year should be able to choose their suppliers.

The concessions made for political reasons in the First Gas Directive hindered the opening of natural gas markets. The Madrid Forum, an annual meeting of natural gas market stakeholders, was established in 1999 with a remit to identify further harmonisation needs. With 9 out of 15 member states planning to open their natural gas markets fully by 2008, the EC moved to adopt the Second Gas Directive in 2003 and complementary regulations in 2005, establishing requirements for cost-of-service calculations and market oversight by regulatory authorities to expedite the creation of an internal energy market. This time, a new electricity sector directive was adopted alongside the Gas Directive.

The main provisions of the Second Gas Directive were:

- **Stronger unbundling**: Mandatory functional and legal unbundling for the transportation and distribution sectors and voluntary functional and legal unbundling for storage and LNG facilities, to address issues associated with lenient account unbundling in the First Gas Directive and sluggish development of the third-party access regime.
- **Stronger third-party access**: A right of third parties to non-discriminatory access to networks and LNG facilities, with the aim of enabling new suppliers to enter the market and providing consumers with a range of suppliers to choose from.

---

[3]EU directives have to be transposed into national law whereas EU regulations apply directly to market participants. Both types of EU law—directives and regulations—govern the natural gas market.

[4]Under negotiated access, network operators are free to negotiate terms of network use in good faith with prospective users. The main commercial conditions of these agreements must be published ex-ante. In contrast, under regulated access the commercial conditions of access are determined by the regulator.

- **National regulatory authorities**: Mandatory establishment of regulatory authorities in all member states, with responsibility for compliance with non-discriminatory network access principles, ensuring appropriate levels of transparency and competition, setting tariffs and the methodology for calculating them, and settlement of disputes. Establishment of the Agency for the Co-operation of Energy Regulators, which co-ordinates the implementation of Gas Directive provisions by national regulators.
- **Principles of network tariff calculation**: The establishment of the principles of tariff calculation for natural monopoly networks. These principles should allow pipeline transmission companies to recover costs and receive a return on investments while at the same time providing incentives to maintain existing infrastructure and construct new infrastructure.
- **Exemptions for new infrastructure**: An exemption for interconnectors, LNG and storage facilities from third-party access requirements and cost regulation, subject to various conditions, in order to reduce and mitigate risks associated with new infrastructure build.

4. The Third Package

The 2009 Gas Directive, known as the Third Package, further developed unbundling and open-access arrangements. It sought to create a competitive, secure and environmentally sustainable market in natural gas. To achieve these goals, the Third Package offered member states a choice between ownership unbundling on the one hand and setting up an independent system operator (ISO) or becoming an independent transmission operator (ITO) on the other. The ISO and ITO models are essentially less stringent forms of unbundling. Unlike ownership unbundling, setting up an ISO or ITO would let the transmission network remain under the ownership of vertically integrated firms. Under an ISO model, the transmission networks remain

under the ownership of vertically integrated energy companies, but an ISO would be responsible for operation and control of day-to-day business. Investment decisions would be jointly made by the ISO and the owner of the infrastructure. Under an ITO model, vertically integrated energy companies retain ownership of their transmission networks. The ITO would be a legally independent joint stock subsidiary operating under its own brand name, with strictly autonomous management and under stringent regulatory control. Investment decisions would be made jointly by the parent company and the regulatory authority.

The directive recommended ownership unbundling as "the most effective tool by which to promote investments in infrastructure in a non-discriminatory way, fair access to the network for new entrants, and transparency in the market". Nonetheless, it also stated that setting up an ISO or ITO that is independent from supply and production interests should enable a vertically integrated company to maintain ownership of network assets while ensuring an effective separation of interest. The Third Package thereby implied that there is more than one way to achieve the necessary alignment of incentives along the gas value chain to deliver gas at least cost while still maintaining investment.

In addition, the Third Package further strengthened the independence of regulators from private or public interests to achieve equally effective regulatory supervision across member states. It set out various roles and responsibilities for regulatory authorities, including setting tariffs, ensuring that tariffs are non-discriminatory and cost-reflective, and issuing binding decisions and penalties in relation to natural gas undertakings.

Finally, the Third Package and complementary regulations focused on providing incentives for investment in natural gas infrastructure and ensuring security of supply. The Third Package upheld the exemption from third-party access requirements and cost regulation in the Second Gas Directive for particularly risky new infrastructure, such as cross-border pipelines and LNG terminals. It also introduced new requirements for long-term infrastructure planning, in terms of 10-year network development plans to be developed by transmission network owners and operators. These plans are to be developed at the national as well as European level. National-level plans are binding, and investments by independent transmission operators over a three-year horizon can be enforced by national regulators. The package also allowed member states to impose public service obligations on entities operating in the natural gas sector, mainly in the areas of security of supply, regularity and quality of service, price, environmental protection and energy efficiency. These entities can refuse access to their systems if it compromises their ability to meet these obligations.

In 2010, EU regulations further strengthened security of supply considerations within the context of the internal market. The aim of these regulations was to ensure trade and supply even under extreme and emergency conditions. The regulations put emphasis on infrastructure development to increase internal natural gas flows and external LNG imports. They provided for increased co-operation among member states, introduced supply obligations for companies and established minimum supply and infrastructure standards.

The European Union gas directives weakened state ownership rights in the natural gas value chain, especially in terms of wholesale and retail, as well as promoting transnational pipeline development. Even though the European Union's natural gas is in the process of being liberalised, and some countries such as The Netherlands and the United Kingdom have already privatised the natural gas wholesale business, nonetheless the majority of infrastructure is still in the hands of the state so as to ensure undifferentiated treatment of third-party access. In contrast to this are the independent company operations by natural gas shippers in The Netherlands and the United Kingdom. In France and other countries, natural gas transmission is operated by vertically unified company subsidiaries, and natural gas transmission companies are very active among the transnational pipelines.

The relationship between the distribution and retail business is becoming increasingly slack. With regard to transmission, the British distribution network is operated by independent companies, while France's is operated by vertically unified company subsidiaries. Vertically unified companies were traditionally the companies building and controlling reserve capabilities, but after market liberalisation took its first steps, all private entities within Europe began developing underground reserve equipment. In addition, Europe's various merchants have begun developing LNG facilities.

At the same time, the European Union's natural gas market liberalisation and natural gas infrastructure planning has been extended, resulting in higher natural gas prices and reduced natural gas consumption volumes. Through strong natural gas network development and the expansion of transmission infrastructure measures and plans, the European Union has stimulated increases in trade with neighbouring regions, as well as the construction of transnational pipelines and LNG terminals.

5. The impact of market liberalisation

Since the 1970s, the European Union's natural gas consumption volumes have grown fivefold. However, in the first 10 years of the 21st century, growth stagnated (Fig. 18.8). The recent drop-off in gas consumption has been as a result of the recession and subsequent low economic growth rates, increasing competition from coal and renewables in power generation, particularly linked to EU climate policies, rising natural gas prices since the onset of liberalisation, and diversion of LNG shipments to Asia as demand rose following the Fukushima disaster.

The European Union has a well-developed natural gas network, with plans to expand the transmission infrastructure to ease increased trade with neighbouring regions, including cross-border pipelines and LNG import terminals (Fig. 18.8). The EC natural gas directives have led to reduced public ownership within the natural gas value chain, particularly at the wholesale and retail levels, and increased cross-border

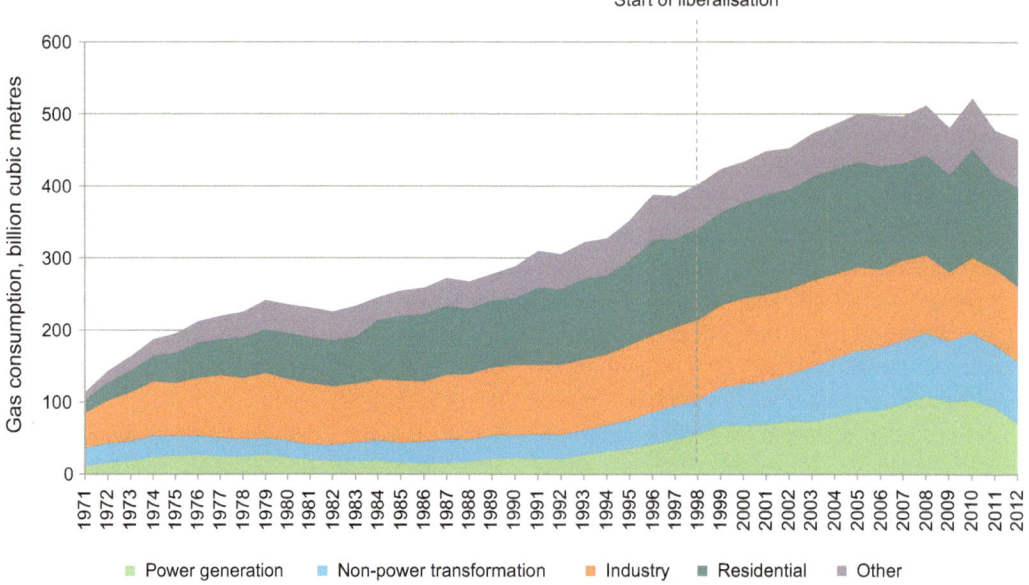

Fig. 18.8 European Union natural gas consumption volume and structure. *Notes* Excludes transport, non-energy use, energy industry own use and losses; non-power transformation includes CHP and natural gas to liquids plants; other includes the services, agriculture and fishing sectors. *Source* Vivid Economics, based on IEA data

pipeline development. The European Union has good natural gas networks, and plans to further expand its gas transmission network to promote natural gas trade development with regions surrounding the European Union. This includes cross-border pipelines and LNG import terminals (see Fig. 18.9). The natural gas directives issued by the European Union have reduced the proportion of state-owned assets in the natural gas industry, especially in the wholesale and retail segments, which has greatly accelerated the development speed of transnational pipelines.

As the European Union's largest natural gas producers, Norway and The Netherlands complemented the liberalisation and increased access to midstream infrastructure with upstream policies to increase competition. At the end of 2013, Norway had confirmed natural gas reserves of 2 trillion m^3, while The Netherlands had 900 billion m^3.

Norway developed its reserves by mandating a 50% stake for publicly owned Statoil in each exploration and production licence. The partnership arrangement enabled Statoil to acquire

exploration and production capabilities to complement its operational capabilities. Statoil was partly privatised in 2001, and the partnerships transferred to Norway through a new state-owned company, Norway Oil and Gas Revenue Management Company. In 2003, Norway began a series of licensing rounds for new producing areas and further exploration in mature areas on its continental shelf. This has introduced a new set of players to the upstream exploration and production segment.

The Netherlands developed its large Groningen natural gas field through a 50:50 public-private partnership. The state expected this ownership structure to lead to rapid development of the reserves, while maximising the benefits to society. The partnership was based on netback pricing, with natural gas prices regulated at 65–85% of the price of a basket of fuel oils. This arrangement provided a steady income stream to encourage new entrants and greater investment in upstream exploration and production, while also ensuring that the investments were competitive compared to alternative fuel sources.

Fig. 18.9 European Union transnational pipelines and LNG terminal distribution. *Source* Euronatural gas

6. **Ancillary policies**

To achieve the energy trilemma goals of energy affordability, security and sustainability at the same time as employing liberalised energy markets, Europe has a relatively well-developed set of climate and environmental policies.

Under the Kyoto Protocol, in 1997 the European Union-15[5] committed to reducing green-house gas emissions by 8% by 2008–2012 from 1990 levels. (Actual reductions were between 12 and 15%.) Since then, it has since adopted a 20% reduction target by 2020 and a 40% target for 2030. In addition, wider environmental policies, such as European clean air policies, have played a significant role in shaping the energy mix in the power sector. The recently proposed air policy package sets ambitious air quality objectives for 2030.

However, the design and implementation of climate policies interacts with electricity markets, affecting the outlook for gas-powered generation. While ambitious policies on climate change and air quality favour gas over coal in electricity

---

[5]In 2004, prior to the eastern expansion, the European Union had 15 member states: Austria, Belgium, Denmark, Finland, France, Germany, Greece, Ireland, Italy, Luxembourg, The Netherlands, Portugal, Spain, Sweden, and the United Kingdom.

generation, the design of these policies could produce some unintended consequences. For example, by setting a separate 2020 target for renewable energy, the EU Renewable Energy Directive has dampened wholesale energy prices, since the marginal cost of renewables has been low, and even negatives at times. This has, in turn, reduced incentives to invest in back-up fossil fuel capacity, which is needed to compensate for the potential of intermittent supply from renewable energy sources. Moreover, flaws in the design and implementation of the EU Emissions Trading Scheme, the European Union's primary mechanism for meeting its greenhouse gas emissions target, combined with recent weak economic conditions and the relatively low price of coal, has led to the "energy paradox": a shift of the energy mix towards coal and renewables at the expense of natural gas.

These unintended interactions between energy market and climate policies have undermined the profitability of power suppliers and destabilised their business models. The problem is illustrated by the challenges faced by companies like E.ON and RWE in Germany, whose marginal cost pricing-based business model has been put at risk as subsidised and distributed sources of renewable energy of significant size enter the energy mix. Going forward, Europe renewables are likely to remain a significant part of the energy mix to 2030 if Europe is to meet its climate policy ambitions, and electricity market regulation will need to evolve alongside climate policies. In turn, this will have knock-on impacts for natural gas demand and markets.

18.3.3   Case Study 3: United Kingdom

From 1986 to 2002, the United Kingdom implemented natural gas market liberalisation, contributing to the "dash for gas" in the power sector in the 1990s. The process comprised three distinct phases: pre-liberalisation, liberalisation and post-liberalisation (Fig. 18.10).

The UK natural gas market took off in the late 1960s, when domestic coal gas started to be replaced by natural gas from North Sea basins.

At that time, the newly formed Conservative government had just come to power and was facing a weakened economy in 1979. They formulated policies to privatise state-owned industries, using this approach to promote investment without requiring actual financial support from the government. By selling assets, the revenues could be used to prop up government finances, to invigorate business and to weaken the influence of unions.

In the 1980s during the asset sales, the first set of sales included a vertically unified company— British Gas. After the implementation of the Gas Act 1986, British Gas began a process of privatisation, opening access to its pipelines for its competitors, and establishing a natural gas regulator, Ofgas. With the new third-party access agreements, the privatised British Gas was forced to abandon a portion of the content of its existing natural gas supply contracts, unbundling the company's business to independent subsidiaries and providing space for entry into the market by competitors. The Gas Act 1995 made further progress by promoting retail competition. In 1996, the Network Code was implemented, fine-tuning third-party access rules while building a system for managing daily settlements and national balancing points. In 2002, the newly established settlement mechanisms entered their final stage. In that same year, British Gas sold its transmission business to achieve complete ownership rights spinoff. As an independent British transmission operator, Britain's National Grid established a national gas transmission network. Following market liberalisation, because of excessive exploration in the North Sea natural gas fields, domestic natural gas production volumes in Britain declined. Further changes after market liberalisation primarily consisted of alterations to the terms of third-party access, improved settlement agreement efficiency and improved supply security.

1. **Before market liberalisation**

In the 1970s, as natural gas supplies continued to increase, the United Kingdom established a state-owned natural gas company. In the late

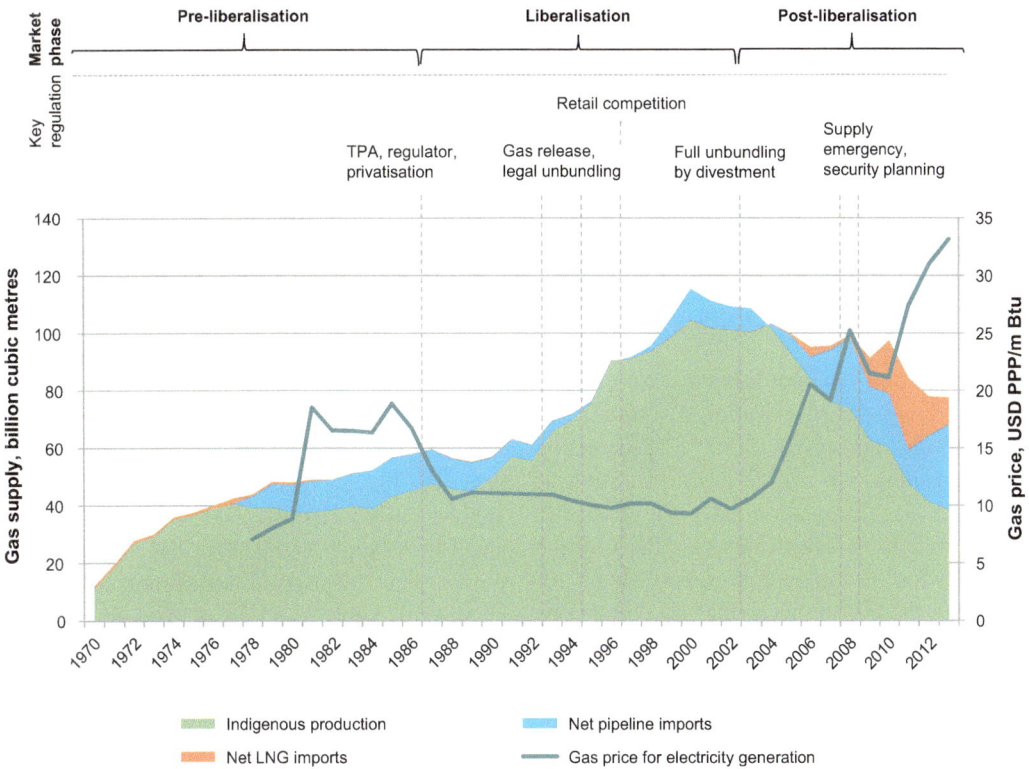

Fig. 18.10 Phases, development and key regulations in the development of the British natural gas market. *Note* Nominal natural gas prices paid at power plant, national currency converted to average GDP purchasing power parity. *Source* Vivid Economics, based on IEA data

1960s, self-produced North Sea Basin natural gas gradually replaced domestic coal gas, and the United Kingdom's natural gas market began to flourish. The Gas Act of 1972 merged the 12 existing area natural gas boards and the Gas Council to create the British Gas Corporation. British Gas was the sole buyer of natural gas produced in and around the United Kingdom. It also held a statutory monopoly on the supply of natural gas to end users. The company relied on flexibility in its long-term natural gas purchasing contracts to provide reliable balancing of input and output volumes on its network. Natural gas purchase prices varied greatly as British Gas negotiated individual contracts with producers. Downstream pricing was based on the weighted average cost of these natural gas purchase prices, a margin to cover transportation and distribution costs, and a profit margin.

In the 1970s, in face of high inflation and unemployment rates, the United Kingdom applied for loan financing from the International Monetary Fund. Poor labour relations and rising oil prices led to the Winter of Discontent in 1978–79, a period of widespread strikes. Margaret Thatcher was elected prime minister of a new Conservative government in 1979. She immediately faced difficulty funding the nationalised energy sectors, which needed to invest to secure supply. The government developed plans for the privatisation of the nationalised industries, to generate income and enable new investment by the new private entities. At the time, privatisation was not driven so much by a desire to increase competition as by the immediate challenge of government finances that could not support the necessary investments in energy infrastructure.

A first move towards privatisation of the natural gas sector was made with the Oil and Gas (Enterprise) Act of 1982, which removed British Gas's statutory right of first refusal on purchases of upstream natural gas. This effectively opened the market to third parties, although in reality it remained challenging for new entrants to purchase, transport and eventually sell natural gas.

2. Market liberalisation

The Gas Act 1986 introduced real competition by sanctioning the privatisation of British Gas and requiring third-party access to transmission and distribution networks. The privatisation of the British Gas Corporation led to the formation of the similarly named British Gas Plc, and the new entity lost its monopoly to supply the largest customers: those using more than 25,000 therms (around 730 MWh) of natural gas a year. The act also obliged the newly incarnated British Gas to grant third-party access to its pipeline infrastructure. Finally, the act established the first natural gas regulator, the Office of Gas Supply or Ofgas.

Competition was slow to develop under the new arrangements. In 1988, a Monopolies and Mergers Commission inquiry found that British Gas's practices were anti-competitive and that price discrimination prevailed. This led to the recommendation that company be barred from purchasing more than 90% of new natural gas on the market. The company was also instructed to publish prices charged to industrial and commercial customers. The government viewed the initial stages of natural gas market liberalisation to be a political success, and used it as a template for the power sector privatisation that followed in 1989 and 1990.

Competition in the natural gas sector was subject to close government scrutiny in subsequent years, leading to British Gas being asked in 1992 to further reduce its market share to 40% of the natural gas market by 1995. The threshold for British Gas's large consumer supply monopoly was further reduced to 2500 therms. British Gas also undertook to release some of its natural gas under existing contracts by selling it to other suppliers. Further inquiry into the company by the Monopolies and Mergers Commission in 1993 resulted in the recommendation that British Gas unbundle its divisions into separate subsidiaries. This took place in 1994. It led to the creation of Transco, which had responsibility for transport and storage.

The Gas Act 1995 started a phased introduction of full competition in the natural gas market, including retail competition, which established complete supplier choice in the market. The act established a new licensing system for designated pipeline operators, wholesale companies or shippers, and retail companies. As a result, British Gas lost its majority market position. In October 1990, British Gas controlled all of small-firm supply and 93% of large-firm supply, but by June 1996 these had shrunk to 43% of small-firm supply and 19% of large-firm supply.

The 1996 Network Code established the rules and procedures for third-party access to pipelines and introduced a regime for daily balancing. Transco gained responsibility for securing the physical balance of the system, capacity planning, the forecasting of demand and distribution arrangements, and the overall operation of the system. The code also created the NBP as a virtual location where the transmission system operator could balance the system on a daily basis. The NBP quickly became the United Kingdom's most liquid natural gas transmission system, and was adopted by traders for nominating their buys and sells on a standardised basis. This was followed by the introduction of natural gas trading arrangements in 1996, which simplified natural gas trading procedures. Further significant reforms in 1999 introduced an on-the-day commodity market to help with balancing and to incentivise to Transco to minimise overall balancing costs. In 1998, an interconnector pipeline between Belgium and the United Kingdom was opened, for the first time allowing natural gas to flow between grids in Britain and continental Europe.

British Gas was restructured several times. In 2000, it broke up into two companies: BG Group, holding upstream assets, and Centrica, encompassing a mixture of upstream and downstream assets. Transco was sold to National Grid in 2002. These combined businesses make up the United Kingdom's independent transmission system owner and operator for natural gas and electricity markets (Fig. 18.11). In addition, National Grid owns and operates four distribution networks, as well as the Grain LNG import terminal and the Avonmouth LNG storage facility.

3. After market liberalisation

The end of the liberalisation phase in 2002 was marked by full unbundling of the natural gas business. It also signalled the successful transition to competitive and liquid wholesale markets: daily balancing by shippers, with penalties for out-of-balance portfolios, and a role for the transmission system operator to maintain safe balance in the transmission system. In 2005, the Uniform Network Code replaced the Network Code, adding clarity and detail to enable shippers and the transmission system operator to balance more effectively. This included identifying types of capacity held along the natural gas value chain and obliging owners of that capacity to provide stock data to the transmission system operator. The Uniform Network Code has continued to be

GAS TRANSMISSION

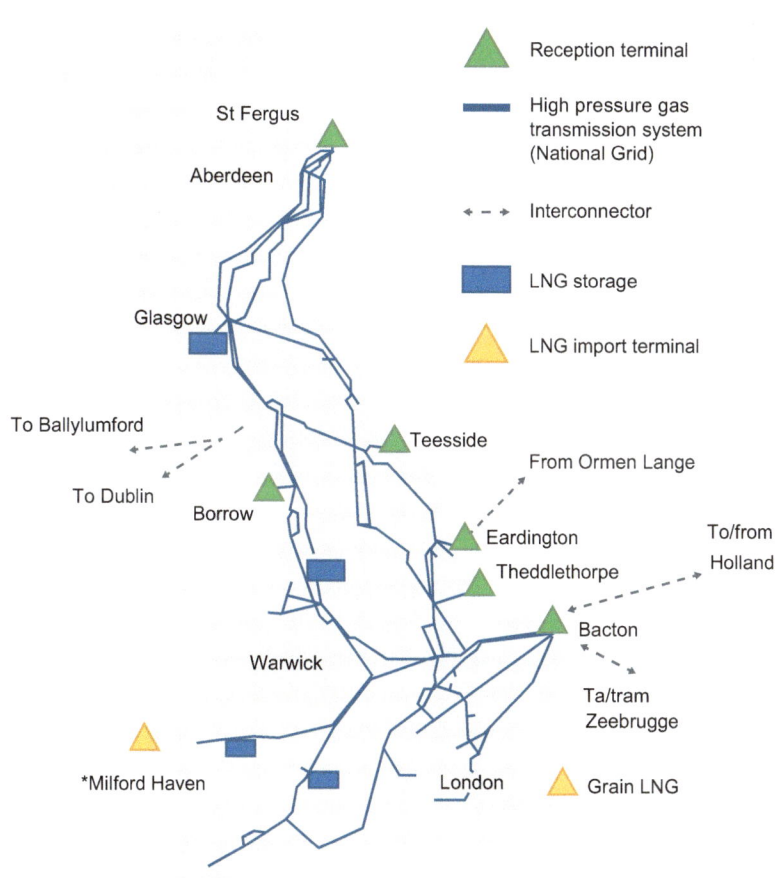

Fig. 18.11 National Grid completely owned and operated British transmission networks. *Source* National Grid

updated and amended over time to improve practices and ensure the safe and efficient functioning of the United Kingdom's natural gas supply.

More recently, security of natural gas supply has emerged as a key priority in the United Kingdom. This is governed by several additional codes, including:

- **the Fuel Security Code** (2007), which grants the Department of Energy and Climate Change the option to instruct power stations to use alternative fuels;
- **the National Emergency Plan for Gas and Electricity**, which sets out the response to emergency situations—such as supply deficits, storage safety breaches, transportation constraints and the loss of supply to more than 50,000 customers—and describes the roles and responsibilities of market participants; and
- **the Transmission Network Planning Code**, which is maintained by the National Grid and identifies weaknesses in the UK natural gas

transportation system, based on long-term demand and supply forecasts. The code requires a security margin to cover uncertainties in forecasts and a design margin for maximum demand, subject to approval to the regulator.

4. The impact of market liberalisation

The liberalisation of the UK natural gas market triggered the "dash for gas" in power generation during the 1990s (Fig. 18.12). The power sector had access to new combined-cycle natural gas turbine technology, which was relatively quick to build. The liberalisation of the natural gas and electricity sectors, along with abundant supply and a lifting of the ban on use of natural gas for power generation, drove a rapid rise in gas consumption as the number of combined-cycle turbines used by power generators increased. The decline in natural gas use in power generation in recent years has been the result market forces which have made natural gas less competitive compared to coal.

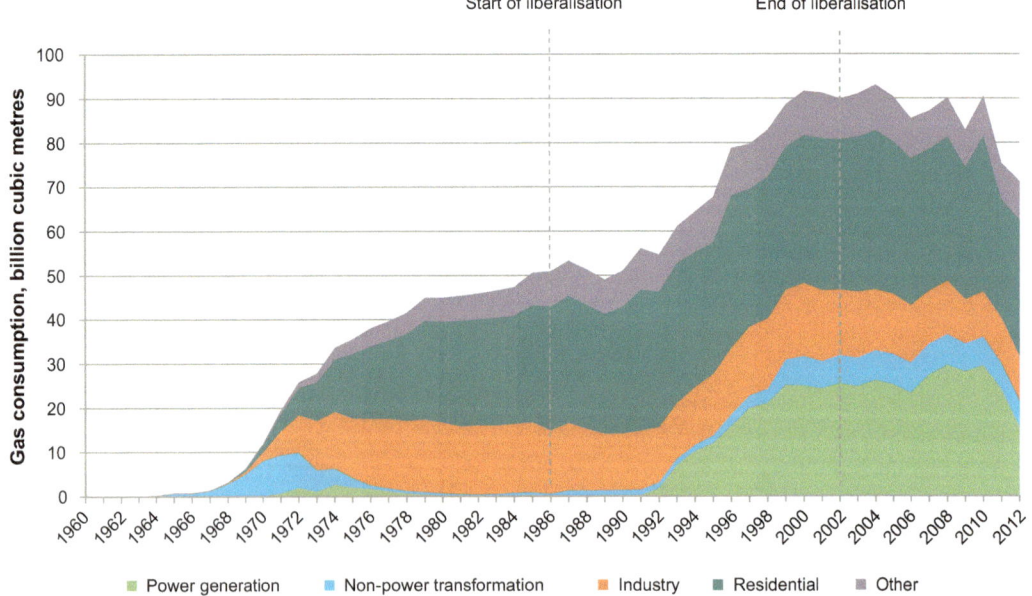

Fig. 18.12 UK natural gas consumption by sector, 1960–2012. *Note* Excludes transport, non-energy use, energy industry own use and losses; non-power transformation includes CHP and natural gas to liquids plants; other includes the services, agriculture and fishing sectors. *Source* Vivid Economics, based on IEA data

Upstream regulation in the United Kingdom is based on principles of transparency and open access. The United Kingdom had 200 billion m^3 of proven natural gas reserves at the end of 2013. The Petroleum Act of 1998 sets out the rules around upstream natural gas exploration and production. Natural gas resources are owned by the state, and the Secretary of State of the Department of Energy and Climate Change is responsible for granting licences for exploration and production rights.

As exploration and production efforts on the United Kingdom's continental shelf have become increasingly difficult, licensing tenures have lengthened. The Department of Energy and Climate Change assesses bids, focusing on policy objectives including environmental risk and infrastructure quality. If the winner of a licence

does not keep up with the agreed work programme or does not meet safety and environmental standards, the department may revoke the licence.

18.3.4   Case Study 4: Japan

Vertically integrated businesses delivered all aspects of Japanese natural gas supply until liberalisation began in 1995. The development of Japan's natural gas market has gone through three distinct phases, with liberalisation still incomplete (Fig. 18.13).

From 1954 to 1995, prior to Japan's natural gas market being liberalised, it was managed through the 1954 "Natural Gas Business Law". Under this law, the natural gas industry was

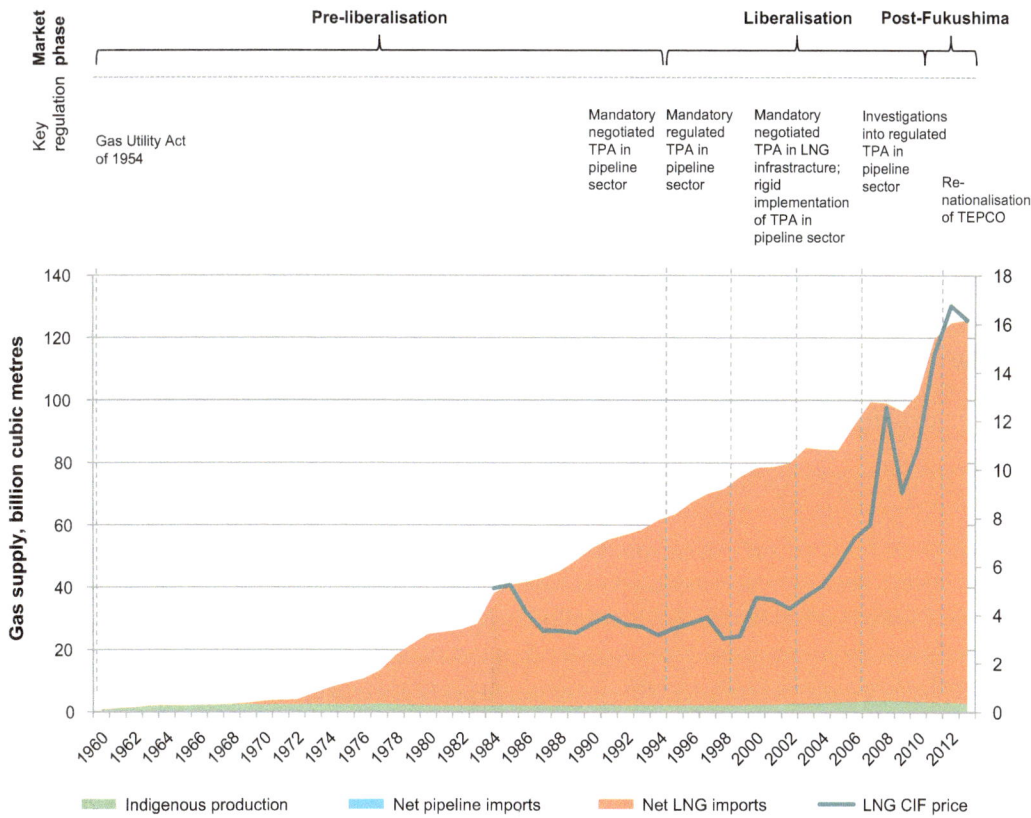

Fig. 18.13 Phases, development and key regulations in the development of the Japanese natural gas market. *Note* Nominal LNG cost, insurance and freight (CIF) import prices converted to US$. *Source* Vivid Economics, based on IEA data and BP Statistical Review of World Energy 2013

monopolised by vertically unified entities. Beginning in 1995, Japan's government began to implement natural gas sector market liberalisation gradually. Powerful opening up of user rights systems in natural gas transmission, including functional spin-off and outside access of controls, permitted large clients to choose suppliers. LNG infrastructure incorporated negotiated third-party access systems.

In 2011, the Fukushima nuclear disaster had a massive impact on Japan's energy systems, causing the re-nationalisation of the nation's major LNG importer, Tokyo Electric Power, in 2012. This marked the end of the Japanese government's liberalisation plans for the energy market. However, in public debate, natural gas market reforms are continually a focus, and the government has been forced by this pressure to add energy sector competition. After the rise in oil prices due to the Fukushima disaster, because demand grew, and LNG price fluctuations were significant, the Japanese government was forced to change policy to weaken controls on long-term contracts linked to oil. The government has also taken other measures weakening the influence of oil markets on natural gas, including proposing the construction in 2015 of a LNG futures market. The futures market would determine natural gas prices based on domestic demand factors.

1. Before market liberalisation

Prior to liberalisation, Japan's natural gas market was governed by the 1954 Gas Utility Act. Under this act, the natural gas industry was regulated as a vertically integrated monopoly. Companies wishing to enter the natural gas market needed licences from the Ministry of Economy, Trade and Industry. The vertically integrated natural gas companies were responsible for importing LNG, and the city natural gas companies were responsible for distribution. Tokyo Gas, Osaka Gas and Toho Gas were major vertically unified natural gas companies responsible for the import of natural gas, operation of LNG terminals, the storage of LNG, operating natural gas transmission distribution networks, as well as retail business. The smaller-scale city

natural gas companies also provided distribution and retail services, reselling natural gas purchased from the vertically integrated companies. Power generators were allowed to import LNG for their own use, including large users such as Tokyo Electric Power (TEPCO) and Kansai Electric Power.

2. Market liberalisation

The Japanese government gradually introduced liberalisation in transmission, distribution and LNG facilities beginning in 1995. Natural gas prices in Japan have long been higher than the average in OECD countries, and the Japanese government sought to reduce prices by introducing competition, with open access to infrastructure and unbundling. The first two phases of liberalisation, in 1995 and 1999, focused on the transportation and distribution sectors.

People believed that these two stages were not successful. Since 2003, oil prices have remained high and, compared to oil products, natural gas has become more attractive, with subsequent rises in natural gas demand. Therefore, beginning in 2003, the government started to implement more unbundling measures. Although there not a complete opening of LNG infrastructure to third-party access, nonetheless this stage began to implement open access.

The Gas Utility Act of 1995 attempted to introduce negotiated third-party access and unbundling, but was unsuccessful. The act introduced mandatory negotiated third-party access, allowing large natural gas consumers with an annual natural gas consumption of more than 2 million m^3 to buy natural gas from non-incumbent natural gas suppliers. Tokyo Gas, Osaka Gas and Toho Gas were required to unbundle their services. However, almost a decade later these three companies still accounted for more than 70% of domestic sales (Fig. 18.14).

This regime was not considered successful in opening the natural gas market as terms and conditions for access were not sufficiently standardised and transparent. The companies using or seeking access to other companies' pipeline

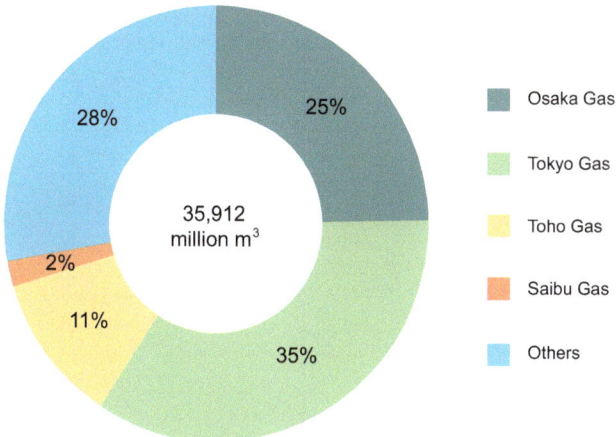

Fig. 18.14 Japan natural gas market sales volume, 2012. *Source* Osaka Gas (2012)

complained of entry barriers and non-transparency in the wheeling system which moves natural gas from one system to another. Many of the complaints related to the lack of information on the available capacity of pipelines, non-transparent standards and procedures for assessing fees, and other compensation mechanisms.

The Gas Utility Act was revised in 1999 to introduce regulated third-party access, expanding from the three largest vertically integrated companies to include a fourth, Saibu Gas. The threshold for eligible consumers was reduced by half, to 1 million m^3 in natural gas consumption a year. Natural gas companies were also required to publish the terms and conditions for use of their transportation and distribution pipelines.

When the first phase of liberalisation did not achieve an effective open-access regime, further unbundling and market transparency measures were adopted from 2003 onwards. High oil prices provided added impetus, making natural gas more attractive compared to oil products and substantially raising demand. In 2003, account and functional unbundling were introduced to separate natural gas transportation businesses from sales businesses by requiring separate filings of accounts and a separation of day-to-day management. The reforms also sought to open LNG terminals. The threshold for customers to be eligible for third-party access was further reduced to 500,000 m^3 a year in 2003 and again

to 100,000 m^3 in 2007 (Fig. 18.15). By 2007, 61% of the natural gas market by volume was deregulated, and this was unchanged in 2013.

In addition, open access was extended to all natural gas companies with eligible customers. Natural gas companies were required to prepare standardised third-party access contracts with terms and conditions for transportation services that were to be approved by the energy authority.

Japan does not have a well-developed long-distance natural gas transportation network. There are no cross-border natural gas pipelines and the high-pressure transmission natural gas pipeline length is only around 2500 km (Fig. 18.16), compared to 500,000 km in the United States. Although there are around 43 main interconnection points between areas, the trunk line networks are not necessarily connected to each other as they have separately developed around LNG terminals. The lack of interconnection between regions may limit the scope for competition through third-party access.

Starting in 2003, there were calls in Japan to delay the development of national high-pressure natural gas pipeline networks, on the grounds that managed third-party access terms could reduce the incentivising effect of natural gas infrastructure investment, and hinder energy security.

In 2003, there were natural gas legal reforms, with LNG infrastructure sectors implementing negotiated third-party access plans.

Annual contracted gas
consumption volume

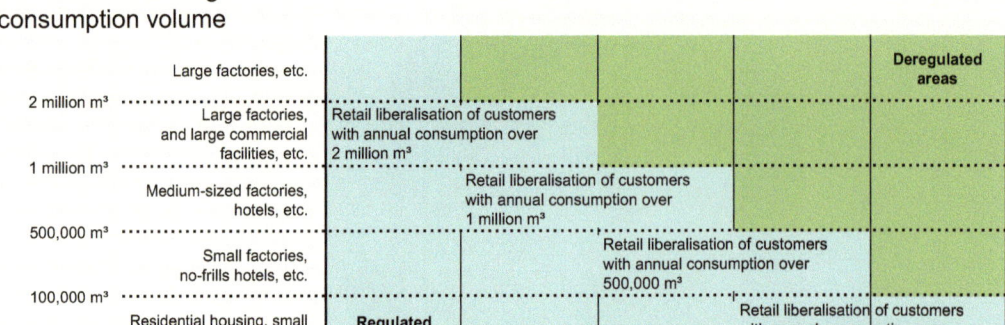

Market deregulation (2013.3)

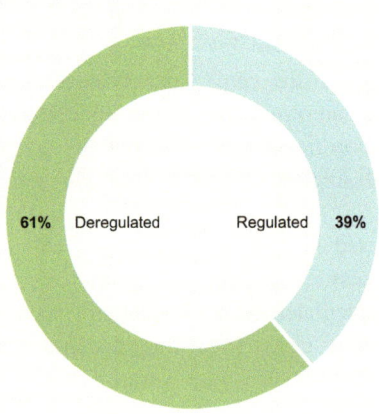

Fig. 18.15 Japan's natural gas market liberalisation progress. *Source* Osaka Gas

The need for open access in LNG infrastructure was included in the 1999 US-Japan Third Joint Status Report under the Enhanced Initiative on Deregulation and Competition Policy, an offshoot of economic co-operation efforts between the two countries. The issue was particularly relevant as LNG terminals are the entry points for imported natural gas, and without free access to LNG terminals and storage facilities, only a limited degree of competition, if any, would be possible in Japan. In response, the Japanese government amended the Gas Utility Act in 2003 and introduced an open-access scheme for LNG facilities. Under this scheme,

natural gas companies were to define preconditions for negotiated access. Further transparency requirements related to the capacity information for LNG storage and facilities were also applied. However, by 2012, no third-party access had been effectively granted to any company at a Japanese LNG terminal.

3. **Prospects for continued liberalisation post-Fukushima**

After the Fukushima nuclear disaster in 2011, the appetite for further liberalisation in natural gas markets in Japan diminished further. The

This document and any map included here are without prejudice to the status of or sovereignty over any territory, to the delimitation of international frontiers and boundaries and to the name of any territory, city or area.

Fig. 18.16 Japan LNG terminals and high-pressure pipelines. *Source* IEA (2013)

renationalisation of TEPCO, a significant LNG importer, in 2012 epitomised the change in momentum regarding energy sector liberalisation that followed the Fukushima disaster. However, in 2013, a plan for reforms to deregulate the electricity market was passed, signalling a renewed interest in energy sector reform. The electricity sector is particularly important to Japan's natural gas market because it accounts for about two-thirds of the country's total natural gas consumption and that share has grown rapidly since the Fukushima disaster (Fig. 18.17).

Japan's 10 regional electricity utilities have sought to increase regulated electricity tariffs paid by end users to help cover their costs. Since 2012, the Ministry of Economy, Trade and Industry has approved tariff increases of between 7 and 11% for six utilities. In 2013, the first of

three electricity market reform packages was passed, to reduce LNG procurement costs through competition between electric utilities. The measure established an Organization for Cross-Regional Co-ordination of Transmission Operators to promote the development of electricity transmission and distribution networks and to enhance the nationwide function of adjusting the supply-demand balance of electricity. The second reform measure would fully liberalise entry to the electricity retail business and the third would require legal unbundling of the transmission and distribution sector.

As Japan depends heavily on LNG imports to meet its electricity generation needs, the government has sought to ensure diversified LNG supply based on cost-efficiency and energy security. The key elements of Japan's overall energy security policy are diversifying its

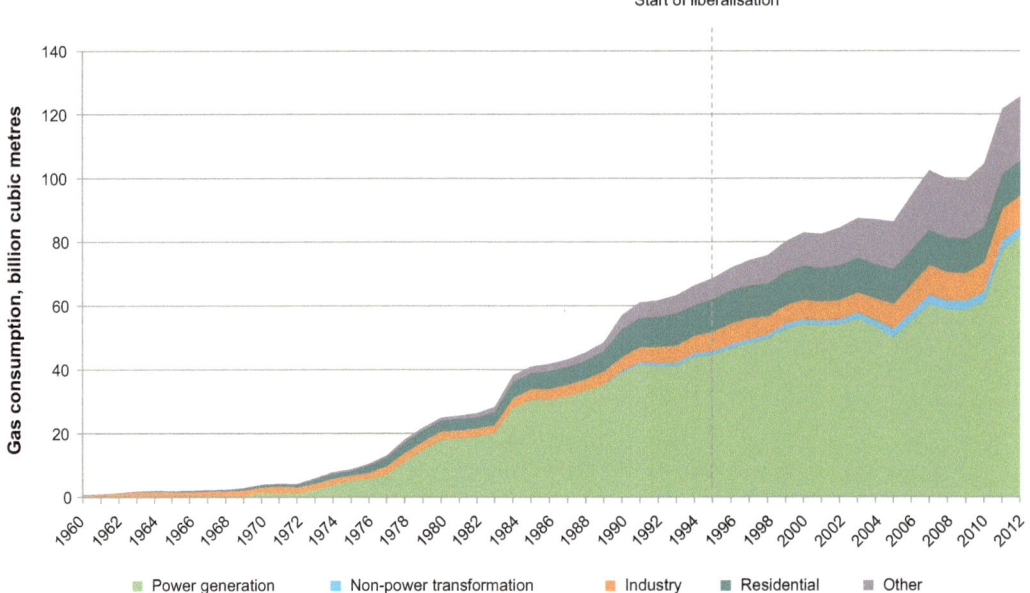

Fig. 18.17   Major components of Japanese natural gas consumption. *Note* Excludes transport, non-energy use, energy industry own use and losses; non-power transformation includes CHP and natural gas to liquids plants; other includes the services, agriculture and fishing sectors. *Source* Vivid Economics, based on IEA data

long-term supply contract portfolio, ensuring contractual flexibility to increase imports in an emergency and using voluntary commercial LNG stocks in industry. Japan's largest natural gas supplier, Australia, represented just 21% of total imports of the country in 2013, illustrating the diversity of Japan's natural gas suppliers (Fig. 18.18).

In recent years, Japan's proportion of imports from Malaysia and Indonesia has dropped, while import proportions from Australia and Qatar have increased.

4. Focus: Japanese natural gas pricing

Traditionally, Japan's natural gas contracts have been linked to oil prices, but the government has already begun to weaken the influence of crude oil prices. In the 1970s and 1980s, long-term LNG import contracts were closely linked to international crude oil prices. These contracts are about to expire, and importers are being forced to renegotiate contracts or else lock into shorter-term supplies. Crude oil price increases have led to dual increases in cost and electricity prices in obtaining LNG.

Japan's METT encourages electricity companies to negotiate in order to obtain LNG at prices the same or lower than previous contracts, as a prerequisite to collecting higher fuel fees from electricity consumers. Japan's natural gas and electricity companies also carry out negotiations for LNG contracts, and these contracts are not restricted by crude oil prices that are lower than natural gas market prices in the United States. For example, Kansai Electric Power came to a long-term agreement in 2012 with British Petroleum, and this agreement was governed by the US Henry Hub price. In 2012, Japan's government proposed establishing a LNG futures market in March 2015, with the futures market determining natural gas prices based on domestic demand factors. Japan released its latest strategic energy plans in April 2014, encouraging natural gas and electricity companies to carry out targeted loosening of LNG long-term contract terms. Beginning in 2012, many companies were already carrying out US upstream liquefaction

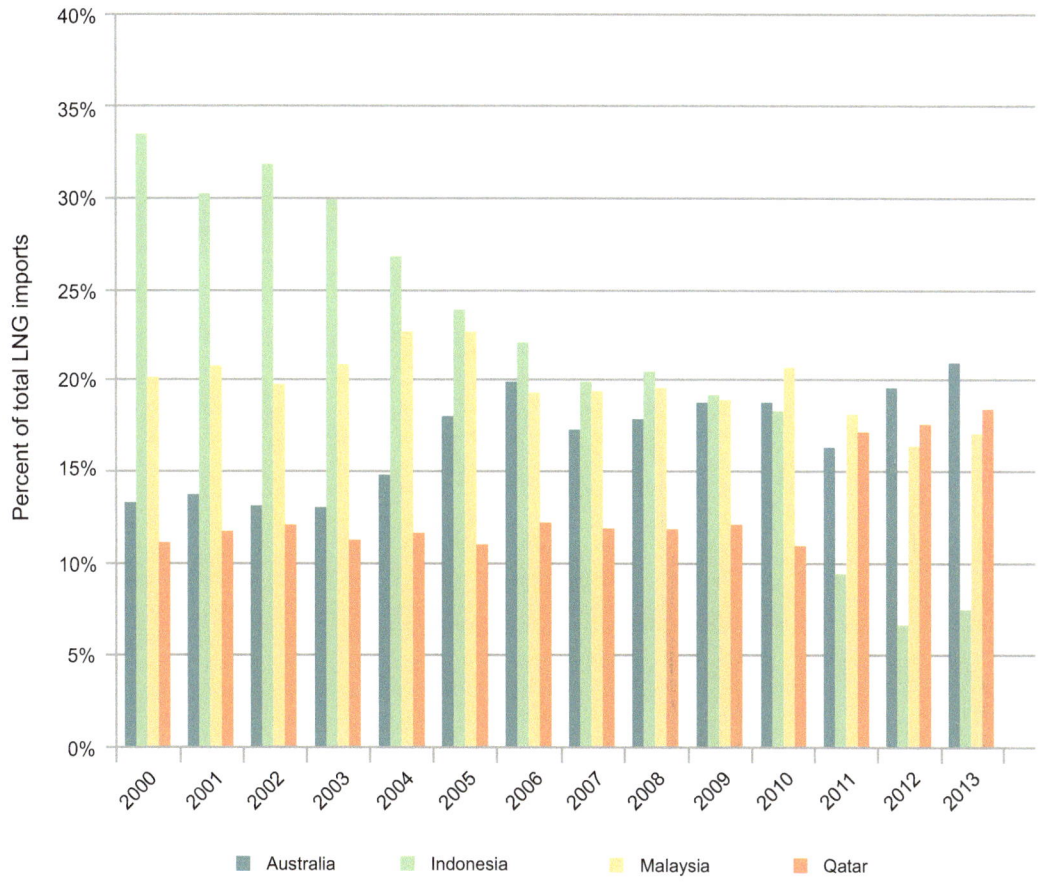

Fig. 18.18 Four major country suppliers of LNG gas imports to Japan since 2000. *Source* Vivid Economics, based on IEA data

projects, and this will account for 20% of Japan's total natural gas demand.

Furthermore, Japan's LNG importers have formed partnership relationships with the region's other LNG purchasers in order to increase negotiating power and reduce purchase prices. For example, Tokyo Electric Power formed a full alliance in October 2014, covering the entire energy supply chain, with Chubu Electric Power. Their co-operation incorporates a wide arrange of activities, whether it be in upstream investments and procurement of fuel for power plants, or carrying out a series of joint projects, including natural gas upstream investment on a global scale, construction of new thermal power plants and procurement of LNG.

18.3.5   Case Study 5: South Korea

Since 1997, South Korea's efforts to liberalise the natural gas sector have led to partial unbundling and open access, but progress has been slow in the wake of political resistance that began in the early 2000s (Fig. 18.19).

Prior to market liberalisation in 1997, KOGAS (a state-owned corporation) was the exclusive owner of LNG infrastructure and natural gas transmission facilities. In 1997, the government attempted to break KOGAS's import monopoly. Even though some large-scale natural gas users could directly import natural gas for their own use, KOGAS still maintained its substantive monopoly in LNG, transmission and

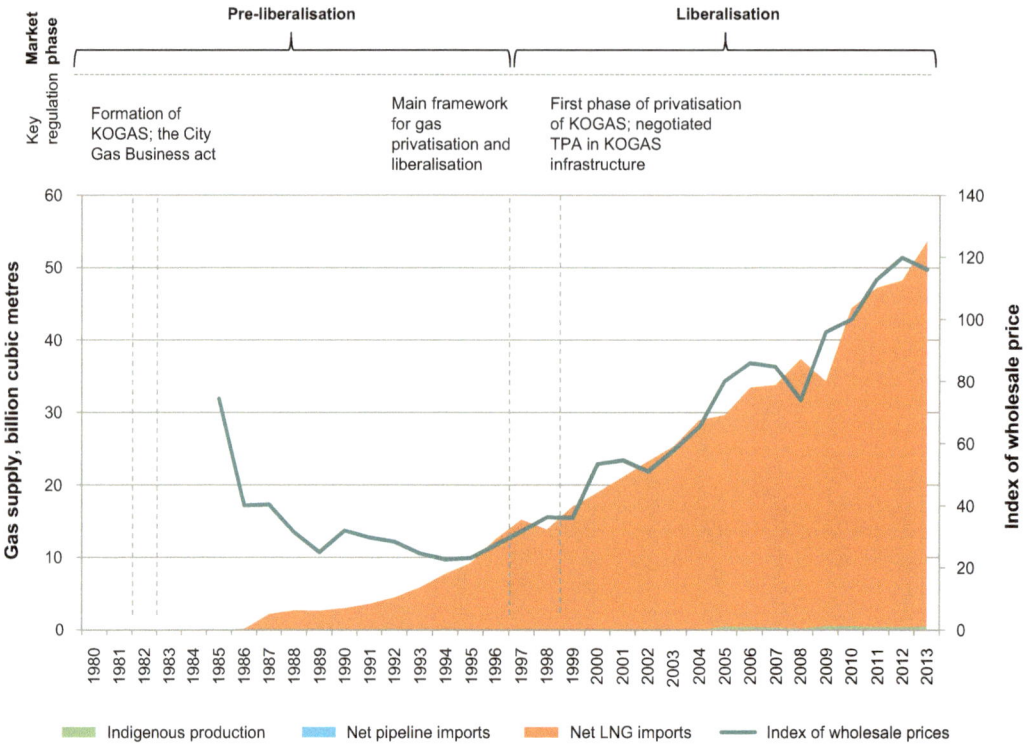

Fig. 18.19 Phases, development and key regulations in the establishment of a liberalised natural gas market in Korea. *Notes* Index of wholesale prices is the natural gas sub-index of the producer price index; 2010 = 100. *Source* Vivid Economics based on IEA data

wholesale supply. In the first 10 years of the 21st century, after the natural gas supply crisis, market liberalisation progress halted. Moreover, California's major power cuts not only increased worries about the compatibility of electricity market liberalisation and energy security, but also weakened political determination. Finally, opposition from trade unions at KOGAS further impeded the implementation of government reform plans.

1. Before market liberalisation

Before liberalisation began in 1997, the state-owned utility, KOGAS, was the sole owner of LNG infrastructure and natural gas transportation facilities. KOGAS is a vertically integrated state-owned natural gas company formed by the Korea Gas Corporation Act in 1982. Similar to the vertically integrated regional

monopolies in Japan, KOGAS was primarily an import-oriented company, given the lack of indigenous natural gas production in South Korea, and it controlled and operated LNG terminals, storage and natural gas transportation pipelines.

The City Gas Business Act of 1983 established city natural gas companies and introduced a regulatory framework comprising licensing requirements for the supply of natural gas, the construction of natural gas facilities and the terms and conditions of natural gas supply. The main business of the city natural gas companies was to buy natural gas from KOGAS or liquefied petroleum natural gas from oil companies and distribute it to customers. This local distribution system has remained largely in place, and there are 30 privately owned city natural gas companies, each of which enjoys exclusive retail sales rights within its areas.

2. Market liberalisation

In response to the 1997 Asian financial crisis, the South Korean government announced plans to liberalise the natural gas sector by introducing competition and privatising KOGAS. The 1997 National Energy Plan set the main framework for natural gas liberalisation and privatisation, but did not specify detailed regulatory changes. In 1999, the South Korean government developed the Basic Plan for Restructuring the Gas Industry. The assets and staff of KOGAS were to be divided into two groups: infrastructure, including receiving terminals, pipelines and storage infrastructure; and import and wholesale functions. KOGAS would retain ownership of the infrastructure group, while the import and wholesale divisions were to be further split into three natural gas supply subsidiaries by the end of 2001. Of these three trading entities, one would remain a subsidiary of KOGAS and the other two would be independent.

The government succeeded in partially privatising KOGAS, and 39% of the equity was sold by the government by the end of 1999. However, the plan to further privatise and unbundle KOGAS was not completed because of fierce opposition from the company's labour union. Additionally, the country faced a domestic natural gas supply crisis at the time when detailed privatisation plans were being discussed. Natural gas demand from the residential, industrial and power sectors rose sharply, raising concerns over energy security (Fig. 18.20). The electricity blackouts in California further fuelled concerns about whether liberalisation and energy security were compatible. These factors together undermined the political appetite for electricity and natural gas sector liberalisation.

Although the 1999 Restructuring Plan stalled, a voluntary service unbundling initiative and a negotiated third-party access regime were introduced for KOGAS infrastructure. In 1999, the City Gas Business Act abolished part of KOGAS's monopoly on importing LNG and operating LNG infrastructure. The act set out a regulated third-party access regime, but it lasted only a short time before being suspended in 2000.

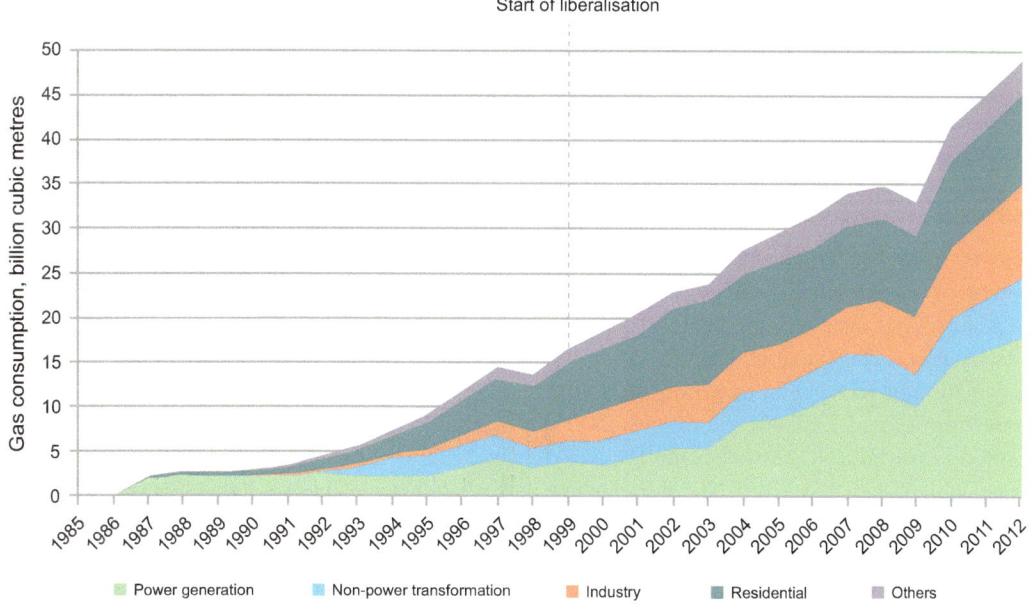

Fig. 18.20 Korean natural gas consumption by sector. *Note* Excludes transport, non-energy use, energy industry own use and losses; non-power transformation includes CHP and natural gas to liquids plants; other includes the services, agriculture and fishing sectors. *Source* Vivid Economics based on IEA data

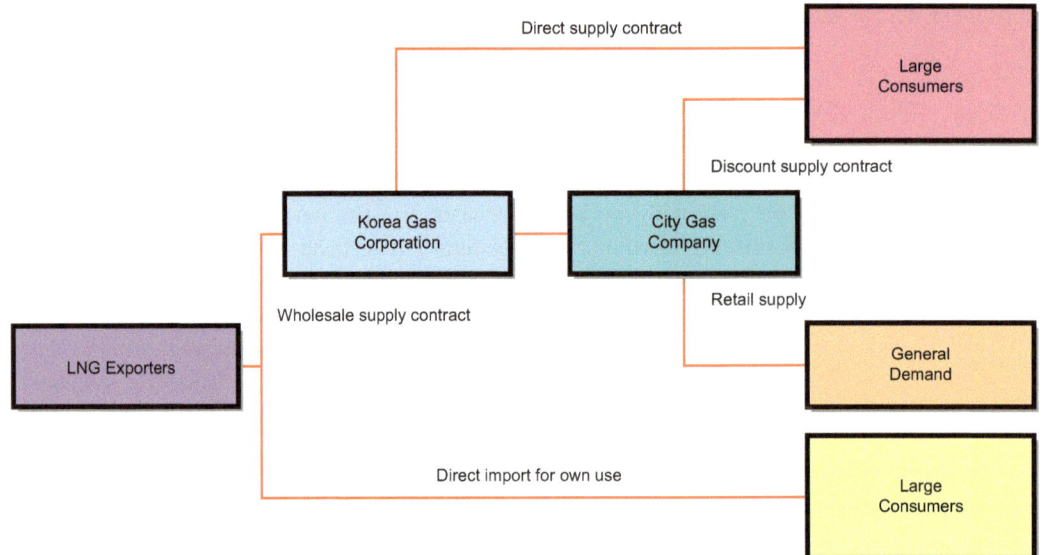

Fig. 18.21 Korean natural gas industry structure. *Source* Asia Pacific Energy Research Centre

Despite the reform efforts, third-party access to LNG storage facilities, the transmission network and LNG terminals owned by KOGAS has been limited. In 2002, KOGAS remained the dominant natural gas player, even while large users like Pohang Iron and Steel Company operated their own LNG import terminals and imported LNG for their own use (Fig. 18.21).

Energy security concerns over LNG market volatility and meeting peak winter demand have hampered progress in natural gas market liberalisation. The importance of diversified LNG supply sources, ensuring LNG supply on the basis of long-term contracts, and expansion of storage capacity to meet high seasonal demand are perceived to outstrip the potential benefits of liberalisation. The government promotes the participation of KOGAS in upstream natural gas production abroad and strives to preserve its position on world LNG markets, where KOGAS is the single largest buyer of LNG. For example, the Tenth Long-Term Gas Supply and Demand Plan, published in 2010, proposed that KOGAS

secure oil-indexed long-term contracts with improved flexibility and conditions from 2015. This makes it unlikely that plans to privatise KOGAS will be revived soon, and the government is likely to continue to exert a direct, commanding influence on South Korea's natural gas sector.

KOGAS owns and operates Korea's national pipeline network and three quarters of the country's LNG terminals. Korea currently has no transnational natural gas pipelines. Its national transmission pipelines stretch for 3588 km (Fig. 18.22). According to plans announced, this will be extended to 4928 km by 2027.

The majority of Korea's contracts are linked to the price of crude oil. KNGC has signed mid-term and long-term contracts linked to crude oil for the import of 80–90% of its LNG, with the remainder being through spot market purchases. In 2012, Qatar was the largest supplier of natural gas to Korea, followed by Indonesia, Oman, Malaysia and Yemen (Fig. 18.23). Among the 10 long-term natural gas supply and demand

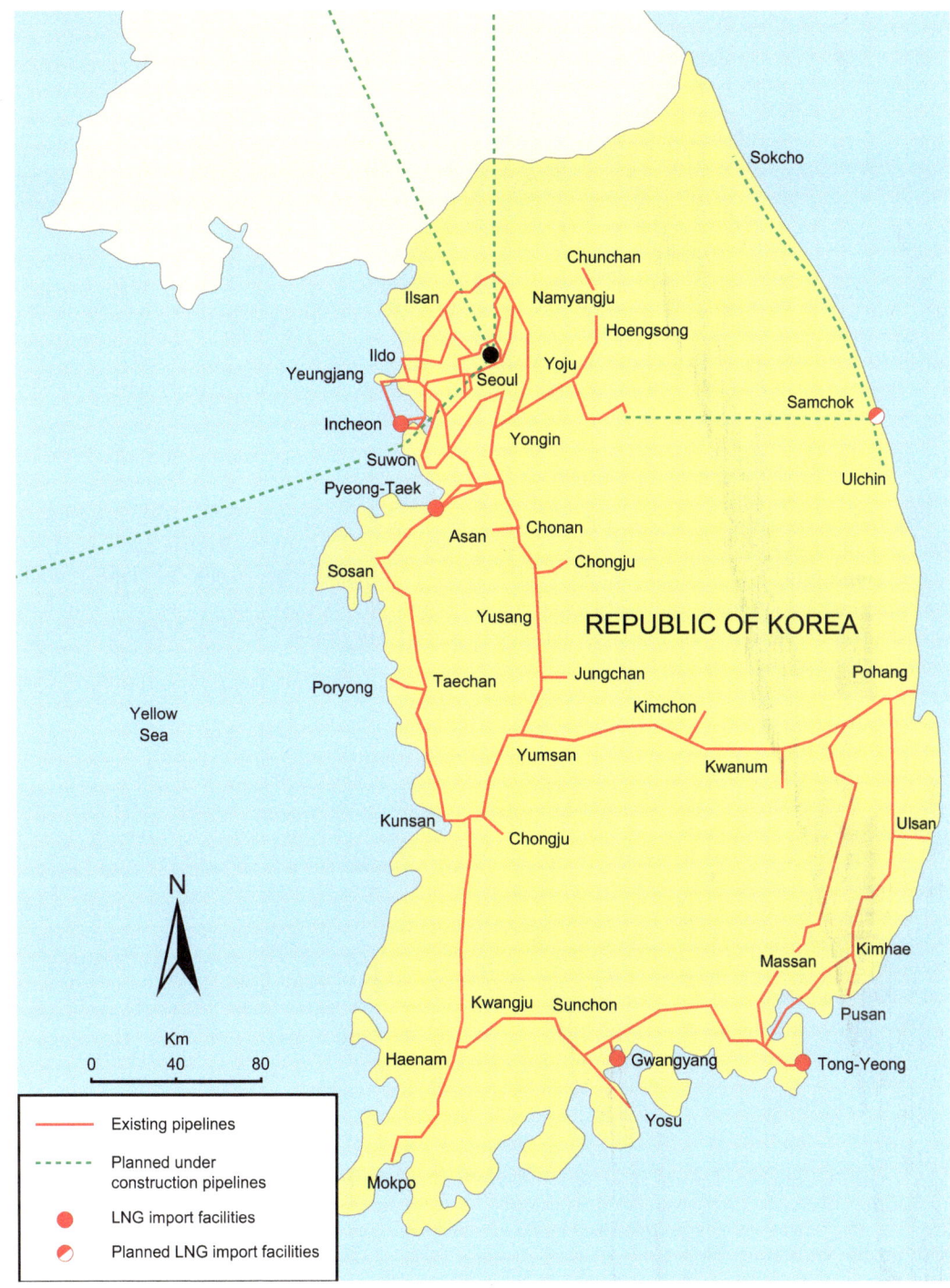

Fig. 18.22 Korean LNG terminals and pipeline networks. *Source* IEA (2011)

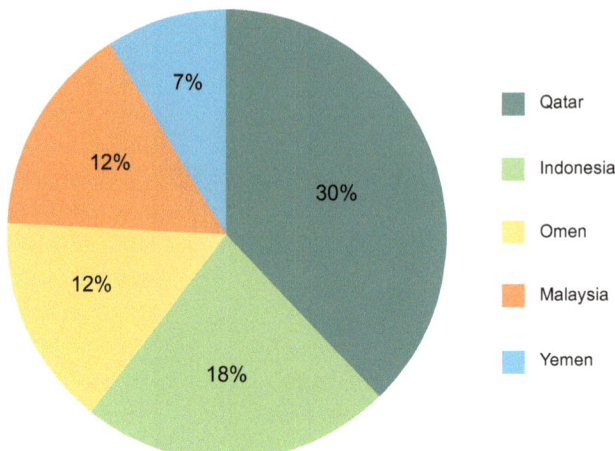

Fig. 18.23 Major importers of natural gas to Korea, 2012. *Source* Vivid Economics, based on IEA data (2012)

policies announced in 2010 was the proposal that, beginning in 2015, KNGC should ensure that oil-linked long-term contracts increase their degree of flexibility and terms. Further co-operation between Korea and its LNG importing neighbours (such as Japan) could improve buying power as part of new long-term contract agreements.

A Close Look at the Natural Gas Industry Chain

19

The previous section set out international experience in the development and evolution of natural gas markets. A key insight from the five country case studies is that liberalisation requires different regulatory approaches for different parts of the natural gas value chain.

This section draws from international experience and takes a closer look at regulatory efforts at specific points of the value chain. In the upstream segment, countries have encouraged exploration and production by increasing competition, including through fiscal and licensing regimes, to encourage new entrants into the industry. In the midstream segment, where natural monopolies exist, third-party access to transmission and distribution infrastructure has been crucial. In addition, some degree of unbundling of companies integrated across the value chain has been essential, both to reduce anti-competitive behaviour with respect to midstream infrastructure and to encourage competition across the natural gas value chain. In the downstream segment, countries have encouraged greater competition in the wholesale and retail

markets supplying end users. Countries with significant domestic natural gas resources have liberalised their wholesale markets by encouraging greater competition and the establishment of natural gas hubs.

19.1 The Upstream Segment: Fiscal Policies and Licensing Systems

In addition to furthering a government's fiscal objectives, these regimes have important implications for the development of a country's oil and natural gas resources, the natural gas regulatory environment, the relationship between various levels of government and various agencies, and the overall nature of the oil and natural gas industry.

Upstream fiscal and licensing regimes play a critical role in the development of a country's oil and natural gas resources. The two of them work in unison to offer benefits by incentivising investments and strengthening market competitiveness. Through an analysis of the experiences of the United States, Australia, Argentina and Mexico, this section carries out an in-depth exploration of natural gas upstream sector financing and taxes, permit system arrangements, and co-ordination and contact between various levels of government, all of which can be useful as references as China formulates similar policies.

* This chapter was overseen by Jigang Wei from the Development Research Center of the State Council and Taoliang Lee from Shell Eastern Trading Corporation. It was jointly completed by Yaodong Shi, Zifeng Song, Ren Miao from the Energy Research Institute of the National Development and Reform Commission and Cindy Wang from Shell China. Other members of the topic group participated in discussions and revisions.

© The Editor(s) and The Author(s) 2017
Shell International and The Development Research Center (Eds.), *China's Gas Development Strategies*, Advances in Oil and Gas Exploration & Production, DOI 10.1007/978-3-319-59734-8_19

19.1.1   International Taxation and Licensing Systems Regulating Upstream Production

Our analysis of international experience in fiscal and licensing regimes to regulate the upstream produced four overarching observations.

- **Unconventional resources**: International experience in crafting fiscal and licensing regimes shows that policy makers need to consider the fundamental differences between the exploration and development of conventional natural gas reserves and that of unconventional reserves. Unconventional natural gas developments have an ongoing need to explore, develop and produce on order to find the most productive areas, the so-called "sweet spots". These are characterised by shorter plateaus and more rapid depletion of reserves than conventional natural gas wells. As a result, unconventional operations require ongoing capital expenditures, whereas the majority of capital costs for conventional operations occur at the start of a development. Moreover, unlike conventional operations, unconventional natural gas developments typically require significant quantities of water, potentially affecting water availability for other users in the area, as well as water quality from the run-off and discharge of water used in processes such as hydraulic fracturing. This requires appropriate policy frameworks and processes to manage and minimise these impacts.
- **Co-ordination across government tiers**: Fiscal and licensing regimes should also take account of the interplay between national and regional authorities. Each level of government has an important role to play in the development of a country's energy resources, in terms of taxation and wider policy development. A review of experiences in the United States, Australia and Argentina shows

that national and regional governments typically share the revenues generated, both directly and indirectly. Co-ordination between national and regional authorities is also required in terms of regulatory policies. For example, in the United States and Australia, access to midstream infrastructure—essential for encouraging competition in the upstream and downstream segments—requires policy co-ordination and alignment between the national and state governments.
- **Looking across the value chain**: Fiscal and licensing regimes have a direct impact on promoting upstream developments, but they are just one part of an overall energy approach. Efforts to stimulate other parts of the value chain, especially demand, can have a profound influence on upstream activities. In Australia, for example, regulatory support for a switch to natural gas in power generation created momentum for greater domestic exploration and production.
- **Balancing the need for foreign investment and expertise with domestic interests**: Investment in upstream exploration and production is often supported by foreign direct investment (FDI). However, at times, increased FDI can also create tensions with national oil companies or concerns that national interests are being neglected. The experience of Mexico and its Bid Round Zero illustrates how countries can balance these interests, providing opportunities and safeguards for domestic companies while bringing in foreign funding and capabilities.

19.1.2   Case Study 1: The United States—Leasing, Taxation and Development of Information Sharing

Under the United States federal system, most regulations for the oil and natural gas industry are enacted and enforced by the individual states.

The US upstream investment regime uniquely is underpinned by ownership of mineral resources by the owner of the land. This means that onshore ownership of the resources is by private individuals, and private leases can be negotiated without a licensing authority. The exception to this norm concerns federal and state land, where the relevant government authority issues leases through tenders. For federal land, the Bureau of Land Management is responsible for issuing leases (Fig. 19.1). For federal waters—typically beyond three miles of the coastline—the Bureau of Ocean Management is responsible for issuing leases. Federal leases typically have a term of 5–10 years and can be renewed as long as production attributable to the lease continues.

With regard to the mining of natural gas, federal and state governments in the United States primarily use a tax and royalty regime. Onshore rates range from 12.5 to 30%, while the offshore rate is 18.75%. Most states also charge a severance tax, with its structure and level varying by state. For example, the Texas severance tax is 7.5% of the market value of the natural gas, while in neighbouring Louisiana, the severance tax is set each year based on NYMEX Henry Hub settled prices and on spot prices in the state. In 2014, this was 16.3 cents per thousand cubic feet.

The regime in the United States is also characterised by significant information sharing around the development of tight and shale natural gas, often cited as a reason behind the country's successful development of unconventional oil and natural gas resources. Data on permits, drilling, testing and production is required to be made available in a timely manner, although the details vary from state to state. Interested parties can obtain the information directly from the state or from consulting companies that collect such data. In addition to legal disclosure requirements,

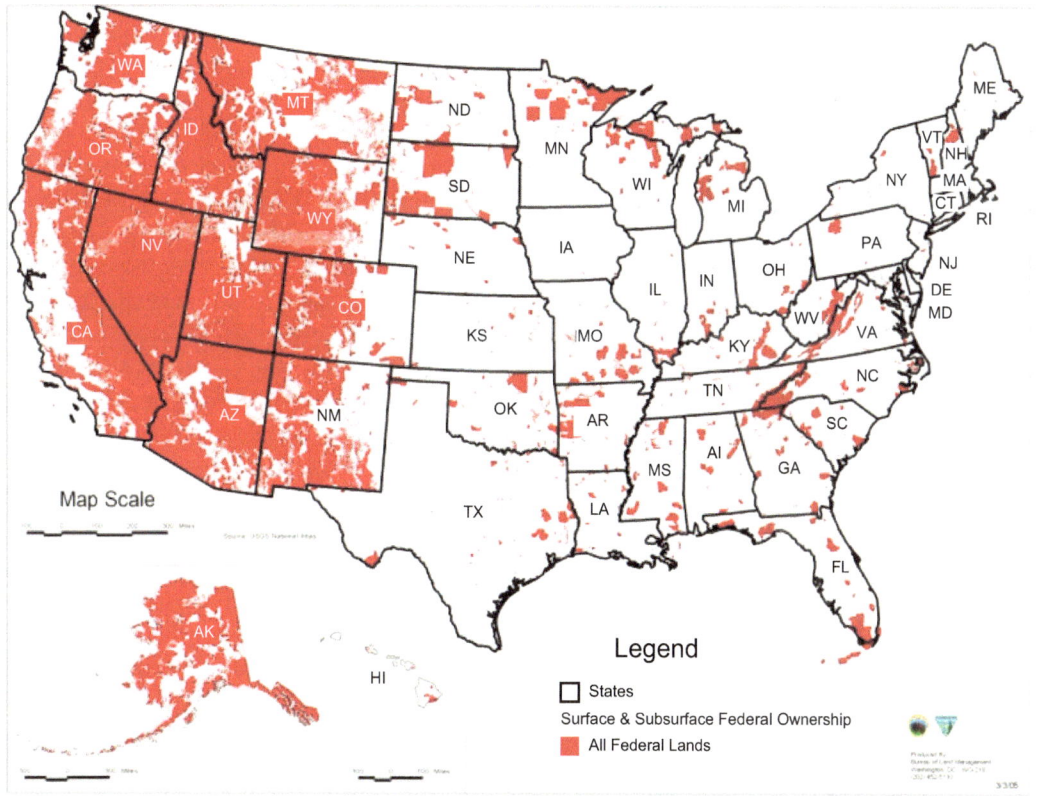

Fig. 19.1 US federal public land surface and underground levels. *Source* US Bureau of Land Management

companies also have incentives to disclose information—occasionally including details on individual wells—to support their share price and their efforts to raise capital. Publicly traded companies are also required to submit detailed quarterly and annual reports to the federal government, which are publicly available. In addition, several federal government agencies, including the US Geological Survey, the Department of Energy and the Energy Information Administration, monitor and publish key statistics weekly, quarterly and annually. Adding to the data flow, industry associations, technical and business journals, investment reports and conferences provide a platform for upstream companies, investment banks, academics and government to exchange information.

19.1.3 Case Study 2: Australia—Licensing, Finance and Taxation, and Third-Party Access

According to forecasts, in the near future Australia will become the largest LNG exporter in the world. It has liquefaction capacity of about 26 million metric tonnes a year in operation, and is expected to add about 62 million metric tonnes in additional capacity over the next several years. The new projects include three projects in Queensland, primarily using coalbed methane to produce LNG. A diverse range of companies are involved in onshore Queensland natural gas production, including PetroChina, Sinopec and the China National Offshore Oil Corp. (CNOOC).

Australia's domestic natural gas system comprises the East Coast market, covering Queensland, New South Wales, South Australia, the Australian Capital Territory and Victoria, and the Western Australian market, covering the rest of the country. Australia does not have an extensive pipeline network, especially compared to Europe or the United States, because demand is highly concentrated in the various state and territory capitals and because supply has

historically come primarily from two basins, Cooper and Gippsland (Fig. 19.2).

Alongside fiscal and licensing measures to stimulate upstream exploration and production, Australia also provided demand incentives to increase the share of natural gas in the energy mix. The Queensland Gas Scheme illustrates the success of such measures in driving coalbed methane production. Introduced by the Queensland government in 2005, the scheme required all electricity retailers to acquire 15% of the power they sold from natural gas-fired generation plants. Accredited power producers generated Gas Electricity Certificates (GEC) for each MWh of eligible natural gas-fired electricity they produced. These GECs are then sold to the electricity retailers, providing an alternative revenue source for power producers to offset the higher cost of natural gas-fired power generation compared to coal. The scheme ended in 2013 when the state government declared that its goals of promoting the state's natural gas industry and reducing greenhouse natural gas emissions had been achieved. During this time, natural gas-fired power generation had gone from 5% of total power in 2005 to 20% in 2013, and the state's unconventional natural gas production had grown from 1 billion m^3 in 2004–05 to 6 billion m^3 in 2010–11 (Fig. 19.3).

Australia has three levels of government—federal, state or territory, and local—and has a concessionary fiscal regime. The federal government and state and territory governments share jurisdiction over petroleum resources, and there are no private ownership petroleum resources.

State and territory governments manage the licences for development onshore and within 3 miles of the coastline. They collect royalties of between 10 and 12.5%, based on the wellhead value of the petroleum resource. Licences in Queensland are issued and managed by the Department of Natural Resources and Mines, and the published royalty rate is 10%.

The federal government manages licences for offshore development beyond 3 miles of the coastline. It levies a Petroleum Resource Rent Tax of 40% for both onshore and offshore oil and

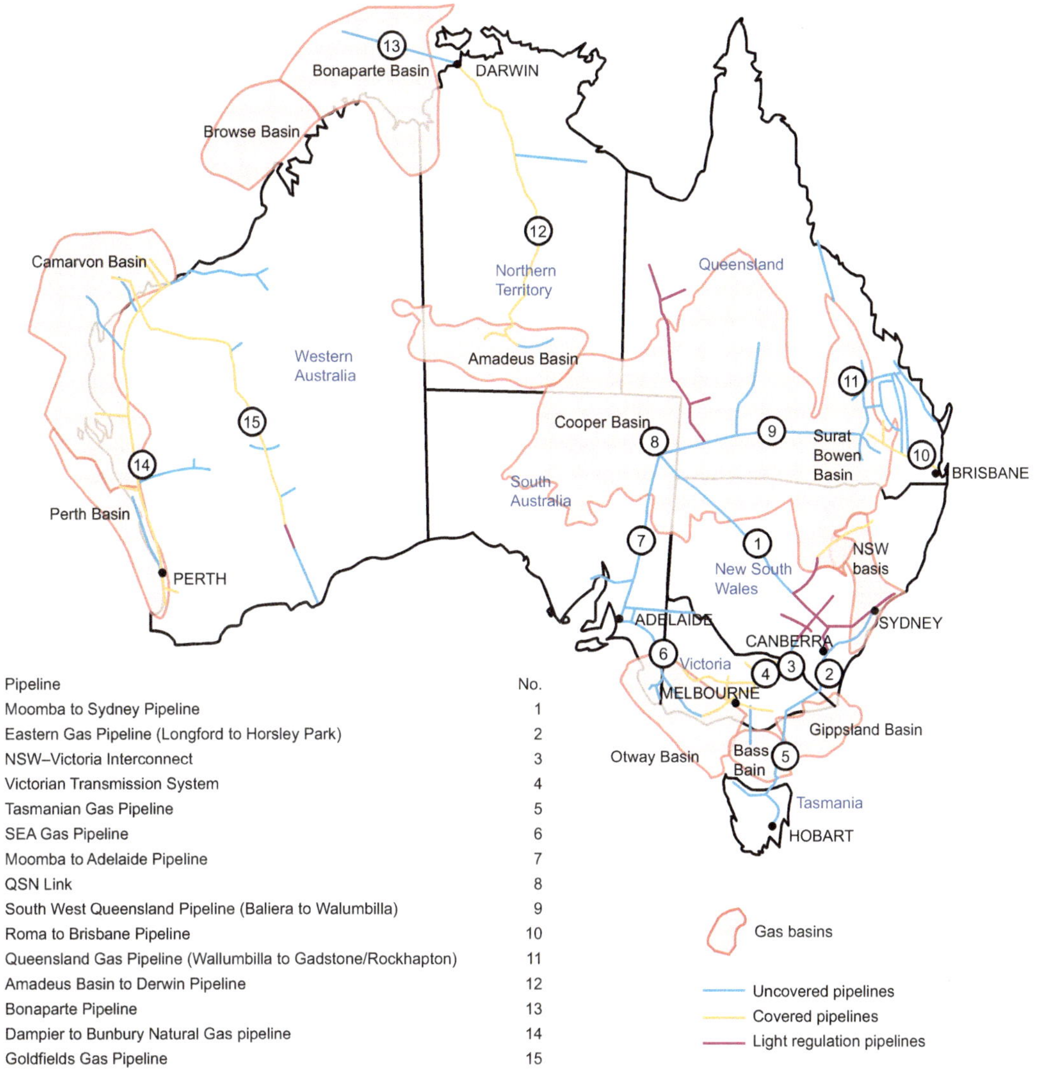

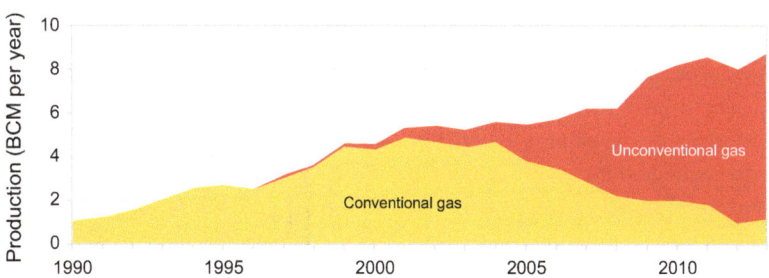

Bonaparte Basin
DARWIN
Browse Basin
Carnarvon Basin
Northern Territory
Queensland
Western Australia
Amadeus Basin
Cooper Basin
Surat Bowen Basin
BRISBANE
South Australia
Perth Basin
PERTH
New South Wales
NSW basis
SYDNEY
ADELAIDE
CANBERRA
Victoria
MELBOURNE
Gippsland Basin
Otway Basin
Bass Bain
Tasmania
HOBART
Pipeline | No.
Moomba to Sydney Pipeline | 1
Eastern Gas Pipeline (Longford to Horsley Park) | 2
NSW–Victoria Interconnect | 3
Victorian Transmission System | 4
Tasmanian Gas Pipeline | 5
SEA Gas Pipeline | 6
Moomba to Adelaide Pipeline | 7
QSN Link | 8
South West Queensland Pipeline (Baliera to Walumbilla) | 9
Roma to Brisbane Pipeline | 10
Queensland Gas Pipeline (Wallumbilla to Gadstone/Rockhapton) | 11
Amadeus Basin to Derwin Pipeline | 12
Bonaparte Pipeline | 13
Dampier to Bunbury Natural Gas pipeline | 14
Goldfields Gas Pipeline | 15
Gas basins
Uncovered pipelines
Covered pipelines
Light regulation pipelines

Fig. 19.2 Australia's natural gas basins and major transmission pipelines. *Source* Australia Energy Regulator

Production (BCM per year)
10
8
6
4
2
0
1990
1995
2000
2005
2010
Unconventional gas
Conventional gas

Fig. 19.3 Natural gas production in Queensland, Australia. *Source* Queensland Department of Natural Resources and Mines

natural gas, with state and territory royalties deductible from the amount due.[1] In addition, the federal government levies a corporate tax of 30%. It also levies a value added tax, known as a Goods and Services Tax, of 10% on most transactions, with receipts redistributed to the state and territory governments.

The three levels of government in Australia co-ordinate to develop natural gas resources and markets. The Council of Australian Governments is the main intergovernmental forum that promotes policy reforms with national relevance or those that require co-ordinated action at all levels. Members of the council are the prime minister, the premiers or chief ministers of each state or territory respectively, and the president of the Australian Local Government Association. In addition, the Standing Council on Energy Resources (SCER) was founded in 2011 and comprises federal, state and territory ministers responsible for energy and resource matters. The council's responsibilities include oversight of the natural gas and electricity markets, energy laws and regulations, energy security and emergency management, and promoting the economic development of Australian resources.[2]

Companies developing new upstream infrastructure may be granted exemptions from third-party access requirements if they satisfy criteria outlined in the legislation.

Access to third-party infrastructure is established under the federal Competition and Consumer Act. The National Competition Council (NCC) is established through consensus by the SCER and is responsible for recommending regulations on third-party access to monopoly infrastructure, including natural gas transmission pipelines. Factors taken into consideration include whether access would increase competition in a market, the economics of developing competing infrastructure, and whether the infrastructure is already subject to an effective access regime.

For example, 15-year exemptions have been granted to cover the pipelines being built to transport natural gas from the three Queensland LNG projects to the liquefaction plants in Gladstone. The NCC concluded that third-party access "would not promote a material increase in competition in any likely dependant market".[3] Even though the developers can obtain a waiver, nonetheless the Australian Competition and Consumer Commission (which also has jurisdiction beyond energy) and the Australian Energy Regulator (AER) both carry out significant oversight, strictly executing energy market rules and providing protection for competition and consumers.

19.1.4 Case Study 3: Argentina—The Special Licensing System and Encouraging Investment

Argentina is a significant producer and consumer of natural gas. According to data from the US Energy Information Administration, the country's technically recoverable shale natural gas reserves are second only to those in China, and its shale oil resources are the world's fourth largest, after Russia, the United States and China. According to forecasts, Argentina's natural gas demand will have been 44 billion m[3] in 2015, accounting for 45.7% of its energy structure. Currently, the domestic shortfall is met by LNG purchased primarily from the spot market and imported through two floating regasification and storage units. Natural gas prices are regulated and subsidised below cost, but prices have begun to be deregulated—mainly through bilateral negotiations—to encourage upstream investment and to raise revenues for the government.

Argentina's concession-based licensing system replaced a system based on risk service contracts. State-controlled Energía Argentina S.A. (ENARSA), which is 35% publicly owned, owns all unlicensed federal offshore exploration acreage more than 12 nautical miles from shore. Any activity in these blocks must be done in partnership with ENARSA. Provincial governments issue tenders for onshore projects.

---

[1]Australian Taxation Office website.

[2]Jurisdictional scope of the Standing Council on Energy and Resources.

[3]NCC website.

All licences are subject to royalties, income tax, provincial sales tax and other signature bonuses and rentals, as well as possible export duties on oil and natural gas. The Oil and Gas Law, passed in 2014, centralises implementation of the royalty system and licensing, but leaves the administration to provincial regulators. Before this law was enacted, provincial governments had jurisdiction on oil and natural gas licensing and operations, and many provincial governments took equity interests in licences. From this it is clear that provincial-level governments can become involved in relevant licensing matters through their various oil and natural gas departments (Figs. 19.4 and 19.5).

Due to growing energy needs, governments have begun to focus on stimulating investment. In October 2014, the Oil and Gas Law was passed. This contained a series of measures providing incentives for investment in unconventional oil and natural gas exploration and production, including:

- distinguishing between conventional and unconventional resources, for example by providing unconventional assets with an exploitation period of 35 years compared to 25 years for conventional resources (exploration periods are 13 years);
- plans to allow higher prices for natural gas produced using unconventional means;
- creating an unconventional production licence, providing five years for a pilot project to become commercially viable, with royalties to be dropped by 25% for unconventional projects after the pilot phase is completed; and
- allowing 20% of production to be exported.

However, obstacles remain to building an unconventional natural gas industry in Argentina, including a shortage of trained labour. The opening of the Unconventional Fields Technology Center in Neuquén Province is a step forward, but more is needed to build capabilities in shale natural gas technology. In addition, current regulations do not address environmental concerns adequately, which could become problematic for unconventional developments near population centres.

19.1.5 Case Study 4: Mexico— Reopening the Market and Round Zero Tender

Mexico's oil industry is one of the world's oldest; oil was first discovered in Mexico in 1904. By 1921, Mexico's oil production had reached 530,000 bod, accounting for 25% of the world's oil production. Following labour disputes, international oil company assets in Mexico were nationalised in 1938, which led to the creation of the national oil company Petróleos Mexicanos (Pemex). Pemex held exclusive rights to the country's oil and natural gas fields.

In the following decades, Pemex largely ran its own oil and natural gas operations, aided by international service companies. As a state monopoly, its budget and financial management were heavily controlled by the Ministry of Finance and Public Credit, leading in part to its gradual stagnation. Its production volume continually dropped, beginning in 2004 (Fig. 19.6). Its production volume peak was between 2004 and 2005, with approximately 2.6 million barrels/day, as well as 3 billion ft^3 of natural gas. Rapid further decline is projected for the coming years.

In an attempt to halt production declines, in 2013 the political party known as PRI began implementing energy reforms after 71 years. President Enrique Peña Nieto proposed an Energy Reform Act in August 2013. The act was eventually passed in December 2013, allowing international oil companies and other investors to invest once again in Mexican oil and natural gas businesses through three possible legal arrangements: service contracts, production- or profit-sharing contracts, or licences.

To protect the national interest in Pemex, the government ran a unique tender in 2014, Bid Round Zero. Under the tender, Pemex had to submit applications for which fields and blocks it wanted to keep and relinquish the rest. The submission had to demonstrate technical,

Fig. 19.4 Argentina's primary basins and LNG receiving stations

financial and operational capabilities to operate productive assets, and, for exploration acreage, Pemex had to show it had either drilled exploration wells, conducted sub-surface studies or both. The Energy Secretary and the Comisión Nacional de Hidrocarburos (National Hydrocarbons Commission) made the final decisions on the applications.

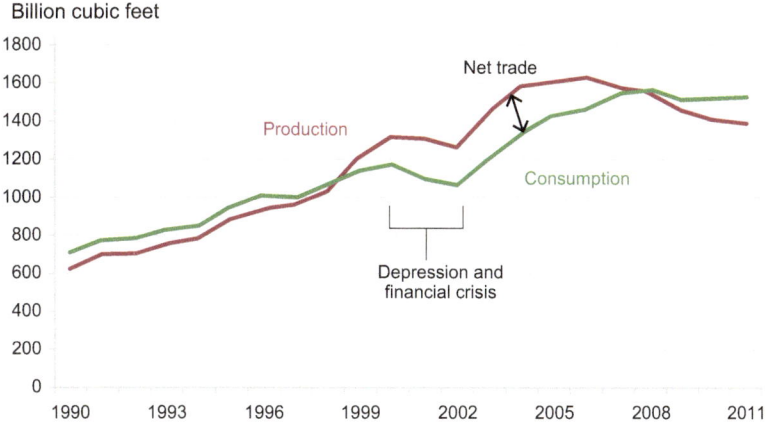

Fig. 19.5 Argentina's natural gas production and consumption. *Source* Energy Information Administration, International Energy Statistics

Fig. 19.6 Oil and natural gas production in Mexico. *Source* Wood Mackenzie

As a result of Bid Round Zero, Pemex retained about 83% of proven and probable oil and natural gas reserves (Fig. 19.7) and 21% of the prospective resources, based on commission estimates (Fig. 19.8). Pemex was also allowed to invite foreign investors to join in the exploration, development and production of assets it retained.

A measure like Bid Round Zero could be an attractive option for China. Such a tender would allow existing state oil companies to nominate

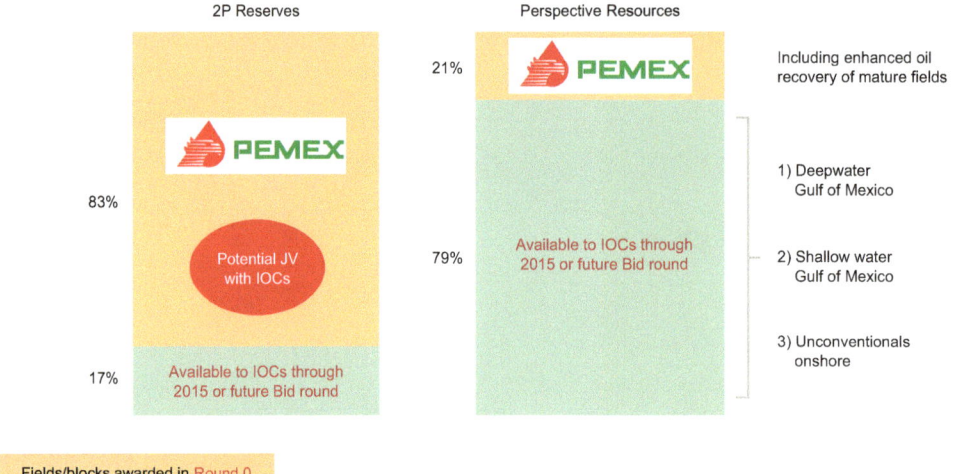

Fig. 19.7 Pemex reserves analysis after bid round zero

assets they would like to retain, freeing up other assets for new entrants. The move could attract greater energy investment and accelerate the development of unconventional shale natural gas production. Increased domestic production would reduce the country's dependence on foreign natural gas, which rose from zero in 2005 to 32% of domestic consumption in 2013.

In late 2014, the National Hydrocarbons Commission began Bid Round One, which will be a staged tender and include shallow water exploration, shallow water development, onshore projects, enhanced production for the Chicontepec field, unconventional oil and natural gas, and deepwater projects (Fig. 19.9).

Without question, the most important factor in these events has been the speed and determination of the Mexican government in taking clear and transparent steps toward energy reform, including the 20%+ of discovered oil/natural gas fields provided to international oil companies following the final Round Zero Tender. It is worth considering which of these measures could suit China's situation, especially in the accelerated exploration and development of unconventional natural gas and shale natural gas, as well as

in attracting suitable international oil companies and private investors to accelerate the trials in this process. It is also worth considering whether Mexico will continue on its path of open policies after the six-year term of President Enrique Peña Nieto.

19.2   The Midstream Segment: Building Infrastructure and Managing Access

Midstream assets—transmission pipelines, LNG regasification terminals and storage facilities, and distribution networks—are natural monopolies. They have high fixed capital costs of investment but low, and decreasing, marginal costs of operation. As a result, it is economically more efficient if fewer of these assets are built and fully utilised, rather than assets being duplicated and dividing volumes among them through competition.

Natural gas pipelines—both transmission and distribution—have more significant natural monopoly characteristics than LNG terminals or storage facilities. The economies of scale for LNG terminals and storage facilities are not as

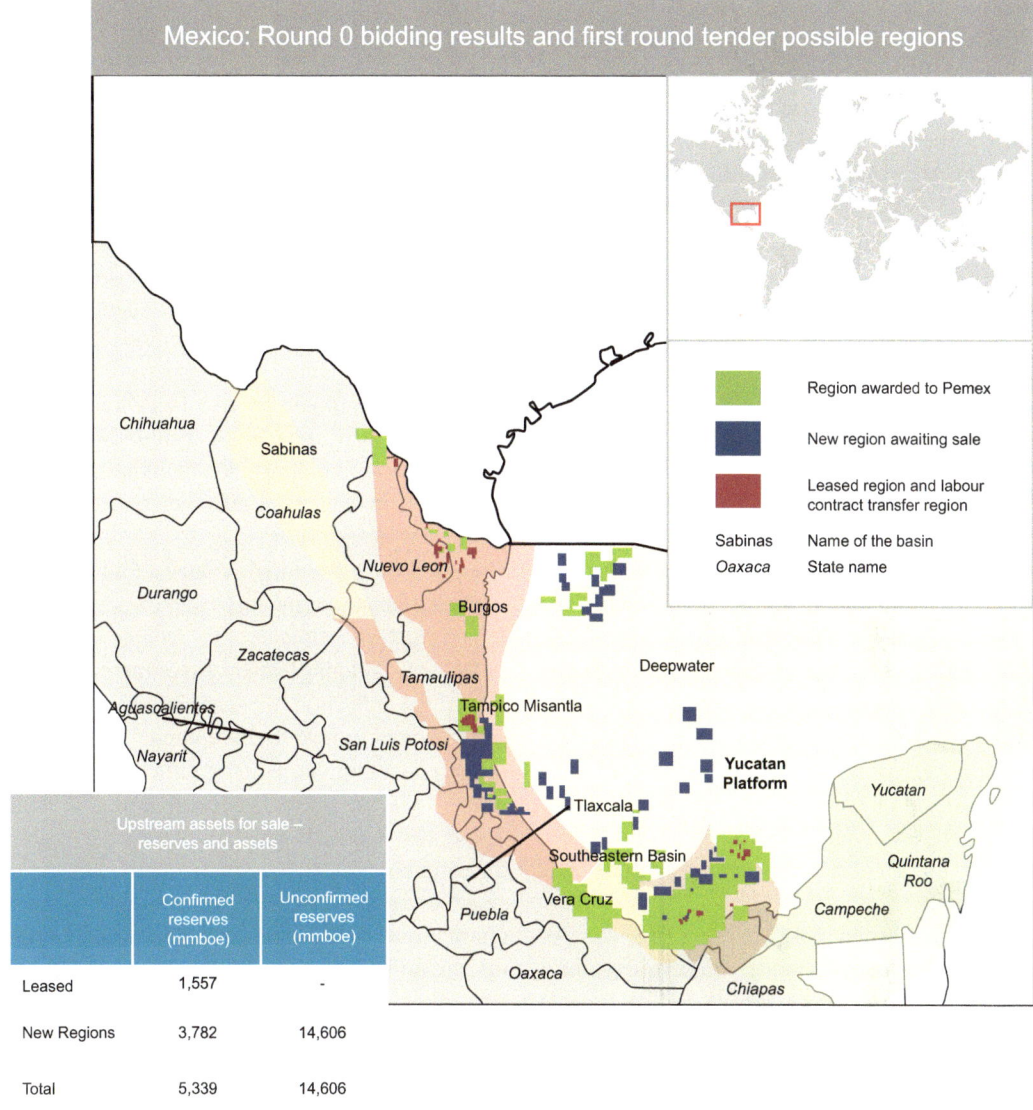

Fig. 19.8 Pemex unconfirmed resources, 2014. *Information source* HIS, PFC Energy

	Shall water: Exploration	Shall water: Development	Onshore	Chicontepec and Unconventionals	Deepwater
Publication of terms	December 2014	January 2015	February 2015	March 2015	April 2015
Opening of data rooms	January 2015	January 2015	March 2015	April 2015	May 2015

Fig. 19.9 Suggested timeline for a new round of tenders

great as those for pipelines. Hence, for a given market size, the optimal number of facilities required will be greater compared to the number of pipelines required to serve a market of similar size. Moreover, LNG terminals and storage facilities can compete with other similar assets once they are connected through a pipeline network.

In the United States, LNG receiving stations are viewed as upstream assets, because they are seen as extensions of upstream wells. However, in Europe, LNG receiving stations are midstream assets. The difference in perception depends on the degree of competition in the market. In the United States, LNG receiving stations compete with domestic products, and thus they seen as comparable to domestic wells. In Europe, the majority of competition comes from imported natural gas, and thus LNG receiving stations are viewed as part of the transmission network. If LNG receiving stations are seen as upstream assets, then regulatory measures will be different, and there will be fewer concerns about the monopolistic nature of the receiving stations. In China, there is uncertainty regarding the character of LNG receiving stations, because development is lacking in competitive natural gas markets.

The natural monopoly characteristics of midstream infrastructure mean that government oversight and regulation is required to prevent access being restricted and monopoly rents being charged. A monopoly owner of midstream infrastructure has an incentive to charge very high prices to maximise its profits, reducing pipeline utilisation in the process. It may therefore be necessary to unbundle midstream assets from vertically unified companies. Examples of the options for accomplishing this can be found by looking at the international experiences seen in the case studies.

Even if midstream assets do not belong to a given vertically unified company, if no oversight is implemented, there is still the possibility of problems arising. On the one hand, independent owners need upstream and midstream participants to use their midstream assets. On the other hand, independent owners also have an incentive

to collect the highest price possible to maximise their profits. Regulation is the solution, to ensure that usage rights to midstream assets are available to all parties (in other words, third-party access).

Regulatory institutions will often need to unbundle and re-establish third-party access. In practice, even if there are rules for third-party access, there are still many means to exclude third parties, such as lack of transparency over pricing. Using regulatory institutions to implement strict regulation of third-party access can prevent such anti-competitive behaviour. At the same time, many regulatory institutions believe that through unbundling such motivations can be eliminated, which is an effective supplement to third-party access rules.

In a growing market in need of investment, such as China, regulatory frameworks should ensure that investments can receive sufficient profits, encouraging capital investment. Regulation of the collection of usage right fees can help asset owners to recover their capital. Regulation of fee collection can employ either a "regulated return" system or a "regulated tariff" system. These two methods are very similar, but have differing risk allocations in terms of midstream asset users and owners. Under a "regulated return" oversight approach, the owner of the midstream is authorised to receive a high enough rate of return to act as a motivation for future pipeline construction. Under the "regulated tariff" regime, such tariffs should likewise be sufficiently high to provide an incentive for future asset investments.

The challenge therefore faced by government departments is to achieve a balance between making usage rights available and encouraging investment in midstream assets. This balance can be achieved through a precisely designed regulatory framework, which will need to include the following key factors: rules for liability and standards, a consensual pricing framework for third-party access and a mechanism to assure returns for investors. In some frameworks, when regulatory institutions solve downstream market structure or legacy issues, these mechanisms can be used to balance out two different goals.

When a country's regulatory mechanisms reflect various domestic factors, such evolutions and influences can be referenced by China. We examine three case studies below: the third-party access arrangements of the United Kingdom in the North Sea; Singapore and Japan's LNG experience; and Shell's IPO of its Shell Midstream Partners enterprise in the United States. These case studies yield five primary lessons:

- if it is to promote usage rights and investment, regulation must strike a balance between the economic interests of midstream asset users and owners;
- regulatory frameworks must be stable, reliable and clear, thereby facilitating general risk minimisation and stimulation of investment;
- mechanisms based on rules can provide a means of resolution for this;
- at the same time as generating revenue, regulations such as mandatory third-party access requirements can ensure that midstream assets are used efficiently;
- even while balancing usage rights and investment, regulatory institutions can also consider market structure in addition to the value chain. Current regulatory policy will influence future economic opportunities.

19.2.1   Balancing Third-Party Access and Investment Incentives

Third-party access needs to be balanced against incentives for further investment in midstream infrastructure, especially in growing markets such as China. While allowing for third-party access, the regulatory framework also needs to provide adequate incentives—and returns—on future investment in the midstream. For example, access charges can be regulated to enable asset owners to recover fixed costs.

Regulated charging regimes can be of two types: either regulating the rate of return achieved by midstream infrastructure owners or regulating the tariff charged for access to midstream infrastructure. Under a regulated return regime, the owner is granted a rate of return on its midstream asset value that is high enough to provide incentives for future investment in maintaining and extending the infrastructure network. Under a regulated tariff regime, the tariff is set at a level that provides incentives for future asset investment. Both approaches are similar, but vary in the distribution of risk among the users and the owners of midstream assets.

While an individual country's regulatory regime reflects a variety of domestic factors, a review of the evolution and impact of these regimes can offer insights for China in achieving a balance between returns to investment and access to midstream assets. The oversight of natural gas infrastructure reflects a country's experience in the realm of natural gas market liberalisation, as well as its systemisation and government and policy priorities. There is no single oversight system that can be transplanted from one jurisdiction to another. In fact, if residual issues, systems and political factors are not taken into consideration, replicating another country's oversight structure will result in failure (United Nations Economic Commission for Europe 2012). However, it is possible to learn by comparing existing oversight frameworks that the key is to balance the attainment of public and private goals.

Regulatory institutions often need to unbundle third-party access. In reality, even if there are third-party access rules, there are still many means by which to exclude third parties—for example, a lack of transparency over pricing of capacity. Regulatory institutions can prevent such anti-competitive behaviour by rigorous oversight of third-party access rules, but many regulatory institutions believe that unbundling is an effective solution that eliminates the incentive for anti-competitive behaviour.

In growing markets in need of investment, such as China, regulatory frameworks should also allow for sufficient profits, thereby providing momentum for future midstream investment. Using a "regulated return" system, owners of midstream assets are allowed sufficiently high rates of return that there is an incentive for future piping construction. Under a regime of "regulated tariffs", such tariffs should likewise be sufficiently high to provide an incentive for future asset investments.

The challenge for government departments is therefore to achieve a balance between midstream asset usage rights and investment, as shown in Fig. 19.10. Within a carefully designed regulatory framework, it is possible to achieve this goal. Many countries have achieved such success, and the regulatory mechanisms that may be of interest to China as precedents include:

- **Rules**: Regulators establish and enforce rules on liability and standards.
- **Prices**: Regulators provide a framework for investors and users to agree on prices in both third-party-access and regulated-return regimes.
- **Mechanisms to reduce risk**: Regulators minimise risk for investors and users.

19.2.2   International Experience of Managing Midstream Asset Access

The case studies provide five overarching insights:

- **Regulation balances the economic interests of users and owners of midstream assets to facilitate access and investment**. Economic opportunities drive demand for access to and investment in infrastructure. However, demand by itself is an insufficient condition for access and investment, as users and owners often cannot agree terms, often because of the asset owner's natural monopoly power. Thus, players require a strong regulatory framework that grants access to the midstream and provides incentives for investment. For example, experience in Singapore and the United Kingdom in establishing third-party access shows that strong regulation can assist users of assets in their negotiations with midstream operators.

The experience of the United States and Japan has shown that in terms of risk/reward for investors, oversight is beneficial in creating investment opportunities and encouraging

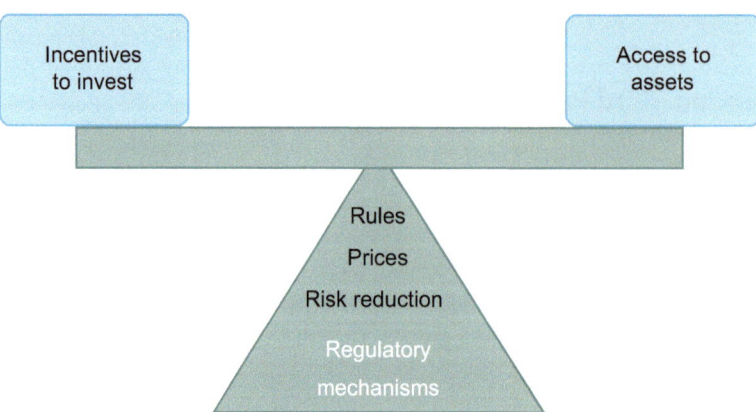

Fig. 19.10 The balance of factors needed in the regulatory framework

more funds to be channelled to the construction of midstream infrastructure.

- **Regulation can ensure that the midstream asset is used efficiently**. Regulation, such as mandatory third-party access requirements, can ensure that players upstream and downstream can access the pipeline and that the fees charged and returns made by the owner of the midstream asset are fair. Regulation can take various forms, such as negotiated or regulated access, but all forms of regulation are aimed at increasing access while ensuring reasonable returns to the asset owner and supporting sufficient investment to maintain energy security. Controls implemented according to specific circumstances can utilise agreement negotiations or controls to achieve third-party access. The British North Sea owner and user negotiations successfully implemented third-party access, with new companies able to develop small-scale gas fields without needing to invest in the construction of new piping. Midstream asset owners can obtain revenue from amortised assets, and the government can obtain relevant oil and natural gas production taxes related to asset development. However, in the rapidly growing market of China, the experience from the North Sea is of limited use. There is very little need for new investment in piping in the North Sea, and this made it easier for midstream facility owners and users come to an agreement. This is not the case in China, and the facility owners tend to look to use their position to collect monopolistic lease amounts. As a result, China's rapidly growing emerging market should give greater consideration to using mandatory third-party access systems to stimulate investment in midstream facilities by promoting recovery of capital and reasonable risk returns.

- **The regulatory framework should be stable, credible and clear to minimise risk and encourage investment**. The success of the master limited partnership in the United States has been in large part the result of government providing a statutory basis for partnership treatment: federal rules on qualifying income and assets gave all parties the necessary certainty to proceed. In the United Kingdom, the combination of a statutory, rules-based framework and the threat of arbitration ensures that when negotiating third-party access, parties have similar expectations and incentives, with it being very rare that events end in disagreement. The regulatory layers collectively help to reduce uncertainty for participants (see Fig. 19.11). In contrast, the absence of enforcement has meant that third-party access rules are not credible, and midstream infrastructure is for the most part not open to third parties.

- **Regulators should take into account market structures along the value chain when balancing access and investment**. The nature of competition along the entire value chain influences regulatory choices. For example, Japan has a fragmented market where all natural gas is imported by LNG shipments and there is limited interconnection

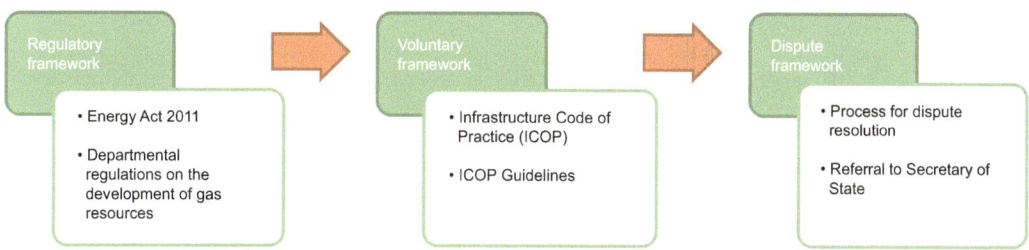

Fig. 19.11 The three layers of agreements supporting North Sea usage rights

between regional markets. This means that competition is hard to achieve and brings few benefits, and as a result third-party access is limited.

In the United Kingdom, the natural gas market was originally created upstream, with competitive downstream markets only starting in the 1980s, and thus its regulatory framework design focus is on the promotion of midstream usage rights.

In Singapore, private funding for an open-access LNG terminal fell through as a result of sufficient pipeline gas supply to meet demand and liberalised downstream markets, which favoured cheaper pipeline gas over more expensive LNG imports. The government stepped into develop the LNG terminal as a means of diversifying sources of supply and enhancing energy security. Ownership is deliberately separated from operation of the multi-user LNG terminal, and the government has developed a framework to open the terminal to third-party access. At the same time, the government has also sought to secure downstream demand for LNG by restricting new pipeline imports (Table 19.1).

- **The regulatory framework a country pursues affects the set of available policy choices**. The regulatory legacy can affect the level of utilisation of infrastructure, the level of activity by new entrants and the form of private contracts. The private sector's capacity to identify and develop economic opportunities is a function of how a country's regulatory regime and its infrastructure have evolved.

In the first years of development of North Sea gas, large companies obtained permits from the government and independently invested to develop the oil and gas fields. In the past 12 years, because of drops in yields, these oil and gas fields have seen reduced rates of return, and there is less interest from large companies. However, the amortised assets that they possess can be transferred to smaller enterprises to develop small-scale oil fields that pay the larger companies for usage of their assets. Therefore, from the 1960s to the 1990s, the United Kingdom's regulation system has increased the opportunities today. In Japan, even though on the surface LNG receiving stations seem to be accessible, regulators have chosen to use negotiated approach a third-party access, rather than making it mandatory. This approach gives facility owners a powerful negotiating position. Combined with a lack of competition upstream and downstream, there is a push for suppliers and upstream facility owners to sign long-term supply agreements. Without reliable access systems, new entrants to the market generally build new LNG receiving stations in order to take advantage of downstream business opportunities.

Table 19.1 The influence of the nature of competition on midstream infrastructure regulation choices

Market	Nature	Challenges for asset owner/user	Consequence for regulation
United Kingdom upstream and downstream	Liberalised markets	Ease of entry, but investment costs high	Framework for asset owners and users to negotiate terms of access
Japan downstream	Monopolistic geographic markets	Lack of interconnection creates concern over security of supply	Facilitate construction of new LNG terminals with limited third-party access
Singapore downstream	Liberalised market but fragmented infrastructure	Uncertainty that downstream can be accessed given pipeline natural gas supplies	Pipeline natural gas moratorium and midstream third-party access

19.2.3   Case Study 1: The UK North Sea—The Framework for Negotiated Third-Party Access

The United Kingdom's North Sea experience illustrates one approach to arranging access to midstream infrastructure. In particular, it focuses on the government's efforts to develop a regulatory framework that grants new entrants access to the pipelines built by earlier North Sea field developers.

Several key lessons emerge from this experience:

- **A liberalised market with clear rules encourages new investment**. Output from the North Sea was declining and costs were rising. However, new firms that specialise in declining fields were able to enter the liberalised market, which has provided asset owners with new business and the government with greater tax revenue.
- **Users of assets have been favoured over owners to some extent**. As midstream assets were already in place as a legacy of past exploration of the North Sea, and faced declining utilisation, new users seeking access had greater bargaining power as asset owners had few alternative customers. In addition, by encouraging access, the government could benefit from greater hydrocarbon production tax revenues associated with resource development.
- **Where interests between midstream asset owners and users are broadly aligned, a voluntary agreement on access provision between the parties is feasible**. The United Kingdom had a layered regulatory framework to access. Parties were encouraged to reach voluntary agreements, with clear rules for dispute resolution and, ultimately, a legal framework if voluntary agreements could not be reached. This system has worked well, with many parties reaching voluntary agreements, because interests are broadly aligned and the rules are clear and credible.

China faces a different gas market context than seen in the North Sea example. China's natural gas market is growing rapidly, whereas output from the North Sea was declining. As a result, the economic pressures felt by UK asset owners that supported negotiated third-party access—specifically, existing assets with declining utilisation—would not be as prevalent in China, where limited midstream capacity gives asset owners more bargaining power. Nevertheless, the North Sea experience demonstrates how a credible regulatory framework can help achieve agreement and avoid disputes.

1. **Background**

The regulatory regime for natural gas exploration in the United Kingdom from the mid-1960s through to the 1980s was characterised by liberalisation and increasingly competitive markets. When the first licences were offered by government, major companies chose to develop oil and natural gas fields as stand-alone installations. Upstream pipelines and offshore processing facilities were usually built by field owners to process and transport output from specific oil and natural gas fields. However, since then, recovery rates—and returns—from these fields have fallen, and larger enterprises have been less interested in developing the remaining resources.

As spare capacity became available in pipelines and terminals, opportunities arose. The existence of their depreciated pipeline assets has allowed smaller players to enter the industry and exploit smaller fields, paying larger companies to use their assets to transport oil and natural gas to the mainland. A liberalised downstream market meant there was a ready market for small-field developers if they could access midstream assets. There were benefits to infrastructure owners, too, in terms of additional revenue from granting third-party access to new users and from deferring the costs of decommissioning these assets. The government saw opportunities in terms of additional job creation, maximising recovery of North Sea oil and gas resources, and continued production tax revenues.

The 1996 Network Code established the rules and procedures for third-party access to pipelines and introduced a regime for daily balancing, while the Petroleum Act of 1998 set out the rules around upstream exploration and production. By the early 2000s, liberalisation of the sector was complete, with the creation of the National Grid as a fully unbundled, independent transmission system owner and operator of the United Kingdom's natural gas and electricity markets in 2002.

However, the legal framework by itself was insufficient and left room for commercial disputes between owners and users of the infrastructure. Through the 1990s, there were concerns that the tariffs to access infrastructure were too high compared to costs and risks of developing small fields. This led to a new process for negotiating access. The legislative and regulatory changes provided a statutory basis for a rules-based framework, including a process for arbitration of disputes, which helped co-ordinate activity and create common expectations.

2. The framework for negotiated third-party access

Working with government, the natural gas industry sought to develop voluntary frameworks for third-party access. The first industry offshore Code of Practice was introduced in 1996. It sought to establish a timely process for seeking, offering and negotiating third-party access to natural gas infrastructure in the North Sea. It also sought to ensure that access was easy and fair, with terms offered on a negotiated, non-discriminatory basis.

In 2004, market participants agreed to a strengthened code, the Infrastructure Code of Practice. The code outlined best practice and expected behaviour in conducting negotiations for access to infrastructure. According to the industry group Oil & Gas UK, the code was intended "to facilitate the utilisation of infrastructure for the development of remaining UKCS [United Kingdom continental shelf] reserves through timely agreements for access on fair and reasonable terms, where risks taken are reflected by rewards". Its main tenets are:

- Parties uphold infrastructure safety and integrity and protect the environment.
- Parties follow the Commercial Code of Practice.
- Parties provide meaningful information to each other before and during negotiations.
- Parties support negotiated access in a timely manner.
- Parties undertake to settle disputes if needed through the Automatic Referral Notice process which involves the UK minister responsible for energy.
- Parties resolve conflicts of interest.
- Infrastructure owners provide transparent and non-discriminatory access.
- Infrastructure owners provide tariffs and terms for unbundled services, where requested and practicable.
- Parties seek to agree on fair and reasonable tariffs and terms, where risks taken are reflected by rewards.
- Parties publish key commercial provisions from the agreements.

Companies seeking access to infrastructure falling within the scope of the third-party access provisions of the Energy Act 2011 must apply first to the owners. However, the act also allows the government minister responsible to take the initiative and set the terms in cases where there is no realistic prospect of reaching an agreement. If the parties are unable to agree to satisfactory terms and conditions, the prospective user may apply to the minister to resolve the dispute. The minister can require access to pipelines, associated offshore production facilities, and onshore natural gas processing facilities, and dictate the appropriate terms. In most cases, these terms will be in line with those offered by infrastructure owners in a competitive environment. In such cases, the legislation stipulates that appropriate notices be issues to the parties to implement the terms. While a dispute resolution procedure exists and the minister can intervene to adjudicate, the third-party access framework is essentially self-enforcing and has rarely resulted in dispute.

A crucial factor in the success of the framework is that it reduces commercial risk by giving stakeholders guidance on tariffs and a variety of other terms. Terms of agreements tend to be quite specific with regard to liability, transparency and technical standards. Terms covered often include:

- the basis for modification of infrastructure, if necessary;
- the duration of service;
- identifying which party has ownership and risk exposure to the natural gas while it is within the infrastructure owner's facility;
- capacity terms and the ability of the owner to change them;
- charges for providing the service;
- rights of both parties to terminate the agreement; and
- liabilities covering damage and loss relating to people, property and pollution.

Within the regulatory framework, both owner and user considerations are important in determining terms of access. Competition and other market forces, the user's ability to pay and contract flexibility are all key considerations to be taken into account when agreeing the terms of access (Fig. 19.12).

Owners and users have flexibility when negotiating access. For example, when agreeing to provide access, the infrastructure owner reserves capacity in its system for the shipper. The shipper gains certainty that the required capacity will be available, subject to normal operational availability, and the infrastructure owner will not market reserved capacity to others. In turn, the infrastructure owner will want to mitigate the risk of the user booking more capacity than required, such as through send-or-pay provisions, which guarantee minimum payment regardless of the extent to which the reserved capacity is used. In declining fields, such as in the North Sea, terms of access are often based on operating costs, plus a risk premium. This contrasts with the approach for newer fields, where the terms of access also tend to include operating costs, as well as capital cost recovery.

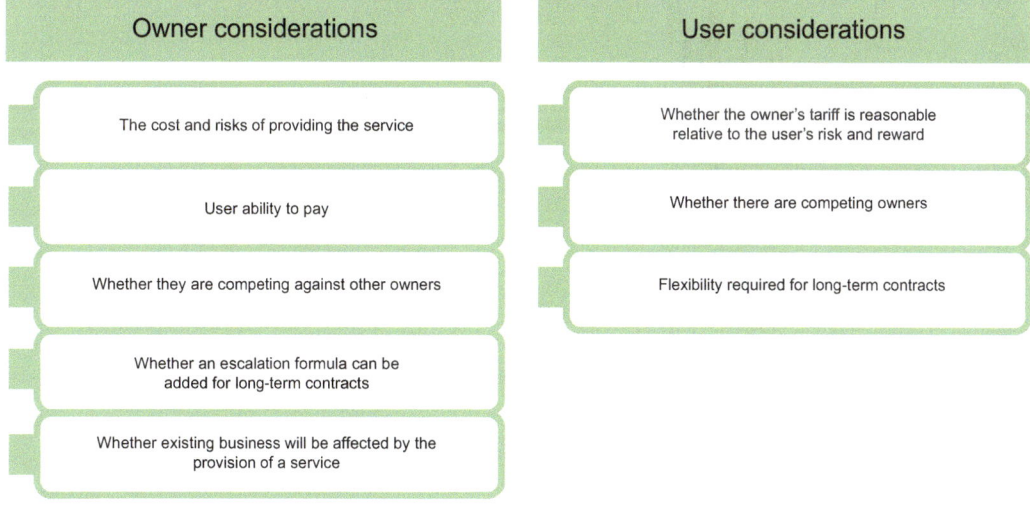

Fig. 19.12 Analysis of regulatory frameworks

19.2.4 Case Study 2: Japan and Singapore LNG— National Power and the Influence of Oligopoly

Singapore and Japan have had entirely opposite experiences in the development and control of midstream infrastructure, especially in terms of LNG regasification receiving stations (Table 19.2). However, the experiences of the two countries highlight that for a balance to be struck between access to and investment in midstream infrastructure, both a streamlined industry chain and consideration of the influence of the market structure are necessary.

1. Singapore LNG: national power involved in influencing construction of infrastructure

Singapore has no domestic natural gas resources and is dependent on imports. Natural gas consumption grew rapidly in the 2000s, from 1.3 billion m^3 in 2000 to 8.7 billion m^3 in 2010.

Into the early 2000s, the island-state's natural gas market was characterised by government ownership along the value chain through Singapore Power Group and SembCorp, limited or no unbundling in commodity and transportation activities, significant price controls on electricity, and generally homogeneous supply through pipelines deliveries from Malaysia and Indonesia. In 2004, the country began progressive

liberalisation, resulting in a competitive midstream and downstream natural gas market, including regulated open-access networks, a significant number of primarily private market participants, unbundled commodity and transportation activities, and oversight entrusted to an independent energy regulator, the Energy Market Authority (Fig. 19.13). The government's role is limited to ensuring fair access to infrastructure.

In 2006, the government decided to develop an LNG regasification terminal to enhance energy security by diversifying sources of supply away from existing pipeline gas. However, meeting the policy objective on energy security was not consistent with Singapore's liberalised and competitive wholesale and retail markets. Singapore had a well-functioning domestic natural gas market, with sufficient supply from pipelines, and there was little market demand for an LNG regasification terminal, despite the energy security benefits it offered.

The country originally granted a licence to a private consortium of PowerGas and GDF Suez to make the necessary investment, but this arrangement collapsed in 2009 because of a funding shortfall. While the problem was linked to the 2008 global financial crisis, it also reflected the competitive nature of the downstream market, which dampened the commitment among potential customers of the LNG facility. The benefits of diversifying supply were outweighed by the risks of being a first mover to a more expensive and untested source, reducing the incentive for energy companies and end users to switch supply sources.

Table 19.2 Differing LNG management systems resulting from different market traits in Singapore and Japan

Influencing factor	Singapore	Japan
Industry organisations	Separation of midstream from downstream assets	Vertically integrated supply chains
Downstream market	Liberalised except for moratorium on piped natural gas	Monopolistic due to lack of geographic integration
Shipping and commercial activities	Functional unbundling	No mandatory functional unbundling
Midstream asset ownership	Public utility (terminal)	Private
Access mechanism	Third-party access	Bilateral access arrangements, but with no credible enforcement
Support mechanisms	Rules-based framework and arbitration	Ease of regulation on construction of new terminals

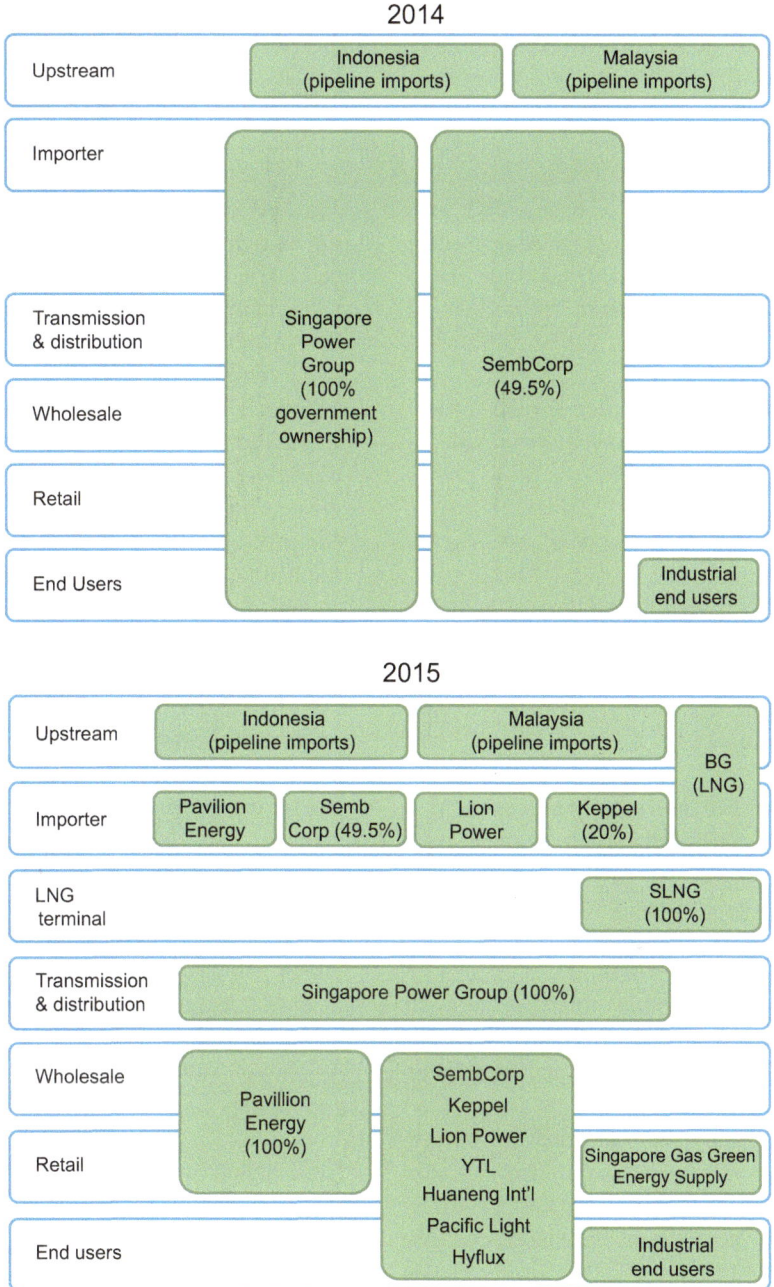

Fig. 19.13 Market diversification in Singapore, 2004 versus 2014

In 2009, the government announced it would take over the development and ownership of the LNG terminal. It established the Singapore LNG Corp. to develop and operate the terminal. The facility, which opened in 2013, was regulated by the Energy Market Authority. It was open access, with third-party access arrangements underpinned by a rules-based regulatory framework, including a dispute-resolution process, and public ownership of the infrastructure was deliberately kept separate from commercial operation.

The regulatory framework established by the government and the Energy Market Authority was designed to guarantee fair and transparent access to midstream infrastructure while reducing investor risk. It comprised three key elements:

- **Aggregator**: A private company, BG, was appointed natural gas aggregator to secure supplies. By pooling demand from large users, Singapore maximised its negotiating power on global markets and secured competitively priced supply.
- **Framework**: Terminal access rules agreed between BG and Singapore LNG provided a commercial framework for other third parties that sought access and enabled a greater degree of forward planning by energy companies.
- **Support**: The government offered a purchase agreement with any power-generation company that committed to LNG supply and a tariff structure that allowed companies to pass through the cost of LNG.

The terminal had the capacity to handle three times as much natural gas as the country consumed. While it initially opened with a 3 million tonnes per year capacity, the opening of its third tank in 2014 doubled capacity to 6 million tonnes. A fourth tank is expected to open by 2017, which would bring total capacity to 9 million tonnes. The terminal could eventually accommodate seven tanks, for a total capacity of 15 million tonnes.

Because capacity was planned to exceed domestic requirements, the government has also sought to secure downstream demand for LNG. To do this, the government introduced controls on new pipeline imports in 2006. These controls allow existing contracts to be honoured, but new contracts were subject to approval from the Energy Market Authority. The decision moved Singapore away from its free market approach and introduced a measure of uncertainty about the evolution of Singapore's natural gas markets. While the moratorium on new pipeline imports is likely to be lifted, it remains unclear whether the

government will continue to limit direct competition between pipeline natural gas and LNG.

Key lessons from Singapore's experience of developing and regulating its midstream LNG infrastructure include:

- **Markets can face challenges when investing to achieve policy objectives**. Singapore had a well-functioning domestic gas market, with sufficient supply from pipelines. Given this, there was little market demand for an LNG terminal, despite the energy security benefits it offered.
- **Regulation is needed to deliver policy objectives**. As the market would not provide the investment, the government had to step in and invest in the LNG terminal itself. At the same time, the government also sought to secure downstream demand for LNG by restricting new pipeline imports. However, in recognition of the efficiency of market forces, the government operated the terminal as a multi-user access terminal to enable competition downstream between importers.
- **Regulators have an important role in setting up the commercial framework for access**. The terms of access were clear and credible, allowing private companies to make long-term plans despite the uncertainty that comes from an open access terminal, such as available docking times and capacity.

2. **Japanese LNG: oligopoly restricting third-party access**

Japan imports most of its natural gas, and its imports are exclusively LNG. The country has 30 operating LNG terminals, with a total capacity of 172 million tonnes a year, a level well above domestic demand. However, Japan remains constrained by how much LNG it can receive, based on berthing, ship size and other infrastructure limitations. Five additional terminals are currently under construction and expected to become operational by 2016, adding capacity of at least 7 million tonnes.

The majority of LNG terminals are near main population centres and manufacturing hubs around

Fig. 19.14 Japanese LNG facilities and their proximity to large cities and manufacturing hubs. *Note* This document and any map included here are without prejudice to the status of or sovereignty over any territory, to the delimitation of international frontiers and boundaries and to the name of any territory, city or area. *Source* IEA

Tokyo, Osaka and Nagoya (Fig. 19.14). They are owned by local power companies, either alone or in partnership with natural gas companies, and the same companies own much of Japan's LNG tanker fleet.

Japan has unique energy fundamentals that make midstream assets particularly important. Japan has no domestic production of natural gas and few other domestic sources of energy.

The country also does not have a well-developed or integrated domestic long-distance natural gas transportation network, partly because its mountainous geography limits the development of a network. Instead, LNG regasification terminals have proliferated along the coastline and take the vital role of a transmission network. However, as a result of these conditions, downstream markets are fragmented and dominated by regional monopolies. Domestic natural gas transport and distribution infrastructure, including LNG terminals, is

owned and operated by vertically integrated private natural gas and power companies. The natural gas market remains largely uncompetitive, with four companies supplying more than 70% of natural gas and almost all LNG terminals owned by a small number of companies and no actual third-party access.

Open access in LNG infrastructure was discussed in the US-Japan Third Joint Status Report under the Enhanced Initiative in 1999, an economic co-operation agreement between the two countries. In response, the Japanese government amended the Gas Utility Act in 2003 and introduced an open-access scheme that obliged natural gas companies to define preconditions for negotiated access. Further transparency requirements related to the capacity information for LNG storage and facilities were also introduced. The changes were intended to act as a guide for infrastructure users, but did not go far enough:

third-party access to terminals was characterised as "desirable" and LNG terminals were not classified as "essential facilities". As a result, while a mechanism for access exists, the regulatory framework does not promote ease of access. Users face a lengthy application process, the timing, flexibility and tariffs are unclear, and there is no arbitration or dispute resolution process.

Since the changes were introduced, no company has effectively negotiated third-party access. Incumbents have successfully argued against third-party access, claiming that regulated third-party access, such as in Singapore, may create disincentives for investment in natural gas infrastructure and hamper energy security. As a result, regulators have tended to promote investment over access. Lack of downstream competition has also led to LNG shippers and LNG terminal owners signing long-term supply contracts, which make it more difficult for newcomers to enter the market.

Natural gas market liberalisation in Japan has been ineffective. The country has rules in place for market liberalisation, such as retail price liberalisation for the non-household sector and third-party access rules. However, natural gas market reform has been piecemeal. Third-party access rules have not been complemented by mandatory unbundling or by other measures to increase competition in downstream retail markets. If the market were to be liberalised, reform is needed on all fronts: while effective third-party access facilitates competition, the lack of downstream competition has prevented third-party access from being effective.

Energy security remains the primary policy priority. Japan's regulatory regime cannot focus on minimising costs as vigorously as that of countries with more favourable fundamentals, such as China, which has greater energy security and can build a national transmission network.

Key lessons from Japan's experience of developing and regulating its midstream LNG infrastructure include:

- **Japan has unique energy fundamentals that make midstream assets very important**. The country had no domestic production of gas and few other domestic sources of energy, and its mountainous geography limited the development of a national transmission network. As a result, energy security was a high priority, and LNG terminals played the role of a transmission network by providing access at points along the coastline.

- **Japan has experienced large investments in midstream assets**. Japan has invested in a large number of LNG terminals, which were warranted from an energy security and geography perspective. Incumbents have successfully argued against third-party access, claiming that regulated third-party access, such as in Singapore, may create disincentives for investment in natural gas infrastructure and hamper energy security. As a result, regulators have tended to promote investment over access.

- **Gas market liberalisation in Japan has been ineffective**. Japan had rules in place for market liberalisation, such as retail price liberalisation for the non-household sector and third-party access rules. However, in practice the market remained uncompetitive, with four companies supplying more than 70% of natural gas and almost all LNG terminals owned by a small number of companies with no meaningful third-party access.

- **The failure of liberalisation demonstrates the importance of regulating across the value chain**. Gas market reform in Japan has been piecemeal. Third-party access rules have not been complemented by mandatory unbundling or by other measures to increase competition in downstream retail markets. If the market were to be liberalised, reform is needed on all fronts: while effective third-party access facilitates competition, the lack of downstream competition has prevented third-party access from being effective.

19.2.5   Case Study 3: The United States—Master Limited Partnerships

In the United States, a unique corporate structure, master limited partnerships, enables ownership of midstream assets to be reallocated to owners with appropriate risk profiles. The structure facilitates the transfer of midstream assets from organisations focused on high-risk, high-return activities, such as oil and gas exploration and development to organisations seeking low-risk, low-return assets, such as pension funds. While the case study focuses on oil pipelines, it is applicable to midstream infrastructure more generally.

Key lessons from the US experience with master limited partnerships include:

- **Midstream assets are low-risk, low-return assets and can attract significant investment**. Once constructed, midstream assets such as pipelines have a relatively simple business model, especially in markets with high demand and regulated charges, such as the United States. Such assets were attractive to investors comfortable with low-risk holdings and a regular return, such as pension funds.
- **Greater capital liquidity enables better allocation of capital**. Many midstream assets were constructed by companies with a greater appetite for risk and higher return expectations, such as domestic and international oil companies with upstream assets, as a way of transporting wellhead natural gas to consumers. However, these companies specialised in higher-risk, higher-return investments. Instruments such as master limited partnerships enabled these companies to sell midstream assets to investors with more appropriate risk profiles and recycle the proceeds into new investment more consistent with their own risk-reward profile.
- **Tax efficiency is important in attracting investor interest**. A master limited partnership is a particular form of business entity that is not liable for US federal income tax.

Allowing midstream assets to be structured into a master limited partnership reduced tax liability and improved the returns, therefore increasing investor interest.

- **Regulation gives clarity and certainty, further enhancing capital liquidity**. Beyond the corporate structure, master limited partnerships in the United States have benefited from a clear, credible and stable regulatory regime, which has created confidence among investors less familiar with the energy industry. Regulated access charges, a history of long-term contracts, and a stable regulatory and tax regime helped lower the risk of an investment and attract new capital.

1. Background

The United States is the world's largest oil consumer. The level of domestic crude oil production has increased over the past few years, reversing a decline that began in 1986. According to the IEA, crude oil production increased from 5 million barrels a day in 2008 to just less than 5.7 million barrels a day in 2011 and 6.5 million barrels a day in 2012. This increase in oil production was largely brought about by new seismic and horizontal drilling technology and hydraulic fracturing, bringing domestic resources that were previously considered non-viable into production.

Pipelines are the common transport mode for shipping crude oil and refined products. In total, the country has more than 275,000 km of crude-gathering and distribution pipelines, operated by 2338 companies. The top 10 operators alone run almost 90,000 km of pipeline. In 2011, this network delivered 514.3 million barrels a day of crude oil between regions. The highest concentration of pipelines is in the Gulf Coast, which is also home to almost half the country's refining capacity.

Because of the capital-intensive nature of pipeline operations, many companies have sought to structure these units as master limited partnerships, publicly traded businesses that are taxed as partnerships. Unlike a corporation, a partnership in the United States is classified as a pass-through entity and is not liable for federal income taxes. Instead, partners pay tax based on their allocated share of the partnership. The structure offers two significant advantages. First, tax is levied only at a single level of federal income tax, which is applied individually on the total income of each member of the partnership. The master limited partnerships do not pay corporation tax. The other advantage is that favourable tax treatment allows a master limited partnership to access lower-cost capital than if it were taxed as a corporation.

These advantages make the master limited partnership approach an especially attractive option for capital-intensive pipeline infrastructure transmission enterprises.

2. Benefits to infrastructure owners

One example of the structure was Shell Midstream Partners, a master limited partnership that held onshore and offshore oil infrastructure. Three of the four pipelines in its portfolio served the Gulf Coast, while the fourth reached into the Northeast (Fig. 19.15).

For the infrastructure owner, a particular advantage of this corporate structure is that the business retains control of the pipelines. Unlike shareholders in a publicly traded company,

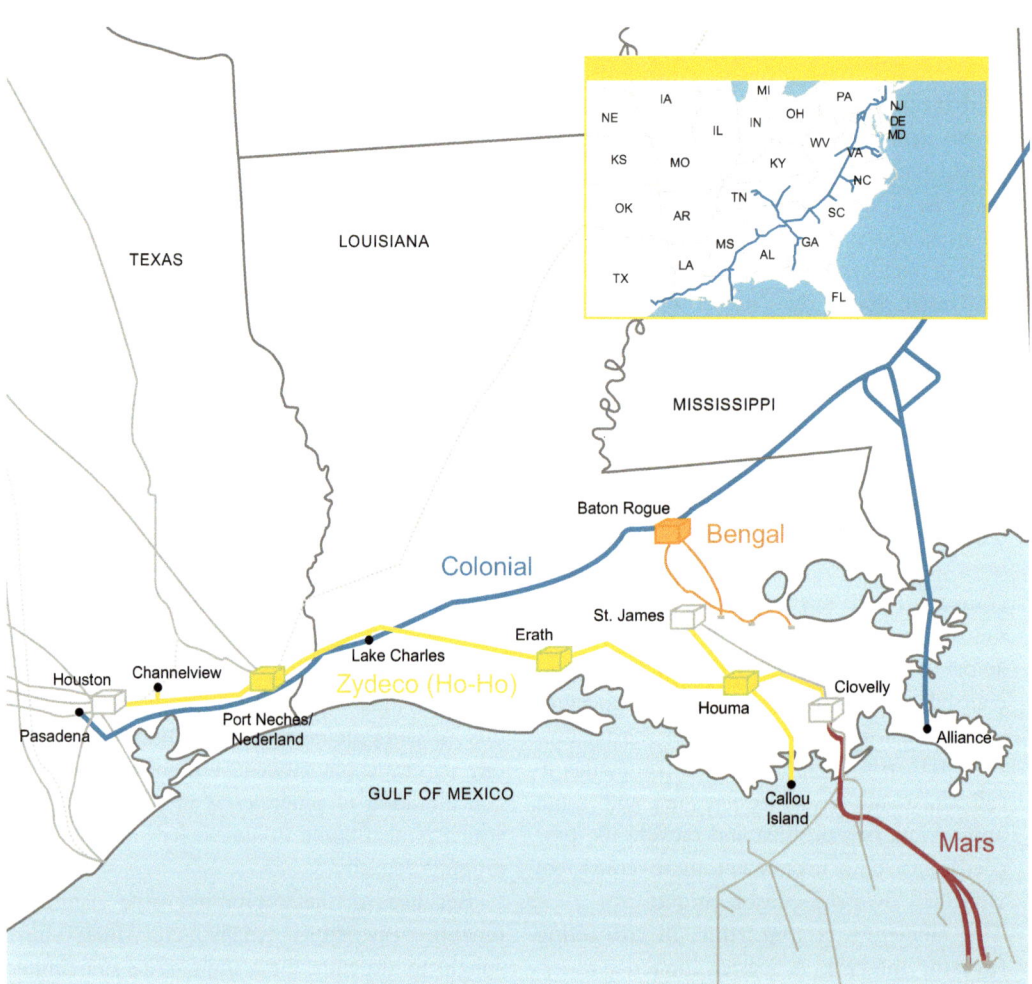

Fig. 19.15 Shell midstream partners pipelines

Table 19.3 Shell Midstream Partners ownership rights allocation

Entity	Shell Midstream Partners (%)	SPLC retained ownership rights (%)	Length (km)	Capacity (kbpd)
Zydeco (Ho–Ho)	43.0	57.0	563	375
Mars	28.6	42.9	262	400
Bengal	49.0	1.0	254	515
Colonial	1.6	14.5	8851	2500

Notes SPLC is Shell Pipeline Company Limited Partners; kbpd is thousand barrels per day; ownership interests may not add to 100%

unitholders—investors holding shares of master limited partnerships—are not entitled to elect the general partner or any directors. Instead, they benefit from a stream of income in accordance with their ownership interest, while the partnership itself is managed by a board of directors and executive officers appointed by a general partner, in this case Shell Pipeline Company LP, an affiliate of Shell. The distribution of ownership interests varied across the pipelines in the portfolio (Table 19.3).

Launching an initial public offering for Shell Midstream Partners allowed Shell to release capital for investment elsewhere in its value chain and optimise its risk profile. It also allowed Shell to divest itself of low-return, non-core assets and to improve its focus on other activities. At the same time, investors in Midstream Partners, such as pension funds, were attracted by its lower risk-return profile.

3. Importance of the regulatory framework

The regulatory framework provided certainty for investment by detailing eligible assets and activities. Master limited partnerships were only possible as a result of the regulatory authorities providing rules for partnership treatment: federal rules on qualifying income and assets gave Shell the necessary certainty to proceed. It also enabled Shell to offer stable and predictable cash flows to investors. Rules on access meant that Midstream Partners' assets generated stable revenue under FERC-based tariffs and long-term transportation agreements. In addition, ship-or-pay contracts mitigated volatility in cash flows by limiting

exposure to changing market dynamics that could reduce production and affect shipper demand. Finally, life-of-lease agreements, some of which have a guaranteed return, reduced cash flow exposure to volume reductions.

19.3   Unbundling Midstream Infrastructure

Midstream infrastructure unbundling is a critical step in natural gas market liberalisation progress and in the promotion of natural gas market competition. This section analyses the effects of unbundling with the background of natural gas market liberalisation as well as related goals of unbundling. In addition, a comparison of five partition models is carried out using case studies from the United Kingdom, the European Union and Japan. These case studies offer experience and a reference point to China in its future unbundling work.

Natural gas markets have historically been dominated by large vertically integrated companies operating across the various segments of the value chain. Vertically integrated pipeline owners face particularly strong incentives to engage in anti-competitive behaviour, charging excessive rates to third-party natural gas shippers for access to midstream infrastructure to protect profits in their own production or retail businesses.

Unbundling seeks to separate the incentives facing midstream operations from those of upstream and downstream operations and to reduce the opportunity and incentives for

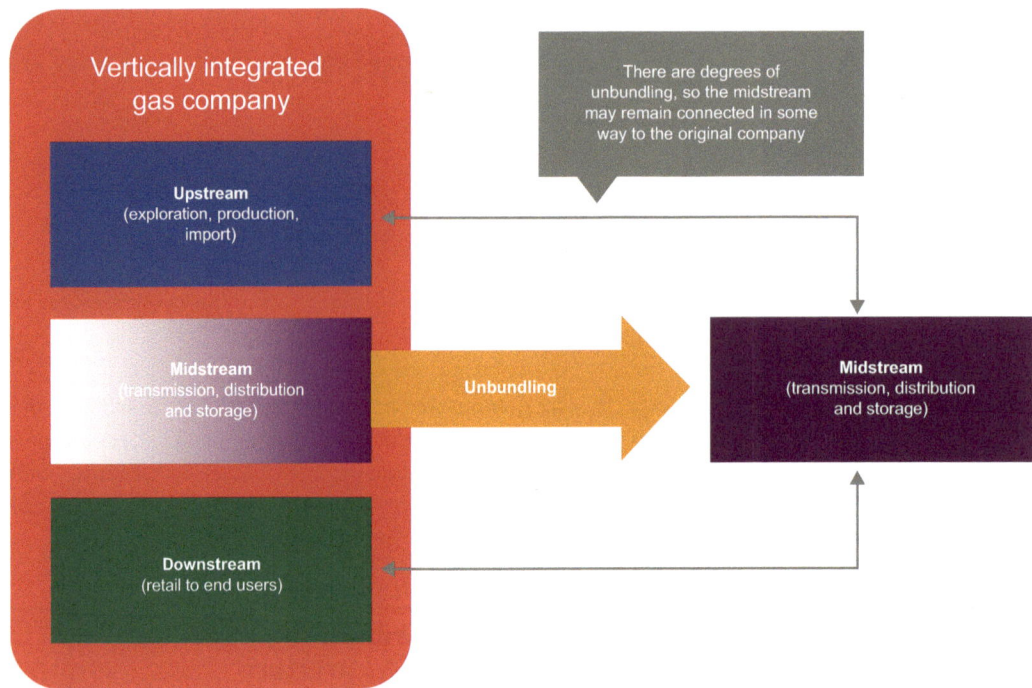

Fig. 19.16 Unbundling is the separation of midstream business from that of upstream and downstream business

anti-competitive behaviour. Without this, owners of midstream assets will retain an incentive to operate in a way that favours their upstream or downstream businesses. Unbundling focuses on the midstream segment because of its natural monopolies and the potential for misusing market power (Fig. 19.16).

Full ownership unbundling is the most complete form of unbundling, although some models of unbundling seek to change these incentives without separation of ownership. For example, if a single vertically integrated firm participates in the upstream segment, transportation and retail sales, but regulation effectively prevents it from operating its midstream assets to the advantage its upstream or downstream businesses, vertical integration does not threaten efficient market operations.

Unbundling is a key step in natural gas market liberalisation, as part of a broader effort to create open access to midstream infrastructure, particularly to pipelines which have stronger characteristics of natural monopolies. Unbundling is required in addition to third-party-access

regimes. Third-party access requirements address the incentive of a natural monopoly owner to charge very high prices to maximise its profits, reducing pipeline utilisation in the process. Unbundling requirements, on the other hand, address the incentive for vertically integrated owners to make monopoly profits by favouring their vertically integrated partners and excluding third parties.

Vertically integrated companies can misuse their market power even if third-party access to midstream assets is allowed, for example through a lack of transparency over available capacity and prices or through onerous contracting or technical studies requirements for third parties. Such anti-competitive behaviour can be remedied by strict policing of third-party access rules, but removing the incentive for this behaviour through unbundling has been seen by many regulators as a part of the solution. Unbundling is not itself the goal, but rather a means to ensure that third-party access is effective. The success of an unbundling regime should be measured by the ultimate effectiveness of an open access reform package in supporting an efficient and competitive gas market.

19.3.1 Models of Unbundling

Five unbundling models are defined and examined using case studies from the United Kingdom, the European Union and Japan. They are:

- **Service unbundling**: Businesses must offer use of their midstream assets as a distinct service separate from the wholesale supply of natural gas.
- **Account unbundling**: Midstream businesses must maintain separate accounts from upstream and downstream businesses to prevent cross-subsidisation and distorting behaviour.
- **Legal unbundling**: Midstream assets are placed into a separate legal entity, such as a wholly-owned subsidiary, to reinforce the operational separation of these assets from other assets within a vertically integrated entity.
- **Structural unbundling**: Strict regulatory requirements are placed on how a midstream business is operated to ensure that it supports the efficient operation of the natural gas market, rather than to the benefit of related upstream or downstream interests.
- **Ownership unbundling**: The legal entity that owns midstream assets does not have common ownership with upstream or downstream interests, fully separating their economic incentives.

Each of these models builds sequentially on each other, alongside an overarching process known as functional unbundling, which can reinforce each of these models. Functional unbundling restrictions can vary from being fairly mild to quite onerous. They could include, for instance, requiring separate management, locations, support units and logos, prohibiting the sharing of information, and imposing additional compliance and reporting requirements. Functional unbundling is a critical element of any

unbundling process. To illustrate, a legally unbundled midstream business that shared management or offices with a related upstream or downstream business could be less effective than, say, a midstream business that legally remained part of a vertically integrated entity, but had more rigorous separation of staff and management.

Each of the five models of unbundling serves different purposes and follows sequential changes in moving from service unbundling to ownership unbundling (Table 19.4).

19.3.2 Key Insights from the Case Studies

Our review of international experience in unbundling midstream assets from upstream and downstream businesses produced three general insights that might be valuable to China's policy discussions.

- **Complete ownership unbundling is not necessary under a strong regulator**. The case studies indicated that there are viable models that don't require full ownership unbundling. In particular, structural unbundling under a strong regulator could capture many of the benefits of full ownership unbundling. Of the country case studies, only the United Kingdom and The Netherlands pursued full ownership unbundling. France and Germany argued for structural unbundling that fell short of full ownership unbundling, and the latest EU Gas Directive included this option along with strong regulatory measures to enforce a well-developed structural unbundling model. Experience in the United States also suggested that a competitive natural gas market can be supported without mandatory ownership unbundling, provided the operations of transmission companies are tightly prescribed. Each of these cases followed a unique path to unbundling (Fig. 19.17).

  In general, structural unbundling can be effective if the regulator is strong. However, anything less than structural unbundling is

Table 19.4 Operational changes resulting from five types of unbundling

Model	Change	Goal	Relation to previous changes
Service partition	Midstream service (primarily management services) must be independent from natural gas wholesale business	If natural gas transmission services are not separated from the wholesale business, then anti-competitive activities will occur, since natural gas suppliers will be unable to sell natural gas through pipeline owners	N/A
Account partition	Midstream business operation business revenue and costs must be separated from upstream and downstream business	The goal is to prevent midstream business from benefiting from overlapping subsidies in upstream or downstream business	If midstream services are not made into independent services having independent fees, independent accounts cannot be possessed
Legal partition	Midstream business becomes an independent legal entity, but is still within the vertically unified corporation, for example as a wholly-owned subsidiary	The subsidiary company is the legal person assuming legal obligations. Using independent legal entities achieves better management separation	Independent legal entities must manage independent accounts and provide separate services
Structural partition	Midstream business formulates strategies to effectively separate its economic incentives from upstream or downstream business incentives	Separating the economic incentives of midstream business from related upstream or downstream economic incentives can support obtaining effective market results by removing incentives to hinder the competition	Midstream business must be an independent legal entity and adhere to existing procedures and assume obligations
Ownership rights partition	Midstream business separates from the vertically unified enterprise and transitions to be an entity with independent ownership rights	Complete ownership separation is the most effective means of separating the interests of midstream business from related upstream and downstream business	For independently owned companies, they must be independent legal entities that have independent accounts and provide independent services

Note Structural unbundling encompasses the independent transmission operator model within the EU's Third Gas Package
Source Vivid Economics, based on Gao

likely to be inadequate in delivering liberalised natural gas markets. The experience of the United Kingdom and Japan indicated that service unbundling alone offered few benefits and that service and account unbundling, combined with meaningful functional unbundling, was the minimum viable unbundling models. However, this combination of unbundling models should be seen as a stepping stone to deeper unbundling, rather than an end point. The European experience also indicated that legal unbundling offered

little without moving to deeper structural unbundling, complemented with functional unbundling.

These observations suggest a streamlined unbundling process consisting of three steps (Fig. 19.18). Functional unbundling supports this process at key junctions, principally in the account, legal and structural unbundling phases.

• **Domestic production can increase the benefits of unbundling**. A key benefit of unbundling was to open upstream activities to

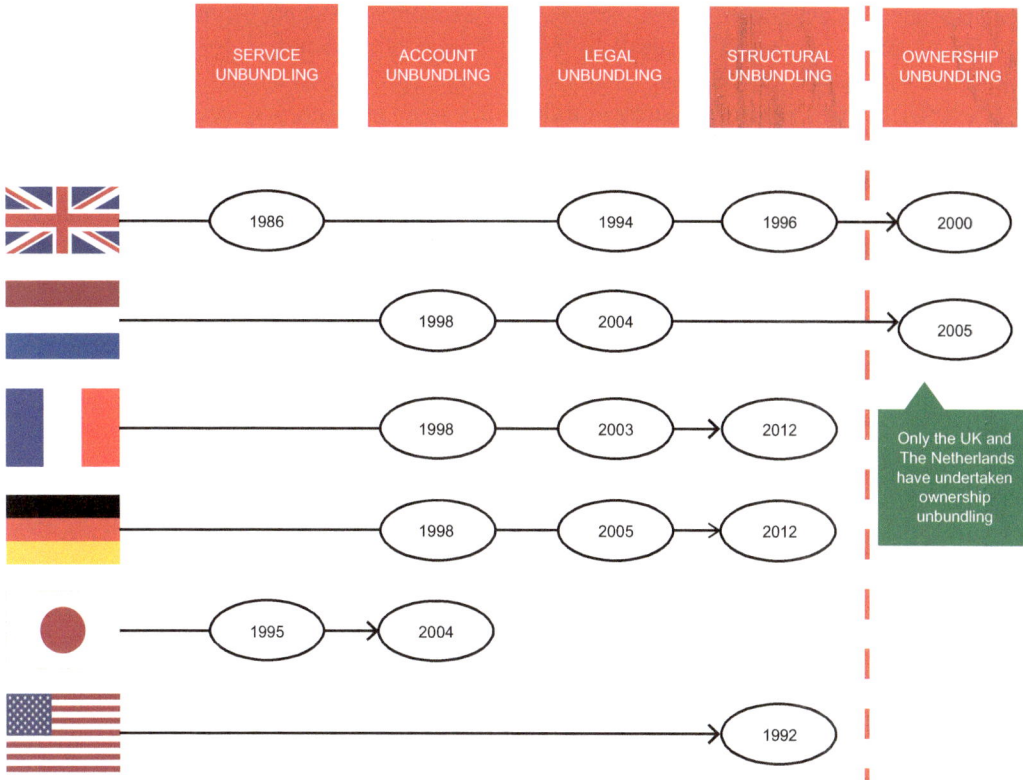

Fig. 19.17   The path of unbundling (ultimate ownership rights unbundling might not be necessary)

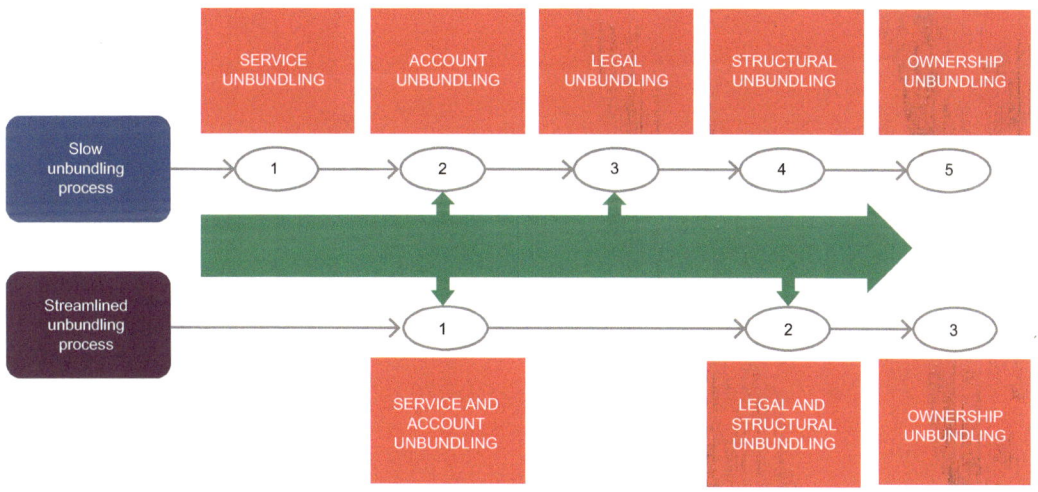

Fig. 19.18   Streamlined unbundling process (three steps only)

competition and efficient production. In markets that had few indigenous natural gas resources and were heavily reliant on imports, unbundling and other open-access measures promoted the efficient delivery of natural gas to market, but had only a limited impact on upstream activity. In contrast, in countries with substantial indigenous resources, unbundling and other open-access measures could promote domestic production by enhancing access to downstream wholesale and retail markets. Indeed, countries with more domestic production have tended to unbundle earlier and more comprehensively (Fig. 19.19).

The Netherlands and United Kingdom, with their indigenous North Sea resources, pursued unbundling more rapidly than nations such as France and Germany with limited domestic reserves. The United States, with its substantial and geographically diverse upstream resources, was the world leader in unbundling efforts, requiring legal and other structural unbundling measures as early as 1992. Japan, with its negligible domestic natural gas resources, pursued unbundling only tentatively. This suggests that countries with significant domestic resources perceive higher economic benefits from unbundling—in terms of promoting a dynamic and competitive upstream sector and unlocking the economic and energy security benefits of domestic production—and are more willing to pursue these reforms.

- **LNG and storage facilities tend to face less strict unbundling regimes than pipelines**. Unbundling requirements for LNG terminals and storage facilities were generally less stringent than for pipelines, reflecting the greater potential for competitive pressure on these facilities, as well as the relatively new development of this type of infrastructure. The potential for LNG regasification terminals or storage facilities to exercise excessive market power depends on the number of alternative providers of similar services. In a physically well-connected market, LNG terminals are one source of supply competing alongside

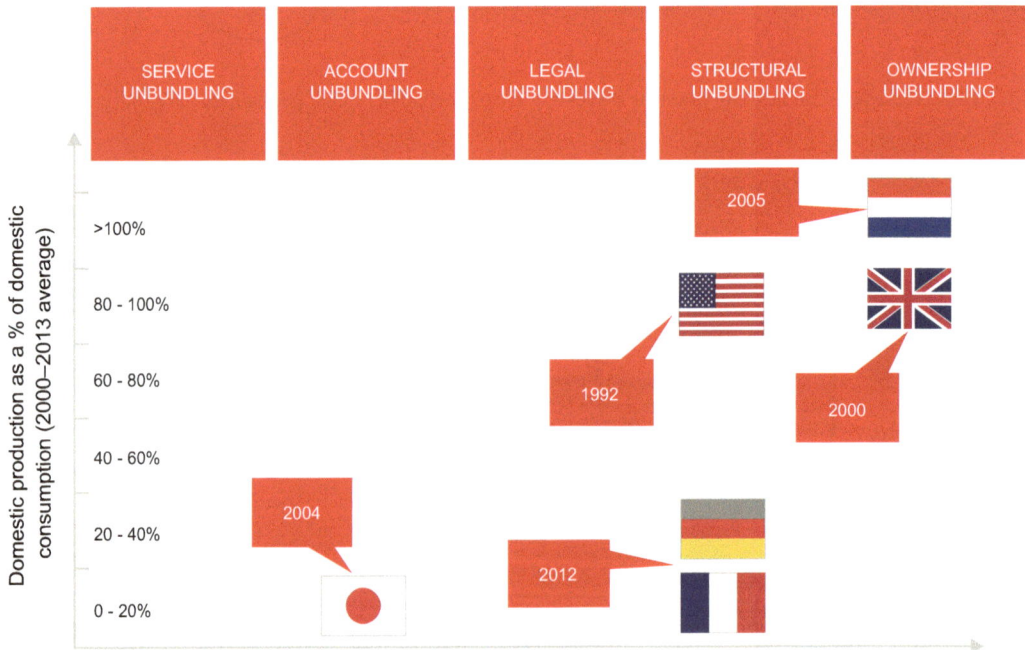

Fig. 19.19 Correlation of domestic natural gas production volumes and levels of and timing of unbundling. *Note* Characterising 1992 reforms in the United States as structural unbundling is based on the strong non-discrimination, transparency and capacity release requirements of FERC Order 636, reinforced with extensive functional unbundling measures. *Source* Vivid Economics, based on Gao and IEA

many others and have limited market power. Similarly, a diversified and well-connected market may have multiple storage providers, as well as other options such as pipeline linepack (in which natural gas is stored in the pipeline), flexible production capacity and demand curtailment. Other than in very small or fragmented markets, the potential for LNG terminals and storage facilities to exercise market power is likely to be limited. As a result, both unbundling and third-party access requirements tended to be more lenient for these facilities.

19.3.3 Case Study 1: The UK—The Long and Difficult Road to Spinoff

Despite the United Kingdom's position today as an exemplary case of full natural gas supply chain unbundling, the process has been far from smooth. Regulatory arrangements were reviewed and amended on numerous occasions before the present market structure was largely reached in 2000. Unbundling was not always synchronised with liberalisation, and with measures to introduce competition in the broader gas market. Full ownership unbundling was only achieved in 2000 when the then British Gas Plc separated its Transco subsidiary into the separately owned Lattice Group Plc, well after the emergence of substantial gas market competition through the early and mid-1990s.

By that time, substantial competition had already emerged in large parts of the UK natural gas market. British Gas had already lost significant market share before major reforms under the Gas Act of 1995 were implemented. The Gas Release Scheme, for instance, had forced British Gas to release natural gas contracts to competing suppliers. A compounding factor was that many of British Gas's legacy North Sea natural gas contracts were priced higher than prevailing market prices, pressuring the company's market position.

The first phase of reform was marked by the Gas Act of 1986. The act privatised the state-owned British Gas Corp. as British Gas Plc and provided a limited form of service unbundling of the company's pipeline operations. The act allowed larger natural gas customers to source natural gas from any authorised supplier and required British Gas to allow these suppliers to use its pipelines. In effect, British Gas was told to offer a standalone pipeline service to competitors seeking to supply large customers. However, this service unbundling was not supported by account or functional unbundling and proved to be largely ineffective. The Monopolies and Mergers Commission found in 1988 that British Gas had been abusing its market position by engaging in price discrimination and sought to address this by requiring the company to publish common carriage terms to support pipeline access. The change, however, was still largely ineffective.

Subsequent reviews by the Office of Fair Trading in 1991 and the Monopolies and Mergers Commission in 1993 suggested that further unbundling was necessary to promote competition. The Office of Fair Trading review addressed a range of other structural elements of the natural gas market other than unbundling. It led to British Gas undertaking to make supply available to competitors under the Gas Release Strategy, which triggered a substantial reduction in its market share with large industrial customers between 1991 and 1993. It was these provisions, rather than its unbundling recommendations, that were instrumental in driving a decline in British Gas's market share.

In response to the critical 1993 Monopolies and Mergers Commission review, British Gas bypassed the account unbundling phase and voluntarily moved directly to legal unbundling in 1994. A transportation and storage subsidiary, Transco, was created and separated from British Gas's trading and sales activities. Despite the commission's recommendation for full ownership unbundling, in response to regulatory and commercial pressures, the unbundling of British Gas progressed in two further phases.

The Gas Act of 1995 required Transco to establish a Network Code to govern the national transmission system, which effectively constituted structural unbundling of the network business. The code was published in 1996, establishing the rules and procedures for third-party access to pipelines and introducing a regime for daily balancing. The code strengthened access to the pipeline system for natural gas market players, supporting competition even though Transco remained a British Gas subsidiary.

However, this business structure did not last, as competitive pressures built around British Gas. The 1995 Gas Act extended the scope of competition to include the smallest, residential customers, and full retail competition was implemented across the United Kingdom by 1997. As competition in both large and small user markets deepened, the relatively high price of British Gas's legacy upstream contracts for North Sea gas weighed on its business. The pressure prompted British Gas to voluntarily divest its trading and sales businesses. British Gas was unbundled into two entities: Centrica plc, which held sales and retail units, as well as upstream assets in Morecombe Bay, and BG plc, which retained Transco, the midstream unit, as well as domestic and global upstream assets.

This structural unbundling model appears to have been effective, even though Transco remained a subsidiary of BG, a major upstream player. Transco held some ownership of upstream assets, and its separation from BG was maintained by the Network Code.

The final step in full ownership unbundling came in 2000, when BG split into the BG Group and the Lattice Group, which took ownership of Transco. The split meant that Transco was entirely separated from all upstream and downstream assets. Further ownership changes saw the Lattice Group merge with the operator of the United Kingdom's electricity grid, National Grid, in 2002 and sell off some local distribution networks in 2005. While Transco initially included storage assets, these were spun off into a storage subsidiary, which was privatised in 2001 and purchased by Centrica in 2002.

19.3.4   Case Study 2: Europe—The Step-by-Step Progression Towards Spinoff and the Multitude of Choices

The EU regulatory regime has evolved towards greater unbundling over the course of three major regulatory packages since 1998. These stages broadly comprise account and functional unbundling in 1998, legal unbundling from 2003 and more detailed structural and ownership bundling approaches since 2009. Some member states have moved to comply with EU directives more swiftly than others, and several have gone beyond the minimum requirements of EU regulation.

Differences on the merits of unbundling have led to a diversity of approaches among EU member states and an EU regulatory approach that offers a menu of unbundling options rather than a single prescriptive model. Development of the Third Gas Directive was contentious and ultimately led to a political and regulatory compromise that offered two main unbundling models: structural unbundling and ownership unbundling.

The characteristics of the member states that took different positions within this debate illuminate the motivations and concerns of those arguing for and against ownership unbundling models. For example, countries with domestic natural gas resources, such as the United Kingdom and The Netherlands, appear to be more enthusiastic proponents of unbundling. The benefits of unbundling are greater in such countries, promoting competition among natural gas producers. Conversely, in countries like Germany, where there are limited domestic natural gas resources and the primary physical assets held by natural gas businesses are pipelines, unbundling was more hesitant. These fundamental differences could, in part, explain varying regulatory attitudes toward unbundling.

The First Gas Directive successfully achieved account unbundling, but does not appear to have had substantial effects on gas market competition. The directive required separate accounts for transmission, distribution and storage activities to support transparency and competition in these activities. A 2003 benchmarking report by the European Commission found that all member states had complied with minimum account unbundling requirements and several had gone further to legal or ownership unbundling. However, the same report noted that prospects for competition in the natural gas market were lagging significantly behind those for electricity in many member states, citing Germany in particular. Around 2001, the European Commission itself began arguing that the first package of measures had been ineffective and that further reforms were needed.

One weakness of the First Gas Directive may have been limited functional unbundling in support of the core model of account unbundling. While the directive included some broad provisions against discriminatory behaviour and a general requirement for transmission providers not to abuse commercially sensitive information obtained from third parties, it did not include detailed and enforceable provisions to separate information and management systems between transmission and other business units. Some member states, including The Netherlands and Ireland, voluntarily adopted functional unbundling measures to separate management of transportation businesses. Others, such as Austria, Belgium, Denmark and Italy, went further, adopting legal unbundling. Meanwhile, Spain and the United Kingdom moved to ownership unbundling. However, some major member states, notably France and Germany, only implemented the directive's minimum requirement of account unbundling and appear to have been slower in progressing natural gas market competition during this period.

The Second Gas Directive, adopted in 2003, mandated the legal separation of transmission businesses, reinforced with functional unbundling measures. The directive required transmission system operators to be "independent at least in terms of legal form, organisation, and decision-making from other activities not relating to transmission". Specific functional unbundling

measures required entirely separate management, effective independent decision-making rights and a compliance programme to maintain independence. Member states adopted elements of the Second Gas Directive at different speeds and to differing degrees. By 2005, for example, The Netherlands had established full ownership unbundling, while Germany had only implemented minimum regulatory requirements for a legal unbundling regime.

Policy-makers and market observers quickly concluded the Second Gas Directive was also ineffective in promoting competition and argued for further reform. A 2007 European Commission report found that "wholesale gas trade has been slow to develop, and the incumbents remain dominant in their traditional markets". It concluded that the level of unbundling in place was inadequate to "resolve the systemic conflict of interest inherent in the vertical integration of supply and network activities". In the same year, an IEA review of German energy policies concluded that legal unbundling was ineffective and that the involvement of traders in pipeline businesses had limited the liquidity of wholesale markets. A similar 2009 IEA review of French energy policies identified the dominance of GDF-Suez, especially in the bilateral trade of wholesale natural gas, as constraining the ability of other market players to supply the French market.

Earlier reform efforts had targeted mandatory ownership unbundling. For example, the European Commission argued in 2007 that "full ownership unbundling is the most effective means to ensure choice for energy users and encourage investment." However, nine member states—including France and Germany—opposed full ownership unbundling, saying that it would not improve competition or reduce prices. They argued that breaking up large, vertically integrated energy companies would weaken their negotiating power in international markets—for example with powerful external suppliers such as Russia's Gazprom—with no benefit to consumers.

The adoption of the Third Gas Directive was contentious, with several large member states

arguing against mandatory ownership unbundling. The compromise captured in the Third
Directive presented ownership unbundling as the
ideal model, but allowed member states to
choose from two alternative structural unbundling models. The directive recommended full
ownership unbundling as "the most effective tool
by which to promote investments in infrastructure in a non-discriminatory way, fair access to
the network for new entrants, and transparency
for the market". It also argued that the two
alternative models, the independent transmission
operator (ITO) model and the independent system operator (ISO) model, "should enable a
vertically integrated undertaking to maintain
ownership of network assets while ensuring an
effective separation of interests, provided that...
extensive regulatory control mechanisms are put
in place".

The ISO and ITO models were essentially less
stringent forms of unbundling. Unlike ownership
unbundling, setting up an ISO or ITO would let
transmission networks remain under the ownership of vertically integrated firms.

EU member states have adopted either full
ownership unbundling or the ITO model. The
United Kingdom, The Netherlands, Denmark and
Spain have ownership unbundling models,
though these regimes were in place before the
Third Gas Directive was adopted. Elsewhere, the
ITO model has been adopted, complemented by
a range of strong functional unbundling
requirements. By mid-2014, 21 natural gas
transmission operators had been certified.

No country had pursued the ISO model by
mid-2014. The core element of the ISO model
was that network ownership remains with a subsidiary of the vertically integrated entity, but
system operation and investment decisions are
made by a legally separate entity with separate
ownership. Vertically integrated companies
seemed unconvinced by an approach that would
require them to fund investments made by entirely
separate entities, while the ITO model allowed
investment decisions to be made jointly by the
parent company and the regulatory authority.

The choice between the full unbundling and
the ITO model may reflect different evolutionary

stages in the development of natural gas markets.
Countries that embraced full ownership
unbundling started their reform processes earlier
and progressed more rapidly along the spectrum
of unbundling models. The progress made in
markets such as the United Kingdom and The
Netherlands may reflect the maturity of their
unbundling efforts, rather than the advantage of
ownership unbundling over alternative models.
Indeed, an effective and mature structural
unbundling model may be sufficient to support
and enable competition in the upstream and
downstream segments of the natural gas value
chain.

19.3.5   Case Study 3: Japan—Market Characteristics Restricting Unbundling

Japan's two tentative attempts at unbundling
appear to have had only a modest effect on its
natural gas market, although the lack of competition may also reflect the fundamentals of the
Japanese markets as much as the merits of its
regulatory efforts. One of the benefits of
unbundling appears to be a dynamic upstream
market, but since Japan has no domestic natural
gas resources, this incentive is less compelling.
Minimal pipeline interconnectivity further limits
the potential for regulation to unlock competition. In general, the limitations of Japan's
unbundling efforts since 1995 may be both a
symptom and a cause of the lack of competition.

Japan began its unbundling measures in 1995
with a service unbundling regime. The 1995
reform required Japan's three largest natural gas
companies to offer pipeline services to competitors seeking to supply large consumers, under-
pinning the emerging third-party access regime.
These requirements were extended to additional
customers in 1999. They were reinforced in 2000
with functional unbundling requirements that
directed companies to separate the transportation
function from other elements of their business.
However, the measures were lenient and did not
materially change the behaviour of incumbent
natural gas suppliers. A second wave of reform in

2003 and 2004 introduced account unbundling supported by new and more stringent functional unbundling requirements. The functional unbundling measures required separate physical office locations, staff functions, and decision-making and prohibited information sharing. Compliance improved, but was not uniform.

The overall reform programme has been only modestly successful, with markets remaining highly concentrated. As of 2012, three large, vertically integrated players together supplied around 70% of the market. Companies wanting to use pipelines owned by other companies have complained of entry barriers and opaque access requirements. In particular, many complained of insufficient information on the available capacity of pipelines and on standards and procedures for assessing fees and other compensation mechanisms.

The Japanese experience suggests that account and functional unbundling can deliver benefits, but its value may have been limited by the fundamentals of the Japanese natural gas market. Service and account unbundling, combined with meaningful functional unbundling, appears to be the minimum viable unbundling model, acting as a stepping stone to more advanced unbundling model.

19.3.6   Unbundling LNG Terminals and Gas Storage Facilities

Unbundling requirements for LNG terminals and storage facilities are generally less stringent than for pipelines, reflecting the greater potential for competitive pressure on these facilities, as well as the relatively new development of this type of infrastructure. The potential for LNG regasification terminals or storage facilities to exercise market power depends on the number of alternative providers of similar services and the degree of interconnection between them. In most markets, there are likely to be a larger number of LNG terminals and storage sites than pipeline routes because of the smaller economies of scale. Moreover, in a physically well-connected

market, LNG terminals will compete with each other and with other sources of supply, limiting their market power. Similarly, a diversified and well-connected market may have multiple storage providers, alongside storage services provided by pipeline linepack, flexible production capacity and demand curtailment.

Except in very small or fragmented markets, the potential for LNG terminals and storage facilities to exercise market power is likely to be limited, and regulations mandating unbundling and third-party access requirements will not tend to be as stringent as for pipelines. For example, Japan requires only account unbundling for LNG facilities and places no requirements on storage facilities. Unbundling requirements in the European Union are stronger than those in Japan for both LNG regasification terminals and storage facilities, with both types of facilities required to have unbundled accounts. In addition, storage operators cannot participate in company structures of vertically integrated natural gas companies, imposing an additional functional unbundling requirement.

A range of European LNG and storage facilities have gone beyond the minimum regulatory requirements for account unbundling, adopting either legal or ownership unbundling. In some cases, particularly for LNG terminals, legal unbundling has been adopted for practical reasons associated with forming joint ventures. Ownership unbundling has been adopted in instances when transportation assets face unbundling requirements, and since storage facilities or LNG terminals are held by transportation owners, they become unbundled from upstream and downstream interests by default. In other cases, legal unbundling through wholly-owned subsidiaries has occurred on a voluntary basis, possibly pre-empting regulatory requirements to unbundle or other regulations limiting the benefits of vertical integration. Finally, new LNG terminals can receive an exemption from third-party requirements if, among other things, they are unbundled.

Examples of LNG facilities and storage facilities and their prevailing level of unbundling show a wide range of approaches (Table 19.5).

Table 19.5 European LNG receiving stations and gas reserve facilities and their voluntary unbundling

Facility	Country	Owner position	Unbundling type	Owner	Operating entity
Dragon LNG receiving station	United Kingdom	Upstream	Legal	BG Group (50%), Petronas (50%)	Dragon LNG (JV)
South Hook LNG receiving station	United Kingdom	Upstream	Legal	Qatar Petroleum (67.5%), Exxon Mobil (24.15%), Total (8.35%)	South Hook LNG (JV)
Montoir and Fos-Tonkin LNG receiving station	United Kingdom	Vertically integrated	Legal	GDF-Suez	Elengy (LNG subsidiary)
Fos Cavaou LNG receiving station	France	Vertically integrated	Legal	GDF-Suez (70%), Total (30%)	Fosmax LNG (JV)
Grain LNG receiving station	United Kingdom	Midstream	Ownership rights	UK National Grid	Grain LNG (LNG subsidiary)
Zeebrugge LNG receiving station	Belgium	Midstream	Ownership rights	Fluxys	Fluxys LNG (LNG subsidiary)
Gate LNG receiving station	The Netherlands	Midstream	Legal	Gasunie (47.5%), Vopak (47.5%), receiving station user holdings 5%	Gate terminal B.V (JV)
Rough gas reserve facilities	United Kingdom	Upstream and downstream	Legal	Centrica	Centrica Storage (gas reserve subsidiary)
6 gas reserve facilities	Germany	Vertically integrated	Legal	VNG AG	VNG Gasspeicher GmbH (gas reserve subsidiary)
14 gas reserve facilities	France	Vertically integrated	Legal	GDF-Suez	Storengy (gas reserve subsidiary)

Note Gate LNG could be characterised primarily as ownership unbundling as 95% of the company is held by parties with no upstream or downstream interests, and, formally, the 5% ownership held by terminal users could justify its classification as legally unbundled

Overall, the range of experience suggests that access to LNG or storage terminals is not a key regulatory concern in the European market and that legal unbundling of these facilities is not an imposition on normal commercial behaviour.

19.4 The Downstream Segment: Natural Gas Trading Hubs and the Liberalisation of Wholesale Natural Gas Markets

The downstream segment of the natural gas value chain includes wholesale and retail markets supplying gas to end users. Greater competition in the downstream segment can increase choice and reduce the price paid by the end users of natural gas, for example by establishing natural gas hubs and increasing competition in wholesale markets. Competitive wholesale markets, in turn, are important for ensuring effective retail competition and delivering benefits for consumers and end users. Whether focused on a physical location, such as Henry Hub in the United States, or a virtual platform, such as Britain's NBP, activity on wholesale markets is the dominant factor in pricing and delivering natural gas to retailers and, eventually, end users.

This section analyses the experience of the United States, the United Kingdom and continental Europe in constructing natural gas trading hubs, providing a reference point to China in promoting its own natural gas downstream market liberalisation.

19.4.1   Pros and Cons of Natural Gas Hubs

There are two basic types of natural gas hub: physical hubs, such Henry Hub in the United States, and virtual hubs, such as the NBP in the United Kingdom. The IEA defines a physical hub as a geographical point in the network where the price is set for natural gas delivered to that specific location. A virtual hub, on the other hand, sets the price for natural gas across a wider geographic area. Unlike a physical hub, the NBP price reflects the price in the entire area without geographic differentials linked to transport costs. In general, a natural gas hub cannot be both physical and virtual, but it can evolve over time, for instance, starting as a physical hub and then expanding to become a wider virtual hub.

A natural gas hub has two core functions. The first is to physically connect buyers and sellers in a natural gas system. Before the creation of a hub, a gas system may already have many or all of the individual elements required for a hub, and some of them may already be connected. A hub strengthens these connections by linking physical flows between buyers and sellers across the entire system (Fig. 19.20). The second is to competitively determination natural gas prices. A gas hub provides a point or area in the gas network where demand and supply are balanced by a price. This balancing price is determined through competition between many buyers and sellers in the marketplace. In this way, hub prices reflect the fundamental market conditions for natural gas rather than prices tied to other commodities such as oil-indexed prices or regulated prices set by the government.

Natural gas hubs also provide a set of institutional rules, standardised contracts and price benchmarks. Hub prices serve as a focal point for market participants, ensuring that natural gas market conditions are widely known. In this way, hubs increase market transparency and lower transaction costs. Hub prices can also be used as references in contracts without delivery through the hub itself.

The creation of a natural gas hub comes with a variety of benefits, but also has potential drawbacks. The benefits include:

- **Market-based price signals raise the economic efficiency of trade and investment decisions**. The competitive trading of natural gas sets a price that reflects the true cost and value of gas in the economy. These competitive prices direct natural gas to where it is most needed from whoever can supply it most

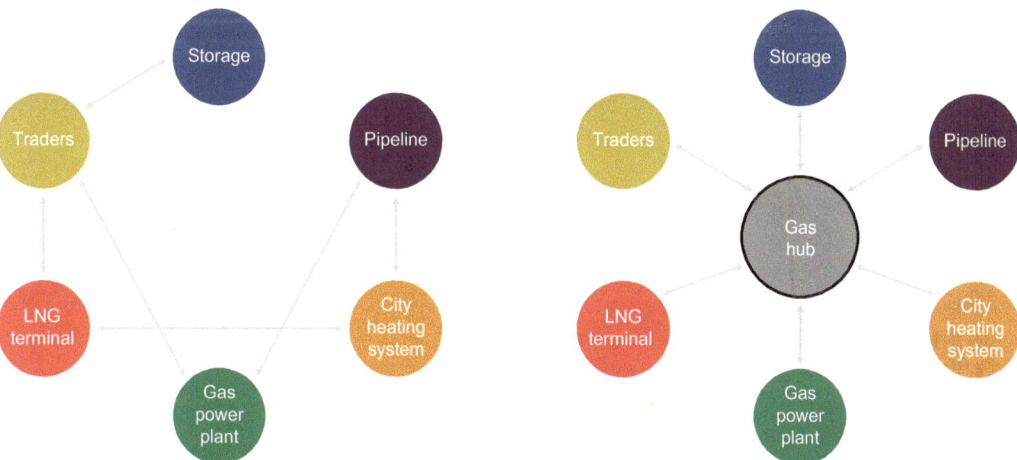

Fig. 19.20   The position of natural gas hubs in connecting buyers and sellers

cheaply. Trade takes place only when it is economically efficient, maximising the overall gains from trade. Such co-ordination is difficult to achieve other than through market mechanisms, especially in large and complex gas systems. In the longer term, the signals from competitive prices raise the quality of capital allocation as they drive natural gas market players to make investments only when they can create real economic value.

- **Hubs benefit from powerful and often self-reinforcing network effects**. Better market co-ordination, contact standardisation and greater transparency facilitate trading by reducing the costs of transactions. A hub lowers market participants' costs in searching for trading partners, who can now be easily found at exchanges or over the counter at the hub. A hub also reduces bargaining over the terms of the trade: a reference price and standardised contracts avoid wasted time and resources in negotiation over the terms of bespoke bilateral contracts. By reducing transaction costs, a hub makes it easier for new players to enter the natural gas market, widening market participation and increasing competition.
- **A natural gas hub can improve a country's security of energy supply**. Hubs help to diversify sources of supply; for example, by connecting regions within a country. Moreover, efficient price signals mean that, when markets are tight, additional supply is encouraged to come to market while also inducing reductions in gas demand. This helps to reduce the likelihood of unanticipated natural gas shortages.

However, natural gas trading hubs also have some potential drawbacks:

- **Hubs can disrupt pre-existing arrangements in a country's natural gas industry**. In particular, hub prices are determined by competitive forces, not by the government or a dominant incumbent. A transition to hub pricing puts pressure on incumbent players and leads to a shift in the structure of natural

gas markets. Hub pricing also puts pressure on regulated prices, as well as any remaining oil-indexed contracts. For example, the international trend since 2005 towards hub pricing, most notably in Europe, has reduced the role of long-term, oil-linked contracts. Natural gas in North America is almost entirely priced on hubs, while half of European natural gas is priced on hubs, rather than through oil-indexed contracts.

- **There is no guarantee that prices will decline with the emergence of hub pricing**. At any particular time, hub prices may be higher or lower than oil-indexed contract prices. However, there appears to be a tendency for hub prices to be lower. For example, NBP prices were below European oil-indexed prices 77% of the time between 2000 and 2014. An increase in prices may have adverse impacts on broader state development objectives, especially in poorer regions, for example putting upward price pressure on domestic heating. However, hubs tend to develop when the oil-indexed contract price fails to clear the market, that is, when supply is available in excess of contracted volumes and demand for this extra supply exists at prices below oil-indexed contract prices. As a result, the hub price paid for non-contracted volumes will be below the oil-indexed contract price, and the potential savings between oil-indexed prices and hub prices encourage buyers to switch from oil-indexed contracts to hubs.
- **Hub prices have tended to be more volatile than oil-indexed contract prices**. Because of averaging and time lags, oil-indexed contracts tend to be less volatile than hub prices. The volatility of hub prices may be perceived as costly by natural gas buyers, although experience suggests that such price risks can be managed, especially in well-developed hub markets with a wide array of financial contracts for natural gas.

A China natural gas hub is likely to create long-lasting benefits, while its drawbacks are likely to be temporary. China's large and complex domestic gas market could benefit from value-based pricing, market co-ordination and lower transaction costs. China's market size means that valuable network effects could also operate more strongly. A natural gas hub could further encourage local gas production, including from shale natural gas, and improve energy import diversity and responsiveness. Consumer gas prices could fall following the establishment of hub pricing, although they could rise again later.

19.4.2   The Important Inspiration of International Experience in the Development of Natural Gas Trading Hubs

A successful natural gas hub can only develop and be successful under a clear set of physical, market, and institutional conditions (Table 19.6). These conditions follow from the functions that a natural gas hub performs, informed by the experience from countries that have developed hubs.

Since the 1990s, the world has accumulated considerable experience in creating and developing natural gas hubs. This experience centres on US and European markets. The US natural gas market was restructured in the 1980s and 1990s, and its Henry Hub price has become the world's leading natural gas price indicator. A similar process of liberalisation began in the United Kingdom in the 1990s, leading to creation of the NBP, the leading hub in Europe. In continental Europe, market liberalisation began later and has proceeded more slowly as part of European Commission's plans for an Internal Energy Market.

Insights from the experience of the United States, continental Europe and the United Kingdom could benefit China. Indeed, China shares some structural features with the United States—in particular, both are geographically vast, with domestic production far removed from centres of demand. These characteristics suggest that, like the US Henry Hub, China may need a physical hub to accommodate significant differences in transmission costs to various dispersed destinations. Essentially, a physical hub prices natural gas at a single physical point of the network, after which transmission costs are added. This contrasts with a virtual hub, such as the NBP, which sets the price for natural gas within a region and imposes a fixed transmission cost for natural gas delivered within that region. A virtual hub is appropriate for a market with a relatively small delivery region and fixed transmission costs. The situation in China may be best served by multiple hubs that follow the pricing of a main hub, a structure similar to that seen in the United States.

The process of building a hub in Europe also has relevance for China. Hub development in

Table 19.6   Necessary preconditions in three areas for the successful development of a natural gas hub

Area	Precondition	Description
Physical	Well-developed natural gas transmission network	Connection of domestic and foreign natural gas supply to demand Open-access to pipeline network by suppliers
Market	Large numbers of participants	Sufficiently large number of market participants, both as buyers and sellers
	Low market concentration	No dominant supplier or consumer Possibility of natural gas imports
	Trading activity	Entrepreneurial activity to arbitrage prices, requiring third-party access, part of supply uncontracted and part of demand contestable; price reporting agencies
Institutional	Liberalisation of price determination	Liberalised prices for wholesale natural gas and along the supply chain
	Stable regulatory framework	Competition policy and market regulation ensures level playing field

continental Europe and the United Kingdom took place against a backdrop of domestic production and imported natural gas, both through pipeline and as LNG. In contrast, US hub development took place without LNG playing an important role. Continental European natural gas markets have also been undergoing a gradual transition from a highly regulated, state-dominated system in the 1990s towards a fully liberalised system. Finally, potentially analogous to Asian markets, Europe has been home to several regional hubs that coexist on the EU natural gas market.

Overall, a robust and credible liberalisation process, along with favourable market conditions, has been instrumental in the development of natural gas hubs. The US experience shows that open access to pipelines is critical for market-based pricing, and that liberalisation can help wholesale markets shift away from long-term contracts and towards market-based pricing in 5–10 years. The experience of continental Europe is of hubs being established as a result of favourable market conditions enabled by the regulatory reform process. In these cases, market forces created a buyers' market in natural gas, which, combined with the liberalisation of natural gas markets, accelerated hub development. The UK experience is one of market and regulatory changes working in parallel to increase competition in wholesale markets, leading to the establishment of the NBP. Standardisation of contracts played a key role in driving hub success in the United Kingdom.

19.4.3 Case Study 1: The United States—Regional Price Balances and Short-Term Pricing

The US natural gas market underwent fundamental reform beginning in the 1980s, with the emergence of ample natural gas supplies and regulatory changes to liberalise the market. In the 1980s and early 1990s, the US regulator, FERC, implemented reforms to decouple the production and trading of natural gas from its transportation by pipeline. The economic argument was that the

production and consumption of natural gas can involve many different buyers and sellers and has the potential to be competitive. Meanwhile, pipeline transport services were often highly concentrated, as a result of their natural monopoly characteristics, and should be regulated. As an alternative, guaranteed open access for all market participants on a non-discriminatory basis could create greater competition among producers and natural gas shippers and better allocation of economic resources.

The empirical evidence shows that reforms in the United States had substantial impact quickly. In the early 1990s, movements of natural gas prices in different regions were weakly correlated, with significant price divergence. By the late 1990s, wholesale price correlations had risen to very high levels with the application of the law of one price, an approach that equalised prices across locations, excluding transport costs. Evidence suggests that regulation to ensure non-discriminatory open access to pipelines played a key role in driving the move to a competitive, national natural gas market.

In parallel, the underlying contract structure of the industry shifted from long-term contracts to short-term pricing. Existing long-term contracts were either renegotiated in light of the changed market environment or terminated. Spot and futures prices emerged as the key co-ordinating mechanisms within the US natural gas sector, with pricing centred on Henry Hub. Market concentration fell, with the 20 largest players accounting for 44% of physical and financial transaction volumes in 2014. More recently, the Henry Hub price, plus a mark-up for transport, has been used in LNG contracts between US exporters and foreign buyers.

19.4.4 Case Study 2: Continental Europe—Market Conditions for Natural Gas Hub Development

European natural gas hubs began developing only in 2008, reflecting the slow progress of EU natural gas market liberalisation. Northern

Europe is more advanced than southern Europe in terms of natural gas trading and hub development. The legacy system of bilaterally negotiated oil-indexed natural gas contracts remains more prevalent in the south.

Several factors drove deepened natural gas trading and hub development in Western Europe after 2008. The 2008 global financial crisis dampened European natural gas demand, in contrast to the boom that had been expected. Increased shale natural gas production in the United States meant that the country no longer imported LNG, with much of the freed supply diverted to Europe. These two factors created a buyers' market in natural gas, and traders and brokers in key European markets—especially Germany and France—began to enjoy access to uncontracted pipeline natural gas. Low natural gas demand and greater LNG supply coupled with high oil prices meant that oil-linked contract prices rose above market-based natural gas prices, creating powerful incentives for buyers to renegotiate long-term contracts with upstream natural gas suppliers, such as Norway and Russia. Within a few years, long-term contracts had either been terminated or renegotiated to reflect market-based hub prices rather than oil prices.

The Dutch Title Transfer Facility (TTF) is widely seen as the most successful of the continental European natural gas trading hubs. The facility was deliberately designed as a virtual hub, based on its physical pipeline and LNG connections, its geographic location and the large domestic natural gas supply from the Dutch Groningen field. Development of the TTF accelerated in the late 2000s, and it has overtaken NBP as Europe's most liquid hub. In contrast, the Belgian Zeebrugge hub is generally seen as a much less successful natural gas trading centre. Conditions at Zeebrugge—its location and relatively early development—once suggested that it could become the leading European natural gas hub, but local trading restrictions limited the number of market participants and slowed the development of its trading platform.

Overall, the move towards more active natural gas trading has increased the correlation of natural gas prices across European countries, similar to the US experience in the 1990s. Spot prices in the European Union now account for about half the natural gas trade.

19.4.5 Case Study 3: The UK—Two Natural Gas Market Reform Bills

Natural gas market liberalisation began in the United Kingdom in the 1990s, following a wave of privatisations in the previous decade. The United Kingdom was the first country in Europe to begin natural gas market reform, and aspects of first-mover advantage remain.

Two pieces of regulation were especially important to the reform effort in the United Kingdom. The Gas Act of 1986 removed the de facto monopoly of British Gas in serving large natural gas customers, while also requiring open access to its pipelines for its competitors. The Gas Act of 1995 set out a statutory timetable for full competition in the British natural gas market, including at the residential level. As a result, the market share held by British Gas declined rapidly, from nearly 100% in 1990 to 29% by 1996. These institutional and regulatory changes led to a surge in the number of market participants and trading activity during the mid-1990s. Among the new entrants to the natural gas market were merchant banks, electricity companies and foreign and domestic trading houses. The number of participants in the wholesale market rose from fewer than 15 in 1995 to more than 50 within about two years.

The 1995 Gas Act also established the Network Code, a key element in the development of natural gas trading. The code is a set of rules and procedures governing third-party access to the UK natural gas pipeline network. The code also enabled a system of daily balancing, which required short-term natural gas trades.

These regulatory developments were accompanied by shifts in the country's demand and supply of natural gas. On the demand side, the "dash for gas" in the British electricity industry saw the growth of generation capacity using flexible combined-cycle natural gas turbines. On the supply side, natural gas was available from domestic production in the North Sea. The regulatory changes introduced by the Gas Acts of 1986 and 1995 meant that new participants were able to access natural gas segments across the value chain.

The regulatory and market factors combined put pressure on the old system of bilateral contracts, which were typically indexed to other commodities, such as gas oil or fuel oil. The NBP was created within the code as a virtual hub to promote the balancing of the natural gas system. From the late 1990s, the NBP evolved quickly as a trading point and was used as the basis for the standardised NBP97 contract, which became the cornerstone of British over-the-counter contracts. The NBP also became the delivery point for the Intercontinental Exchange natural gas futures contract.

With the deepening of UK natural gas trade since the late 1990s, NBP has become a leading natural gas hub in Europe. A number of factors helped its ascension, including the existence of domestic natural gas reserves, periods of high demand for natural gas and a stable regulatory framework that enabled open-access market participation via open access and offered a clear set of rules and standardised contracts.

19.4.6   Establishing Natural Gas Hubs in China

Natural gas hubs are market institutions, and thus to create a successful hub in China, the market structure of the natural gas industry in China will need to be changed. This will generate efficiency and generate costs, while also creating winners and losers. However, the long-term benefits of China establishing a hub are clear, as it will raise domestic natural gas production, raise energy diversification and also have major significance to response capabilities. Construction of natural gas hubs in China is therefore critical to the development of natural gas. This section will discuss the overall prospects for construction of natural gas hubs in China and the factors that must be considered in hub development, and analyse the potential steps.

1. **The overall outlook for construction of natural gas trading hubs in China**

Insights gained from international experience in establishing natural gas hubs, combined with an assessment of China's circumstances, lead to five key conclusions.

- China is in a good position to develop more market-based natural gas trading with hub pricing. Key positive factors include domestic gas production, exposure to both pipeline and LNG imports, and its size within Asian energy markets.
- China's large and complex domestic natural gas market would benefit from value-based pricing, market co-ordination and lower transaction costs. It is likely that these efficiency gains from hub pricing will bring large and long-lasting benefits, while any adverse effects, particularly disruption of current market arrangements, would be short-lived.
- A successful hub would require significant further liberalisation of the natural gas sector in China, which would reduce the influence of state-owned incumbents, improve market transparency, and create a level playing field for new market entrants through third-party access arrangements.
- China can move towards hub pricing in a series of small steps, backed by government policy. Beginning with pilot natural gas trading in a particular region, these steps may gradually satisfy the pre-conditions for a successful hub and ease the transition for incumbents.
- China is likely to develop the leading hub in Asia. Even though international experience suggests that first-mover advantages exist in the development of gas hubs, it is likely that a China hub would become the Asian market leader even if it emerges later than hubs in other countries in the region.

2. **Factors for consideration when developing hubs**

Based on international experience, the primary conditions for creating a successful hub are a well-developed physical natural gas transmission network that market participants can access under non-discriminatory conditions; a large number of independent buyers and sellers actively engaged in arbitrage, but without significant market power; and a government commitment to liberalising wholesale gas markets and a stable, transparent and credible regulatory framework.

Many of these preconditions for a successful hub are not currently met in Asian countries, including China. For instance, in China transmission infrastructure remains under development, markets are concentrated, prices are widely regulated and regulations such as third-party access rules that support the smooth operation of a market with many buyers and sellers are not yet in place.

A plan to develop a Chinese natural gas hub could begin by securing the preconditions through liberalisation of the natural gas sector in China. A hub is a market institution and requires market liberalisation to function well. Liberalisation facilitates a hub by enabling prices to be set competitively, reducing the influence of state-owned incumbents, improving market transparency and creating a level playing field for new market entrants through third-party access arrangements.

Hub pricing creates winners and losers along the natural gas value chain. Some players, protected under the previous regulatory regime, could lose revenues. Improving market signals could undermine the main rationale for vertical integration and for co-ordinating action across a supply chain. Other players, on the other hand, could benefit from greater market connectivity and the ability to optimise asset portfolios more actively.

Who wins and who loses, and by how much, is also affected by the oil price. Historically, natural gas prices have been driven by oil indexation. A higher oil price would mean higher natural gas contract prices, and moving to hub pricing would create strong downward pressure on natural gas prices. Midstream utilities that are locked into long-term oil-indexed contracts but unable to pass through their higher costs to consumers would be among the losers. However, in a world with lower oil prices, as seen since late 2014, legacy oil-linked natural gas contract prices would lie below the market-determined price for natural gas, thus potentially reversing the roles of winners and losers were a hub established, and slowing the adoption of hub pricing.

International experience shows that natural gas hubs are formed in response to market pressure coming from domestic sources or international market developments. In the past, the development of hubs has often been driven by large additional supplies of natural gas seeking access to consumers. This supply has either been domestically produced, as in the case of the UK's NBP hub development in the 1990s, or available from global LNG markets, as in the deepening of continental European hub markets from the late 2000s.

At least three supply-side factors are exerting market pressure to form a hub in China: expanding domestic gas production, including from unconventional sources; supplies from the global LNG market not already contractually committed to particular export markets; and additional gas imports from Russia from existing or planned pipelines.

The development of natural gas hubs elsewhere in Asia could also challenge China's domestic natural gas pricing arrangements and create momentum for a Chinese natural gas hub. For example, Singapore has put itself forward as an LNG hub, with limited trading since 2014. Japan has begun to publish spot LNG prices and has proposed the creation of a futures market. Although the European experience suggests that there are first-mover advantages in hub creation, a potential Chinese hub is likely to become the principal Asian hub even if it is not the first. China's economy is much larger, with large industrial consumers of natural gas. China also has significant domestic production and has begun to develop a national pipeline network. China's natural gas imports come over pipelines as well as LNG, unlike the rest of Asia, which relies heavily on LNG. These factors suggest that, supported by the appropriate regulatory environment, a Chinese hub is likely to emerge as the leader in Asia.

3. **Potential steps for developing hubs**

International experience shows that successful hub development takes time—at least 5–10 years —even in countries with energy sectors that were much more liberalised than that in China.

Taking an incremental and phased approach would allow China to address the preconditions for full hub pricing, within the broader market liberalisation process. Rushing to establish a full hub by government mandate in the early stages of liberalisation is likely to fail because of a shortage of participants and institutions. A phased approach, on the other hand, would allow market participants to learn and build institutions concurrently with a liberalisation programme. A phased approach would also allow for the gradual creation of a support network among participants, as various players have time to adjust to the new liberalised markets and the benefits of market-based pricing become more apparent.

The evolution of China's natural gas sector towards hub pricing could be phased over 5–10 years and cover six specific aspects (Fig. 19.21).

- **Pilot trading market**: Within an individual region, such as Shanghai, the government could create a pilot trading market with contestable natural gas demand and some uncontracted natural gas supply. The market would be connected through a natural gas network with third-party access. Entrepreneurs would also be encouraged to trade the uncontracted natural gas supply to meet demand among the customers, who are free to choose their supplier.
- **Price transparency**: The government could require publication of prices of natural gas sold under all types of contracts. The mandate would help reveal opportunities for arbitrage to entrepreneurial buyers and sellers, vital components of a successful market-based arrangement.
- **Government recognition**: Once significant trading activity develops within the pilot region—for example, a shift to standardised over-the-counter trading—the government could explicitly endorse the market by establishing a regulatory body. This body would standardise contracts further and seek to reduce transaction costs.

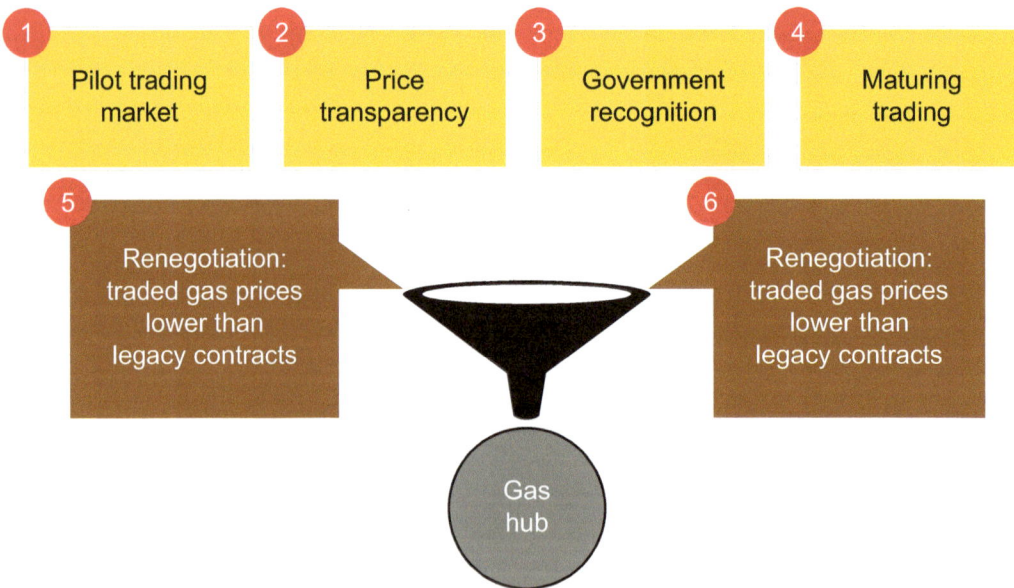

Fig. 19.21 Hub development and its support from regulatory measures and their market effects

- **Maturing trading**: As exchange-based trading develops, the government would increase the scope of the pilot region to cover a larger geographic area or more types of buyers and sellers. At this stage, market regulation may have to become more stringent to reflect the growing system-wide role of natural gas trade.

Following these steps, a natural gas hub could then become self-sustaining with two final changes:

- **Renegotiation**: If traded natural gas prices are lower than legacy oil-indexed natural gas contract prices, companies paying legacy prices would be put at a disadvantage and may incur losses if they cannot pass along the price difference to their customers. With access to the liberalised trading on the natural gas hub, such companies would have a strong incentive to renegotiate legacy contracts, recalibrating prices or volumes, and buy traded natural gas instead.

- **Network effects**. As more and different organisations join hub trading, the benefits of the natural gas hub would increase as a result of network effect. The use and scope of hub pricing would tend to grow until it reaches the limits of either incomplete liberalisation or of geography.

Bibliography

BP (2014) BP statistical review of world energy 2014

Chen M (2014) The development of Chinese gas pricing: drivers, challenges and implications for demand. The Oxford Institute for Energy Studies, OIES paper, NG89

Cuddington JT, Wang Z (2006) American natural gas spot market integration assessment: From daily price data proof. J Regul Econ 29:195–210

Development Research Center of the State Council and the World Bank (2013) The China 2030 report: building a modern, harmonious, and creative society

Doane MJ, Spulber DF (1994) American natural gas spot market access and development. J Law Econ 37: 477–517

European Environment Agency (2014) 2008 to 2012 European industrial facility air pollution costs

Financial Times (2015) With link to Russia smashed, EU seeks new sources. February 25

Forster C (2013) Australian LNG: busting the budget. Platts. Available at: www.platts.com/IM.Platts.Content%5Caboutplatts%5Cmediacenter%5Cpdf%5Cinsight-oct13-australia.pdf. Accessed 30 Sept 2016)

Global Commission on the Economy and Climate. Faster growth, safe climate: the new climate economy report. Available at: http://static.newclimateeconomy.report/wp-content/uploads/2014/08/NCE_GlobalReport.pdf. Accessed 31 May 2015

IEA. IEA current policies outlook. Available at: www.iea.org

Institute for Health Metrics and Evaluation (2014) Global burden of disease. Available at: http://www.healthdata.org/gbd. Accessed 31 May 2015

International Gas Union (2014) World LNG report. Available at: http://www.igu.org/sites/default/files/node-page-field_file/IGU%20-%20World%20LNG%20Report%20-%202014%20Edition.pdf. Accessed 31 May 2015

Jiaofeng G et al (2014) Utilizing optimization of market net return on value as a breakthrough point to promote natural gas price reforms. State Council Development Research Center Investigative Research Report No. 148 (Z No. 4647)

Matus K, Nam K-M, Selin NE, Lamsal LN, Reilly JM, Paltsev S (2011) Health damages from air pollution in China. Global Environmental Change 22:55–66

Natural Gas Daily (2014) TTF is now the "European Yardstick" after increases

Nielson CP, Ho MS (2013) Clearer skies over China. MIT Press

Nulle G (2014) Shale development prospects beyond the United States. Colorado School of Mines

Shell (2014) New lens scenarios

World Bank and Chinese State Environmental Protection Administration (2007) Cost of pollution in China: economic estimates of physical damages. World Bank

Xu W (2015) What can China learn from US regulatory experiences? Dev Res 3:4–14

© The Editor(s) and The Author(s) 2017
Shell International and The Development Research Center (Eds.), *China's Gas Development Strategies*,
Advances in Oil and Gas Exploration & Production, DOI 10.1007/978-3-319-59734-8

Further Reading

American Supreme Court. Phillips Petroleum Co. v. Wisconsin (347 U.S. 672 (1954)). Available at: https://supreme.justia.com/cases/federal/us/347/672/case.html. Accessed 31 May 2015

APERC (2002) Gas storage in the APEC region. Available at: http://aperc.ieej.or.jp/file/2010/9/26/Gas_Storage_in_the_APEC_Region_2002.pdf. Accessed 31 May 2015

Bloomberg (2014) Singapore bids for role as LNG hub with second terminal. Available at: www.bloomberg.com/news/articles/2014-02-25/singapore-plans-to-build-second-lng-terminal-in-country-s-east. Accessed 31 May 2015

BP (2011) BP world energy statistical yearbook 2011

BP (2013) BP world energy statistical yearbook 2013

Canqi WU (2012) Next 10-year natural gas usage trend in China. International Petrol Economics Z1

Carlisle LE, Hagan DA, Rueger JE (2013) American pipeline infrastructure investment: Can the proposed "Proprietor Limited Partnership Equality Act" stimulate a round of investment? Harvard Law Rev 4:32–37

Department of Energy and Climate Change (2012) Upstream oil and natural gas infrastructure third party access dispute resolution

Disavino S, Krishnan B (2014) King of American natural gas trade—Henry Hub, Marcellus. Available at: http://www.reuters.com/article/2014/09/25/us-natnaturalgas-henryhub-marcellus-analysis-idUSKCN0HK17E20140925. Accessed 25 Sept 2014

European Commission (2003) Natural gas international market directive 2003/55/EC—overview. Available at: http://eur-lex.europa.eu/legal-content/EN/TXT/?uri=uriserv:l27077. Accessed 31 May 2015

Eikeland PO (2004) Internal energy market slow and circuitous roads—European policy consistencies and contradictions. FNI report

EurActiv (2007) EU states reject breaking up energy firms. Available at: www.euractiv.com/energy/eu-states-reject-breaking-energy-firms/article-164398. Accessed 31 May 2015

European Natural Gas (2013) Annual statistical report 2014

European Commission (2003) Second comparison report regarding internal power and natural gas markets. 32:54–55. Based on (European Commission) 1/2003

Rule No. 17 regarding EU natural gas and power industry investigations [Z], 2007, pp 1–15

European Commission (commencing March 2011) Natural gas internal market directives 2009/73/EC. Available at: http://eur-lex.europa.eu/legal-content/EN/TXT/?qid=1413403818553&uri=URISERV:en0017. Accessed 31 May 2015

European Commission (2014) Internal energy market optimization progress. ITO model report

European Environment Agency (2013) EMEP/EEA air pollution emissions list guidelines. Available at: http://www.eea.europa.eu/publications/emep-eea-guidebook-2013. Accessed 31 May 2015

Federal Energy Regulatory Commission (2014) Characteristics of American natural gas trading: Federal Energy Regulatory Commission. May 16. Form 552 Letter of opinion

Financial Post. (2013) Korean companies to invest more in Canadian LNG resources: analyst. Available at: http://business.financialpost.com/2013/03/27/korean-companies-to-invest-more-in-canadian-lng-resources-analyst/?__lsa=880e-847b. Accessed 27 Mar2013

Gao AM-Z (2010) Natural gas market regulation: United States, EU, Japan, South Korea, and Taiwan partitioning and market access system comparative research

GIIGNL Commercial Study Group (2010) Liquefied natural gas receiving station third party access. Available at: www.giignl.org/system/files/publication/130516_giignl_2010_tpa_lng_giignl.pdf. Accessed 31 May 2015

Gilbert S (2009) Russia's EU natural gas politics. J Polit Sci Int Aff 5:126–138

Heather P (2012) Are Europe's continental natural gas hubs truly useful?

Heather P (2010) The role and development of UK natural gas trade markets. Oxford Institute for Energy Studies

Helm D (2004a) Energy, nations, and markets: UK energy policy since 1979 (revised). Oxford University Press

IEA. Current policy prospects

IEA (2002) Work energy prospects 2002—natural gas. Available at: www.eia.gov/oiaf/archive/ieo02/nat_naturalgas.html. Accessed 31 May 2015

© The Editor(s) and The Author(s) 2017
Shell International and The Development Research Center (Eds.), *China's Gas Development Strategies*, Advances in Oil and Gas Exploration & Production, DOI 10.1007/978-3-319-59734-8

IEA (2007) IEA national energy policy: Germany. OECD/IEA, Paris France

IEA (2009) IEA national energy policy: France. OECD/IEA, Paris France

IEA (2011) Oil and natural gas security: IEA national emergency response—South Korea

IEA (2012a) IEA national energy policy: South Korea

IEA (2012b) Natural gas pricing and regulation: challenges and international experiences facing China. IEA Partner Nation Series

IEA (2013a) Developing Asian natural gas trade hubs: challenges and opportunities

IEA (2013b) Oil and natural gas security: Japan. Available at: www.iea.org/publications/freepublications/publication/2013_OSS_Japan.pdf. Accessed 31 May 2015

IEA (2014) Energy supply security. Available at: www.iea.org/media/freepublications/security/EnergySupplySecurity2014_US.pdf. Accessed 31 May 2015

International Gas Union (2014) Natural gas wholesale price survey 2014. 2005–2013 global price formation mechanism survey

Japan Gas Association (2013) Japan natural gas. Available at: www.naturalgas.or.jp/naturalgasfacts_e/#page_num=0. Accessed 31 May 2015

Japan Ministry of Foreign Affairs (2000) Third joint status report on the U.S.-Japan enhanced initiative on deregulation and competition policy. Available at: www.mofa.go.jp/region/n-america/us/report0007.html. Accessed 15 May 2015

Japanese Natural Resources and Energy Agency (2013) Japan electricity market reforms. Available at: www.meti.go.jp/english/policy/energy_environment/electricity_system_reform/pdf/201311EMR_in_Japan.pdf. Accessed 31 May 2015

Jun H, Shouying L (2014) Rural organic waste production of natural gas effects and policies. Development Research Center of the State Council Survey Research Report Summary Extracts 2014.12

Kern F (2012) UK electricity industry combined-cycle gas turbine development and "rush to gas" policy (1987–2000). UK Energy Research Center project work package 2's final case study report: "Does CCS technology have potential?"

Kumins L, Bamberger R (2005) Oil and gas disruption from hurricanes Katrina and Rita. Congressional Research Service. Available at: http://fpc.state.gov/documents/organization/55824.pdf

MacAvoy PW (2000) Natural gas market: six years of strict oversight and deregulation. Yale University Press

Makholm JD (2012) The political economy of pipelines: a century of comparative system development. Chicago University Press

Mergers and Monopolies Commission (1997) BG: 1986 report on restrictions caused under the "Natural Gas Act" to gas shipping and reserve service prices

Ministry of Environmental Protection. (2015) Take stringent measures to curb environmental harmful actions: Full interpretation of the four supplementary measures in the new Environmental Protection Act. Available at: http://english.mep.gov.cn/News_service/infocus/201502/t20150209_295637.htm. Accessed 31 May 2015

National Grid (2014) Natural gas and electricity network routes. Available at: http://www2.nationalgrid.com/uk/services/land-and-development/planning-authority/naturalgas-and-electricity-network-routes/. Accessed 6 Oct 2014

Natural Gas Supply Association of America (XXXX) Regulated markets. Available at: http://naturalnaturalgas.org/regulation/market/. Accessed 31 May 2015

Norske Veritas—Comar and Coway (2013) LT-ST market natural gas research. Project under the European Commission DG Energy Technical Support Framework Service Contract TREN/R1/350-2008.

OECD (2013) Cleaner, healthier environmental reforms: OECD Economic Survey: China

Oil & Gas UK. Code of practice on access to upstream oil and gas infrastructure on the UK continental shelf. Available at: www.ogel.org/article.asp?key=1671. Accessed 31 May 2015

Osaka Gas Co Ltd (2012) Annual report. Available at: www.osakanaturalgas.co.jp/en/ir/library/ar/pdf/2012/12_04.pdf. Accessed 31 May 2015

Osaka Gas Co Ltd (2013). Annual report. Available at: www.osakanaturalgas.co.jp/en/ir/library/ar/pdf/2013/ar2013e.pdf. Accessed 31 May 2015

Rogers HV, Stern JP (2014) European natural gas market dynamics: hub price determination factors and major roles and risks. Oxford Institute for Energy Studies

Securities and Exchange Commission (2014) Shell midstream partners prospectus. Available at: www.sec.gov/Archives/edgar/data/1610466/000119312514387075/d738367d424b4.htm#toc. Accessed 31 May 2015

Smith CE (2014) Oil pipeline product pipeline completions lead planned construction lower. Available at: www.ogj.com/articles/print/volume-112/issue-2/special-report/worldwide-pipeline-construction/product-pipeline-completions-lead-planned-construction-lower.html. Accessed 17 Oct 2014

Stern J, Bradshaw M, Flower A, Fridley D, Jung N, Joshi S, Paik K-W et al (2008) Asian natural gas: China, India, Japan, and South Korea growth challenges

Stern J (ed) (2012) Pricing for international natural gas trade. Oxford University Press

Tobin J (2014) Natural gas market centers and hubs: 2003 TEPCO and Chubu Electric to form comprehensive alliance—Creation of a global energy company competing worldwide. Available at: www.tepco.co.jp/en/press/corp-com/release/2014/1242656_5892.html. Accessed 31 May 2015.

Tokyo Gas Company (2011) Creating joint value and realising growth. Available at: www.tokyo-naturalgas.co.jp/IR/english/library/pdf/anual/11e06.pdf. Accessed 31 May 2015

UNECE (2012) European economic zone liquefied natural gas current and future prospects research

UNEP (2014) Asia Pacific development—clean fuels and vehicle partners

US Energy Information Administration. Natural Gas Legislation and Regulatory Measures (1935–2008).

Available at: www.eia.gov/oil_naturalgas/natural_naturalgas/analysis_publications/ngmajorleg/ngmajorleg.html. Accessed 1 Oct 2014

US Energy Information Administration (2009) Natural gas residential options as of December 2009. Available at: www.eia.gov/oil_naturalgas/natural_naturalgas/restructure/restructure.html. Accessed 10 Oct 2014

US Energy Information Administration (2015) Japan. Available at: www.eia.gov/countries/cab.cfm?fips=ja. Accessed 31 May 2015

US Energy Information Administration (2015) National overview—Japan. Available at: www.eia.gov/countries/cab.cfm?fips=ja. Accessed 31 May 2015

United States Environmental Protection Agency (2014) Clean energy plan proposed rules.

Vivoda V (2014) Asian natural gas: trade, markets, and regional institutions

Wang Z, Krupnick A (2013) American Shale gas development review survey: what has catalyzed the flourishing of Shale gas? Future Resources Research Center Discussion Draft.